T0099954

Cell Fusion

METHODS IN MOLECULAR BIOLOGY™

John M. Walker, SERIES EDITOR

475. **Cell Fusion:** *Overviews and Methods,* edited by *Elizabeth H. Chen, 2008*
474. **Nanostructure Design:** *Methods and Protocols,* edited by *Ehud Gazit and Ruth Nussinov, 2008*
473. **Clinical Epidemiology:** *Practice and Methods,* edited by *Patrick Parfrey and Brendon Barrett, 2008*
472. **Cancer Epidemiology, Volume:** *2 Modifiable Factors,* edited by *Mukesh Verma, 2008*
471. **Cancer Epidemiology, Volume 1:** *Host Susceptibility Factors,* edited by *Mukesh Verma, 2008*
470. **Host-Pathogen Interactions:** *Methods and Protocols,* edited by *Steffen Rupp and Kai Sohn, 2008*
469. **Wnt Signaling, Volume 2:** *Pathway Models,* edited by *Elizabeth Vincan, 2008*
468. **Wnt Signaling, Volume 1:** *Pathway Methods and Mammalian Models,* edited by Elizabeth Vincan, 2008
467. **Angiogenesis Protocols:** *Second Edition,* edited by *Stewart Martin and Cliff Murray,* 2008
466. **Kidney Research:** *Experimental Protocols,* edited by *Tim D. Hewitson and Gavin J. Becker,* 2008.
465. **Mycobacteria,** *Second Edition,* edited by *Tanya Parish and Amanda Claire Brown,* 2008
464. **The Nucleus, Volume 2:** *Physical Properties and Imaging Methods,* edited by *Ronald Hancock,* 2008
463. **The Nucleus, Volume 1:** *Nuclei and Subnuclear Components,* edited by *Ronald Hancock,* 2008
462. **Lipid Signaling Protocols,** edited by *Banafshe Larijani, Rudiger Woscholski, and Colin A. Rosser,* 2008
461. **Molecular Embryology:** *Methods and Protocols, Second Edition,* edited by *Paul Sharpe and Ivor Mason,* 2008
460. **Essential Concepts in Toxicogenomics,** edited by *Donna L. Mendrick and William B. Mattes,* 2008
459. **Prion Protein Protocols,** edited by *Andrew F. Hill,* 2008
458. **Artificial Neural Networks:** *Methods and Applications,* edited by *David S. Livingstone,* 2008
457. **Membrane Trafficking,** edited by *Ales Vancura,* 2008
456. **Adipose Tissue Protocols,** *Second Edition,* edited by *Kaiping Yang,* 2008
455. **Osteoporosis,** edited by *Jennifer J. Westendorf,* 2008
454. **SARS- and Other Coronaviruses:** *Laboratory Protocols,* edited by *Dave Cavanagh,* 2008
453. **Bioinformatics, Volume 2:** *Structure, Function, and Applications,* edited by *Jonathan M. Keith,* 2008
452. **Bioinformatics, Volume 1:** *Data, Sequence Analysis, and Evolution,* edited *by Jonathan M. Keith,* 2008
451. **Plant Virology Protocols:** *From Viral Sequence to Protein Function,* edited by *Gary Foster, Elisabeth Johansen, Yiguo Hong, and Peter Nagy,* 2008
450. **Germline Stem Cells,** edited by *Steven X. Hou and Shree Ram Singh,* 2008
449. **Mesenchymal Stem Cells:** *Methods and Protocols, edited by Darwin J. Prockop, Douglas G. Phinney, and Bruce A. Brunnell,* 2008
448. **Pharmacogenomics in Drug Discovery and Development,** edited by *Qing Yan,* 2008.
447. **Alcohol:** *Methods and Protocols,* edited by *Laura E. Nagy,* 2008
446. **Post-translational Modifications of Proteins:** *Tools for Functional Proteomics, Second Edition,* edited by *Christoph Kannicht,* 2008.
445. **Autophagosome and Phagosome,** edited by *Vojo Deretic,* 2008
444. **Prenatal Diagnosis,** edited by *Sinhue Hahn and Laird G. Jackson,* 2008.
443. **Molecular Modeling of Proteins,** edited by *Andreas Kukol,* 2008.
442. **RNAi:** *Design and Application,* edited by *Sailen Barik,* 2008.
441. **Tissue Proteomics:** *Pathways, Biomarkers, and Drug Discovery,* edited by *Brian Liu,* 2008
440. **Exocytosis and Endocytosis,** edited by *Andrei I. Ivanov,* 2008
439. **Genomics Protocols,** *Second Edition,* edited by *Mike Starkey and Ramnanth Elaswarapu,* 2008
438. **Neural Stem Cells:** *Methods and Protocols, Second Edition,* edited by *Leslie P. Weiner,* 2008
437. **Drug Delivery Systems,** edited by *Kewal K. Jain,* 2008
436. **Avian Influenza Virus,** edited by *Erica Spackman,* 2008
435. **Chromosomal Mutagenesis,** edited by *Greg Davis and Kevin J. Kayser,* 2008
434. **Gene Therapy Protocols: Volume 2:** *Design and Characterization of Gene Transfer Vectors,* edited by *Joseph M. LeDoux,* 2008
433. **Gene Therapy Protocols: Volume 1:** *Production and In Vivo Applications of Gene Transfer Vectors,* edited by *Joseph M. LeDoux,* 2008
432. **Organelle Proteomics,** edited by *Delphine Pflieger and Jean Rossier,* 2008
431. **Bacterial Pathogenesis:** *Methods and Protocols,* edited by *Frank DeLeo and Michael Otto,* 2008
430. **Hematopoietic Stem Cell Protocols,** edited by *Kevin D. Bunting,* 2008
429. **Molecular Beacons:** *Signalling Nucleic Acid Probes, Methods and Protocols,* edited by *Andreas Marx and Oliver Seitz,* 2008
428. **Clinical Proteomics:** *Methods and Protocols,* edited by *Antonia Vlahou,* 2008
427. **Plant Embryogenesis,** edited by *Maria Fernanda Suarez and Peter Bozhkov,* 2008
426. **Structural Proteomics:** *High-Throughput Methods,* edited by *Bostjan Kobe, Mitchell Guss, and Thomas Huber,* 2008
425. **2D PAGE:** *Sample Preparation and Fractionation, Volume 2,* edited by *Anton Posch,* 2008
424. **2D PAGE:** *Sample Preparation and Fractionation, Volume 1,* edited by *Anton Posch,* 2008
423. **Electroporation Protocols:** *Preclinical and Clinical Gene Medicine,* edited by *Shulin Li,* 2008
422. **Phylogenomics,** edited by *William J. Murphy,* 2008
421. **Affinity Chromatography:** *Methods and Protocols, Second Edition,* edited by *Michael Zachariou,* 2008
420. **Drosophila:** *Methods and Protocols,* edited by *Christian Dahmann,* 2008
419. **Post-Transcriptional Gene Regulation,** edited by *Jeffrey Wilusz,* 2008
418. **Avidin–Biotin Interactions:** *Methods and Applications,* edited by *Robert J. McMahon,* 2008
417. **Tissue Engineering,** *Second Edition,* edited by *Hannsjörg Hauser and Martin Fussenegger,* 2007
416. **Gene Essentiality:** *Protocols and Bioinformatics,* edited by *Svetlana Gerdes and Andrei L. Osterman,* 2008
415. **Innate Immunity,** edited by *Jonathan Ewbank and Eric Vivier,* 2007
414. **Apoptosis in Cancer:** *Methods and Protocols,* edited by *Gil Mor and Ayesha Alvero,* 2008
413. **Protein Structure Prediction,** *Second Edition,* edited by *Mohammed Zaki and Chris Bystroff,* 2008

METHODS IN MOLECULAR BIOLOGY™

Cell Fusion

Overviews and Methods

Edited by

Elizabeth H. Chen

Department of Molecular Biology and Genetics, Johns Hopkins University School of Medicine, Baltimore, MD

Editor
Elizabeth H. Chen
Department of Molecular Biology and Genetics
Johns Hopkins University School of Medicine
Baltimore, MD

Series Editor
John M. Walker
School of Life Sciences
University of Hertfordshire
Hatfield, Hertfordshire AL10 9AB
UK

ISBN: 978-1-58829-911-6 e-ISBN: 978-1-59745-250-2
ISSN: 1064-3745 e-ISSN: 1940-6029
DOI: 10.1007/978-1-59745-250-2

Library of Congress Control Number: 2008921782

Cover Illustration: Fig. 2, chapter 22, "Methods to Fuse Macrophages In Vitro," by Agnès Vignery.

Printed on acid-free paper

9 8 7 6 5 4 3 2 1

springer.com

Preface

Cell fusion is a specialized cellular event that is critical for the conception, development and physiology of a multicellular organism. Known as a phenomenon for over a hundred years, cell fusion took center stage in the early analysis of gene expression, chromosomal mapping, monoclonal antibody production, and cancer therapy. It is only recently, however, that the molecular mechanisms of cell fusion have begun to come to light, thanks to the application of new technologies in genetics, cell biology, molecular biology, biochemistry, and genomics.

Exciting work in the past decade has revealed commonalities and differences among individual cell fusion events. The aim of *Cell Fusion: Overviews and Methods* is to bring together a collection of overviews that outline our current understanding of cell fusion and methods that present classic and state-of-the-art experimental approaches in a variety of systems. The first half of this volume consists of nine overviews that describe different cell fusion events from yeast to mammals. The second half consists of thirteen chapters illustrating commonly used methods to assay cell fusion in different systems.

The overall goal for this book is to serve as a comprehensive resource for anyone who is interested in this fascinating biological problem. It is intended for both newcomers and active researchers in the field to either acquire basic knowledge on cell fusion or to compare and contrast different cell fusion events. The user-friendly format of the method chapters should enable beginning students and experienced researchers to conduct assays in a variety of cell fusion systems.

The completion of *Cell Fusion: Overviews and Methods* would not have been possible without the enthusiastic support and outstanding contributions from all authors, to each of whom I owe a big and hearty thanks. My deep gratitude also goes to Chelsea Newhouse for her excellent assistance in proofreading all the chapters, communicating with the authors, and organizing the manuscripts. Finally, I would like to thank the series editor, Dr. John Walker, for his encouragement and guidance throughout the editing process.

Elizabeth H. Chen

Contents

Contributors

SUSAN M. ABMAYR • *The Stowers Institute for Medical Research, Kansas City, MO*
SCOTT ALPER • *Laboratory of Respiratory Biology, NIEHS, NIH, Department of Medicine, Duke University Medical Center, Durham, NC*
DORA BACZYK • *Womens and Infants Health, Samuel Lunenfeld Research Institute, Department of Obstetrics and Gynaecology, Mount Sinai Hospital, University of Toronto, Ontario, Canada*
MARY K. BAYLIES • *Program in Developmental Biology, Memorial Sloan Kettering Institute and Weill Graduate School at Cornell Medical School, New York, NY*
KAREN BECKETT • *Program in Developmental Biology, Memorial Sloan Kettering Institute, New York, NY*
MARCUS BORGES • *Department of Obstetrics, Paulista Medicine School, UNIFESP-Federal University of São Paulo, São Paulo, Brazil*
ELIZABETH H. CHEN • *Department of Molecular Biology and Genetics, Johns Hopkins University School of Medicine, Baltimore, MD*
SASCHA DREWLO • *Womens and Infants Health, Samuel Lunenfeld Research Institute, Department of Obstetrics and Gynaecology, Mount Sinai Hospital, University of Toronto, Ontario, Canada*
CAROLINE DUNK • *Womens and Infants Health, Samuel Lunenfeld Research Institute, Department of Obstetrics and Gynaecology, Mount Sinai Hospital, University of Toronto, Ontario, Canada*
STAR EMS • *Department of Genetics and Developmental Biology, University of Connecticut Health Center, Farmington, CT*
BEATRIZ ESTRADA • *Division of Genetics, Department of Medicine, Brigham and Women's Hospital and Harvard Medical School, Boston, MA*
CHARLES A. ETTENSOHN • *Department of Biological Sciences, Carnegie Mellon University, Pittsburgh, PA*
ANDRÉ FLEIßNER • *Department of Plant and Microbial Biology, The University of California, Berkeley, CA*
ALISON E. GAMMIE • *Department of Molecular Biology, Princeton University, Princeton, NJ*
ERIKA R. GEISBRECHT • *The Stowers Institute for Medical Research, Kansas City, MO*
N. LOUISE GLASS • *Department of Plant and Microbial Biology, The University of California, Berkeley, CA*

ERIC GROTE • *Department of Biochemistry and Molecular Biology, Johns Hopkins Bloomberg School of Public Health, Baltimore, MD*
PAUL G. HODOR • *Department of Biological Sciences, Carnegie Mellon University, Pittsburgh, PA*
BERTHOLD HUPPERTZ • *Institute of Cell Biology, Histology and Embryology, Center of Molecular Medicine, Medical University of Graz, Graz, Austria*
NAOKAZU INOUE • *Genome Information Research Center, Research Institute for Microbial Diseases, Osaka University, Osaka, Japan*
KATIE M. JANSEN • *Graduate Program in Biochemistry, Cell and Developmental Biology, Department of Pharmacology, Emory University, Atlanta, GA*
JOHN KINGDOM • *Womens and Infants Health, Samuel Lunenfeld Research Institute, Department of Obstetrics and Gynaecology, Mount Sinai Hospital, University of Toronto, Ontario, Canada*
EVELINE S. LITSCHER • *Department of Molecular, Cell and Developmental Biology, Mount Sinai School of Medicine, New York, NY*
ALAN M. MICHELSON • *Division of Genetics, Department of Medicine, Brigham and Women's Hospital and Harvard Medical School, Boston, MA*
WILLIAM A. MOHLER • *Department of Genetics and Developmental Biology, University of Connecticut Health Center, Farmington, CT*
MASARU OKABE • *Genome Information Research Center, Research Institute for Microbial Diseases, Osaka University, Osaka, Japan*
GRACE K. PAVLATH • *Department of Pharmacology, Emory University, Atlanta, GA*
BENJAMIN PODBILEWICZ • *Department of Biology, Technion-Israel Institute of Technology, Haifa, Israel*
BRIAN E. RICHARDSON, • *Program in Developmental Biology, Memorial Sloan Kettering Institute and Weill Graduate School at Cornell Medical School, New York, NY*
MARK D. ROSE • *Department of Molecular Biology, Princeton University, Princeton, NJ*
VICTORIA L. SCRANTON • *Department of Genetics and Developmental Biology, University of Connecticut Health Center, Farmington, CT*
JESSICA H. SHINN-THOMAS • *Department of Genetics and Developmental Biology, University of Connecticut Health Center, Farmington, CT*
ANNA R. SIMONIN • *Department of Plant and Microbial Biology, The University of California, Berkeley, CA*
AGNÈS VIGNERY • *Department of Orthopaedics, Yale University, New Haven, CT*
PAUL M. WASSARMAN • *Department of Molecular, Cell and Developmental Biology, Mount Sinai School of Medicine, New York, NY*

Nedra F. Wilson • *Department of Anatomy and Cell Biology, Oklahoma State University Center for Health Sciences, Tulsa, OK*
Casey A. Ydenberg • *Department of Molecular Biology, Princeton University, Princeton, NJ*
Shiliang Zhang • *Department of Molecular Biology and Genetics, The Johns Hopkins University School of Medicine, Baltimore, MD*
Shufei Zhuang • *The Stowers Institute for Medical Research, Kansas City, MO*

I

Cell Fusion

1

Yeast Mating

A Model System for Studying Cell and Nuclear Fusion

Casey A. Ydenberg and Mark D. Rose

Summary

Haploid yeast cells mate to form a zygote, whose progeny are diploid cells. A fundamentally sexual event, related to fertilization, yeast mating nevertheless exhibits cytological properties that appear similar to somatic cell fusion. A large collection of mutations that lead to defects in various stages of mating, including cell fusion, has allowed a detailed dissection of the overall pathway. Recent advances in imaging methods, together with powerful methods of genetic analysis, make yeast mating a superb platform for investigation of cell fusion. An understanding of yeast cell fusion will provide insight into fundamental mechanisms of cell signaling, cell polarization, and membrane fusion.

Key Words: Conjugation; mating; *Saccharomyces cerevisiae*; pheromone; cell polarity; karyogamy.

1. Introduction

Like other eukaryotes, the baker's/brewer's yeast *Saccharomyces cerevisiae* has a true sexual phase; cells of different mating types mate (conjugate) to form a diploid zygotic cell. The diploid cells can be propagated asexually, or in response to nutrient limitation, enter meiosis to form four haploid spores. Exposed to favorable conditions, the haploid spores germinate and reenter the mitotic cycle. With commonly used laboratory strains, haploid cells are stable and can be grown asexually indefinitely. When haploid cells of different mating types (called **a** and α) encounter each other, either because they were neighboring spores or were placed together in the laboratory, the two cells begin the complex process of conjugation. In wild strains of yeast, haploid cells are able to switch their mating types, allowing any cell to find a mate. Several detailed

From: *Methods in Molecular Biology, vol. 475: Cell Fusion: Overviews and Methods*
Edited by: E. H. Chen © Humana Press, Totowa, NJ

reviews of the process of conjugation have been published ***(1–5);*** below we describe the overall process, highlighting key areas.

1.1. Description of Yeast Mating

Haploid yeast cells detect their partner's presence by smell; each mating type secretes a specific peptide pheromone that is detected by a specific receptor in the membrane of the opposite mating type. Detection of the pheromone initiates a complex series of events, including arrest at the next G_1 phase of the cell cycle, a broad transcriptional response, and morphological change ***(1).*** Because yeast cells are not motile, they can only approach their partners by growing toward them ***(6)***. The pheromone gradient surrounding each cell thereby provides the critical directional cue that determines the axis of cell growth. The production of a projection, oriented toward the high end of the pheromone gradient, changes the cell's shape from oval to pear shape, often referred to as a *shmoo*. The new axis of cell growth overrides the preexisting haploid mitotic growth pattern, which specifies that new buds preferentially arise adjacent to the previous bud site ***(7,8)***. The dramatic reorganization of the cell growth axis in response to pheromone affects all of the major cellular pathways, including the actin cytoskeleton, the secretory pathway, the cytoplasmic microtubules, and nuclear migration.

Once in contact, pheromone-stimulated cells adhere to each other tightly, via the elaboration of mating-specific adhesion molecules (agglutinins; **refs. *9,10)***. Adhesion occurs at the tips of the shmoo projections (**Fig. 1A**), because these are the sites that are closest together after growth and because the shmoo tip will have the highest concentrations of the induced secreted agglutinins. Subsequent cell fusion requires the removal of the cell wall separating the mating partners (*see* **Fig. 1B,C**). As cells are normally under positive osmotic pressure, premature removal or removal at inappropriate sites would likely be lethal. Hence it is thought that tight association is important to help seal the cells together as they fuse.

After cell wall breakdown, the plasma membranes can come close enough to fuse, resulting in a mixed zygotic cytoplasm ***(11,12)***. The two haploid nuclei are positioned close to the zone of cell fusion, having migrated there via the action of cytoplasmic microtubules interacting with the cortex at the shmoo tip ***(3,13,14)***. After plasma membrane fusion, the nuclei move together, ultimately fusing to produce a diploid nucleus (*see* **Fig. 1C–E**). The diploid zygote downregulates genes involved in mating and reenters the mitotic pathway, producing diploid buds (*see* **Fig. 1F**). Diploids cells are refractory to mating and reproduce asexually until stimulated to enter meiosis.

Although formally a pathway of fertilization, the relationship of yeast cell fusion to mammalian fertilization is not clear. Aspects of morphology bear

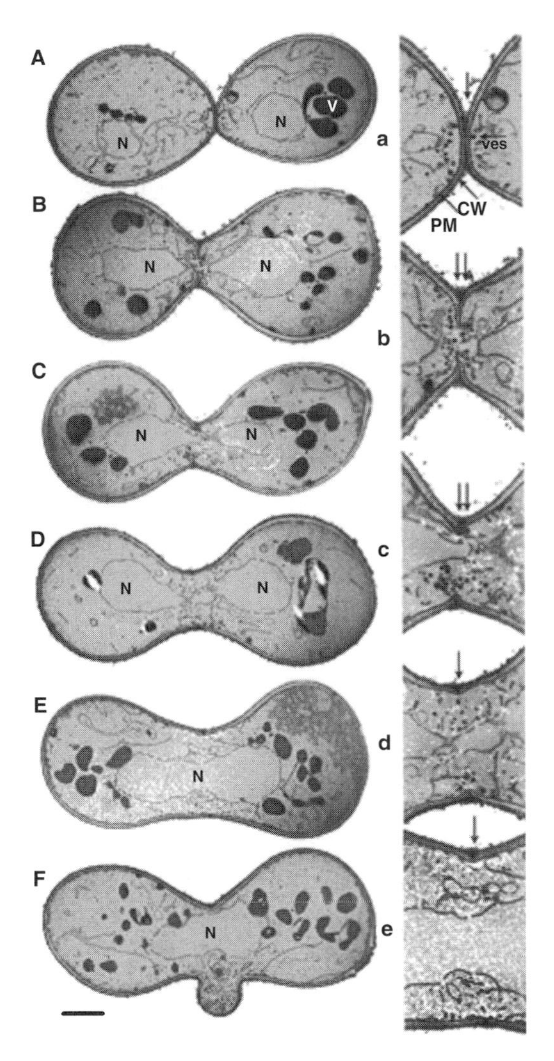

Fig. 1. Cell fusion and karyogamy in wild-type mating. **(A)** Prezygote. **(B)** Early zygote before karyogamy. **(C)** Early zygote during karyogamy. **(D)** Later zygote during karyogamy. **(E)** Mature unbudded zygote. **(F)** Mature budding zygote. **(a–e)** Twofold enlargements of respective zones of cell fusion in A–E. At least one example of each salient cellular structure is noted as follows: CW, cell wall; N, nucleus; PM, plasma membrane; V, vacuole; ves; vesicles. Arrows in a denote the zone of cell fusion. Double arrows in b and c point to regions containing cell wall remnants after fusion. Single arrows in d and e mark the cell fusion scar. Bar = 1 μm. (Reproduced with permission from **ref. *11*.**)

at least superficial resemblance to somatic cell fusion such as the formation of muscle fibers by myoblast fusion ***(15,16)***. However, clearly homologous pathways have yet to be described. Instead, the importance of yeast cell fusion lies with its utility as a model system for studying problems of broadly conserved basic cell biology. These include issues of cell–cell signaling, cell polarization, membrane fusion, and microtubule dynamics. With its sophisticated classic and molecular genetics, yeast provides an ideal system for performing studies on complex biological pathways. Furthermore, because yeasts cells are normally propagated asexually, a wide variety of mutations have been isolated that block mating at almost every step in the pathway, without necessitating the use of conditional mutations. Strengthening its utility, recent improvements in optical and electron microscopic methods have allowed a detailed description of the critical events.

1.2. Mutations Affecting Yeast Mating

Mutations that block mating have been isolated in a variety of selections and screens ***(17–20)***. Without delving into the details of their isolation, the mutants fall into four broad categories. Mutations that result in defects in pheromone signaling (either failure to make or to respond to the pheromone) result in profound defects in mating (up to 10^6-fold decrease in mating). Many of the affected genes are called *STE* genes for sterility. Mutations that interfere with cell fusion form a diverse group but generally arise from defects in cell polarization, cell signaling, cell wall removal, and plasma membrane fusion ***(2,5)***. The three genes first identified as being required for cell fusion were called *FUS*; however, because of the diversity of functions and identification in prior genetic screens, several other gene names have been used (e.g., *PRM1, RVS161, SPA2,* etc.).

Mutations that interfere with nuclear fusion (karyogamy, *kar*) are of two types. The first type affects cytoplasmic microtubules and interferes with nuclear congression. The second type affects components of the nuclear envelope and interferes with nuclear membrane fusion. For both the cell fusion and the nuclear fusion mutants, the severity of the defect is generally less than the sterile mutants, decreasing mating by as little as 50% to as much as 90%–95%. The requirements for different proteins are further confused by the existence of partial genetic redundancy between different pathways required for cell fusion, as well as the observation that for many mutations both parents must be defective to significantly compromise cell fusion.

1.3. Phenotypes of Cell and Nuclear Fusion Mutants

In wild-type zygotes (and all nuclear fusion mutant zygotes) the intervening cell wall is completely removed as the zone of cell fusion widens, leading to a smooth internal cell wall (*see* **Fig. 1D,E**; **ref.** ***11***) The only indication of the site of cell fusion is a small "scar" visible by electron microscopy (*see* **Fig. 1D,E**). Defects in cell fusion are broadly manifest as zygotes in which remnant cell wall material and/or plasma membrane continue to separate the two haploid cells. Some cell fusion mutants allow partial cell fusion, in which a portion of the intervening cell wall is removed, allowing plasma membrane fusion and even nuclear fusion. Zygotes that fail to accomplish any cytoplasmic mixing may reinitiate mating with other partners, leading to extended chains of cells of alternating mating type ***(21)***.

Like wild-type zygotes, mutants in which nuclear fusion has failed reenter the mitotic pathway but with two haploid nuclei instead of one. Typically, both haploid nuclei undergo mitosis, resulting in four nuclei within the zygote. Most frequently, one of the haploid nuclei will enter the first zygotic bud, resulting in a haploid bud with mixed cytoplasm from both parents (called a *cytoductant*), and a tri-nuclear zygote. If one of the parents contains a mitochondrial mutation, the cytoductants can be easily detected by their unique genetic constitution ***(22)***. Zygotes can undergo repeated rounds of mitosis, but there is no evidence that multiple nuclei within the zygote can fuse, consistent with the observation that certain functions required for nuclear fusion are induced only during mating ***(23)***.

2. Pheromone Signaling

Treatment of cells with mating pheromone induces a signal transduction cascade that results in a transcriptional response, cell cycle arrest in G_1, and polarization along the pheromone gradient. This subject has been extensively reviewed elsewhere ***(1,24–26)*** and will be only briefly outlined here, with areas of ongoing research interest highlighted. Haploid cells sense pheromone produced by cells of the opposite mating type by virtue of a seven-transmembrane G-protein-coupled receptor on the plasma membrane. These receptors are distinct in **a** and α cells; the rest of the pathway is shared between the two cell types. Activation of the receptor stimulates guanine nucleotide exchange on the α-subunit of a heterotrimeric G protein, which causes it to release the β- and γ-subunits. Free G-βγ can then recruit the scaffolding protein Ste5p and the mitogen-activated protein (MAP) kinase kinase kinase Ste11p to the plasma membrane, where Ste11p can be phosphorylated by the p21-activated kinase kinase Ste20p. Ste20p is active only in the presence of guanosine triphosphate (GTP)-bound Cdc42p, and this is important for creating cell polarity (*see* **Section 3**). Ste11p then phosphorylates the MAP kinase kinase Ste7p, which in turn activates the MAP kinase Fus3p.

In contrast to other members of the pathway, Fus3p is not absolutely required for response to pheromone. Subsets of its functions are shared with Kss1p, a related MAP kinase involved in the filamentation response to starvation ***(26)***. In particular, this includes the activation of gene expression by the transcription factor Ste12p ***(26)***. How the signaling specificity of the pheromone and filamentation responses is achieved is a subject of ongoing research ***(27)***. Part of the answer lies in the fact that Ste12p activates filamentation genes only when complexed with Tec1p ***(28)***, and Tec1p is degraded in response to phosphorylation by Fus3p ***(29,30)***. During the pheromone response, Ste12p activates genes required for agglutination (*see* **Section 4**), cell fusion (*see* **Sections 5** and **6**), nuclear congression and nuclear fusion (*see* **Section 7**). Genes required for later events in the pathway additionally require the transcription factor Kar4p, the gene for which is itself a Ste12p target ***(31)***. This is believed to create a temporal delay by which genes for later events in the pathway are expressed after genes for earlier events. Other targets of Ste12p include *SST2*, encoding a negative regulator of pheromone signaling ***(32)***. This creates a negative feedback loop by which cells that do not mate eventually recover from pheromone and resume normal growth.

In addition to the activation of Ste12p-dependent transcription, pheromone signaling leads to cell cycle arrest in the subsequent G_1 phase of the cell cycle. Cell-cycle arrest requires the phosphorylation of Far1p by Fus3p ***(33,34)***. Activated Far1p acts as an inhibitor of the G_1-cyclin/cyclin-dependent kinase complex, although how it carries out this function is as yet unclear ***(35)***. Although Kss1p can activate the transcriptional response to pheromone, it apparently cannot phosphorylate Far1p ***(27,36)***. Consequently, both *fus3* and *far1* mutants are defective for pheromone-induced cell-cycle arrest and activate the pheromone response in cells that are otherwise progressing through mitosis ***(33,37)***.

3. Cell Polarization

To conjugate, yeast cells must grow toward and contact their partner and subsequently remove their cell wall in a spatially restricted manner. The only known spatial information is the gradient of pheromone surrounding the mating partner. Hence the spatial information must be encoded by the increased level of receptor signaling occurring on the side of the cell facing the high end of the pheromone gradient ***(6,38,39)***. Two pathways are known that couple the receptor signaling to cell polarization. The first entails the recruitment of proteins required for bud emergence away from the incipient bud site. In mitosis, Far1p is nuclear, where it sequesters Cdc24p, a guanine-nucleotide exchange factor for Cdc42p. At the onset of G_1, Far1p is degraded, freeing Cdc24p for nuclear export and the activation of bud emergence by Cdc42p ***(40,41)***. During mating, as cells arrest in G_1, Far1p is phosphorylated by Fus3p, preventing its degradation. Phospho-Far1p then relocalizes from the nucleus together with Cdc24p, where it associates with free G-βγ at the cell cortex ***(8,40–42)***. The localization of Cdc24p presumably recruits Cdc42p and Bem1p away from the previously established incipient bud site. Mutants lacking Far1p form shmoo projections, but these are mislocalized to the site of bud emergence, indicating Far1p's role in orientation. Cells lacking both the marker for the incipient bud site *(bud1)* and Far1p-dependent orientation lose the ability to form shmoo projections. Nevertheless, such mutants still exhibit polarized growth in response to pheromone ***(43)***, reflecting the presence of a second underlying mechanism for cell polarization.

The second pathway of polarization acts through Fus3p-mediated activation of the actin-nucleating formin protein Bni1p ***(44)***. Fus3p phosphorylates Bni1p in vitro, and both the phosphorylation of Bni1p and its localization during mating are dependent on Fus3p. Like *bni1* mutants, *fus3* mutants become completely depolarized upon pheromone treatment. In addition, overproduction of Bni1p partially suppresses the Fus3p requirement for pheromone-induced polarization ***(44)***. Phosphorylated Fus3p binds to free G-α, and G-α mutants defective for binding Fus3p partially recapitulate the polarization defects of *fus3* mutants ***(45)***. Thus, in principle, the binding of phospho-Fus3p to free G-α provides a second spatial cue encoding the pheromone gradient, helping to activate actin assembly and cell growth toward the mating partner. However, the G-α/Fus3p interaction, by itself, is not sufficient to overcome the preexisting cell growth axis determined by the bud site, as shown by the mislocalization of the shmoo projection in *far1* mutants ***(8,42)***.

4. Adhesion/Agglutination

In response to mating pheromone, a family of glycophosphatidylinositol-anchored cell surface proteins called *agglutinins* are expressed and cause adhesion between the **a** and α cell types ***(9)***. *FIG2* and *AGA1* are expressed in **a** cells, whereas *FIG2*, *AGA1*, and *SAG1* are expressed in α cells. Deletion of individual agglutinins produces a mating defect only in liquid culture; however, synthetic deletion of all agglutinins expressed in a given cell type results in a strong mating defect on solid medium ***(46)***. The primary defect associated with these mutations is at the level of adhesion ***(9)***, although morphogenetic and cell fusion defects have also been reported for *fig2* ***(47,48)***.

5. Cell Wall Degradation

Yeast cells are surrounded by an ~200 nm thick cell wall that provides osmotic stability and protects the cell from physical damage (for reviews, *see* **refs.** ***49–51***). This structure consists primarily of two kinds of β-glucan polymer and an outer layer of mannoprotein. There is also a small amount of chitin (β-1,4-GlcNAc), which is concentrated subapically in shmoos. For cell fusion to occur, the cell wall must be degraded in a spatially and temporally restricted manner so as to prevent lysis. Electron micrographs of wild-type mating partners reveal a large number of dark-staining vesicles tightly clustered at the zone of cell fusion (*see* **Fig. 1A,B; ref.** ***11***) The contents of these vesicles have not been determined; however, it is tempting to speculate that they contain hydrolytic enzymes. Consistent with this view, a pair of closely related, secreted glucanases, *SCW4* and *SCW10*, produce a synthetic cell fusion defect when deleted in both partners of a mating pair ***(52)***.

Even prior to cell fusion, wild-type shmoos show reduced levels of β-1,3-glucan at the tip of the shmoo where cell fusion would occur ***(53)***. Not surprisingly, overproduction of β-1,3-glucan can inhibit cell fusion. For example, deletion of the GTPase-activating protein Lrg1p, a negative regulator of the Rho1p-GTPase, produces a cell fusion defect ***(53)***. Rho1p, in turn, promotes β-1,3-glucan synthesis ***(54–56)***. The *lrg1* mutants overproduce β-1,3-glucan and lack the spatial restriction at the shmoo tip ***(53)***. Similarly, *fus2* mutants show increased levels of β-1,3-glucan at the shmoo tip, which can be suppressed by overexpression of Lrg1p ***(53)***.

The phenotypes of several mutants defective for cell fusion suggest that failure in cell wall removal may underlie part of their defect. In these mutants, *spa2, fus2, fus1,* and *rvs161*, the majority of mating pairs fail to fuse and contain intact cell wall between the two partners ***(11)***. Fus1p, Fus2p, and Rvs161p are all induced by pheromone and localize to the shmoo tip ***(20,21,47,57,58)***. In the *spa2* and *fus1* mutants, vesicles either fail to cluster or are reduced in number, whereas in *rvs161* and *fus2* mutants the vesicles cluster normally ***(11)***. These observations, together with epistasis experiments, suggest that Spa2p and Fus1p act early in the pathway, whereas Fus2p and Rvs161p act later ***(11,58)***. This view is consistent with Spa2p playing a role in actin-dependent cell polarization (*see* below, this Section). However, Fus1p also negatively regulates the high-osmolarity glycerol response ***(59)***, activation of which is inhibitory to cell fusion ***(60)***. How these functions are related, whether Fus1p has additional biochemical roles, and what the functions of Rvs161p and Fus2p are, remain to be determined.

In contrast with the view that removal of cell wall material is critical for cell fusion, mutants affecting chitin synthesis also show reduced efficiency of mating ***(61–63)***. Loss of a putative chitin synthase catalytic subunit (Chs3p) results in a modest two- to threefold decrease in mating efficiency ***(61)***. Given its subapical location, it is unclear how loss of chitin would directly impact cell fusion. It is tempting to speculate that diminished mating may reflect the effects of a cell wall-related stress response. Mutation of a targeting protein for chitin synthase, Chs5p, does result in a profound cell fusion defect, which is most likely due to a defect in Fus1p localization ***(63)***.

Efficient cell polarization is also required for efficient cell wall degradation. *BNI1*, *SPA2*, and *PEA2* are all required for proper polarization of actin cables in response to pheromone; mutations in these genes result in a large number of unfused zygotes ***(64)***. Indeed, *spa2* mutant zygotes contain broad zones of cell fusion with unclustered vesicles ***(11)***, suggesting that, when polarization is compromised, the cell fusion machinery is not localized and is thus not effective. *CDC42* plays yet another role in mating, because *cdc42* mutant alleles have been recovered that block cell fusion in a small number of zygotes ***(65,66)***. These mutants fail to localize Spa2p and Fus1p to the shmoo tip. Polarization of the plasma membrane itself also occurs during mating, because Fus1p is associated with sterol and sphingolipid-rich detergent-resistant membranes, which appear to be enriched at the shmoo tip ***(67,68)***. In mutants compromised for sterol or sphingolipid synthesis, Fus1p fails to localize and mating efficiency is reduced. These studies highlight the importance of spatial restriction of cell fusion and suggest that several pathways overlap to ensure that the machinery is active only at the zone of cell fusion.

Finally, mutants that produce reduced levels of **a**-factor or α-factor also form unfused zygotes, but this effect is suppressed by mating to a strain sensitized to pheromone ***(20,69)***. This indicates that sensing high levels of pheromone is a prerequisite for carrying out cell wall degradation. Whether high levels of pheromone are also sufficient to activate the cell fusion machinery in the absence of a mating partner or whether there are additional cues transmitted by cells that have made contact remains unclear.

6. Plasma Membrane Fusion

Traditional genetic screens have failed to identify any proteins involved in the membrane fusion step of mating. With this in mind, Heiman and Walter ***(12)*** undertook an in silico approach to identify candidates. By mining available data, they looked for proteins that were induced by pheromone, and, by analogy with viral membrane fusion proteins and SNAREs, contained at least one predicted transmembrane domain. The strongest candidate that emerged, Prm1p, is induced by pheromone and localizes to the shmoo tip. Unlike SNAREs and some viral fusion proteins, Prm1p contains five predicted transmembrane domains. Approximately half of mating pairs in which both partners lack Prm1p fail to fuse. Unlike other cell fusion mutants, in the *prm1* mutants the cell wall is removed, and the two plasma membranes become closely juxtaposed.

Because half of the *prm1* mating pairs do fuse, it remains unclear whether Prm1p is indeed a membrane fusion protein that is partially redundant with other proteins that catalyze fusion or if it acts more indirectly, contributing to the efficiency of fusion. Further work has complicated the role of *PRM1* without resolving the issue. Deletion of *KEX2*, encoding a Golgi protease that processes plasma membrane-bound cargo, enhances the *prm1* defect ***(70)***. It seems likely that some of this cargo cooperates with Prm1p to carry out cell fusion, but no candidates have emerged. Deletion of *FIG1*, encoding a transmembrane protein required for Ca^{2+} influx during the pheromone response, also enhances the *prm1* defect, and this effect is independent of *KEX2* (P. Aguilar, personal communication; **refs.** ***71,72***). Interestingly, a subset of *prm1* mating pairs lyse after cell wall degradation ***(73)***, and the penetrance of this phenotype is dramatically increased by removing Ca^{2+} from the media ***(72)***. Because lysis occurs in both cells simultaneously and is independent of both cell wall integrity and programmed cell death ***(73)***, an attractive possibility is that it is the result of a fusion event gone awry. If this is the case, then *prm1* would be required for wild-type levels of membrane fusion, whereas *prm1* and Ca^{2+} would be jointly required to prevent lysis during fusion ***(72)***.

In contrast to *prm1* mutants, membrane fusion in wild-type cells is certainly initiated before cell wall breakdown is complete ***(11)***. Fusion appears to occur within a small region within the zone of cell contact. The fused membranes then form a pore that expands outward as the wall continues to be degraded. Because remodeling of the two organelles is coupled, it is possible that *FUS1, FUS2*, *RVS161,* and other genes may have additional roles in membrane fusion. Indeed, the expansion of the fusion pore is significantly delayed in *fus1* mating pairs ***(74)***.

7. Karyogamy: Nuclear Congression and Fusion

After cells have fused, the two haploid nuclei move together (congression) and fuse to form a single diploid nucleus (nuclear fusion; **refs.** ***2–4,20***). Mutations that block nuclear congression (*bik1, bim1, cik1*, *kar1, kar3, kar4, kar9, mps2, mps3,* and *tub2,* among others; **refs.** ***13,14,75–80***) generally affect the spindle pole body (SPB; the microtubule organizing center) or the cytoplasmic microtubules. Prior to cell fusion, the nuclei become oriented toward and migrate to the tip of the shmoo projection, dependent on movement generated by the cytoplasmic microtubules ***(14,81–83)***. To orient the nuclei, Bim1p (the yeast homolog of EB1) and Kar9p (a putative yeast homolog of APC1) at the tips of the cytoplasmic microtubules ***(84–86)*** interact with cortical determinants dependent on the actin cytoskeleton, including Myo2 and Bni1p ***(87,88)***. Force is generated on the microtubule, from the cortex, by the kinesin-like motor protein Kar3p, in concert with its associated light chain Cik1p ***(13,82,83)***. In addition, Bik1p, a CLIP-170 ortholog, plays a critical role in cortical microtubule attachment ***(83)*** and possibly also the loading of Kar9p on the microtubules ***(89)***.

Remarkably, during mating the region of the SPB responsible for nucleating the cytoplasmic microtubules changes from the central plaque to the membranous "half-bridge" ***(90)***. The shift is due to a pheromone-induced association between the γ-tubulin complex and Kar1p, an essential protein embedded in the half-bridge ***(91)***.

After cell fusion, the cytoplasmic microtubules become attached to each other at or near their plus ends ***(83)***. Cytoplasmic microtubule attachment and movement are dependent on Kar3p ***(13,83)***. Movement is clearly dependent on depolymerization of the microtubules, via Kar3p-dependent plus-end–specific depolymerization ***(83,92)***. One attractive model is that depolymerization is coupled to Kar3p's minus-end–directed movement, effectively removing the microtubules as the nuclei move together ***(92,93)***.

Ultimately the double nuclear envelopes fuse, catalyzed by proteins resident in the nuclear envelopes (Kar2p, Kar5p, Kar8/Jem1p, Prm3p, Sec72p, and others). Early electron microscopy suggested that nuclear envelope fusion was initiated at the SPB ***(90)***, but more recent work suggests that fusion initiates elsewhere and that SPB fusion occurs later in the pathway ***(94)***.

After nuclear fusion, the zygote usually reenters the mitotic pathway. The mechanism by which newly formed zygotes stop responding to pheromone, downregulate their existing response, and reactivate the cell cycle is not well understood. Recent work indicates that two pathways are important. First, the coexpression of the α cell–specific receptor for **a**-factor and Asg7p, a protein normally found only in **a** cells, leads to the rapid downregulation of the pheromone response, in part by internalization of Ste4p, the β-subunit of the trimeric G protein ***(95)***. Sst2p, a GTPase activating protein for the G-α-subunit, previously identified for its role in adaptation to chronic stimulation by pheromone, is also likely to be required ***(96)***.

8. Unanswered Questions

Successful yeast mating requires the complex interplay of multiple cell biological pathways, including cell polarization, cell–cell and intracellular signaling, microtubule dynamics, and plasma and nuclear membrane fusion. Their elucidation in this exquisite system will likely continue to provide insight into these conserved cell biological processes.

Although much is known about the pathway of cell fusion, large questions remain. For example, although the overall circuitry of the intracellular response to pheromone is relatively well understood, one of the major questions concerns the details of how the pheromone response incorporates the positional information from the gradient. Cells are able to reliably distinguish gradients in which the pheromone concentration on either side of the cell differs by as little as 1% ***(6)***. Presumably positive feedback pathways, integrated over time, allow this remarkable precision. The differential spatial activation of the receptors then leads to activation of at least two intracellular pathways regulating polarization. Whether these are the only pathways and how they are integrated remain important questions.

Even less well understood is how the response is insulated from and/or integrated with other regulatory pathways, some of which share components with the pheromone response. How the pheromone signaling leads to cell-cycle arrest is also not well understood, nor is how the cell reestablishes the mitotic cycle after mating is complete.

Although one candidate for the fusogen for plasma membrane fusion has been identified, it seems clear that other pathways and/or proteins must also play a significant role. The hunt for the remaining fusogens is a very active area of investigation. One clue is that it is likely that the fusogen is processed by the Kex2 protease during its transport to the plasma membrane ***(70)***.

Like other fertilization events, yeast cell mating shows remarkable fidelity. The frequency of matings in which more than two partners are genetic donors is less than 10^{-7} (our unpublished observation). How yeast cells prevent the equivalent of "polyspermy" is completely unknown. Furthermore, we are only beginning to appreciate the mechanisms by which the cells protect themselves from increased environmental vulnerability during mating. It seems likely that mating is carefully regulated, both positively and negatively, to ensure efficient mating while preventing lysis. It remains to be determined whether the temporal regulation of mating is as complex as that governing the cell cycle. However, the requirement that two cells coordinate their activities raises the possibility that significantly different biological mechanisms will be found.

References

1. Bardwell, L. (2005) A walk-through of the yeast mating pheromone response pathway. *Peptides* **26,** 339–350.
2. Marsh, L. and Rose, M. D. (1997) The pathway of cell and nuclear fusion during mating in *S. cerevisiae*, in *The Molecular and Cellular Biology of the Yeast Saccharomyces* (J. R. Pringle, J. R. Broach, and E. W. Jones, eds.), vol. 3. Cold Spring Harbor Laboratory Press, Cold Spring Harbor, NY, pp. 827–888.
3. Molk, J. N. and Bloom, K. (2006) Microtubule dynamics in the budding yeast mating pathway. *J. Cell Sci.* **119,** 3485–3490.
4. Rose, M. D. (1996) Nuclear fusion in the yeast *Saccharomyces cerevisiae. Annu. Rev. Cell Dev. Biol.* **12,** 663–695.
5. White, J. M. and Rose, M. D. (2001) Yeast mating: getting close to membrane merger. *Curr. Biol.* **11,** R16–R20.
6. Segall, J. E. (1993) Polarization of yeast cells in spatial gradients of alpha mating factor. *Proc. Natl. Acad. Sci. U.S.A.* **90,** 8332–8336.
7. Casamayor, A. and Snyder, M. (2002) Bud-site selection and cell polarity in budding yeast. *Curr. Opin. Microbiol.* **5,** 179–186.
8. Butty, A. C., Pryciak, P. M., Huang, L. S., Herskowitz, I., and Peter, M. (1998) The role of Far1p in linking the heterotrimeric G protein to polarity establishment proteins during yeast mating. *Science* **282,** 1511–1516.
9. Lipke, P. N. and Kurjan, J. (1992) Sexual agglutination in budding yeasts: structure, function, and regulation of adhesion glycoproteins. *Microbiol. Rev.* **56,** 180–194.

10. Zhao, H., Shen, Z. M., Kahn, P. C., and Lipke, P. N. (2001) Interaction of alpha-agglutinin and a-agglutinin, *Saccharomyces cerevisiae* sexual cell adhesion molecules. *J. Bacteriol.* **183,** 2874–2880.
11. Gammie, A. E. Brizzio, V., and Rose, M. D. (1998) Distinct morphological phenotypes of cell fusion mutants. *Mol. Biol. Cell* **9,** 1395–1410.
12. Heiman, M. G. and Walter, P. (2000) Prm1p, a pheromone-regulated multispanning membrane protein, facilitates plasma membrane fusion during yeast mating. *J. Cell Biol.* **151,** 719–730.
13. Meluh, P. B. and Rose, M. D. (1990) KAR3, a kinesin-related gene required for yeast nuclear fusion. *Cell* **60,** 1029–1041.
14. Miller, R. K. and Rose, M. D. (1998) Kar9p is a novel cortical protein required for cytoplasmic microtubule orientation in yeast. *J. Cell Biol.* **140,** 377–390.
15. Chen, E. H. and Olson, E. N. (2004) Towards a molecular pathway for myoblast fusion in *Drosophila. Trends Cell Biol.* **14,** 452–460.
16. Doberstein, S. K., Fetter, R. D., Mehta, A. Y., and Goodman, C. S. (1997) Genetic analysis of myoblast fusion: blown fuse is required for progression beyond the prefusion complex. *J. Cell Biol.* **136,** 1249–1261.
17. Mackay, V. and Manney, T. R. (1974) Mutations affecting sexual conjugation and related processes in *Saccharomyces cerevisiae.* I. Isolation and phenotypic characterization of nonmating mutants. *Genetics* **76,** 255–271.
18. Wilson, K. L. and Herskowitz, I. (1987) STE16, a new gene required for pheromone production by a cell of *Saccharomyces cerevisiae. Genetics* **115,** 441–449.
19. Berlin, V., Brill, J. A., Truchcart, J., Boeke, J. D., and Fink, G. R. (1991) Genetic screens and selections for cell and nuclear fusion mutants. *Methods Enzymol.* **194,** 774–792.
20. Kurihara, L. J., Beh, C. T., Latterich, M., Schekman, R., and Rose, M. D. (1994) Nuclear congression and membrane fusion: two distinct events in the yeast karyogamy pathway. *J. Cell Biol.* **126,** 911–923.
21. Trueheart, J., Boeke, J. D., and Fink, G. R. (1987) Two genes required for cell fusion during yeast conjugation: evidence for a pheromone-induced surface protein. *Mol. Cell. Biol.* **7,** 2316–2328.
22. Conde, J. and Fink, G. R. (1976) A mutant of *Saccharomyces cerevisiae* defective for nuclear fusion. *Proc. Natl. Acad. Sci. U.S.A.* **73,** 3651–3655.
23. Rose, M. D., Price, B. R., and Fink, G. R. (1986) *Saccharomyces cerevisiae* nuclear fusion requires prior activation by alpha factor. *Mol. Cell. Biol.* **6,** 3490–3497.
24. Elion, E. A. (2000) Pheromone response, mating and cell biology. *Curr. Opin. Microbiol.* **3,** 573–581.
25. Naider, F. and Becker, J. M. (2004) The alpha-factor mating pheromone of *Saccharomyces cerevisiae*: a model for studying the interaction of peptide hormones and G protein–coupled receptors. *Peptides* **25,** 1441–1463.
26. Gustin, M. C., Albertyn, J., Alexander, M., and Davenport, K. (1998) MAP kinase pathways in the yeast *Saccharomyces cerevisiae. Microbiol. Mol. Biol. Rev.* **62,** 1264–1300.
27. Breitkreutz, A., Boucher, L., and Tyers, M. (2001) MAPK specificity in the yeast pheromone response independent of transcriptional activation. *Curr. Biol.* **11,** 1266–1271.

28. Madhani, H. D. and Fink, G. R. (1997) Combinatorial control required for the specificity of yeast MAPK signaling. *Science* **275,** 1314–1317.
29. Bao, M. Z., Schwartz, M. A., Cantin, G. T., Yates, J. R., 3rd, and Madhani, H. D. (2004) Pheromone-dependent destruction of the Tec1 transcription factor is required for MAP kinase signaling specificity in yeast. *Cell* **119,** 991–1000.
30. Chou, S., Huang, L., and Liu, H. (2004) Fus3-regulated Tec1 degradation through SCF^{Cdc4} determines MAPK signaling specificity during mating in yeast. *Cell* **119,** 981–990.
31. Lahav, R., Gammie, A., Tavazoie, S., and Rose, M. D. (2007) Role of transcription factor Kar4 in regulating downstream events in the *Saccharomyces cerevisiae* pheromone response pathway. *Mol. Cell. Biol.* **27,** 818–829.
32. Dohlman, H. G., Song, J., Ma, D., Courchesne, W. E., and Thorner, J. (1996) Sst2, a negative regulator of pheromone signaling in the yeast *Saccharomyces cerevisiae:* expression, localization, and genetic interaction and physical association with Gpa1 (the G-protein alpha subunit). *Mol. Cell. Biol.* **16,** 5194–5209.
33. Chang, F. and Herskowitz, I. (1990) Identification of a gene necessary for cell cycle arrest by a negative growth factor of yeast: FAR1 is an inhibitor of a G1 cyclin, CLN2. *Cell* **63,** 999–1011.
34. Peter, M., Gartner, A., Horecka, J., Ammerer, G., and Herskowitz, I. (1993) FAR1 links the signal transduction pathway to the cell cycle machinery in yeast. *Cell* **73,** 747–760.
35. Gartner, A., Jovanovic, A., Jeoung, D. I., Bourlat, S., Cross, F. R., and Ammerer, G. (1998) Pheromone-dependent G1 cell cycle arrest requires Far1 phosphorylation, but may not involve inhibition of Cdc28-Cln2 kinase, in vivo. *Mol. Cell. Biol.* **18,** 3681–3691.
36. Elion, E. A., Brill, J. A., and Fink, G. R. (1991) FUS3 represses CLN1 and CLN2 and in concert with KSS1 promotes signal transduction. *Proc. Natl. Acad. Sci. U.S.A.* **88,** 9392–9396.
37. Elion, E. A., Grisafi, P. L., and Fink, G. R. (1990) FUS3 encodes a cdc2+/CDC28-related kinase required for the transition from mitosis into conjugation. *Cell* **60,** 649–664.
38. Barkai, N., Rose, M. D., and Wingreen, N. S. (1998) Protease helps yeast find mating partners. *Nature* **396,** 422–423.
39. Vallier, L. G., Segall, J. E., and Snyder, M. (2002) The alpha-factor receptor C-terminus is important for mating projection formation and orientation in *Saccharomyces cerevisiae. Cell Motil. Cytoskeleton* **53,** 251–266.
40. Nern, A. and Arkowitz, R. A. (2000) Nucleocytoplasmic shuttling of the Cdc42p exchange factor Cdc24p. *J. Cell Biol.* **148,** 1115–1122.
41. Shimada, Y., Gulli, M. P., and Peter, M. (2000) Nuclear sequestration of the exchange factor Cdc24 by Far1 regulates cell polarity during yeast mating. *Nat. Cell Biol.* **2,** 117–124.
42. Nern, A. and Arkowitz, R. A. (1999) A Cdc24p-Far1p-Gbetagamma protein complex required for yeast orientation during mating. *J. Cell Biol.* **144,** 1187–1202.
43. Nern, A. and Arkowitz, R. A. (2000) G proteins mediate changes in cell shape by stabilizing the axis of polarity. *Mol. Cell* **5,** 853–864.

44. Matheos, D., Metodiev, M., Muller, E., Stone, D., and Rose, M. D. (2004) Pheromone-induced polarization is dependent on the Fus3p MAPK acting through the formin Bni1p. *J. Cell Biol.* **165,** 99–109.
45. Metodiev, M. V., Matheos, D., Rose, M. D., and Stone, D. E. (2002) Regulation of MAPK function by direct interaction with the mating-specific Galpha in yeast. *Science* **296,** 1483–1486.
46. Guo, B., Styles, C. A., Feng, Q., and Fink, G. R. (2000) A *Saccharomyces* gene family involved in invasive growth, cell-cell adhesion, and mating. *Proc. Natl. Acad. Sci. U.S.A.* **97,** 12158–12163.
47. Erdman, S., Lin, L., Malczynski, M., and Snyder, M. (1998) Pheromone-regulated genes required for yeast mating differentiation. *J. Cell Biol.* **140,** 461–483.
48. Zhang, M., Bennett, D., and Erdman, S. E. (2002) Maintenance of mating cell integrity requires the adhesin Fig2p. *Eukaryot. Cell* **1,** 811–822.
49. Cid, V. J., Duran, A., del Rey, F., Snyder, M. P., Nombela, C., and Sanchez, M. (1995) Molecular basis of cell integrity and morphogenesis in *Saccharomyces cerevisiae. Microbiol. Rev.* **59,** 345–386.
50. Lesage, G. and Bussey, H. (2006) Cell wall assembly in *Saccharomyces cerevisiae. Microbiol. Mol. Biol. Rev.* **70,** 317–343.
51. Orlean, P. (1997) Biogenesis of yeast wall and surface components, in *The Molecular and Cellular Biology of the Yeast Saccharomyces* (J. R. Pringle, J. R. Broach, and E. W. Jones, eds.), vol. 3. Cold Spring Harbor Laboratory Press, Cold Spring Harbor, NY. pp. 229–362.
52. Cappellaro, C., Mrsa, V., and Tanner, W. (1998) New potential cell wall glucanases of *Saccharomyces cerevisiae* and their involvement in mating. *J. Bacteriol.* **180,** 5030–5037.
53. Fitch, P. G., Gammie, A. E., Lee, D. J., de Candal, V. B., and Rose, M. D. (2004) Lrg1p Is a Rho1 GTPase-activating protein required for efficient cell fusion in yeast? *Genetics* **168,** 733–746.
54. Drgonova, J., Drgon, T., Tanaka, K., Kollar, R., Chen, G. C., Ford, R. A., Chan, C. S., Takai, Y., and Cabib, E. (1996) Rho1p, a yeast protein at the interface between cell polarization and morphogenesis. *Science* **272,** 277–279.
55. Qadota, H., Python, C. P., Inoue, S. B., Arisawa, M., Anraku, Y., Zheng, Y., Watanabe, T., Levin, D. E., and Ohya, Y. (1996) Identification of yeast Rho1p GTPase as a regulatory subunit of 1,3-beta-glucan synthase. *Science* **272,** 279–281.
56. Watanabe, D., Abe, M., and Ohya, Y. (2001) Yeast Lrg1p acts as a specialized RhoGAP regulating 1,3-beta-glucan synthesis. *Yeast* **18,** 943–951.
57. McCaffrey, G., Clay, F. J., Kelsay, K., and Sprague, G. F. Jr. (1987) Identification and regulation of a gene required for cell fusion during mating of the yeast *Saccharomyces cerevisiae. Mol. Cell. Biol.* **7,** 2680–2690.
58. Brizzio, V., Gammie, A. E., and Rose, M. D. (1998) Rvs161p interacts with Fus2p to promote cell fusion in *Saccharomyces cerevisiae. J. Cell Biol.* **141,** 567–584.
59. Nelson, B., Parsons, A. B., Evangelista, M., Schaefer, K., Kennedy, K., Ritchie, S., Petryshen, T. L., and Boone, C. (2004) Fus1p interacts with components of the Hog1p mitogen-activated protein kinase and Cdc42p morphogenesis signaling pathways to control cell fusion during yeast mating. *Genetics* **166,** 67–77.

60. Philips, J. and Herskowitz, I. (1997) Osmotic balance regulates cell fusion during mating in *Saccharomyces cerevisiae. J. Cell Biol.* **138,** 961–974.
61. Santos, B., Duran, A., and Valdivieso, M. H. (1997) *CHS5*, a gene involved in chitin synthesis and mating in *Saccharomyces cerevisiae. Mol. Cell. Biol.* **17,** 2485–2496.
62. Santos, B. and Snyder, M. (1997) Targeting of chitin synthase 3 to polarized growth sites in yeast requires Chs5p and Myo2p. *J. Cell Biol.* **136,** 95–110.
63. Santos, B. and Snyder, M. (2003) Specific protein targeting during cell differentiation: polarized localization of Fus1p during mating depends on Chs5p in *Saccharomyces cerevisiae. Eukaryot. Cell* **2,** 821–825.
64. Dorer, R., Boone, C., Kimbrough, T., Kim, J., and Hartwell, L. H. (1997) Genetic analysis of default mating behavior in *Saccharomyces cerevisiae. Genetics* **146,** 39–55.
65. Barale, S., McCusker, D., and Arkowitz, R. A. (2004) The exchange factor Cdc24 is required for cell fusion during yeast mating. *Eukaryot. Cell* **3,** 1049–1061.
66. Barale, S., McCusker, D., and Arkowitz, R. A. (2006) Cdc42p GDP/GTP cycling is necessary for efficient cell fusion during yeast mating. *Mol. Biol. Cell* **17,** 2824–2838.
67. Bagnat, M. and Simons, K. (2002) Cell surface polarization during yeast mating. *Proc. Natl. Acad. Sci. U.S.A.* **99,** 14183–14188.
68. Proszynski, T. J., Klemm, R., Bagnat, M., Gaus, K., and Simons, K. (2006) Plasma membrane polarization during mating in yeast cells. *J. Cell Biol.* **173,** 861–866.
69. Brizzio, V., Gammie, A. E., Nijbroek, G., Michaelis, S., and Rose, M. D. (1996) Cell fusion during yeast mating requires high levels of a-factor mating pheromone. *J. Cell Biol.* **135,** 1727–1739.
70. Heiman, M. G., Engel, A., and Walter, P. (2007) The Golgi-resident protease Kex2 acts in conjunction with Prm1 to facilitate cell fusion during yeast mating. *J. Cell Biol.* **176,** 209–222.
71. Muller, E. M., Mackin, N. A., Erdman, S. E., and Cunningham, K. W. (2003) Fig1p facilitates Ca^{2+} influx and cell fusion during mating of *Saccharomyces cerevisiae. J. Biol. Chem.* **278,** 38461–38469.
72. Aguilar, P. S., Engel, A., and Walter, P. (2007) The plasma membrane proteins Prm1 and Fig1 ascertain fidelity of membrane fusion during yeast mating. *Mol. Biol. Cell.* **18,** 547–556.
73. Jin, H., Carlile, C., Nolan, S., and Grote, E. (2004) Prm1 prevents contact-dependent lysis of yeast mating pairs. *Eukaryot. Cell* **3,** 1664–1673.
74. Nolan, S., Cowan, A. E., Koppel, D. E., Jin, H., and Grote, E. (2006) FUS1 regulates the opening and expansion of fusion pores between mating yeast. *Mol. Biol. Cell* **17,** 2439–2450.
75. Berlin, V., Styles, C. A., and Fink, G. R. (1990) BIK1, a protein required for microtubule function during mating and mitosis in *Saccharomyces cerevisiae*, colocalizes with tubulin. *J. Cell Biol.* **111,** 2573–2586.
76. Huffaker, T. C., Thomas, J. H., and Botstein, D. (1988) Diverse effects of beta-tubulin mutations on microtubule formation and function. *J. Cell Biol.* **106,** 1997–2010.

77. Page, B. D., and Snyder, M. (1992) CIK1: a developmentally regulated spindle pole body-associated protein important for microtubule functions in *Saccharomyces cerevisiae*. *Genes Dev.* **6,** 1414–1429.
78. Schwartz, K., Richards, K., and Botstein, D. (1997) BIM1 encodes a microtubule-binding protein in yeast. *Mol. Biol. Cell* **8,** 2677–2691.
79. Kurihara, L. J., Stewart, B. G., Gammie, A. E., and Rose, M. D. (1996) Kar4p, a karyogamy-specific component of the yeast pheromone response pathway. *Mol. Cell. Biol.* **16,** 3990–4002.
80. Vallen, E. A., Hiller, M. A., Scherson, T. Y., and Rose, M. D. (1992) Separate domains of KAR1 mediate distinct functions in mitosis and nuclear fusion. *J. Cell Biol.* **117,** 1277–1287.
81. Maddox, P., Chin, E., Mallavarapu, A., Yeh, E., Salmon, E. D., and Bloom, K. (1999) Microtubule dynamics from mating through the first zygotic division in the budding yeast *Saccharomyces cerevisiae*. *J. Cell Biol.* **144,** 977–987.
82. Maddox, P. S., Stemple, J. K., Satterwhite, L., Salmon, E. D., and Bloom, K. (2003) The minus end–directed motor Kar3 is required for coupling dynamic microtubule plus ends to the cortical shmoo tip in budding yeast. *Curr. Biol.* **13,** 1423–1428.
83. Molk, J. N., Salmon, E. D., and Bloom, K. (2006) Nuclear congression is driven by cytoplasmic microtubule plus end interactions in *S. cerevisiae*. *J. Cell Biol.* **172,** 27–39.
84. Korinek, W. S., Copeland, M. J., Chaudhuri, A., and Chant, J. (2000) Molecular linkage underlying microtubule orientation toward cortical sites in yeast. *Science* **287,** 2257–2259.
85. Lee, L., Tirnauer, J. S., Li, J., Schuyler, S. C., Liu, J. Y., and Pellman, D. (2000) Positioning of the mitotic spindle by a cortical-microtubule capture mechanism. *Science* **287,** 2260–2262.
86. Miller, R. K., Cheng, S. C., and Rose, M. D. (2000) Bim1p/Yeb1p mediates the Kar9p-dependent cortical attachment of cytoplasmic microtubules. *Mol. Biol. Cell* **11,** 2949–2959.
87. Hwang, E., Kusch, J., Barral, Y., and Huffaker, T. C. (2003) Spindle orientation in *Saccharomyces cerevisiae* depends on the transport of microtubule ends along polarized actin cables. *J. Cell Biol.* **161,** 483–488.
88. Miller, R. K., Matheos, D., and Rose, M. D. (1999) The cortical localization of the microtubule orientation protein, Kar9p, is dependent upon actin and proteins required for polarization. *J. Cell Biol.* **144,** 963–975.
89. Moore, J. K., D'Silva, S., and Miller, R. K. (2006) The CLIP-170 homologue Bik1p promotes the phosphorylation and asymmetric localization of Kar9p. *Mol. Biol. Cell* **17,** 178–191.
90. Byers, B. and Goetsch, L. (1975) Behavior of spindles and spindle plaques in the cell cycle and conjugation of *Saccharomyces cerevisiae*. *J. Bacteriol.* **124,** 511–523.
91. Pereira, G., Grueneberg, U., Knop, M., and Schiebel, E. (1999) Interaction of the yeast gamma-tubulin complex-binding protein Spc72p with Kar1p is essential for microtubule function during karyogamy. *EMBO J.* **18,** 4180–4195.

92. Sproul, L. R., Anderson, D. J., Mackey, A. T., Saunders, W. S., and Gilbert, S. P. (2005) Cik1 targets the minus-end kinesin depolymerase kar3 to microtubule plus ends. *Curr. Biol.* **15,** 1420–1427.
93. Endow, S. A., Kang, S. J., Satterwhite, L. L., Rose, M. D., Skeen, V. P., and Salmon, E. D. (1994) Yeast Kar3 is a minus-end microtubule motor protein that destabilizes microtubules preferentially at the minus ends. *EMBO J.* **13,** 2708–2713.
94. Melloy, P., Shen, S., White, E., McIntosh, J. R., and Rose, M. D. (2007) Nuclear fusion during yeast mating occurs by a three-step pathway. *J. Cell. Biol.* **179,** 695–670.
95. Kim, J., Bortz, E., Zhong, H., Leeuw, T., Leberer, E., Vershon, A. K., and Hirsch, J. P. (2000) Localization and signaling of G(beta) subunit Ste4p are controlled by a-factor receptor and the a-specific protein Asg7p. *Mol. Cell. Biol.* **20,** 8826–8835.
96. Rivers, D. M. and Sprague, G. F. Jr. (2003) Autocrine activation of the pheromone response pathway in matalpha2-cells is attenuated by SST2- and ASG7-dependent mechanisms. *Mol. Genet. Genomics* **270,** 225–233.

2

Cell Fusion in the Filamentous Fungus, *Neurospora crassa*

André Fleißner, Anna R. Simonin, and N. Louise Glass

Summary

Hyphal fusion occurs at different stages in the vegetative and sexual life cycle of filamentous fungi. Similar to cell fusion in other organisms, the process of hyphal fusion requires cell recognition, adhesion, and membrane merger. Analysis of the hyphal fusion process in the model organism *Neurospora crassa* using fluorescence and live cell imaging as well as cell and molecular biological techniques has begun to reveal its complex cellular regulation. Several genes required for hyphal fusion have been identified in recent years. While some of these genes are conserved in other eukaryotic species, other genes encode fungal-specific proteins. Analysis of fusion mutants in *N. crassa* has revealed that genes previously identified as having nonfusion-related functions in other systems have novel hyphal fusion functions in *N. crassa*. Understanding the molecular basis of cell fusion in filamentous fungi provides a paradigm for cell communication and fusion in eukaryotic organisms. Furthermore, the physiological and developmental roles of hyphal fusion are not understood in these organisms; identifying these mechanisms will provide insight into environmental adaptation.

Key Words: Cell fusion; anastomosis; filamentous fungi; signal transduction; hyphal fusion.

1. Introduction

Filamentous ascomycete fungi, such as *Neurospora crassa,* typically form mycelial colonies consisting of a network of interconnected, multinucleate hyphae. Colonies grow by hyphal tip extension, branching, and fusion ***(1,2)***. In filamentous ascomycete species, hyphal cross-walls or septa are incomplete and contain a single central pore. Septal pores allow cytoplasm and organelles, including nuclei, to move between hyphal compartments, thus making the fungal colony a syncytium. The syncytial, interconnected, organization of a fungal colony enables translocation of cellular contents, such as organelles, metabolites,

From: *Methods in Molecular Biology, vol. 475: Cell Fusion: Overviews and Methods*
Edited by: E. H. Chen © Humana Press, Totowa, NJ

nutrients, or signaling compounds, throughout the colony, presumably facilitating growth and reproduction.

Cell fusion events occur during all stages of the filamentous fungal life cycle *(3)*. These fusion events serve different purposes during the establishment and development of fungal colonies. During vegetative growth, germling fusion events between germinating, and even apparently ungerminated, asexual spores (conidia) are correlated with faster colony establishment (**Fig. 1A; refs.** ***4,5***). Fusion between hyphal branches within a mature fungal colony results in the formation of a network of interlinked hyphae (*see* **Fig. 1B; refs.** ***1,6***). Germling or hyphal fusion between genetically different but heterokaryon-compatible individuals leads to the formation of colonies containing genetically different nuclei (heterokaryon). Within heterokaryons, nonmeiotic or parasexual recombination can result in the formation of new genotypes *(7)*, which possibly contribute to the high adaptability of fungal species that lack sexual reproduction. In the sexual phase of the life cycle, cell fusion between male and female reproductive structures is essential for mating in out-breeding species (*see* **Fig. 1C; ref.** ***8***). After mating, cell fusion is associated with ascus formation (*see*

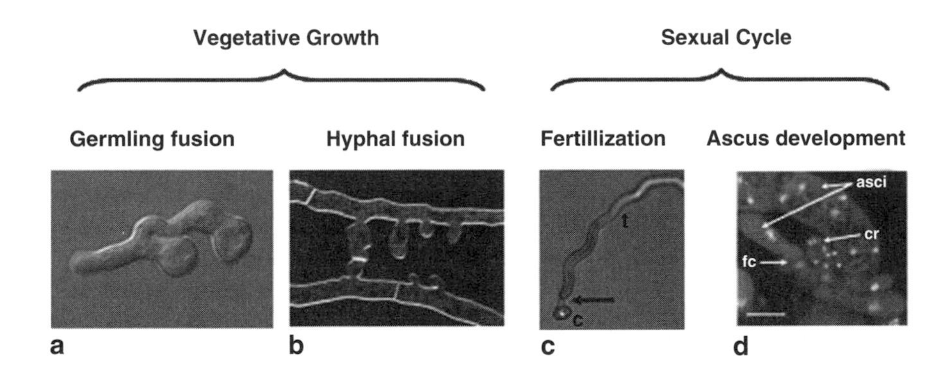

Fig. 1. Stages in the life cycle of *Neurospora crassa* in which fusion occurs. **(A)** Conidia at sufficient cell density undergo fusion between germlings. **(B)** Hyphae within the interior of the colony show chemotropism and hyphal fusion. **(C)** The sexual cycle is initiated by cell fusion between a fertile receptive hyphae (trichogyne [t]) emanating from a female reproductive structure, the protoperithecia (out of view). The trichogyne shows chemotropism toward a conidium of the opposite mating type (c). Arrow indicates fusion point. **(D)** Following fertilization, nuclei of opposite mating type (*mat A* and *mat a*) proliferate in ascogenous hyphae. Opposite mating-type nuclei pair off and migrate into the crozier (cr). In *N. crassa*, karyogamy occurs in the penultimate cell of the crozier. Hyphal and nuclear fusion occurs between the terminal cell and the subtending cell of the crozier (fc). Karyogamy, meiosis, and an additional mitotic division occurs in the ascus resulting in an eight-spored ascus. Asci, ascogenous hyphae and croziers treated with DAPI, a nuclear stain. Bar = 20 μm.

Fig. 1D), the cell in which karyogamy and meiosis occur *(9)*. Whether common cell fusion machinery is involved in both sexual and vegetative fusion events in filamentous fungi remains a question.

Fusion processes in filamentous fungi are comparable to cell fusion events in other eukaryotic organisms. Examples include fertilization events between egg and sperm or somatic cell fusion that result in syncytia (e.g., between myoblasts during muscle differentiation, between macrophages in osteoclast and giant cell formation, and during placental development; **refs.** *10–14*). Although cell fusion events occur in a diversity of species and cell types, they require very similar cellular processes, such as cell recognition, adhesion, and membrane merger. Although in many cases cell types involved in fusion are genetically or physiologically different, such as cell fusion during mating in *N. crassa*, vegetative hyphal fusion occurs between genetically and probably physiologically identical cells. Understanding the molecular basis of hyphal fusion provides a paradigm for self-signaling in eukaryotic cells and provides a useful comparative model for somatic cell fusion events in other eukaryotes. The model organism *N. crassa* is methodically tractable *(15–17)*, thus allowing the direct comparison of the molecular basis of hyphal and cell fusion events during its life cycle.

2. Vegetative Cell Fusion

2.1. Germling Fusion

The life of a fungal individual often begins with the germination of an asexual spore, termed *conidium*. When multiple conidia are placed close to one another, numerous germling fusion events are observed *(5,18)*. As a result, numerous individual germlings become one functional unit, which subsequently develops into a mycelial colony. Germinating conidia can fuse by germ tube fusion (*see* **Fig. 1A**) or by the formation of small hyphal bridges (*fusionshyphen* or *conidial anastomosis tubes*), which are significantly narrower than germ tubes *(4,5,19)*. Fusion events among conidia show a density- and nutrient-dependent function; fusion is suppressed on nutrient-rich media. The merger of initially individual cells into functional units in response to environmental cues is found not only in fungi but also in other species such as the social amoeba of dictyostelid slime molds *(20)*.

2.2. Hyphal Fusion

After germlings create a fused hyphal network, hyphal exploration extends outward from the conidia, thus taking on the morphological aspects of a typical fungal colony *(1,2)*. In *N. crassa* and other filamentous ascomycete species, the frequency of hyphal fusion within a vegetative colony varies from the periphery to the interior of the colony *(21)*. At the periphery, hyphae grow straight out from the colony and exhibit avoidance (negative autotropism), presumably to maximize the outward growth of the colony *(22)*. In the inner portion of a colony, hyphae show a different behavior. Instead of avoidance, certain hyphae or hyphal branches show attraction, directed growth, and hyphal fusion (**Fig. 2; refs.** ***1,21,23***). Similar to germling fusion, the frequency of hyphal fusion events depends on the availability of nutrients. Generally speaking the following rule applies: the fewer the nutrients, the more the fusion events ***(4,24,25)***. For example,

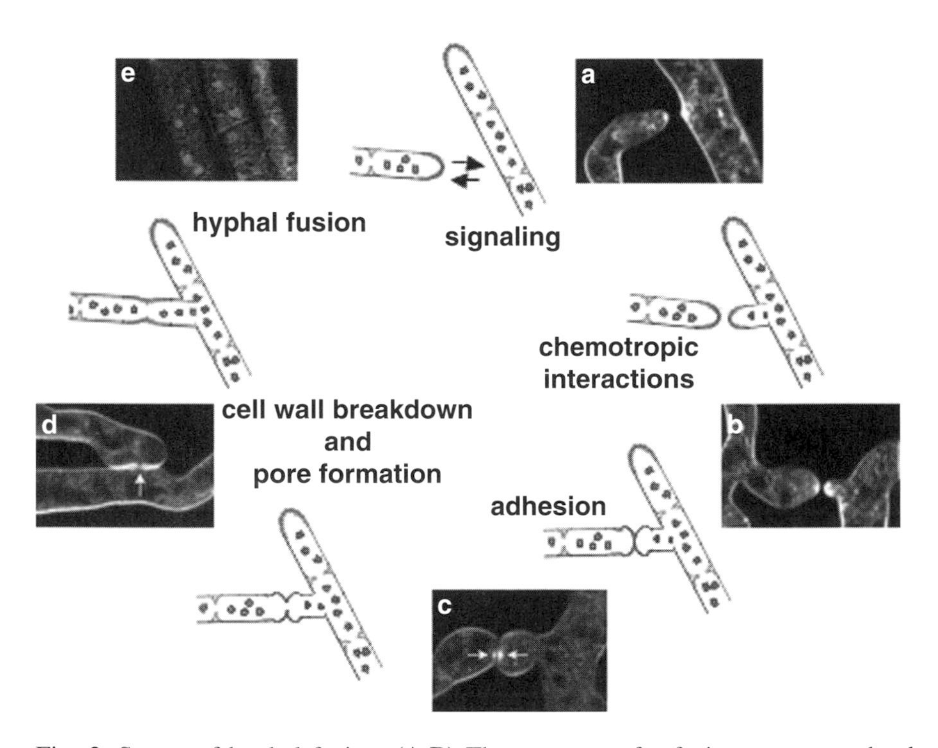

Fig. 2. Stages of hyphal fusion. **(A,B)** The presence of a fusion-competent hypha often results in the formation of a peg in the receptive hypha. Peg formation is associated with the formation of a Spitzenkörper at the tip of the new peg. **(C)** Contact between fusion hyphae is associated with a switch from polar to nonpolar growth, resulting in a swelling of the fusion hyphae at the point of contact. The Spitzenkörper is associated with the site of the future pore in both fusion hyphae (arrows). **(D)** Pore formation (arrow) is associated with cytoplasmic flow. Organelles such as nuclei and mitochondria pass through the fusion pore. Septation is also often associated with hyphal fusion events. **(E)** Fusion results in cytoplasmic mixing. Fusion between one hypha labeled with cytoplasmic GFP and one carrying dsRED-labeled nuclei result in hyphae exhibiting red nuclei in green cytoplasm. Hyphae stained with FM4-64. Bar = 10 µm. (A–D, adapted from **ref.** ***21.***)

addition of nitrogen to nutrient-poor media led to the largest decrease in hyphal fusion frequency in *Rhizoctonia solani* ***(25)***.

2.3. Mechanistic Aspects of Germling and Hyphal Fusion

Mechanistically, the process of germling and hyphal fusion can be divided into three steps: (1) precontact; (2) contact, adhesion, and cell wall breakdown; and (3) pore formation and cytoplasmic flow (*see* **Fig. 2; refs.** ***2,21,26***).

2.3.1. Precontact

The observed attraction between conidial germlings or fusion hyphae suggests chemotropic interactions between the fusion partners. When the relative position of two germlings showing mutual attraction is changed by micromanipulation using optical tweezers, both individuals readjust their growth toward each other to make contact and undergo fusion ***(18,27)***. During hyphal fusion, the presence of a fusion-competent hypha often results in either the alteration of growth trajectory or the formation of fusion branches in a receptive hypha (*see* **Fig. 2; refs.** ***1,4,21***).

The secretion of signaling molecules is a common theme in chemotactic and chemotropic cellular interactions. Instances of cell–cell communication by diffusible substances leading to cell fusion include mating in the unicellular yeast species, *Saccharomyces cerevisiae*, pollen tube growth to the ovary in plant species, or egg–sperm interaction in animals. Mating in *S. cerevisiae* requires two cells of opposite mating types. Haploid cells secrete mating-type specific pheromones, which bind to their cognate plasma membrane receptors in a partner of the opposite mating type ***(28)***. Germinating pollen tubes are also thought to be guided by diffusible chemotropic substances, such as Ca^{2+} or small heat-stable molecules secreted by the style ***(29,30)***. Another diffusible substance that is released by synergid cells guides the pollen tube into the ovule ***(31)***. The eggs of many aquatic animal species also release chemotactic substances to attract sperm; for example, *Xenopus* egg jelly releases a cysteine-rich secretory protein, allurin, to attract sperm ***(32)***. In these examples, the fusion partners are genetically and/or physiologically distinct and either secrete different signaling molecules (such as mating-type–specific pheromones) or only one partner secretes a signal that results in the attraction of the other partner. Although the involvement of secreted signals is not clear in other systems, such as in myoblast fusion during muscle development, in most cases the fusing cells are also different, such that one partner presents an extracellular or surface attractant and the other grows or migrates toward it. In *N. crassa,* there is no evidence that cells that undergo germling and hyphal fusion are genetically or physiologically different. Both cells show chemotropic interactions, indicating that both are secreting and responding to a chemotropic signal. This scenario is somewhat similar to cyclic adenosine monophosphate signaling in *Dictyostelium discoideum*, where a gradient of cyclic adenosine monophosphate mediates attraction of individual cells during the initiation of asexual sporulation ***(20)***. However, in *D. discoideum*, all cells responds to the chemotactic signal, whereas in *N. crassa*, only cells/hyphae destined to fuse do so. The identity of the molecules that mediate chemotropic interactions during germling/hyphal fusion in any filamentous fungus, including *N. crassa*, remains enigmatic.

The chemotropic reorientation of hyphae destined to fuse is associated with alterations in the position of the Spitzenkörper or with the formation of a new Spitzenkörper associated with branch formation in the receptive hypha (*see* **Fig. 2A,B; ref.** ***21***). The Spitzenkörper is a vesicle-rich structure found in growing hyphal tips or at sites of branch initiation ***(21,33)***. Localization of the Spitzenkörper in the hyphal apex has been associated with directionality of growth. In hyphae showing chemotropic interactions prior to hyphal fusion, the Spitzenkörper in the two partner hyphae continually reorient toward each other until the point of contact (*see* **Fig. 2B,C**). Reorientation of the Spitzenkörper and polar hyphal extension toward the fusion partner requires cellular mechanisms linking reception of the fusion signal to reorganization of the cytoskeleton. Adjustment of hyphal growth toward the fusion partner is comparable to cell polarization and shmoo formation during yeast mating ***(28)***, directed pollen tube growth toward the ovary ***(31)***, or the extension and/or stabilization of filopodia during myoblast fusion ***(10)***.

2.3.2. Contact, Adhesion, and Cell Wall Breakdown

After making contact, hyphae involved in fusion switch from polar to isotropic growth, resulting in swelling of hyphae at the fusion point. The two Spitzenkörper of the fusion hyphae are juxtaposed at the point of contact (*see* **Fig. 2C; ref.** ***21***). During chemotropic interactions, vesicles targeted to the Spitzenkörper are associated with hyphal growth. However, once contact occurs, vesicles secreted to the hyphal tips via the Spitzenkörper must be involved in the cell wall degradation at the site of fusion. The localization of the two Spitzenkörper in the fusion hyphae resembles the prefusion complexes found during myoblast fusion in which vesicles line up at the sites of cell contact, forming pairs across the apposing plasma membranes ***(12)***. Interpretation of Spitzenkörper behavior during hyphal fusion as a component of the prefusion complex offers an interesting working hypothesis for further analysis.

Germlings and hyphae involved in fusion events tightly adhere to one another ***(18,21)***, and extracellular electron-dense material associated with fusing hyphae ***(34)*** may be involved in adhesion of participating hyphae. Interaction between adhesive molecules during mating in *S. cerevisiae*, termed *agglutinins*, is required to hold mating pairs together during cell wall breakdown and plasma membrane fusion ***(35)***. During prefusion complex formation in myoblast fusion, extracellular electron-dense material is also found in the area between two aligning vesicles, but not at nonpaired vesicles, suggesting a role for this extracellular material in aligning vesicles during fusion events ***(36)***.

2.3.3. Pore Formation and Cytoplasmic Flow

After fusion of plasma membranes, the cytoplasms of the two participating hyphae mix. In *N. crassa,* the Spitzenkörper remains associated with the fusion pore as it enlarges (*see* **Fig. 2D,E; ref. *21***). Dramatic changes in cytoplasmic flow are often associated with hyphal fusion. Organelles, such as mitochondria, vacuoles, and nuclei, are transferred between hyphae as a result of fusion (*see* **Fig. 2E**). Septum formation near the site of hyphal fusion is also often observed. Physiological changes associated with cytoplasmic mixing upon hyphal/germling fusion are unclear but are presumed to occur; hyphae participating in fusion may be in different developmental states or be exposed to different nutritional conditions.

3. Sexual Fusion

Fusion is also essential for fertilization during mating in filamentous ascomycete species, such as *N. crassa* (*see* **Fig. 1C,D**). Mating requires the production of a specialized female reproductive structure, termed a *protoperithecium*. Reproductive hyphae, called *trichogynes*, protrude from the protoperithecia. Trichogynes are attracted by mating-type-specific pheromones secreted by male cells (microconidia or macroconidia) of the opposite mating type ***(8,37)***. After making physical contact, the tip of the female trichogyne fuses with the male cell (*see* **Fig. 1C**). Following fusion, the nucleus from the male cell migrates through the trichogyne and into the protoperithecium. Following this fertilization event, opposite mating-type nuclei proliferate in a common cytoplasm within the developing perithecium. Opposite mating-type nuclei pair off and migrate into a hook-shaped structure called a *crozier* (*see* **Fig. 1D; ref. *9***). In *N. crassa,* karyogamy occurs in the penultimate cell of the crozier, while hyphal fusion occurs between the terminal cell and the hyphal compartment nearest to the penultimate cell (*see* **Fig. 1D**). Although fusion events occur during both vegetative growth and sexual reproduction in filamentous ascomycete species, it is unclear whether signaling mechanisms and/or hyphal fusion machinery are common to both processes.

4. Identification of Fusion Mutants

Chemotropic interactions observed during hyphal and germling fusion suggest that receptors and signal transduction mechanism are involved. During mating in *S. cerevisiae*, binding of mating-type-specific pheromones to their cognate receptors results in activation of the pheromone response mitogen-activated protein (MAP) kinase (MAPK) pathway. Activation of this signaling pathway results in G_1 growth arrest and transcriptional activation of genes associated with mating, such as *FUS1* and *PRM1* ***(38,39)***. Components of the MAPK pathway, such as the MAPK Fus3p, interact with proteins associated with cytoskeleton rearrangement and cell polarization, such as the formin Bni1p ***(40)***. In *N. crassa,* mutations in homologs of components of the *S. cerevisiae* pheromone response pathway result in strains that cannot perform germling or hyphal fusion (**Fig. 3; ref. *19***). Strains containing mutations in the MAPK gene *mak-2,* the MAPK kinase (MAPKK) gene *NCU04612.3*, or the MAPKK kinase (MAPKKK) gene *nrc-1* show similar phenotypes. In addition to a failure to undergo hyphal or germling fusion, these mutants show reduced growth rates, shortened aerial hyphae, and failure to form female reproductive structures (protoperithecia; **refs. *19,41,42***). Similarly, an *Aspergillus nidulans* mutant disrupted in a MAPKKK *STE11* homolog, *steC*, fails to form heterokaryons (indicating a

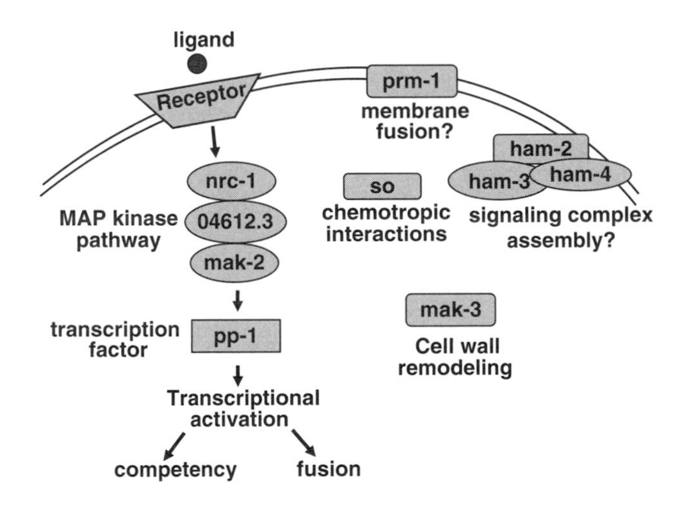

Fig. 3. In *Neurospora crassa,* mutations in the mitogen-activated protein kinase pathway components *nrc-1, NCU04612.3, mak-2,* and *pp-1* result in mutants unable to undergo germling or hyphal fusion ***(19)***. In addition, mutations in *ham-2*, encoding a putative plasma membrane protein, *ham-3,* and *ham-4* result in strains unable to undergo both hyphal and germling fusion ***(57,68)***. Mutations in *so* result in germling/hyphal fusion-deficient strains ***(27)***. In *Fusarium graminearum,* a strain containing a mutation in the ortholog of *SLT2* fails to form a heterokaryon ***(45)***; the *N. crassa* ortholog of *SLT2* is called *mak-3*. The natures of the receptor and ligand involved in anastomosis are unknown. Prm1p mediates membrane fusion in *Saccharomyces cerevisiae* ***(38)***. Preliminary data indicate a similar role of the *N. crassa prm-1* homolog in germling/hyphal fusion. (A. Fleißner, S. Diamond, and N. L. Glass, unpublished data.)

defect in hyphal fusion) and is also affected in formation of sexual reproductive structures ***(43)***. Because mutations in this MAPK pathway affect formation of sexual reproductive structures in filamentous fungi, its role in mating cell fusion has not been addressed.

Phosphorylation of MAK-2 is temporally associated with germling fusion events and is dependent on functional NRC-1 ***(19)***. In *S. cerevisiae*, activation of the pheromone response pathway leads to activation of the transcription factor Ste12p. In *N. crassa,* a strain containing a mutation in the *Ste12* ortholog, *pp-1,* is very similar in phenotype to a *mak-2* mutant and is defective in hyphal and germling fusion (D. J. Jacobson, A. Fleißner, and N. L. Glass, unpublished results; **ref.** ***42***). Live cell imaging and microscopic observations of the *N. crassa nrc-1/mak-2/pp-1* mutants indicate that they are blind to self (mutants neither attract nor are attracted to hyphae/conidia in cases where germling fusion is common in wild-type strains). Furthermore, the *nrc-1* and *mak-2* mutants do not form conidial anastomosis tubes ***(18)***. These data suggest that this MAPK pathway either is involved in early communication between the fusion partners or is required for rendering conidia and hyphae competent to undergo fusion.

In *S. cerevisiae,* the *SLT2* locus encodes an MAPK that is involved in cell wall integrity. The *SLT2* MAPK pathway is downstream of the *FUS3* MAPK pathway and is required for remodeling the cell wall during shmoo formation during mating ***(44)***. Initial data show that a mutant of the *SLT2* homolog in *N. crassa*, *mak-3*, is also hyphal fusion defective (A. Fleißner and N. L. Glass, unpublished results). Mutations in the *SLT2* ortholog in *Fusarium graminearum, MGV1,* resulted in a mutant that is female sterile, fails to form heterokaryons by hyphal fusion, and is substantially reduced in virulence ***(45)***.

In numerous plant pathogenic filamentous fungi, homologs of components of the mating or cell wall integrity MAPK pathways are essential for pathogenic development despite their distinct infection strategies. For example, in *Magnaporthe grisea*, *Colletotrichum lagenarium,* and *Cochliobolus heterostrophus*, strains containing mutations in *FUS3* homologs are defective in appressoria formation and fail to colonize host plants ***(46–48)***. Mutations in the *FUS3* homolog of the biotrophic, nonappressorium-forming grass pathogen *Claviceps purpurea* result in the inability of the fungus to colonize rye ovaries ***(49)***. Possible defects in hyphal fusion have not been addressed in most of these cases. Thus, a role for germling and hyphal fusion for colony development during invasion and growth within host tissue remains unanswered.

Cells recognize extracellular signaling molecules by different types of receptors. All eukaryotes use G-protein-coupled receptors (GPCR) for cell–cell communication and sensing of environmental stimuli. Examples are the mating pheromone receptors in *S. cerevisiae* (**28,50**), cyclic adenosine monophosphate receptors involved in cell–cell communication in *D. discoideum* **(20)**, or GPCRs involved in neuron guidance by extracellular chemical cues (*reviewed in* **ref. 51**). Genome sequence analysis of the *N. crassa* genome has revealed at least 10 seven-transmembrane receptors within the GPCR family **(52)**. The two mating-type-specific pheromone receptors share homology with the *S. cerevisiae* pheromone receptors Ste2p and Ste3p **(53)**. In *N. crassa*, mutations in the putative pheromone receptor gene *pre-1* result in female sterility. Female *pre-1* trichogynes are unable to detect and contact male cells of the opposite mating type, indicating a role of the PRE-1 receptor in pheromone signaling between mating partners. However, heterokaryon formation between two *pre-1* strains was comparable to wild-type strains, indicating that hyphal fusion in the *pre-1* mutant is normal **(53)**.

Binding of ligands to GPCRs results in the disassociation of an intracellular heterotrimeric G protein (Gα, Gβ, and Gγ) and subsequent activation of downstream processes **(50)**. In the *N. crassa* genome, three Gα, one Gβ, and one Gγ genes are present **(52,54)**. *gna-1* and *gnb-1* mutants do not show chemotropic interactions between a trichogyne and conidium, which is required for the initiation of the sexual cycle (*see* **Fig. 1C; ref. 53**). However, G-protein mutants show no defects in vegetative germling or hyphal fusion (A. Fleißner and N. L. Glass, unpublished results), suggesting that GPCRs are not involved in signaling vegetative fusion events. However, there is growing evidence to suggest that GPCRs could function in a G-protein-independent manner **(55)**. Together, these data indicate that signaling molecules and their receptors involved in mating cell fusion in *N. crassa* are different from those involved in vegetative germling/hyphal fusion.

4.1. Proteins Mediating Membrane Fusion

Although cell fusion events are essential for the development of most eukaryotic organisms, the molecular basis of the final step of this process, the fusion of plasma membranes, is only poorly understood. In *S. cerevisiae,* one of the few proteins predicted to be involved in this process is Prm1p. *PRM1* encodes a plasma membrane protein and is found only in fungal species. *prm1* mutants show a significant fusion defect during mating, resulting in the accumulation of prezygotes. Preliminary data indicate that mutations in the *N. crassa prm-1* ortholog also results a fusion defect during germling and hyphal fusion (A. Fleißner, S. Diamond, and N. L. Glass, unpublished results). Further studies will evaluate if *prm-1* is required for fusion of the female trichogyne with the male cell during sexual development. These experiments will reveal whether the different cell fusion events during the *N. crassa* life cycle, which are initiated by different cell–cell communication mechanisms, share the same membrane fusion machinery.

4.2. Genes of Unknown Function: so and ham-2, ham-3, and ham-4

The *N. crassa so* mutant (allelic to *ham-1*) is deficient in both germling and hyphal fusion *(27)* and exhibits an altered conidiation pattern and shortened aerial hyphae. The *so* locus encodes a protein of unknown function, which contains a WW domain predicted to be involved in protein–protein interactions. Homologs of *so* are present in the genomes of filamentous ascomycete fungi but are absent in other eukaryotic species. These data indicate that some aspects of tip growth, polarization, and germling/hyphal fusion require functions that are specific to filamentous fungi. Interestingly, the SO protein accumulates at septal plugs of injured hyphae *(56)*; SO is not essential for wound sealing but contributes to the speed of septal plugging. A possible connection between its function in germling/hyphal fusion and wound sealing is unclear.

The *so* mutant forms female reproductive structures (protoperithecia), and mating cell fusion between the *so* trichogynes and male cells is unimpaired. However, fertilization by a male cell does not result in entry into sexual reproduction *(27)*. Thus, the block in sexual reproduction in the *so* mutant occurs postfertilization. It is possible that *so* may be required for development of the ascogenous hyphae and for the second sexual fusion event during ascus formation (*see* **Fig. 1D**).

In *N. crassa,* the *ham-2* (hyphal anastomosis) locus encodes a putative transmembrane protein *(57)*. *ham-2* mutants show a pleiotropic phenotype, including slow growth, female sterility, and homozygous lethality in sexual crosses. In addition, *ham-2* mutants fail to undergo both hyphal and germling fusion. Laser tweezer experiments showed that *ham-2* mutants are blind to self (fail to attract or be attracted to a wild type during germling fusion events; **ref.** ***18***), similar to the *mak-2* mutants described earlier. Subsequently, a function for a homolog of *ham-2* in *S. cerevisiae,* termed *FAR11,* was reported. Mutations in *FAR11* result in a mutant that prematurely recovers from G_1 growth arrest following exposure to pheromone *(58)*. Far11p was shown to interact with five other proteins (Far3p, Far7p, Far8p, Far9p, and Far10p). Mutations in any of these other genes give an identical phenotype as *far11* mutants. It was proposed that the Far11 complex is part of a checkpoint that monitors mating cell fusion in coordination with G_1 cell-cycle arrest. Apparent homologs of genes encoding several of the proteins that form a complex with Far11p in *S. cerevisiae* are lacking in *N. crassa*, including *FAR3* and *FAR7* (**2**). Preliminary data show that mutations in the homologs that are present in *N. crassa, FAR8* and *FAR9/10* (*ham-3* and *ham-4*, respectively), result in phenotypes similar to the *ham-2* mutant, including defects during germling/hyphal fusion (C. Rasmussen, A. Fleißner, A. Simonin, M. Yang, and N. L. Glass, unpublished results). These data indicate that homologs of proteins of the *S. cerevisiae* Far11 complex might also physically interact in *N. crassa* and that this interaction is essential for vegetative germling/hyphal fusion. Interestingly, HAM-3 shows significant similarity to proteins of the striatin family. In mammals, genes belonging to the striatin family are principally expressed in neurons ***(59)***. Striatin proteins accumulate in dendritic spines in neurons at the point of cell–cell contact or synapses. Striatin family proteins act as scaffolding proteins that organize signaling complexes; for example, formation of a complex between striatin and the estrogen receptor is required for estrogen-induced activation of a MAPK signal transduction pathway ***(60)***. In *Sordaria macrospora,* a species related to *N. crassa,* mutations in the *ham-3* homolog (*pro11*) result in a mutant unable to complete sexual development, but full fertility was restored by expression of a striatin cDNA from mouse ***(61)***. These data indicate that the homologous proteins carry out similar cellular functions in fungi and animals. Further characterization of the function of HAM-3 in *N. crassa* will allow interesting comparisons between neuronal and hyphal signaling and might reveal conserved cellular mechanisms.

5. Physiological and Morphogenetic Consequences of Fusion

There are many advantages associated with hyphal fusion within a colony and between colonies, including increased resource sharing and translocation, increased colony cooperation, hyphal healing, and exchange of genetic material. For example, in *Colletotrichum lindemuthianum*, conidia that undergo fusion exhibit a higher rate of germination compared with single unfused conidia ***(5)***. Also, conidia grown in low-nutrient environments show an increased rate of fusion ***(4)***. These observations suggest that fusion between conidial germlings may serve to increase or pool resources that are important for colony establishment.

It is widely assumed that vegetative hyphal fusion within an established colony is important for intrahyphal communication, cooperation, translocation of water and nutrients, and general homeostasis within a colony. Fusing to create a hyphal network could be important in influencing hyphal patterns of growth and morphogenesis in filamentous fungi. Formation of a connected network may also facilitate signaling within a colony (by molecules, proteins, or perhaps electric fields; *reviewed in* **ref.** ***62***), which may also affect behavior and development of a filamentous fungal colony. In nature, fungal colonies exploit diverse environments with unequal distributions and types of nutrient sources. The ability to form a hyphal network may be needed for coordinated behavior between the different parts of a fungal colony and nutrient transport from sources to sinks ***(6)***.

As well as facilitation of long-distance nutrient transport, vegetative hyphal fusion can function as a healing mechanism to repair hyphal connections when the fungal network has been damaged. Hyphal tips growing out from either side of damaged compartments will eventually find each other, fuse, and reestablish the hyphal network *(1)*. In fungi that are asexual, fusion between different individuals can be a means of exchanging genetic material through the formation of a heterokaryon *(7,63)*.

Fusion between different colonies has potential disadvantages. Hyphal fusion between individuals increases the risk of transfer of deleterious infectious elements, parasitism, or resource plundering, such as the competitive acquisition of resources by one colony from another (*reviewed in* **ref.** *64)*. Many fungi have developed mechanisms for nonself-recognition that result in programmed cell death, which is assumed to minimize the amount of exchange between individual colonies *(64–66)*. Many of the nonself-recognition mechanisms occur following hyphal fusion between genetically different colonies, raising the intriguing possibility that a link is present between the hyphal fusion and programmed cell death machinery in filamentous fungi.

6. Conclusion

Cell fusion events are essential for the vegetative and sexual development of filamentous ascomycete fungi. Live cell imaging has revealed that the processes of cell–cell communication and cell fusion are complex and highly regulated. Characterization of the components required for sexual fertilization versus vegetative fusion indicates that upstream components of the signaling pathways differ. Future analyses will reveal if the machinery associated with fusion processes are similar between sexual and vegetative fusion events. *Neurospora crassa* is an attractive model system with which to study the molecular basis of cell–cell communication and cell fusion in eukaryotes and to dissect similarities and differences in the processes of sexual and vegetative fusion: the genome has been sequenced and annotated *(54)*, well-established molecular and cell biology techniques are available (**ref.** *52; see also* http://www.nih.gov/science/models/neurospora/); and whole-genome microarrays *(67)* and knockout mutants for every single gene are in progress *(17)*. It is currently unclear what the adaptive role of germling and hyphal fusion is in filamentous fungi and what selective advantages it provides. Further analysis of these issues will provide significant insight into environmental adaptation and the evolution of form and function in multicellular microorganisms.

Acknowledgment

The work on germling/hyphal fusion in the N.L.G. Laboratory is funded by a grant from the National Science Foundation (MCB-0131355/0517660). We thank Drs. Denise Schichnes and Steve Ruzin (CNR Biological Imaging Facility) for help with microscopy and Dr. Carolyn Rasmussen for helpful suggestions on the manuscript.

References

1. Buller, A. H. R. (1933) *Researches on Fungi,* vol. 5. Longman, London.
2. Glass, N. L., Rasmussen, C., Roca, M. G., and Read, N. D. (2004) Hyphal homing, fusion and mycelial interconnectedness. *Trends Microbiol.* **12,** 135–141.
3. Glass, N. L. and Fleißner, A. (2006) Re-wiring the network: understanding the mechanism and function of anastomosis in filamentous ascomycete fungi, in *The Mycota* (Kues, Fischer, eds.). Springer-Verlag, Berlin, pp. 123–139.
4. Köhler, E. (1930) Zur Kenntnis der vegetativen Anastomosen der Pilze (II. Mitteilung). *Planta* **10,** 495–522.
5. Roca, M. G., Davide, L. C., Mendes-Costa, M. C., and Wheals, A. (2003) Conidial anastomosis tubes in *Colletotrichum. Fungal Genet. Biol.* **40,** 138–145.
6. Rayner, A. D. M. (1996) Interconnectedness and individualism in fungal mycelia, in *A Century of Mycology* (B.C. Sutton, ed.). University of Cambridge Press, Cambridge, England, pp. 193–232.
7. Pontecorvo, G.(1956) The parasexual cycle in fungi. *Annu. Rev. Microbiol.* **10,** 393–400.
8. Bistis, G. N. (1981) Chemotropic interactions between trichogynes and conidia of opposite mating-type in *Neurospora crassa. Mycologia* **73,** 959–975.
9. Raju, N. B. (1980) Meiosis and ascospore genesis in *Neurospora. Eur. J. Cell Biol.* **23,** 208–223.
10. Chen, E. H. and Olson, E. N. (2004) Towards a molecular pathway for myoblast fusion in *Drosophila. Trends Cell Biol.* **14,** 452–460.
11. Primakoff, P. and Myles, D. G. (2007) Cell–cell membrane fusion during mammalian fertilization. *FEBS Lett.* (in press, corrected proof).
12. Dworak, H. A. and Sink, H. (2002) Myoblast fusion in *Drosophila. Bioessays* **24,** 591–601.
13. Vignery, A. (2005) Macrophage fusion: the making of osteoclasts and giant cells. *J. Exp. Med.* **202,** 337–340.
14. Potgens, A. J. G., Schmitz, U., Bose, P., Versmold, A., Kaufmann, P., and Frank, H. G. (2002) Mechanisms of syncytial fusion: a review. *Placenta* **23** (Suppl A, *Trophoblast Res.* **16**), 107–113.
15. Davis, R. H. (2000) *Neurospora*: *Contributions of a Model Organism.* Oxford University Press, New York.
16. Perkins, D. D., Radford, A., and Sachs, M. S. (2001) *The Neurospora Compendium: Chromosomal Loci.* Academic Press, San Diego.
17. Colot, H. V., Park, G., Turner, G. E., Ringelberg, C., Crew, C. M., Litvinkova, L., Weiss, R. L., Borkovich, K. A., and Dunlap J. C. (2006) A high-throughput gene knockout procedure for *Neurospora* reveals functions for multiple transcription factors. *Proc. Natl. Acad. Sci. U.S.A.* **103,** 10352–10357.
18. Roca, M. G., Arlt, J., Jeffree, C. E., and Read, N. D. (2005) Cell biology of conidial anastomosis tubes in *Neurospora crassa. Eukaryot. Cell* **4,** 911–919.
19. Pandey, A., Roca, M. G., Read, N. D., and Glass, N. L. (2004) Role of a MAP kinase during conidial germination and hyphal fusion in *Neurospora crassa. Eukaryot. Cell* **3,** 348–358.

20. Manahan, C. L., Iglesias, P. A., Long, Y., and Devreotes, P. N. (2004) Chemoattractant signaling in *Dictyostelium discoideum. Annu. Rev. Cell Dev. Biol.* **20,** 223–253.
21. Hickey, P. C., Jacobson, D. J., Read, N. D., and Glass, N. L. (2002) Live-cell imaging of vegetative hyphal fusion in *Neurospora crassa. Fungal Genet. Biol.* **37,** 109–119.
22. Trinci, A.P.J. (1984) Regulation of hyphal branching and hyphal orientation, in *The Ecology and Physiology of the Fungal Mycelium* (D.H. Jennings and A.D.M. Rayner, eds.). Cambridge University Press: Cambridge, England, pp. 23–52.
23. Köhler, E. (1929) Beiträge zur Kenntnis der vegetativen Anastomosen der Pilze I. *Planta* **8,** 140–153.
24. Ahmad, S. S. and Miles, P. G. (1970) Hyphal fusions in *Schizophyllum commune*. 2. Effects on environmental and chemical factors. *Mycologia* **62,** 1008–1017.
25. Yokoyama, K. and Ogoshi, A. (1988) Studies on hyphal anastomosis of *Rhizoctonia solani* V. Nutritional conditions for anastomosis. *Trans. Mycol. Soc. Jpn.* **29,** 125–132.
26. Glass, N. L., Jacobson, D. J., and Shiu, P. K. (2000) The genetics of hyphal fusion and vegetative incompatibility in filamentous ascomycete fungi. *Annu. Rev. Genet.* **34,** 165–186.
27. Fleißner, A., Sarkar, S., Jacobson, D. J., Roca, M. G., Read, N. D., and Glass N. L. (2005) The *so* locus is required for vegetative cell fusion and post-fertilization events in *Neurospora crassa. Eukaryot. Cell* **4,** 920–930.
28. Kurjan, J. (1993) The pheromone response pathway in *Saccharomyces cerevisiae. Annu. Rev. Genet.* **27(4),** 147–179.
29. Lord, E. M. (2003) Adhesion and guidance in compatible pollination. *J. Exp. Bot.* **54,** 47–54.
30. Mascarenhas, J. P. (1993) Molecular mechanisms of pollen tube growth and differentiation. *Plant Cell* **5,** 1303–1314.
31. Higashiyama, T., Kuroiwa, H., and Kuroiwa, T. (2003) Pollen-tube guidance: beacons from the female gametophyte. *Curr. Opin. Plant Biol.* **6,** 36–41.
32. Al-Anzi, B. and Chandler, D. E. (1998) A sperm chemoattractant is released from *Xenopus* egg jelly during spawning. *Dev. Biol.* **198,** 366–375.
33. Riquelme, M., Reynaga-Peña, C. G., Gierz, G., and Bartnicki-García, S. (1998) What determines growth direction in fungal hyphae? *Fungal Genet. Biol.* **24,** 101–109.
34. Newhouse, J. R. and MacDonald, W. L. (1991) The ultrastructure of hyphal anastomoses between vegetatively compatible and incompatible virulent and hypovirulent strains of *Cryphonectria parasitica. Can. J. Bot.* **69,** 602–614.
35. Suzuki, K. (2003) Roles of sexual cell agglutination in yeast mass mating. *Genes Genet. Syst.* **78,** 211–219.
36. Doberstein, S. K., Fetter, R. D., Mehta, A. Y., and Goodman, C. S. (1997). Genetic analysis of myoblast fusion: blown fuse is required for progression beyond the prefusion complex. *J. Cell Biol.* **136,** 1249–1261.
37. Kim, H. and Borkovich, K. A. (2006) Pheromones are essential for male fertility and sufficient to direct chemotropic polarized growth of trichogynes during mating in *Neurospora crassa. Eukaryot. Cell* **5,** 544–554.
38. Heiman, M. G. and Walter, P. (2000) Prm1p, a pheromone-regulated multispanning membrane protein, facilitates plasma membrane fusion during yeast mating. *J. Cell Biol.* **151,** 719–730.

39. McCaffrey, G., Clay, F. J., Kelsay, K. and Sprague, G. F. (1987) Identification and regulation of a gene required for cell fusion during mating of the yeast *Saccharomyces cerevisiae. Mol. Cell. Biol.* **7,** 2680–2690.
40. Matheos, D., Metodiev, M., Muller, E., Stone, D., and Rose, M. D. (2004) Pheromone-induced polarization is dependent on the Fus3p MAPK acting through the formin Bni1p. *J. Cell Biol.* **165,** 99–109.
41. Kothe, G. O. and Free, S. J. (1998) The isolation and characterization of *nrc-1* and *nrc-2*, two genes encoding protein kinases that control growth and development in *Neurospora crassa. Genetics* **149,** 117–130.
42. Li, D., Bobrowicz, P., Wilkinson, H. H., and Ebbole, D. J. (2005) A mitogen-activated protein kinase pathway essential for mating and contributing to vegetative growth in *Neurospora crassa. Genetics* **170,** 1091–1104.
43. Wei, H. J., Requena, N., and Fischer, R. (2003) The MAPKK kinase SteC regulates conidiophore morphology and is essential for heterokaryon formation and sexual development in the homothallic fungus *Aspergillus nidulans. Mol. Microbiol.* **47,** 1577–1588.
44. Buehrer, B. M. and Errede, B. (1997) Coordination of the mating and cell integrity mitogen-activated protein kinase pathways in *Saccharomyces cerevisiae. Mol. Cell. Biol.* **17,** 6517–6525.
45. Hou, Z., Xue, C., Peng, Y., Katan, T., Kistler, H. C., and Xu, J. R. (2002) A mitogen-activated protein kinase gene (MGV1) in *Fusarium graminearum* is required for female fertility, heterokaryon formation, and plant infection. *Mol. Plant–Microbe Interact.* **15,** 1119–1127.
46. Xu, J. R. and Hamer, J. E. (1996) MAP kinase and cAMP signaling regulate infection structure formation and pathogenic growth in the rice blast fungus *Magnaporthe grisea. Genes Dev.* **10,** 2696–2706.
47. Lev, S., Sharon, A., Hadar, R., Ma, H., and Horwitz, B. A. (1999) A mitogen-activated protein kinase of the corn leaf pathogen *Cochliobolus heterostrophus* is involved in conidiation, appressorium formation, and pathogenicity: diverse roles for mitogen-activated protein kinase homologs in foliar pathogens. *Proc. Natl. Acad. Sci. U.S.A.* **96,** 13542–13547.
48. Takano, Y., Kikuchi, T., Kubo, Y., Hamer, J. E., Mise, K., and Furusawa, I. (2000) The *Colletotrichum lagenarium* MAP kinase gene CMK1 regulates diverse aspects of fungal pathogenesis. *Mol. Plant–Microbe Interact.* **13,** 374–383.
49. Mey, G., Oeser, B., Lebrun, M. H,. and Tudzynski, P. (2002) The biotrophic, non-appressorium–forming grass pathogen *Claviceps purpurea* needs a Fus3/Pmk1 homologous mitogen-activated protein kinase for colonization of rye ovarian tissue. *Mol. Plant–Microbe Interact.* **15,** 303–312.
50. Dohlman, H. G., and Thorner, J. (2001) Regulation of G protein–initiated signal transduction in yeast: paradigms and principles. *Annu. Rev. Biochem.* **70,** 703–754.
51. Xiang, Y., Li, Y., Zhang, Z., Cui, K., Wang, S., Yuan, X. B., Wu, C. P., Poo, M. M., and Duan, S. (2002) Nerve growth cone guidance mediated by G protein–coupled receptors. *Nat. Neurosci.* **5,** 843–848.
52. Borkovich, K. A., Alex, L. A., Yarden, O., Freitag, M., Turner, G. E., Read, N. D., Seiler, S., Bell-Pedersen, D., Paietta, J., Plesofsky, N., Plamann, M., Goodrich-

Tanrikulu, M., Schulte, U., Mannhaupt, G., Nargang, F. E., Radford, A., Selitrennikoff, C., Galagan, J. E., Dunlap, J. C., Loros, J. J., Catcheside, D., Inoue, H., Aramayo, R., Polymenis, M., Selker, E. U., Sachs, M. S., Marzluf, G. A., Paulsen, I., Davis, R., Ebbole, D. J., Zelter, A., Kalkman, E. R., O'Rourke, R., Bowring, F., Yeadon, J., Ishii, C., Suzuki, K., Sakai, W., and Pratt, R. (2004) Lessons from the genome sequence of *Neurospora crassa*: tracing the path from genomic blueprint to multicellular organism. *Microbiol. Mol. Biol. Rev*. **68,** 1–108.

53. Kim, H. and Borkovich, K. A. (2004) A pheromone receptor gene, *pre-1*, is essential for mating type-specific directional growth and fusion of trichogynes and female fertility in *Neurospora crassa. Mol. Microbiol*. **52,** 1781–1798.
54. Galagan, J. E., Calvo, S. E., Borkovich, K. A., Selker, E. U., Read, N. D., Jaffe. D., FitzHugh, W., Ma, L. J., Smirnov, S., Purcell, S., Rehman, B., Elkins, T., Engels, R., Wang, S., Nielsen, C. B., Butler, J., Endrizzi, M., Qui, D., Ianakiev, P., Bell-Pedersen, D., Nelson, M. A., Werner-Washburne, M., Selitrennikoff, C. P., Kinsey, J. A., Braun, E. L., Zelter, A., Schulte, U., Kothe, G.O., Jedd, G., Mewes, W., Staben, C., Marcotte, E., Greenberg, D., Roy, A., Foley, K., Naylor, J., Stange-Thomann, N., Barrett, R., Gnerre, S., Kamal, M., Kamvysselis, M., Mauceli, E., Bielke, C., Rudd, S., Frishman, D., Krystofova, S., Rasmussen, C., Metzenberg, R. L., Perkins, D. D., Kroken, S., Cogoni, C., Macino, G., Catcheside, D., Li, W., Pratt, R. J., Osmani, S. A., DeSouza, C. P., Glass, L., Orbach, M. J., Berglund, J. A., Voelker, R., Yarden, O., Plamann, M., Seiler, S., Dunlap, J., Radford, A., Aramayo, R., Natvig, D. O., Alex, L. A., Mannhaupt, G., Ebbole, D. J., Freitag, M., Paulsen, I., Sachs, M. S., Lander, E. S., Nusbaum, C., and Birren, B. (2003) The genome sequence of the filamentous fungus *Neurospora crassa. Nature* **422,** 859–868.
55. Brzostowski, J. A. and Kimmel, A. R. (2001) Signaling at zero G: G-protein–independent functions for 7-TM receptors. *Trends Biochem. Sci*. **26,** 291–297.
56. Fleißner, A. and Glass, N. L. (2007) SO, a protein involved in hyphal fusion in *Neurospora crassa*, localizes to septal plugs. *Eukaryot. Cell* **6,** 84–94.
57. Xiang, Q., Rasmussen, C., and Glass, N. L. (2002) The *ham-2* locus, encoding a putative transmembrane protein, is required for hyphal fusion in *Neurospora crassa. Genetics* **160,** 169–180.
58. Kemp, H. A. and Sprague, G. F. (2003) Far3 and five interacting proteins prevent premature recovery from pheromone arrest in the budding yeast *Saccharomyces cerevisiae. Mol. Cell. Biol*. **23,** 1750–1763.
59. Benoist, M., Gaillard, S. and Castets, F. (2006) The striatin family: a new signaling platform in dendritic spines. *J. Physiol. Paris* **99,** 146–153.
60. Lu, Q., Pallas, D. C., Surks, H. K., Baur, W. E. Mendelsohn, M. E., and Karas R. H. (2004) Striatin assembles a membrane signaling complex necessary for rapid, non-genomic activation of endothelial NO synthase by estrogen receptor alpha. *Proc. Natl. Acad. Sci. U.S.A*. **101,** 17126–17131.
61. Pöggeler, S. and Kück, U. (2004) A WD40 repeat protein regulates fungal cell differentiation and can be replaced functionally by the mammalian homologue striatin. *Eukaryot. Cell* **3,** 232–240.
62. Gow, N. A. R. and Morris, B. M. (1995) The electric fungus. *Bot. J. Scotl.* **47,** 263–277.

63. Giovannetti, M., Fortuna, P., Citernesia, A. S., Morini, S., and Nuti, M. P. (2001) The occurrence of anastomosis formation and nuclear exchange in intact arbuscular mycorrhizal networks. *New Phytol.* **151,** 717–724.
64. Glass, N. L. and Dementhon, K. (2006) Nonself recognition and programmed cell death in filamentous fungi. *Curr. Opin. Microbiol.* **9,** 553–558.
65. Glass, N. L. and Kaneko, I. (2003) Fatal attraction: nonself recognition and heterokaryon incompatibility in filamentous fungi. *Eukaryot. Cell* **2,** 1–8.
66. Saupe, S. J. (2000) Molecular genetics of heterokaryon incompatibility in filamentous ascomycetes. *Microbiol. Mol. Biol. Rev.* **64,** 489–502.
67. Kasuga, T., Townsend, J. P., Tian, C., Gilbert, L. B., Mannhaupt, G., Taylor, J. W., and Glass, N. L. (2005) Long-oligomer microarray profiling in Neurospora crassa reveals the transcriptional program underlying biochemical and physiological events of conidial germination. *Nucleic Acids Res.* **33,** 6469–6485.
68. Wilson, J. F. and Dempsey, J. A. (1999) A hyphal fusion mutant in *Neurospora crassa. Fungal Genet. Newslett.* **46,** 31.

3

Gametic Cell Adhesion and Fusion in the Unicellular Alga *Chlamydomonas*

Nedra F. Wilson

Summary

Differentiation of vegetative cells of the haploid eukaryote *Chlamydomonas* is dependent on environmental conditions. Upon depletion of nitrogen and exposure to light, vegetative cells undergo a mitotic division, generating gametes that are either mating-type plus (mt[+]) or mating-type minus (mt[−]). As gametes of opposite mating type encounter one another, an initial adhesive interaction mediated by flagella induces a signal transduction pathway that results in activation of gametes. Gametic activation results in the exposure of previously cryptic regions of the plasma membrane (mating structures) that contain the molecules required for gametic cell adhesion and fusion. Recent studies have identified new steps in this signal transduction pathway, including the tyrosine phosphorylation of a cyclic guanosine monophosphate–dependent protein kinase, a requirement for a novel microtubular motility known as *intraflagellar transport*, and a mt(+)-specific molecule that mediates adhesion between mating structures.

Key Words: *Chlamydomonas;* gamete; cell fusion; fertilization tubule; flagella; signal transduction.

1. Introduction

The unicellular eukaryote *Chlamydomonas reinhardtii* has become an important model organism for delineating the steps involved in fertilization. *Chlamydomonas* can reproduce both asexually, by mitotic division as haploid vegetative cells (allowing the clonal expansion of cells; **Fig. 1A**), and sexually, through the fusion of haploid mating-type plus (mt[+]) and mating-type minus (mt[−]) gametes to form diploid zygotes. Importantly, a number of mutants exist that are defective at various steps in the fertilization process. Moreover, because *Chlamydomonas* can reproduce asexually, it is relatively straightforward to generate additional mutants defective in fertilization. Mutants can be maintained as vegetative cells with a synchronized cell cycle by exposure to a

From: *Methods in Molecular Biology, vol. 475: Cell Fusion: Overviews and Methods*
Edited by: E. H. Chen © Humana Press, Totowa, NJ

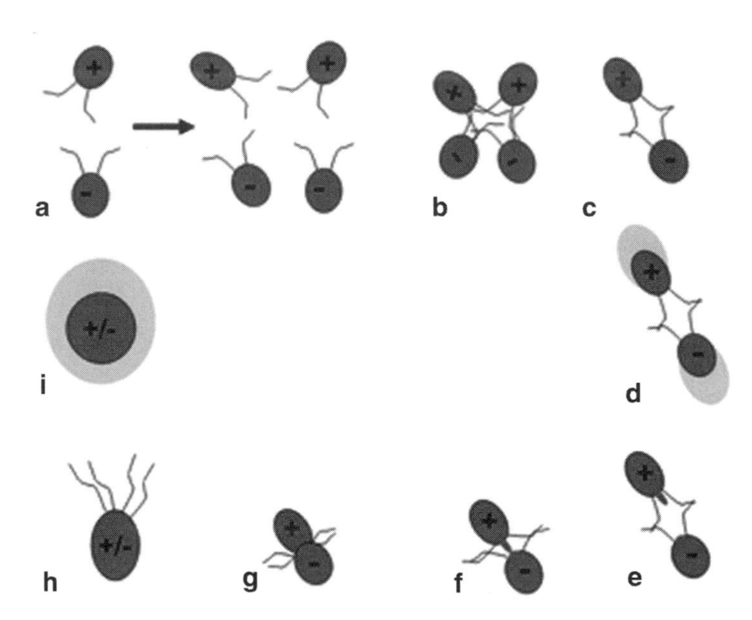

Fig. 1. Diagrammatic representation of fertilization in Chlamydomonas. **(A)** mt(+) and mt(–) vegetative cells divide asexually. **(B)** Gametogenesis is induced upon depletion of nitrogen from the medium and exposure to constant light *(2,3)*. This differentiation process culminates in a mitotic division that generates a homogeneous population of gametes *(3)*. In the first step of fertilization, varying numbers of gametes of opposite mating types interact (agglutinate) via an adhesion molecule, agglutinin, localized on flagella. **(C)** As agglutination continues, the flagellar adhesion molecule, agglutinin, is translocated to the tips of flagella. In addition, the number of gametes undergoing flagellar adhesion decreases such that only one mt(+) and one mt(–) gamete continue agglutinating. **(D)** Agglutination initiates a signal transduction cascade that induces the release of cell walls and activation of mating structures. **(E)** Activated mt(–) gametes form a slightly raised, dome-shaped mating structure, while mt(+) gametes form an actin-filled fertilization tubule. **(F)** Adhesion and subsequent fusion occur between the apex of the activated mt(–) mating structure and the tip of the mt(+) fertilization tubule. Flagellar adhesion of mt(+) and mt(–) gametes orients the cells in such a manner as to facilitate the interaction of the tip of the mt(+) fertilization tubule with the apex of the activated mt(–) mating structure. **(G)** Gametic cell body fusion occurs. **(H)** A quadriflagellated zygote is formed. **(I)** Approximately 4 h postfusion, the flagella are reabsorbed and the zygote secretes a new, thick, highly impenetrable wall.

light–dark cycle *(1)*. These fertilization-defective mutants can then be studied by inducing their differentiation into gametes. Similar to mammalian systems, fertilization in *Chlamydomonas* involves an initial adhesive interaction between gametes of opposite mating types, mt(+) and mt(–). This adhesive interaction activates a signal transduction pathway that induces release of the extracellular matrix (cell wall) that surrounds the plasma membrane of these cells and exposes membrane domains containing the molecules essential for plasma membrane adhesion and fusion.

2. Agglutinins Mediate Flagellar Adhesion

Gametes of *Chlamydomonas* utilize flagella to propel them through their environment to find gametes of the opposite mating type. During differentiation into gametes, flagella are modified by the insertion of sex-specific adhesion molecules (agglutinins) into the plasma membrane ***(4)***. Agglutinins are high molecular weight glycoproteins (>1,000 kDa) rich in hydroxyproline ***(4–7)***. Examination of the amino acid sequence of mt(+) and mt(−) agglutinin revealed that they are only 23% identical and ~30% conserved to each other ***(8)***.

When gametes of opposite mating type encounter each other, the cells initially interact along the length of their flagella, forming large clumps of rapidly agitating or "twitching" cells in a process called *agglutination* (*see* **Fig. 1B,C**). This interaction between agglutinins induces the translocation and turnover of these molecules. Examination of gametes undergoing agglutination reveals that the initial site of adhesion can occur anywhere along the length of the flagella. Once flagellar adhesion has occurred, however, the site of adhesion migrates to the tips of the interacting flagella. This adhesion-induced translocation of the agglutinins to the tips of flagella occurs via a process aptly named "tipping" ***(9, 10)***. Flagellar tipping in gametes can be induced by treatment with antibodies that recognize common flagellar epitopes ***(9)***. This observation suggests that, during flagellar adhesion, the aggregation of agglutinin molecules results in flagellar tipping.

Another consequence of the interactions between agglutinins is the adhesion-induced inactivation and loss of these molecules from flagella. Using isolated mt(+) flagella and intact mt(−) gametes, Snell and Roseman ***(11)*** observed an adhesion-dependent inactivation of agglutinins. As adhesion-competent agglutinins are lost from the flagella they are replaced with agglutinins from the cell body. This cell body pool of agglutinins (which represents 90% of the total cellular agglutinin) is not present intracellularly but instead is found on the plasma membrane of gametic cell bodies ***(12)***. Intriguingly, agglutinins on cell bodies are maintained in an inactive form; cell body agglutinins are not competent to adhere to either cell bodies or isolated flagella of the opposite mating type. The mechanism that restricts the majority of agglutinins to the plasma membrane of cell bodies is not understood. Flagellar adhesion, however, activates a signal transduction pathway that recruits the inactive agglutinins from their storage site on the plasma membrane of cell bodies out onto flagella where they become competent for flagellar adhesion ***(12,13)***. In addition, little is known about the changes that render agglutinins competent for adhesion. It is possible that interactions with one or more protein(s) sequestered to either cell bodies or flagella determine the functional status of the agglutinins.

3. Flagellar Adhesion Activates a Signal Transduction Pathway

Similar to sperm–egg interactions in mammalian organisms, the initial flagellar adhesion between gametes of opposite mating types activates a signal transduction pathway. Analysis of the levels of cyclic nucleotides during mating revealed a transient 10-fold increase in the levels of cyclic adenosine monophosphate (cAMP) and a prolonged 2-fold increase in cyclic guanosine monophosphate (cGMP; **ref.** ***14***). This increase in cAMP levels was coincident with flagellar adhesion. Consistent with a role for cAMP during mating, adenylyl cyclase activity was detected in both cell bodies and flagella of gametes ***(14)***. The kinetics for activation of adenylyl cyclase in flagella is much more rapid than that observed for the cell body form ***(15)***.

Regulation of the flagellar adenylyl cyclase does not occur through interactions with G proteins; no stimulation of adenylyl cyclase activity was observed when assayed in the presence of known activators of G proteins ***(14)***. Instead, the activity of the adenylyl cyclase located in gametic flagella is regulated by phosphorylation ***(16,17)***. When flagella isolated from mt(+) and mt(−) gametes were mixed together, a two- to threefold increase in adenylyl cyclase activity was observed ***(18)***. This adhesion-induced activation of adenylyl cyclase could be blocked by the inclusion of 50 n*M* staurosporine. Interestingly, treatment with higher concentrations of staurosporine (500 n*M* and above) stimulates adenylyl cyclase activity in the absence of flagellar adhesion ***(18)***. These observations suggest two different staurosporine-sensitive steps in the activation of flagellar adenylyl cyclase. In the absence of flagellar adhesion, adenylyl cyclase activity in the flagella is kept at a low level by a protein kinase sensitive to 500 n*M* staurosporine. Upon flagellar adhesion, the activation of a second protein kinase (sensitive to 50 n*M* staurosporine) results in the activation of the flagellar adenylyl cyclase.

Consistent with changes in the activity of protein kinases upon flagellar adhesion, the inclusion of genistein (an inhibitor of protein tyrosine kinases) during mating prevented gametic cell fusion and zygote formation. This inhibition by genistein could be overcome by the addition of dibutrylyl cAMP, suggesting that tyrosine phosphorylation is upstream of activation of adenylyl cyclase. Immunoblot analysis using an antiphosphotyrosine antibody identified a single polypeptide of 105 kDa that was phosphorylated only in adhering flagella ***(19)***. This 105-kDa protein was immunopurified with the antiphosphotyrosine antibody and identified as a cGMP-dependent protein kinase (PKG) by tandem mass spectrometry *(20)*. Consistent with a role in gametic cell activation, the kinase activity of PKG was stimulated severalfold upon flagellar adhesion. Moreover, when the protein level of PKG was significantly decreased by RNA interference, cells were able to undergo flagellar adhesion similar to wild-type cells, but gametic cell fusion was significantly inhibited.

The tyrosine phosphorylation of PKG requires the activity of the microtubule motor protein FLA10 *(19)*. FLA10 is a heterotrimeric kinesin that functions as the anterograde motor for a ciliary transport process known as *intraflagellar transport* (IFT; **ref.** *21*). Intraflagellar transport carries proteins destined for incorporation or removal into and out of these organelles *(22)*. In addition, it has recently been shown that IFT is required for formation of signal transduction complexes during development (*for review, see* **ref.** *23*). Analysis of matings between wild-type gametes and *fla10* gametes, which carry a temperature-sensitive lesion, revealed an inability to undergo fertilization at the restrictive temperature *(24)*. The addition of exogenous cAMP, however, resulted in wild-type levels of gametic cell fusion and zygote formation *(25)*. This observation suggests that IFT activity is required at a step between agglutinin-mediated flagellar adhesion and activation of adenylyl cyclase *(25)*. Recently, Wang and Snell *(19)* made the intriguing observation that, upon flagellar adhesion, PKG moves into a freeze–thaw–labile, particulate compartment in an IFT-dependent manner.

Taken together, these results suggest the following model for gametic cell activation. Flagellar adhesion mediated by agglutinins on gametes of opposite mating types activates a protein tyrosine kinase. This as yet unidentified protein tyrosine kinase in turn phosphorylates and activates PKG. Whether the requirement for IFT in the phosphorylation of PKG is upstream or downstream of the protein tyrosine kinase has not yet been determined. It is intriguing to speculate that, similar to the role of IFT in other ciliary signaling pathways *(23)*, IFT acts as a mobile scaffold to bring signaling molecules to their site of action in flagella. Once activated, PKG, directly or indirectly, activates the flagellar adenylyl cyclase, resulting in an increase in cAMP and the generation of fusion-competent gametes.

4. Cyclic Adenosine Monophosphate Renders Gametes Competent to Undergo Cell–Cell Fusion

The central role that cAMP plays in the activation of gametes prior to cell–cell fusion is illustrated by the observation that exogenous application of dibutrylyl-cAMP is sufficient for inducing all of the morphological changes required for gametes to be competent for cell fusion *(14)*. For example, treatment of gametes with dibutrylyl cAMP induces the translocation of agglutinins to flagellar tips *(10)*. Another early event in gametic cell activation mediated by cAMP is loss of cell walls (*see* **Fig. 1D**). Both vegetative and gametic cells are surrounded by a multilayered cell wall that derives its shape from a protein framework insoluble to boiling in dithiothreitol and sodium dodecyl sulfate ***(26)***. This cell wall is removed during mating by the action of a metalloprotease known as *lysin* or *autolysin* present in the periplasm ***(26–29)***. Upon flagellar adhesion and the subsequent increase in cAMP, a serine protease (p-lysinase) is secreted into the periplasmic space that cleaves prolysin into the enzymatically active lysin ***(30)***. The release of cell walls by lysin exposes the site (mating structures) where gametic cell adhesion and fusion subsequently occur (**Figs. 1E, 2; ref.** ***18***).

5. The Mating Structures

Ultrastructural studies have provided information on the morphology of the mating structures as well as the developmental changes leading to their activation ***(31,32)***. In both mt(+) and mt(−) gametes, the mating structures are associated with the cytoplasmic face of the apical plasma membrane and located between and slightly basal to the two flagella. Studies using thin-section electron microscopy identified an electron-dense region termed the *membrane*

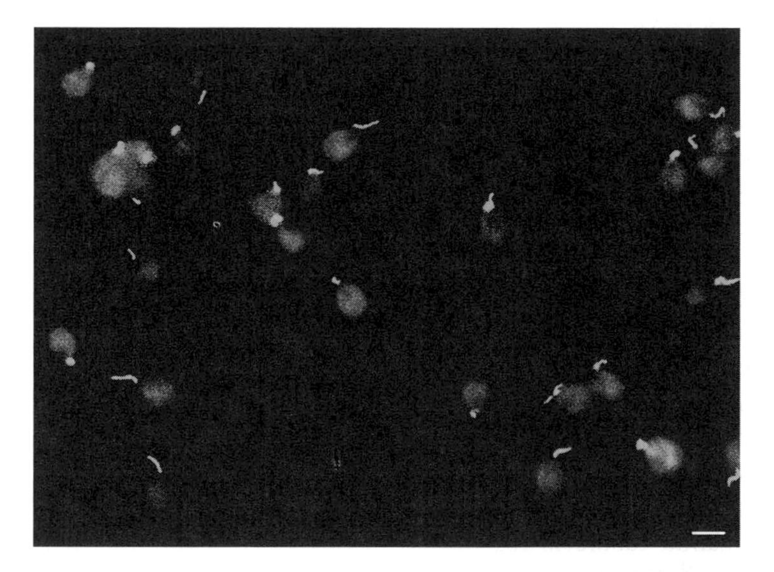

Fig. 2. The mt(+) fertilization tubule. mt(+) gametes were activated by incubation with dibutrylyl cAMP and papaverine for 20 min. Fertilization tubules (green) were visualized by fluorescent microscopy after staining actin filaments with BODIPY phallacidin. Following capture of the image in the fluorescein channel, the residual chlorophyll present in cell bodies (red) was imaged by examination in the rhodamine channel. Bar = 5 μm. (*See* Color Plates)

zone that defines the mating structures. In addition, overlying both mating structures is an extracellular coat, designated *fringe (31,32)*. The unactivated mt(–) mating structure appears cup shaped in three dimensions. Activation of the mt(–) mating structure occurs in two stages. During the "budding stage," the mt(–) mating structure assumes a bud-shaped morphology. The majority of the membrane zone remains at the base of the budded mating structure, which has been shown to contain vesicles, particulate material, and patchy regions of electron-dense submembranous material.

In the second "domed stage" of activation, the mating structure undergoes a morphological change into a domelike structure concomitant with a decrease in its content of particulate material. Extracellularly, the activated mating structure contains an apical tuft of fringe, and intracellularly a submembranous dense material termed the *central zone* is present at the apex of the domed mating structure *(32)*.

In contrast to the domed appearance of the mt(–) mating structure, the activated mt(+) mating structure is a ~3 μm long, microvillous-like fusion organelle (**Figs. 2, 3**). The unactivated mt(+) mating structure differs from its mt(–) counterpart in the presence of an electron-dense structure, the *doublet zone*, that underlies the membrane zone and resembles a lopsided collar or doughnut (**Fig. 3A; ref.** *31*).

In the first step of erection of a fertilization tubule, the unactivated mating structure buds in a manner analogous to the activation of the mt(–) mating structure. Unlike the mt(–) mating structure, however, in the mt(+) bud, the membrane zone remains at the apex of the forming fertilization tubule (*see* **Fig. 3B**). The increase in membrane required for this budding process most likely comes from the fusion of intracellular vesicles present in this region of the cell with the plasma membrane of the forming fertilization tubule (*see* **Fig. 3B**). In addition to the rapid recruitment of membrane, the polymerization of actin promotes the elongation of the fertilization tubule (*see* **Fig. 3C; refs.** *31,33,34*). Closer scrutiny of the ultrastructure of the fertilization tubule demonstrated an intimate interaction between the basal aspect of the doublet zone and the underlying nucleus–basal body connector via a connective lattice system *(35)*.

Once formed, this fusion organelle is a dynamic structure. In the absence of gametic fusion and loss of the flagellar adhesion-induced increase in cAMP, the fertilization tubule is reabsorbed. This process is characterized by both actin filament depolymerization and endocytosis of the plasma membrane as evidenced by numerous small membrane vesicles present within the collapsing fertilization tubule *(33)*.

Further studies employing thin-section and scanning electron microscopy have indicated that binding and subsequent fusion occur between the tip of the mt(+) fertilization tubule and the apex of the activated mt(–) mating structure

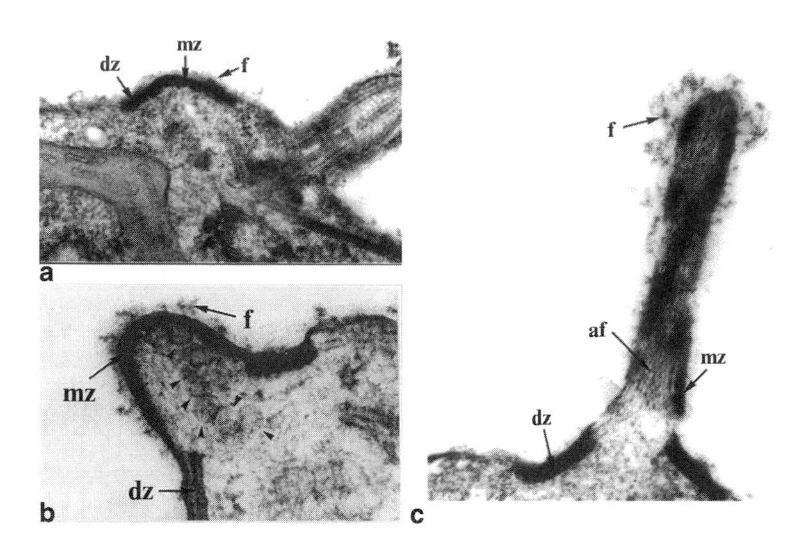

Fig. 3. Elongation of the mt(+) fertilization tubule. **(A)** Unactivated mating structure. The region comprising the mt(+) mating structure is defined by an electron-dense region, the membrane zone (mz) that lies adjacent to the overlying plasma membrane. An additional electron-dense structure, the doublet zone (dz) underlies the membrane zone. Extracellularly, the mating structure is covered by fringe (f), which mediates adhesion to activated mt(–) mating structures. In this section, a flagellum with its associated cytoskeletal elements is visible on the upper right part of the cell. (×60,500; **ref.** ***39***). **(B)** Partially activated fertilization tubule. The fertilization tubule is beginning to elongate as a result of the polymerization of actin into filaments. It is likely that fusion of vesicles visible within the forming fertilization tubule provides the additional membrane required for formation of the fully elongated fertilization tubule. In contrast to the membrane zone (mz), which remains apically localized, the doublet zone (dz) remains at the base of the forming fertilization tubule. Fringe (f) coats the exterior of this elongating fertilization tubule. (×96,700; **ref.** ***39***). **(C)** Fully activated fertilization tubule. Actin filaments (af) fill the erect fertilization tubule, which has fringe (f) located primarily at its apex. dz, Doublet zone; mz, membrane zone. This tubule is approximately 1 μm long ***(39)***.

(*see* **Fig. 1F; refs.** ***32,33,36)***. Consistent with this observation, freeze-fracture electron microscopic studies of the activated mating structures revealed a cluster of intramembranous particles in the region where gametic fusion occurs ***(37)***. These observations suggest that the molecules responsible for gametic adhesion and fusion are sequestered to specific regions of the mating structures.

The microvillous-like nature of the fertilization tubule allowed the development of methods for its isolation and purification ***(38,39)***. With these methods, a 360-fold enrichment of fertilization tubules was obtained, allowing the identification of a number of proteins enriched with these fusion organelles. The ability to isolate fertilization tubules led to the development of an in vitro binding assay that demonstrated the isolated fertilization tubules bound to activated mt(–) mating structures in a trypsin-sensitive manner ***(38,39)***. No binding was observed using unactivated mt(–) gametes that had been treated with lysin to remove their cell walls. These results demonstrated that the isolated fertilization tubules were competent to undergo adhesion and provided evidence that a protein component on the surface of fertilization tubules was responsible for adhesion between mating structures.

6. FUS1 Is the Adhesion Molecule on mt(+) Fertilization Tubules

Although first identified as a structural component of mating structures, subsequent studies on gametic cell fusion using impotent (*imp*) mt(+) mutants suggested a functional role for fringe as a gametic adhesion molecule. One of these mutants, *imp-1,* undergoes normal flagellar adhesion but is unable to fuse with mt(–) gametes ***(32)***. Ultrastructural studies of the fertilization tubule formed by *imp-1* revealed the absence of fringe. The gene affected in *imp-1* was subsequently identified as *FUS1* ***(40)***. As expected, fringe expression was restored in *imp-1* mutants transformed with a wild-type *FUS1* transgene. The ability of these *FUS1*-rescued transformants to fuse with wild-type mt(–) gametes provides compelling evidence that *FUS1* encodes the mt(+) fringe. Recent analysis of the FUS1 sequence detected a 22% identity and 35% similarity to the invasin/intimin family of bacterial proteins, which are utilized by bacteria for adhesion to their mammalian host cells ***(41,42)***. This similarity between FUS1 and invasin includes five immunoglobulinlike repeats consisting of ~90 amino acids.

Further evidence for an adhesive function for fringe stems from studies on the docking interactions between mating structures. Using a mixture of live gametes and activated, glutaraldehyde-fixed gametes of the opposite mating type, Misamore et al. ***(41)*** visualized the docking interactions between mating structures. Only mt(+) and mt(–) gametes that had undergone gametic activation by treatment with dibutrylyl cAMP were competent to undergo docking. Moreover, analysis of *imp-1* gametes with this assay revealed their inability to dock with wild-type mt(–) gametes. With an antibody directed against FUS1, a 95-kDa protein was identified that was present only in mt(+) gametes and enriched in isolated fertilization tubules ***(41)***. Consistent with a role in adhesion of mating structures, FUS1 was immunolocalized along the length of fertilization tubules. Interestingly, fringe and FUS1 are present on unactivated mt(+) mating structures ***(32,41)***. Moreover, treatment of dewalled, unactivated mt(+) gametes with trypsin resulted in the loss of FUS1 as detected by both immunoblot and indirect immunofluorescence.

This result suggests that all of FUS1 is accessible on the surface of the unactivated mt(+) mating structure. The inability of unactivated mt(+) gametes to dock with activated mt(–) gametes suggest that FUS1 is inactive in the absence of sexual signaling. Whether the requirement for cAMP signaling reflects a posttranslational modification of FUS1, an association with accessory proteins, or perhaps the dissociation of an inhibitory protein has yet to be determined. These observations, however, provide compelling evidence for FUS1 as an mt(+)-specific molecule that mediates adhesion between mating structures.

7. The mt(–) Localization of Fusion Protein(s)

Gametic fusion in *Chlamydomonas* has been proposed to occur in a manner analogous to that of virus cell fusion. In viral fusion, an adhesion/fusion protein (present either in a single protein or in different proteins) exclusively present on the virus interacts with the membrane of the host cell to induce destabilization of the membrane and fusion. Similarly, fusion of *Chlamydomonas* gametes has been proposed to occur through the action of an adhesion molecule localized to mt(+) gametes and a fusion protein present only on mt(–) gametes ***(43)***.

This model is based on studies using mt(–) mutants defective in gametic cell fusion. In these studies, an apparent adhesive interaction between mating structures was observed in matings between wild-type mt(+) gametes and mt(–) *gam10* or *gam11* gametes. In contrast, no adhesive interaction between mating structures was observed in matings between *imp-1* mt(+) gametes and wild-type mt(–) gametes ***(36,43)***. Although this is an intriguing model, the interpretation of these results is complicated by the observation that these *gam* mutants undergo normal flagellar adhesion. Additional studies will be required to determine whether the adhesion observed in this assay was due to interactions between mating structures or flagella. The cloning and identification of the gene(s) affected in these *gam* mutants will be required for a complete understanding of mating structure adhesion and fusion. Moreover, what role, if any, the mt(–) fringe plays in the interaction between activated mating structures is unknown. It is possible that the protein encoding mt(–) fringe could represent the receptor for FUS1.

8. Conclusions and Future Directions

The remarkable similarity between fertilization in multicellular organisms and fertilization in *Chlamydomonas* highlights the usefulness of this model organism in understanding the molecular mechanisms that mediate adhesion and fusion of gametes. *Chlamydomonas* has an advantage over other model systems for fertilization in that it is amenable to genetic manipulations as well as cell and molecular biological approaches. Moreover, the ease of culturing *Chlamydomonas* makes it an ideal system for biochemical studies. Although *Chlamydomonas* has provided a wealth of information regarding gametic adhesion and activation, the molecules involved in fusion have remained elusive. Continued studies on the role of PKG in the activation of adenylyl cyclase in gametic flagella, the identification of the FUS1 receptor on mt(−) mating structures, and identification of the genes affected in the fusion-defective mt(−) mutants should provide important new insights into this complex process.

References

1. Lien, T. and Knutsen, G. (1979) Synchronous growth of *Chlamydomonas reinhardtii* (Chlorophyceae): a review of optimal conditions. *J. Phycol.* **15,** 191–200.
2. Sager, R. and Granick, S. (1954) Nutritional control of sexuality in *Chlamydomonas reinhardi. J. Gen. Physiol.* **37,** 729–742.
3. Pan, J-M, Haring, M. A., and Beck, C. F. (1996) Dissection of the blue-light dependent signal-transduction pathway involved in gametic differentiation of *Chlamydomonas reinhardtii. Plant Physiol.* **112,** 303–309.
4. Adair, W. S., Monk, B. C., Cohen, R., Hwang, C., and Goodenough, U. W. (1982) Sexual agglutinins from the *Chlamydomonas* flagellar membrane. Partial purification and characterization. *J. Biol. Chem.* **257,** 4593–4602.
5. Adair, W. S., Hwang, C., and Goodenough, U. W. (1983) Identification and visualization of the sexual agglutinin from the mating-type plus flagellar membrane of *Chlamydomonas. Cell* **33,** 183–193.
6. Cooper, J. B., Adair, W. S., Mecham, R. P., Heuser, J. E., and Goodenough, U. W. (1983) *Chlamydomonas* agglutinin is a hydroxyproline-rich glycoprotein. *Proc. Natl. Acad. Sci. U.S.A.* **80,** 5898–5901.
7. Collin-Osdoby, P., and Adair, W. S. (1985) Characterization of the purified *Chlamydomonas* minus agglutinin. *J. Cell Biol.* **101,** 1144–1152.
8. Ferris, P. J., Waffenschmidt, S., Umen, J. G., Lin, H., Lee, J-H., Ishida, K., Kubo, T., Lau, J., and Goodenough, U. W. (2005) Plus and minus sexual agglutinins from *Chlamydomonas reinhardtii. Plant Cell* **17,** 597–615.
9. Goodenough, U. W. and Jurivich, D. (1978) Tipping and mating-structure activation induced in *Chlamydomonas* gametes by flagellar membrane antisera. *J. Cell Biol.* **79,** 680–693.
10. Goodenough, U. W. (1993) Tipping of flagellar agglutinins by gametes of *Chlamydomonas reinhardtii. Cell Motil. Cytoskeleton* **25,** 179–189.
11. Snell, W. J. and Roseman, S. (1979) Kinetics of adhesion and de-adhesion of *Chlamydomonas* gametes. *J. Biol. Chem.* **254**, 10820–10829.
12. Hunnicutt, G. R., Kosfiszer, M. G., and Snell, W. J. (1990) Cell body and flagellar agglutinins in *Chlamydomonas reinhardtii*: the cell body plasma membrane is a reservoir for agglutinins whose migration to the flagella is regulated by a functional barrier. *J. Cell Biol.* **111,** 1605–1616.
13. Goodenough, U. W. (1989) Cyclic AMP enhances the sexual agglutinability of *Chlamydomonas* flagella. *J. Cell Biol.* **109,** 247–252.
14. Pasquale, S. M. and Goodenough, U. W. (1987) Cyclic AMP functions as a primary sexual signal in gametes of *Chlamydomonas reinhardtii. J. Cell Biol.* **105,** 2279–2292.

15. Saito, T., Small, L., and Goodenough, U. W. (1993) Activation of adenylyl cyclase in *Chlamydomonas reinhardtii* by adhesion and heat. *J. Cell Biol.* **122**, 137–147.
16. Zhang, Y., Ross, E. M., and Snell, W. J. (1991) ATP-dependent regulation of flagellar adenylylcyclase in gametes of *Chlamydomonas reinhardtii. J. Biol. Chem.* **266,** 22954–22959.
17. Zhang, Y. and Snell, W. J. (1993) Differential regulation of adenylylcyclases in vegetative and gametic flagella of *Chlamydomonas. J. Biol. Chem.* **268,** 1786–1791.
18. Zhang, Y. and Snell, W. J. (1994) Flagellar adhesion-dependent regulation of *Chlamydomonas* adenylyl cyclase in vitro: a possible role for protein kinases during sexual signaling. *J. Cell Biol.* **125,** 617–624.
19. Wang, Q. and Snell, W. J. (2003) Flagellar adhesion between mating type plus and mating type minus gametes activates a flagellar protein–tyrosine kinase during fertilization in *Chlamydomonas. J. Biol. Chem.* **278,** 32936–32942.
20. Wang, Q., Pan, J., and Snell, W. J. (2006) Intraflagellar transport particles participate directly in cilium-generated signaling in *Chlamydomonas. Cell* **125,** 549–562.
21. Kozminski, K. G., Johnson, K. A., Forscher, P., and Rosenbaum, J. L. (1993) A motility in the eukaryotic flagellum unrelated to flagellar beating. *Proc. Natl. Acad. Sci. U.S.A.* **90,** 5519–5523.
22. Qin, H., Diener, D. R., Geimer, S., Cole, D. G., and Rosenbaum, J. L. (2004) Intraflagellar transport (IFT) cargo: IFT transports flagellar precursors to the tip and turnover products to the cell body. *J. Cell Biol.* **164,** 255–266.
23. Singla, V. and Reiter, J. F. (2006) The primary cilium as the cell's antenna: signaling at a sensory organelle. *Science* **313,** 629–633.
24. Piperno, G., Mead, K., and Henderson, S. (1996) Inner dynein arms but not outer dynein arms require the activity of kinesin homologue protein KHP1 (FLA10) to reach the distal part of flagella in *Chlamydomonas. J. Cell Biol.* **133,** 371–379.
25. Pan, J. and Snell, W. J. (2002) Kinesin-II is required for flagellar sensory transduction during fertilization in *Chlamydomonas. Mol. Biol. Cell* **13,** 1417–1426.
26. Imam, S. H. and Snell, W. J. (1988) The *Chlamydomonas* cell wall degrading enzyme, lysin, acts on two substrates within the framework of the wall. *J. Cell Biol.* **106,** 2211–2221.
27. Matsuda, Y., Saito, T., Yamaguchi, T., and Kawase, H. (1985) Cell wall lytic enzyme released by mating gametes of *Chlamydomonas reinhardtii* is a metalloprotease and digests the sodium perchlorate-insoluble component of cell wall. *J. Biol. Chem.* **260,** 6373–6377.
28. Matsuda, Y., Saito, T., Yamaguchi, T., Koseki, M., and Hayashi, K. (1987) Topography of cell wall lytic enzyme in *Chlamydomonas reinhardtii*: form and location of the stormed enzyme in vegetative cell and gamete. *J. Cell Biol.* **104,** 321–329.
29. Buchanan, M. J., Imam, S. H., Eskue, W. A., and Snell, W. J. (1989) Activation of the cell wall degrading protease, lysin, during sexual signalling in *Chlamydomonas*. The enzyme is stored as an inactive, higher relative molecular mass precursor in the periplasm. *J. Cell Biol.* **108,** 199–207.
30. Snell, W. J., Eskue, W. A., and Buchanan, M. J. (1989) Regulated secretion of a serine protease that activates an extracellular matrix–degrading metalloprotease during fertilization in *Chlamydomonas. J. Cell Biol.* **109,** 1689–1694.

31. Goodenough, U. W. and Weiss, R. L. (1975) Gametic differentiation in *Chlamydomonas reinhardtii* III. Cell wall lysis and microfilament-associated mating structure activation in wild-type and mutant strains. *J. Cell Biol.* **67,** 623–637.
32. Goodenough, U. W., Detmers, P. A., and Hwang, C. (1982) Activation for cell fusion in *Chlamydomonas*: analysis of wild-type gametes and nonfusing mutants. *J. Cell Biol.* **92,** 378–386.
33. Detmers, P. A., Goodenough, U. W., and Condeelis, J. (1983) Elongation of the fertilization tubule in *Chlamydomonas*: new observations on the core microfilaments and the effect of transient intracellular signals on their structural integrity. *J. Cell Biol.* **97,** 522–532.
34. Detmers, P. A., Carboni, J. M., and Condeelis, J. (1985) Localization of actin in *Chlamydomonas* using antiactin and NBD-phallacidin. *Cell Motil.* **5,** 415–430.
35. Goodenough, U. W. and Weiss, R. L. (1978) Interrelationships between microtubules, a striated fiber, and the gametic mating structures of *Chlamydomonas reinhardtii. J. Cell Biol.* **76,** 430–438.
36. Forest, C. L. (1983) Specific contact between mating structure membranes observed in conditional fusion-defective *Chlamydomonas* mutants. *Exp. Cell Res.* **148,** 143–154.
37. Weiss, R. L., Goodenough, D. A., and Goodenough, U. W. (1977) Membrane differentiations at sites specialized for cell fusion. *J. Cell Biol.* **72,** 144–160.
38. Wilson, N. F., Foglesong, M. J., and Snell, W. J. (1997) The *Chlamydomonas* mating type plus fertilization tubule, a prototypic cell fusion organelle: isolation, characterization, and in vitro adhesion to mating type minus gametes. *J. Cell Biol.* **30,** 1537–1553.
39. Wilson, N. F. (1996) The *Chlamydomonas* mt+ fertilization tubule: a model system for studying the role of cell fusion organelles in gametic cell fusion [Ph.D. thesis]. UT Southwestern Medical Center, Dallas.
40. Ferris, P. J., Woessner, J. P., and Goodenough, U. W. (1996) A sex recognition glycoprotein is encoded by the plus mating-type gene fus1 of *Chlamydomonas reinhardtii. Mol. Biol. Cell* **7,** 1235–1248.
41. Misamore, M. J., Gupta, S., and Snell, W. J. (2003) The *Chlamydomonas* Fus1 protein is present on the mating type plus fusion organelle and required for a critical membrane adhesion event during fusion with minus gametes. *Mol. Biol. Cell* **14,** 2530–2542.
42. Vallance, B. A. and Finlay, B. B. (2000) Exploitation of host cells by enteropathogenic *Escherichia coli. Proc. Natl. Acad. Sci. U.S.A.* **97,** 8799–8806.
43. Forest, C. L. (1987) Genetic control of plasma membrane adhesion and fusion in *Chlamydomonas* gametes. *J. Cell Sci.* **88,** 613–621.

4

Cell Fusion in *Caenorhabditis elegans*

Scott Alper and Benjamin Podbilewicz

Summary

In the nematode *Caenorhabditis elegans*, 300 of the 959 somatic nuclei present in the adult hermaphrodite are located in syncytia. These syncytia are formed by the fusion of mononucleate cells throughout embryonic and postembryonic development. These cell fusions occur in a well-characterized stereotypical pattern, allowing investigators to study many cell fusion events at the molecular and cellular levels. Using tools that allow visualization of cell membranes, cell junctions, and cell cytoplasm during fusion, genetic screens have identified many *C. elegans* cell fusion genes, including those that regulate the fusion cell fate decision and two genes that encode components of the cell fusion machinery.

Key Words: *Caenorhabditis elegans*; nematode; cell fusion; epidermis; uterus.

1. Introduction

Cell–cell fusion is a common process that occurs during animal development, with essential cell fusions occurring between sperm and egg during fertilization; cells of muscle, bone, and placenta; and stem cells. Cell fusion also occurs during the pathogenesis of certain diseases, including the formation of giant cells in cancer, tuberculosis, and viral infections. *Caenorhabditis elegans* has become an excellent model system with which to study cell fusion both because of the resources available to study fusion at the genetic and molecular levels and because of the large number of cells that fuse in a stereotypical fashion. Nearly one third of all somatic cells in *C. elegans* undergo cell fusion during development *(1,2)*. Almost every epidermal (hypodermal) cell in *C. elegans* fuses with a set of epidermal syncytia that grow in size throughout development (**Fig. 1A**). Cell fusion also occurs in several other *C. elegans* organs during development, including the pharynx, uterus, and vulva (**Fig. 1B**). Interestingly, unlike in other animals, the body wall ("skeletal") muscle remains unfused in *C. elegans*.

From: *Methods in Molecular Biology, vol. 475: Cell Fusion: Overviews and Methods*
Edited by: E. H. Chen © Humana Press, Totowa, NJ

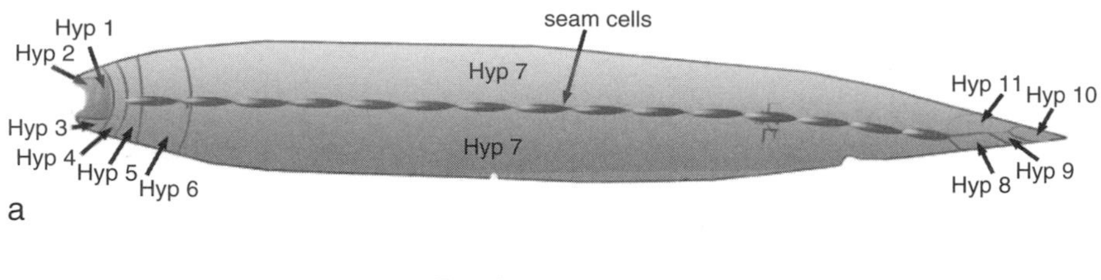

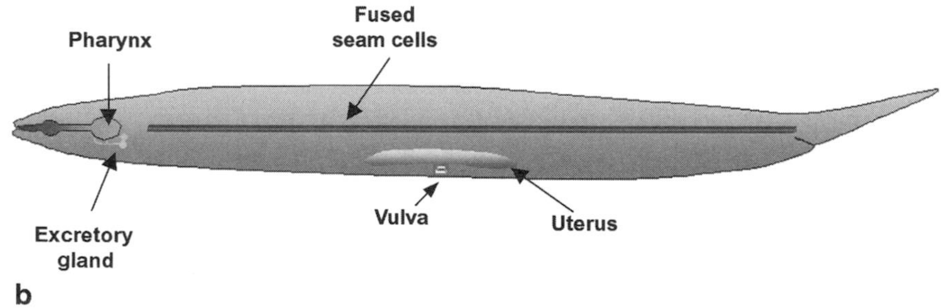

Fig. 1. Overview of syncytia present in *Caenorhabditis elegans*. **(A)** Many of the epidermal syncytia that comprise the *C. elegans* epidermis (hyp1–hyp12) as well as the unfused epidermal seam cells. **(B)** Other organs or tissues that contain syncytia in *C. elegans*. (A, reprinted from *Wormatlas* [www.wormatlas.org], with permission; B, reprinted from Shemer and Podbilewicz *[71]*.)

In addition to the classic genetic and genomic tools that make *C. elegans* an attractive model organism to study various aspects of development, several molecular tools have been developed that allow a detailed analysis of the cell biology of membrane fusion. These resources include green fluorescent protein (GFP) fusions that label the cytoplasm, cell membrane, and apical junctions (S. Alper, unpublished data; **refs. *3–6***). Because *C. elegans* is transparent, these GFP fusions can be visualized using fluorescence microscopy during development in live organisms, allowing monitoring of cell fusion in real time. Membrane-labeling fluorescent dyes and a monoclonal antibody that recognizes apical junctions have also been used in conjunction with fluorescence and electron microscopy to study cell fusion in the nematode ***(5,7–9)***. Forward genetic screens have identified many genes that regulate the fusion cell fate decision ***(1)*** and two membrane fusion effector proteins or fusogens ***(4,10,11)***. These *C. elegans* fusogens are the only such proteins with demonstrated fusogenic activity other than syncytin, which is involved in primate placental morphogenesis ***(12)***. The study of cell fusion in *C. elegans* has focused in large part on four fundamental biological questions: *How*, *when*, *where*, and *why* do cells fuse?

How cells fuse—a question of cell biology. The process of *C. elegans* cell fusion typically involves several steps. First, short cell migrations or cell shape changes bring cells destined to fuse into close proximity. These cells must then recognize each other and adhere, and, finally, the membrane fusion event occurs. All these steps are complicated biological processes, and *C. elegans* is an attractive system to understand this morphogenesis in cellular and molecular detail.

When and where cells fuse—questions of developmental biology. Cells that fuse in the epidermis originate from different anteroposterior (AP) and dorsoventral (DV) positions in the nematode and from different cell lineages. The pattern of fusion is also slightly different between the two sexes. Moreover, the decision to fuse is controlled in a carefully choreographed temporal fashion. Thus, much complex developmental information must be integrated to control morphogenesis and cell fusion.

Why do cells fuse? Is there some advantage to having a fused epidermis or other fused organs? The identification of mutants that block cell fusion in *C. elegans* allows us to begin to address this question.

In this chapter, we focus on several exciting developments in the *C. elegans* cell fusion field in an attempt to answer these questions. This progress is best exemplified by our understanding of epidermal cell fusion in the nematode, so we begin with the epidermis. (We will not consider *C. elegans* sperm–egg fusion in this chapter, but some recent excellent reviews are available ***[13–15]***.)

2. Overview of Epidermal Cell Fusion

The epidermis in the adult hermaphrodite *C. elegans* is composed of 12 cells (hyp1 to hyp12) containing a total of 157 nuclei (13 cells containing 159 nuclei in males) (*see* **Fig. 1A; ref. *2***). Eight of these hyp cells are syncytial, formed by the fusion of mononucleate cells throughout embryonic and postembryonic development. Although smaller syncytia are present in the head and tail, most of the epidermal nuclei reside in the hyp7 syncytium, which spans most of the length of the worm and which contains 139 nuclei in the adult ***(9,16,17)***. Thus, the hyp7 cell in the adult contains almost one sixth of all somatic nuclei in the nematode.

Caenorhabditis elegans initially develops as an egg-encased embryo. Upon hatching, the larvae progress through four larval stages, L1 through L4 ***(18)***. In the embryo, the epidermis is initially formed as six rows of mononucleate epithelial cells that envelop the nematode: two rows of cells on the dorsal surface, two rows of cells on the ventral surface, and a row of laterally located cells (called *seam cells*) on each side of the nematode (**Fig. 2; refs. *9,16***). Prior to embryonic elongation, the two rows of dorsal cells extend their membranes laterally across the dorsal midline and interdigitate, resulting in a single row of

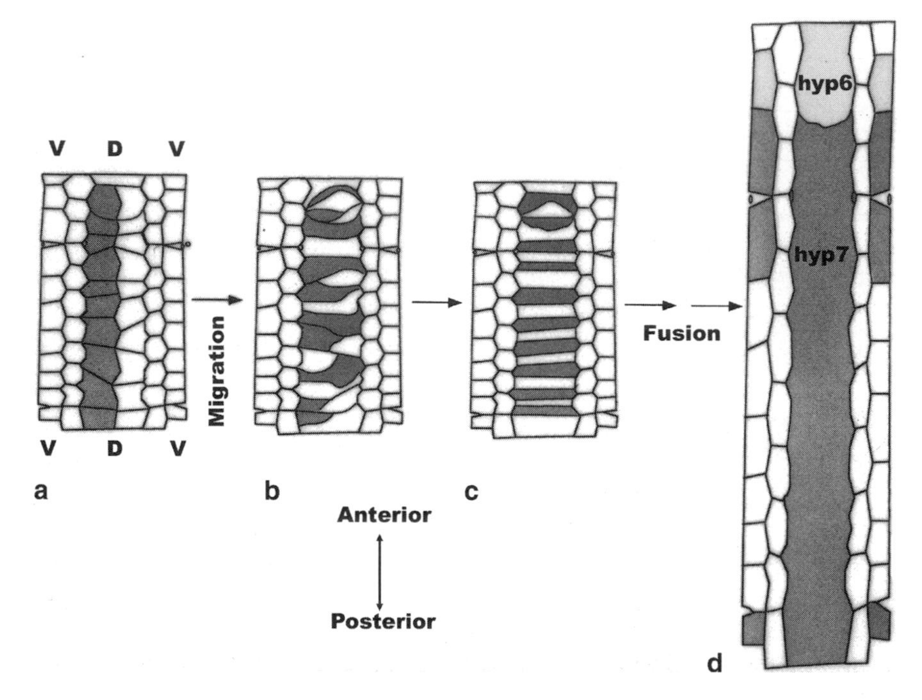

Fig. 2. Schematic depicting epidermal cell shape changes and fusions in the embryo. Cylindrical projection of the midbody region of an embryo that was cut along the ventral midline and viewed from outside the animal. **(A)** The six rows of epidermal cells. **(B,C)** The dorsal cells migrate and interdigitate. **(D)** As the embryo elongates, the 23 cells that make up embryonic hyp7 (dark grey) fuse with each other, as do the 6 cells that make up hyp6 (light grey). D, dorsal cells; V, ventral cells. (Adapted from Podbilewicz and White *[9]*, with permission of Elsevier.)

elongated epithelial cells (*see* **Fig. 2; refs. *9,16,19***). During embryonic elongation, these 17 dorsal epithelial cells and 6 ventrally located cells (4 anterior, 2 posterior) fuse to form the hyp7 cell, which by the end of embryogenesis contains 23 nuclei (*see* **Fig. 2; ref. *9***).

During larval development, cells on the lateral and ventral surfaces of the nematode fuse with hyp7, causing hyp7 to grow in size and nuclear count as the nematode grows. Each lateral surface of a newly hatched larva is composed of a row of 10 epidermal seam cells ***(16)***. These seam cells proliferate during each of the four larval stages of development; typically, after each A/P oriented division, one daughter cell fuses with hyp7 and one cell remains unfused (**Fig. 3; refs. *9,20,21***). Thus, in a sense, the lateral seam cells can be thought of as stem cells that continue to proliferate throughout larval development, generating cells that fuse with hyp7 and regenerating the proliferating stem cell. Usually,

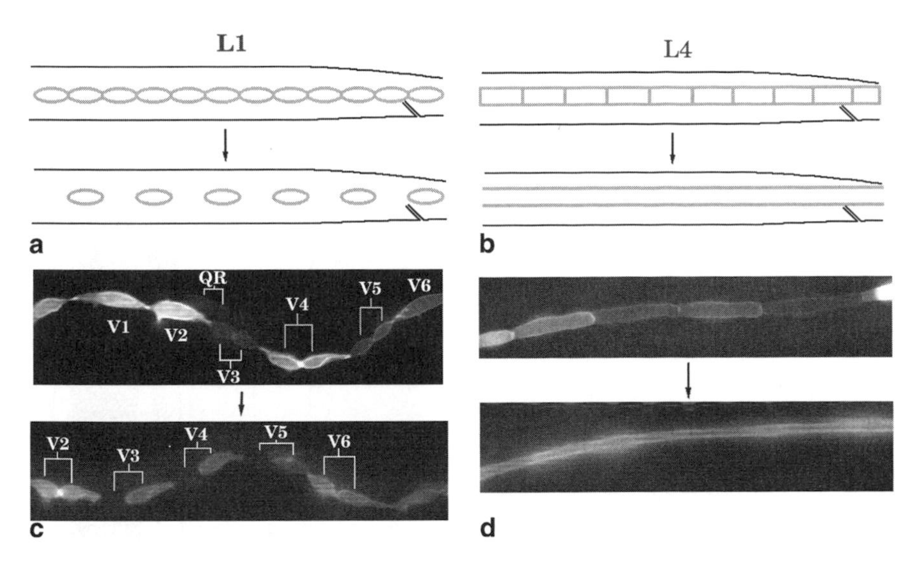

Fig. 3. Overview of lateral epidermal cell fusion in larvae. **(A,B)** Schematics of lateral epidermal fusions in L1 and L4, respectively. **(C,D)** The same cell fusions visualized with a seam cell–specific membrane-GFP strain (S. Alper, unpublished data). During L1 **(A,C)**, one cell from each seam cell division fuses with hyp7 (which is located mostly dorsally); as fusion occurs, the membrane-GFP signal disappears. During L4 **(B,D)**, the seam cells fuse with each other, resulting in a stripe of syncytium on either side of the animal. At this stage hyp7 surrounds most of the animal. Anterior is to the left in all panels.

but not exclusively, it is the anterior daughter from each AP-oriented division that fuses with hyp7 ***(20)***. Additionally, certain seam cells divide and produce neuronal progeny ***(20)***. Thus, these lateral seam cells produce the majority of the more than 100 additional cells that fuse with hyp7 during larval development. Late in the fourth larval stage, the seam cells on each side of the worm fuse with each other, generating two stripes of syncytia, one on either side of the animal (*see* **Fig. 3; ref.** ***9)***.

Most of the ventral surface of the newly hatched nematode larva is composed of two rows of six pairs of P blast cells (with smaller hyp syncytia in the head and tail; **Fig. 4A; refs.** ***9, 20)***. During L1, the six pairs of P cells rotate around each other so that they form a single row of 12 cells along the ventral midline (*see* **Fig. 4A; refs.** ***9, 20)***. Following this short cell migration, the P cells divide. The anterior daughters (the Pn.a cells) become neuroblasts ***(20)***. The posterior daughters (the Pn.p cells) remain epidermal, and late in L1 some of the Pn.p cells fuse with the hyp7 syncytium (*see* **Fig. 4B,C; ref.** ***9)***. However, certain Pn.p cells remain unfused; it is critical that the appropriate cells remain unfused

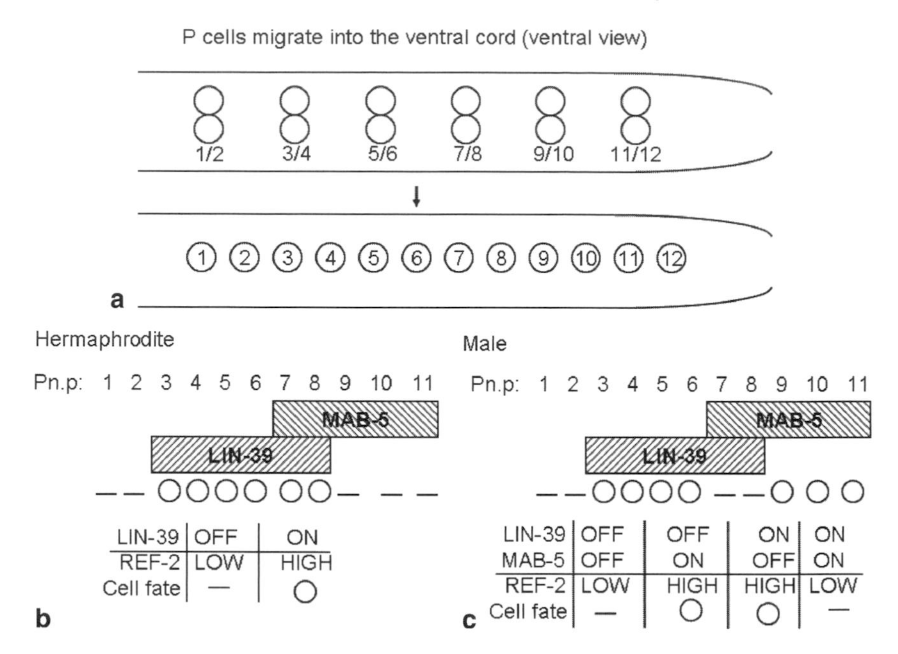

Fig. 4. Overview of ventral epidermal cell migration and fusion. **(A)** During L1, the 6 pairs of P cells migrate around each other, resulting in a single row of 12 P cells (ventral view). Those P cells divide and the posterior daughters remain part of the epidermis (P1.p – P12.p). **(B,C)** At the end of L1, some of the Pn.p cells fuse with the hyp7 syncytium in the patterns depicted. The pattern of fusion is controlled as outlined in B and C by the homeobox genes *lin-39* and *mab-5,* which are expressed in the indicated cells. P12.p behaves differently and is not pictured. The tables explain how homeobox proteins control *ref-2* expression and therefore cell fusion. An open circle indicates a cell that remains unfused; a dash indicates a cell that has fused with the hyp7 syncytium ***(28,29,45,73)***.

because their descendents generate sex-specific mating structures (the vulva, located in the midbody region in hermaphrodites, or the hook, located more posteriorly in males). The Pn.p cells that remain unfused in hermaphrodites are known as the vulval precursor cells; the biology of how these cells generate the vulva has been extensively studied ***(22)***.

Several studies have examined the membrane fusion event in the epidermis, particularly in the embryo. These cell fusions have been characterized using techniques such as fluorescence microscopy in live embryos in conjunction with the fluorescent dye FM 4-64, which stains cell membranes; the AJM-1–GFP fusion, which labels apical junctions; and several soluble GFP fusions that allow visualization of cytoplasm mixing during cell fusion ***(3–7)***. Fixed animals have also been examined using fluorescence microscopy or electron microscopy in conjunction with the MH27 monoclonal antibody, which recognizes the AJM-1 antigen and therefore stains apical junctions *(6–9,23,24)*. Loss of MH27 staining has proved to be a reliable indicator of cell fusion.

With these reagents, it was found that cell fusion typically initiates between two fusing cells with the formation of one or more fusion pores (**Fig. 5; refs.** *5,7*). These pores are formed at the apical surface or very near the apical junction (**5**). The fusion pores then begin to expand (*see* **Fig. 5**). As the fusion pores expand basally, the AJM-1 antigen and the cell junction both move in conjunction with the expanding pores from the apical to the basal surface *(5)*. In fact, the AJM-1 antigen is present briefly on the basal surface once most of the membrane has been removed; shortly afterward, the AJM-1 antigen disappears *(5)*. The entire process takes approximately 30 min at 23°C *(7)*. What happens to the cell membranes during fusion is unclear, although small membranous vesicles have been visualized near the fusion site with fluorescence microscopy and electron microscopy *(5,8)*. Unlike the small vesicles that are present prior to fusion during *Drosophila* myoblast fusion *(25)*, the *C. elegans* vesicles appear during the fusion process and are hypothesized to remove the cell membrane from the fusion site *(5,8)*.

Kinetic, ultrastructural, and mutant studies suggest that cell fusion in the nematode epidermis can be divided into three kinetic steps *(7)*. Fusion pore formation involves two steps, and subsequent pore expansion/resolution is an additional step. One possibility is that the two initial steps could involve the formation of a hemifusion intermediate and subsequent fusion pore formation (**Fig. 6;** *see also* **Heading 3**). The identification of several genes intimately involved in epidermal cell fusion will allow further dissection of this complex process.

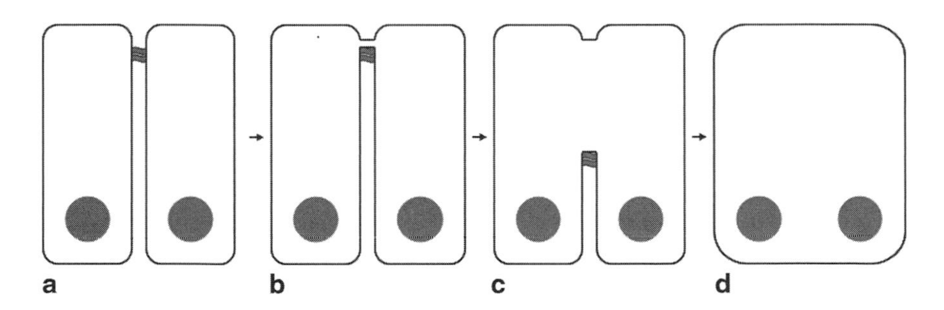

Fig. 5. Schematic depicting epidermal cell fusion. **(A)** Adjacent epidermal cells (apical surface at top, nuclei depicted as grey circles) prior to fusion linked by apical junction (grey squiggly bars). **(B)** Initial fusion pore formation occurs at or near the apical junction. **(C)** As the fusion pore expands, the apical junction sinks basally. **(D)** membrane fusion is complete and the apical junction has disappeared. EFF-1 is required for initial pore formation and subsequent expansion of the pore.

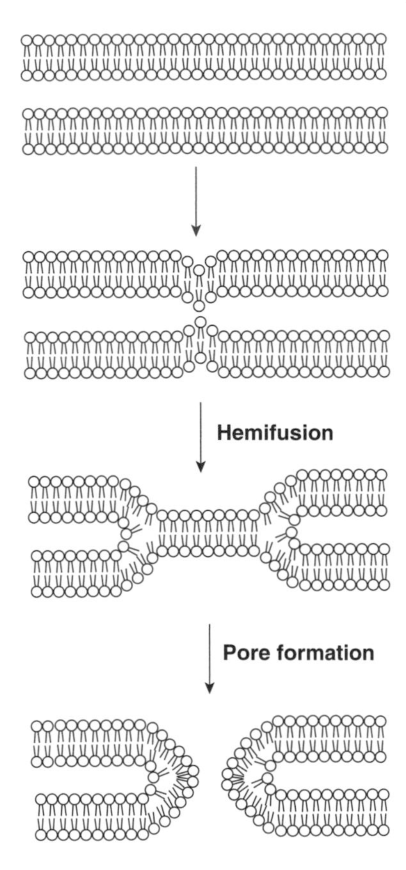

Fig. 6. Schematic depicting membrane fusion. EFF-1 drives pore formation through a hemifusion intermediate ***(10)***.

3. EFF-1 Protein Is Required for Most Epithelial Cell Fusions in *C. elegans*

Numerous genetic screens have been performed to identify mutations that affect cell fusion in *C. elegans* ***(2–4,6,26–40)***. To date, these screens have identified many genes that regulate the fusion cell fate decision but only two genes that are clearly involved in the membrane fusion process itself: *eff-1*, which is required for cell fusion in the epidermis, vulva, and pharyngeal muscle (this section); and *aff-1*, which is required for cell fusion in the uterus, the epidermal seam, and certain vulval cells (*see* **Heading 6**).

The *eff-1* gene was initially identified in two independent screens for mutations that affect epidermal cell fusion ***(4)***. One screen was a direct fluorescence microscopy screen, in which worms carrying the AJM-1–GFP apical junction marker were visually inspected for epidermal fusion defects. Loss of AJM-1–GFP-labeled junctions has been a reliable indicator of cell fusion in many studies. The second screen identified nematodes that failed to lay eggs, indicative of vulval differentiation defects; these mutations were then examined using AJM-1–GFP to visualize cell fusion directly. These *eff-1* alleles and those identified subsequently demonstrate that EFF-1 protein is required for most epidermal, vulval, and pharyngeal cell fusions examined ***(2–4,6,7,10,26,30,33)***. *eff-1* mutant nematodes were viable but had many morphogenetic defects, including egg-laying defects, dumpy body shape, and uncoordinated movement, suggesting that cell fusion is critical to proper organ function ***(4)***. Although several other alleles of *eff-1* have since been identified in additional screens, no other genes that are required for most epidermal cell fusion have been identified, suggesting either that no other nonredundant cell fusion components exist or that all other cell fusion components are required for viability in other processes during development ***(3,6,7,10)***, although *aff-1* is required for the fusion of a very small subset of epidermal cells (*see* **Heading 6; ref.** ***11***).

Judged by several criteria, *eff-1* mutations block membrane fusion completely. In an *eff-1* mutant background, no fusion was observed when cell junctions were visualized using AJM-1–GFP ***(4)***. Similarly, no fusion in the *eff-1* mutant nematode was observed by electron microscopy ***(6)***. Moreover, cytoplasmically localized GFP did not diffuse from one cell to another in *eff-1* mutant animals ***(4)***. Thus, the phenotype of the *eff-1* mutant suggested that EFF-1 was an excellent candidate for a membrane fusogen. EFF-1 exists in several isoforms, including two membrane-bound forms and two secreted forms ***(3,41)***. Consistent with its proposed role as a membrane fusogen, the membrane-bound form of EFF-1–GFP localizes to the site of cell fusion as cells fuse ***(3)***.

EFF-1 contains several domains that suggest a role in the membrane fusion process, including a hydrophobic stretch reminiscent of viral fusion peptides and a phospholipase domain that could be involved in disruption or bending of the outer leaflets of cell membranes ***(4)***. However, mutation of several amino acids in the phospholipase domain did not affect cell fusion, suggesting that this domain is not required for the membrane fusion event ***(3)***. Mutation of the putative hydrophobic peptide domain in EFF-1 blocked cell fusion but also blocked proper EFF-1 protein localization, so it is unclear whether this domain is required for membrane fusion or simply for proper protein folding/subcellular localization of EFF-1 ***(3)***. The putative fusion peptide domain in *C. elegans* EFF-1 is only moderately conserved in EFF-1 in other nematode species and also in AFF-1 in *C. elegans* (see **Heading 6**), suggesting that this domain may not be essential for cell fusion. EFF-1 does contain a domain with a transforming growth factor-β (TGFβ) type I receptor-like (TGFβ-RI) fold that is conserved in EFF-1 and AFF-1 in multiple nematode species. This domain overlaps with the putative phospholipase domain; the identification of an *eff-1* allele with a mutation in a conserved cysteine within these domains suggests that the TGFβ-RI fold is essential for membrane fusion (N. Assaf and B. Podbilewicz, unpublished data; **ref.** ***11***).

eff-1 is clearly necessary for cell fusion. Experiments using a heat shock–inducible promoter demonstrate that *eff-1* is also sufficient to promote cell fusion in *C. elegans* in cells and cell types that normally do not fuse during development ***(3,6)***. These experiments culminated in the demonstration that EFF-1 could induce cell fusion in a heterologous insect cell culture system ***(10)***, demonstrating that EFF-1 is a bona fide fusogen and suggesting that no other nematode components are required for membrane fusion.

How does EFF-1 mediate membrane fusion? Although *eff-1* mutations normally prevent even the formation of the initial fusion pore, electron microscopic analysis of cell fusions in pharynx and epidermis in one temperature-sensitive *eff-1* allele led to the discovery of some fusion pores that are formed but that do not expand ***(6,7)***. This suggests that EFF-1 functions in at least two steps during the fusion process: initial pore formation and subsequent expansion of the fusion pore to the entire membrane (*see* **Fig. 5**). In cell culture experiments examining EFF-1–mediated fusion in heterologous cells, it was found that EFF-1 could stimulate membrane mixing without cytoplasmic content mixing ***(10)***. These experiments and others suggest that EFF-1–mediated fusion proceeds through a hemifusion intermediate similar to that found in viral-mediated fusion (*see* **Fig. 6; refs.** ***42,43***). However, unlike viral fusion, in which the fusogen is required in only one fusing membrane, EFF-1 is required in the membranes of both fusing cells in the nematode and in the cell culture system ***(10)***.

Thus, cell fusion in *C. elegans* proceeds through at least three steps. Initial pore formation involves both a hemifusion intermediate and subsequent production of fusion pores. These pores then expand to encompass all of the membrane between the fusing cells. Both the pore formation and subsequent expansion processes require EFF-1 activity. Although there are no obvious *eff-1* homologs outside of nematodes, many similarities exist between the mechanism of developmental cell fusion in *C. elegans* and the mechanism of viral and intracellular fusion, suggesting that the general mechanisms by which membranes fuse may be highly conserved, even though the individual protein components are not.

4. Homeobox and Zinc Finger Transcription Factors Regulate Cell Fusion in the Lateral and Ventral Epidermis

EFF-1 is a membrane fusogen involved in hundreds of fusion events in *C. elegans*. These cell fusions occur between cells located at different positions along the AP and DV body axes, between cells from different cell lineages, in a

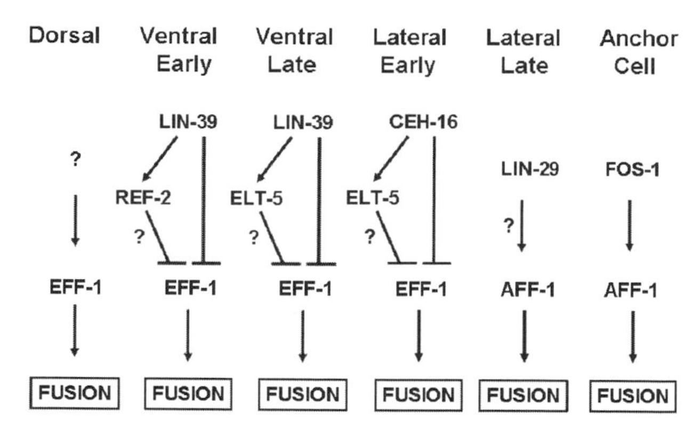

Fig. 7. Genetic pathways that control cell fusion in the epidermis and uterus. Depicted are the pathways that regulate *eff-1* and *aff-1* expression and therefore cell fusion in the indicated epidermal domain (first five pathways) or uterine domain (last pathway) as outlined in the text. LIN-39 and CEH-16 are homeobox-containing transcription factors that regulate the expression of the zinc finger transcription factors ELT-5 and REF-2 as indicated in the different epidermal domains. The zinc finger and homeobox proteins then together inhibit *eff-1*, thereby preventing cell fusion. In cells that do not express *lin-39* or *ceh-16*, EFF-1 is produced and the cells fuse. *aff-1* is required late in the lateral epidermis for cell fusion and may be regulated by the heterochronic gene *lin-29. aff-1* also is needed for the anchor cell to fuse with the utse; *aff-1* expression depends on the FOS-1 transcription factor.

sex-specific pattern and only between cells at the appropriate time. How is this positional, lineage, sex, and temporal information integrated to control *eff-1* and the decision to fuse? What controls this intricate fusion choreography?

Many genes have been identified that affect fusion of lateral and ventral epidermal cells during development. The default state in these cells appears to be to fuse; thus, active repression of *eff-1* is required to prevent cell fusion in the epidermis until the appropriate time and place. A common feature shared by lateral and ventral epidermal cells is that homeobox and zinc finger transcription factors promote epidermal cell fate and repress inappropriate cell fusion (**Fig. 7**). These homeobox-containing transcription factors induce the expression of a set of zinc finger transcription factors. Together, these zinc finger and homeobox proteins repress *eff-1* expression, thereby preventing inappropriate cell fusion.

4.1. elt-5 *and* ceh-16 *Regulate Lateral Epidermal Cell Fusion*

In the lateral epidermis, two GATA-type zinc finger transcription factors, ELT-5/EGL-18 and ELT-6, promote the seam cell fate and prevent these cells from fusing with the growing epidermal syncytium (*see* **Fig. 7; ref. *35***). ELT-5 and ELT-6 may function partly or completely redundantly. Inhibition of *elt-5* and *elt-6* leads to the loss of seam cell dependent structures, because most of the seam cells fuse inappropriately with the epidermal syncytium *(35)*. *elt-5* prevents fusion by repressing *eff-1* transcription (either directly or indirectly), because RNA interference (RNAi)–mediated inhibition of *elt-5* leads to increased *eff-1::gfp* expression (*see* **Fig. 7; ref.** ***36***).

elt-5 expression is regulated by the homeobox-containing transcription factor *ceh-16/engrailed* (*see* **Fig. 7; ref.** ***36***). When *ceh-16* is mutated or inhibited by RNAi, the seam cells in the embryo fuse inappropriately and the embryo dies. CEH-16 prevents seam cell fusion by inhibiting *eff-1* expression in these cells. The effects of *ceh-16* on *eff-1* may be mediated by *elt-5*. *elt-5::gfp* expression decreased in the seam cells when *ceh-16* was inhibited by RNAi *(36)*. Moreover, when *ceh-16* was ectopically expressed, *elt-5::gfp* was inappropriately expressed in dorsal epidermal cells *(36)*. Thus, *ceh-16* induces *elt-5* expression in the seam cells, and *elt-5,* either alone or in conjunction with *ceh-16,* represses *eff-1* expression, thereby keeping the seam stem cells unfused. How this inhibition is overcome in fusing progeny of seam cells remains unclear. One possibility is that transcription factors and microRNAs of the heterochronic pathway regulate the temporal fusion fate by activating *eff-1* (*see* **Heading 5**).

4.2. ref-2 *and the Homeobox Genes Regulate Ventral Epidermal Cell Fusion in Larval Stage 1*

While *elt-5* acts throughout development in both the embryo and the larvae to regulate fusion in the lateral seam cells, an overlapping but distinct set of zinc finger transcription factors acts to control development and cell fusion on the ventral epidermis: *ref-2* during the first larval stage and *elt-5* and *elt-6* during later stages of development (*see* **Fig. 7**).

In a newly hatched nematode, most of the ventral midbody is composed of two rows of six P blast cells (*see* **Fig. 4A; ref.** ***16***). These 12 P cells undergo short cell migrations and divide, and some of their posterior progeny cells (the Pn.p cells) fuse with hyp7 (*see* **Fig. 4; refs.** ***9,20***). Inhibition of the zinc finger transcription factor *ref-2* by RNAi caused multiple defects in the P lineage during L1, including defects in cell migration and cell division of the P cells *(28)*. *ref-2* inhibition also leads to inappropriate fusion of all Pn.p cells with hyp7 *(28)*. Thus, *ref-2* acts as the master regulator of P/Pn.p cell development during L1, controlling all aspects of development in the ventral epidermis and preventing inappropriate fusion.

Similar to the case with lateral cells, homeobox-containing transcription factors control the expression of the zinc finger gene *ref-2* in the ventral epidermis *(28)*. The pattern of Pn.p cell fusion along the AP body axis during L1 is controlled by two genes of the *C. elegans* homeobox gene cluster, *lin-39/Sex combs reduced* and *mab-5/Antennapedia* (*see* **Fig. 4B,C; refs. *44–46***). LIN-39 and MAB-5 are homeobox-containing transcription factors that are expressed in and control midbody and posterior specific cell fates, respectively ***(44–46)***. The homeobox genes are regulated at the levels of both expression and protein activity to control Pn.p cell fusion. *lin-39* is expressed in P(3–8).p in both sexes; likewise, *mab-5* is expressed in P(7–11).p in both sexes (*see* **Fig. 4B,C**). In hermaphrodites, *lin-39* prevents fusion of Pn.p cells in which it is expressed (P[3–8].p); *mab-5* does not affect Pn.p cell fusion in the hermaphrodite because it is inactive in hermaphrodite Pn.p cells during L1 (at least as far as this one cell fate decision is concerned; *see* **Fig. 4B; ref. *45***).

In males, both homeobox genes affect Pn.p fusion; they each individually prevent cell fusion where they are expressed alone: P(3–6).p for *lin-39* and P(9–11).p for *mab-5* (*see* **Fig. 4C; ref. *45***). However, in males, in P7.p and P8.p, where both homeobox genes are expressed, the two homeobox proteins inhibit each other's activities and those cells fuse ***(45)***. This complicated regulation ensures that midbody cells, which generate the vulva, remain unfused in hermaphrodites and that posterior cells, which generate male copulatory structures, remain unfused in males.

The complex interactions between the homeobox genes *lin-39* and *mab-5* ultimately affect cell fusion by regulating the expression of *ref-2* (*see* **Fig. 4B,C. ref. *28***). In hermaphrodites, LIN-39 stimulates *ref-2* expression in the Pn.p cells, thereby preventing fusion of those cells. Because MAB-5 is inactive in these cells in hermaphrodites, MAB-5 does not affect *ref-2* expression in hermaphrodites (*see* **Fig. 4B**). In males, either homeobox protein can induce *ref-2* expression, and thus cells expressing one protein or the other remain unfused (*see* **Fig. 4C**); however, when both homeobox proteins are present in a single Pn.p cell, the two homeobox proteins neutralize each other, *ref-2* expression is not stimulated, and the cells fuse with hyp7 (*see* **Fig. 4C; ref. *28***).

Thus, *ref-2* is the primary "readout" for the positional information provided by the homeobox genes and also for sex-specific information. Consistent with this, a putative enhancer mutation that lies downstream of the *ref-2* coding sequence has been identified that bypasses the interaction between the two homeobox proteins in males; in the presence of this mutation, *ref-2* is expressed strongly even in Pn.p cells containing both homeobox proteins, and those Pn.p cells remain unfused ***(28)***.

Several experiments indicate that transcription of the *eff-1* membrane fusogen is regulated by *lin-39* (***30***). The complex genetic interactions described earlier suggest that both *lin-39* and *ref-2* are required to prevent Pn.p fusion. One possibility consistent with all the data is that LIN-39 directs the transcription of *ref-2* and that REF-2 and LIN-39 both bind to the *eff-1* promoter and function together to repress the *eff-1* gene (*see* **Fig. 7**). In males, MAB-5 also controls *ref-2* expression and probably *eff-1* expression as well.

4.3. elt-5 *and the Homeobox Genes Regulate Ventral Epidermal Fusion Later in Development*

Later in development, *elt-5* and *elt-6*, which control seam cell fate on the lateral surface, also control Pn.p cell fate on the ventral surface and prevent inappropriate Pn.p cell fusion (*see* **Fig. 7; refs. *34,47***). *elt-5* and *elt-6* are expressed in the unfused P and Pn.p cells ***(34,47)***. Mutation or RNAi-mediated inhibition of *elt-5* or *elt-6* results in defects in vulval development, cell division, and inappropriate fusion of the vulval precursor cells ***(34,47)***.

The homeobox proteins also control later development in the Pn.p cells that remain unfused during L1 (*see* **Fig. 7**). At these later times, *lin-39* acts to promote cell proliferation and differentiation and subsequent vulva formation and to prevent Pn.p cell fusion ***(30,48,49)***. LIN-39 and the homeobox cofactor CEH-20/extradenticle are direct transcriptional regulators of the expression of at least one isoform of *elt-5* in the Pn.p cell lineage (although a second characterized isoform is expressed independently of *lin-39*; **refs. *34,47***). *elt-5* and *elt-6* act redundantly in these ventral lineages to prevent Pn.p cell fusion and promote vulval cell fates during the L2 to L4 stages. Consistent with this, *elt-5* is expressed in these cells during this time frame ***(34,47)***. Thus *elt-5* and *elt-6* play similar roles in preventing fusion and promoting cell identity in both the lateral and ventral epidermis, and in both cases they are regulated by homeobox-containing transcription factors (*see* **Fig. 7**). *elt-5* and *elt-6* act after *ref-2*, as mutation of *elt-5* does not affect the Pn.p fusion pattern in L1 (which is regulated by *ref-2*) but does affect fusion at later time points ***(47)***.

5. Other Epidermal Cell Fusion Regulators

Many other genes also affect epidermal cell fusion in *C. elegans*, likely by regulating *eff-1*. All the genes that regulate cell fusion in *C. elegans* are listed in Table 1 in Podbilewicz ***(1)***. We highlight a few of these genes and pathways here.

Numerous genes function with the homeobox genes and *ref-2* to regulate Pn.p cell fusion in L1. The homeobox genes provide AP positional information to *ref-2*. REF-1, a bHLH transcription factor, is regulated by the nematode sex-determination pathway; REF-1 provides sex-specific information to *ref-2* to control cell fusion ***(29,50)***. EGL-27 is a component of a chromatin remodeling complex that regulates many aspects of homeobox gene expression and activity in the Pn.p cells; it too regulates *ref-2* expression ***(51)***.

On the lateral surface of the nematode, the seam cells undergo repeated asymmetric cell divisions in which they regenerate the seam cell and also generate a cell destined to fuse with hyp7 (usually the anterior daughter). The polarity of these cell divisions (and thus of which cells express *eff-1)* is controlled by several *wnt* pathway members, by several components of the mediator complex, and by a receptor tyrosine kinase ***(52–55)***.

Heterochronic genes control development temporally. Many heterochronic genes have been described that cause either precocious or retarded development during *C. elegans* development, which causes developmental stages to be skipped or repeated, respectively. These global developmental changes ultimately impinge on cell fusion. For example, the *lin-29* mutation prevents the transition from the L4 to the adult stage and prevents fusion of the lateral seam cells with each other late in L4 ***(56)***. In contrast, the *dre-1* mutation causes premature fusion of the seam cells with each other ***(57)***. Another heterochronic gene, *lin-41*, affects cell fusion in the nematode male tail ***(58)***. MicroRNAs including *lin-4* and *let-7* also control developmental timing and therefore control cell fusion during development ***(59–61).***

idf-1 mutant nematodes were identified in a screen for mutants blocked in embryonic elongation ***(7)***. In addition to this function, *idf-1* mutants also exhibit a defect in dorsal cell fusion in the embryo. As yet, the molecular identity of *idf-1* is undetermined.

fus-1, which encodes a subunit of the vacuolar ATPase complex, is the only gene identified so far that may affect EFF-1 activity rather than *eff-1* expression ***(31)***. How *fus-1* affects fusion is unclear. Two possibilities are that acidification of the extracellular environment by FUS-1 could help the membrane fusion process or that *fus-1* indirectly regulates fusion by regulating intracellular transport ***(31,62)***.

Thus, many genes that control cell fusion in *C. elegans* have been identified, all of which affect fusion by controlling *eff-1*, directly or indirectly. Thus, understanding the integration of this complex developmental information can effectively be reduced to a problem of understanding the multifaceted, coordinated regulation of *eff-1*.

6. Cell Fusion in Other Tissues and the Role of *aff-1*

While *eff-1* facilitates membrane fusion in many tissues, including the epidermis, pharynx, and most of the vulva, several cell fusions do occur in the *eff-1* null mutant background ***(11)***. These include a few cell fusions in the vulva, the fusion of the seam cells with themselves in L4, and the fusion of the AC cell with the utse cell in the uterus (**Fig. 8**). Many of these fusions occur late in nematode development, suggesting that a different gene may control fusion at this late stage.

Several cells that comprise the uterus are formed by cell fusion during L4. One of these cells, the utse, is formed by the fusion of eight daughters of π cells,

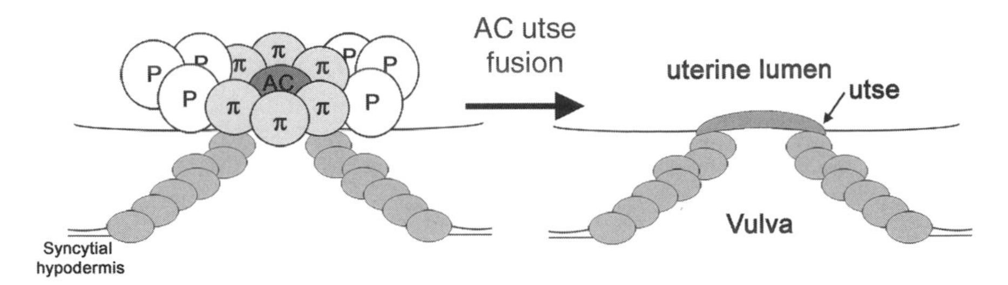

Fig. 8. Schematic depicting anchor cell (AC)–utse fusion during development. The uterine precursor π cells lie dorsal to the vulva. During L4, the AC cell invades the vulva ventrally and fuses with the utse. (Reprinted from Sapir et al. ***(11)***, with permission of Elsevier.)

which subsequently fuse with the anchor cell (AC; *see* **Fig. 8; ref. *63***). The utse is an H-shaped cell that connects the uterus to the seam and that also forms part of a channel between the gonad and vulva through which eggs are passed ***(63)***. The AC cell serves several functions prior to fusion, including providing a signal for vulva and uterus differentiation ***(64–66)***. The migration and fusion of AC into and with the utse has also become a model for cell invasion, because the AC must cross two basement membranes to connect the vulva and uterus before it fuses with the utse ***(67)***.

The *aff-1* mutation was isolated in a screen for mutants that failed to lay eggs; in subsequent experiments, it was found that the AC fails to fuse with the utse in an *aff-1* mutant background ***(11)***. *aff-1* is also required for two of the cell fusions that occur during vulval development, for the fusion of hyp5, and for the fusion of the seam cells with themselves during L4. *aff-1* encodes a protein with moderate homology but with a similar overall structure to EFF-1. Several experiments analogous to those with *eff-1* demonstrate that AFF-1 is a bona fide membrane fusogen ***(11)***. These include experiments in which *aff-1* overexpression induces fusion of cells that normally do not fuse. Moreover, *aff-1*-mediated cell fusion occurs independently of *eff-1*. Additionally, expression of *aff-1* induced cell fusion in a heterologous insect cell culture system.

Although *aff-1* controls AC fusion to the utse, it does not affect the initial fusion of the eight π cell daughters to generate the utse or formation of other syncytial uterine cells ***(11)***. Because the utse cell fuses in *eff-1* and *aff-1* mutant nematodes, it is possible that other fusogens exist in *C. elegans* that have yet to be discovered. The *aff-1eff-1* double mutant is very unhealthy, suggesting that cell fusion is essential for nematode development ***(11)***. The fact that EFF-1 and AFF-1 control the formation of distinct but neighboring syncytia suggests that having independent fusogens may provide some specificity to fusion by allowing fusion between cells expressing *aff-1* but not neighboring cells expressing *eff-1*.

As with the regulation of *eff-1*, the regulation of *aff-1* is likely to be quite complex. FOS-1, a transcription factor that is required for AC invasion into the uterus and fusion with the utse, is required for *aff-1* expression ***(11,68)***. Other regulators of AC fusion include *lin-11*, *cog-2/egl-13*, *smo-1*, and *nsf-1* ***(38,40,69,70)***. NSF-1, or *N*-ethylmaleimide–sensitive factor, which likely functions in vesicular fusion, also is required for fusion of the AC with the utse ***(38)***. It functions in the AC cell to promote fusion.

AFF-1 and EFF-1 have clear homologs in other nematode species but no clear homologs in higher organisms. However, several of the domains in AFF-1 and EFF-1 are conserved throughout the animal kingdom. In particular, the TGFβ-RI domain is likely required for cell fusion and is conserved in higher organisms, suggesting that this domain could serve a similar function in other proteins in other species.

7. Conclusion

The study of cell fusion in *C. elegans* has allowed us to start to answer four fundamental questions: *How*, *when, where*, and *why* do cells fuse?

The answer to the last question remains the most unclear. Several hypotheses have been suggested to explain why cells fuse in the nematode. These include the possibility that cell fusion may be a mechanism for cells to exit developmental pathways (analogous to apoptosis in other systems) or that membrane derived from the fusing cells is recycled to other locations where it can drive global shape changes in the animal such as those that occur during embryonic elongation ***(2,71)***. While many nematodes have a syncytial epidermis, some marine nematodes have cellular skin, indicating that fusion is not strictly necessary for survival of these organisms ***(26,72)***. Single mutants in *C. elegans* have now been identified that exhibit major cell fusion defects; these have large effects on morphogenesis but a more limited effect on viability *per se*. However, the double mutant *eff-1 aff-1* is unviable, suggesting that cell fusion is essential for *C. elegans* development.

The second and third questions are fundamental ones in developmental biology—how is a complicated organ like the epidermis, uterus, or vulva coordinately generated? Many genes have now been identified that coordinate cell lineage and positional, temporal, and sex-specific information to control the carefully choreographed development and fusion that occurs in the epidermis. Most of these genes likely impinge on regulation of the *eff-1* and *aff-1* fusogens.

Perhaps some of the most exciting recent results in the *C. elegans* cell fusion field have been the identification of *eff-1* and *aff-1*, two nematode membrane fusogens. Many components of the cell fusion machinery have been identified and studied in viral cell fusion and intracellular membrane trafficking.

In contrast, although many regulators of developmental cell fusion have been identified, few demonstrated components of the cell fusion machinery have been identified. Although EFF-1 and AFF-1 have some intriguing conserved domains, they do not share obvious homologs in nonnematode species. Yet the mechanism by which they promote cell fusion in vivo and in vitro is quite similar to that for viral and intracellular fusion. Perhaps these different membrane fusion components have evolved in convergent fashion to overcome the similar energetic constraints of protein-mediated membrane fusion. If so, the ability to study these genes in an in vivo system such as *C. elegans*, with all its molecular and genetic resources, will continue to reap benefits and will continue to shed light on the mechanism by which membranes fuse.

Acknowledgments

We thank Queelim Ch'ng and Gidi Shemer for critical reading of the manuscript. B.P. is supported by grants from the Israel Science Foundation. This article was co-written by S.A. in his private capacity; the views expressed here do not necessarily represent the views of the NIH, DHHS, or the United States.

References

1. Podbilewicz, B. (2006) Cell fusion, in *WormBook* (The *C. elegans* Research Community, ed.). Available at: http://www.wormbook.org.
2. Shemer, G. and Podbilewicz, B. (2000) Fusomorphogenesis: cell fusion in organ formation. *Dev. Dyn.* **218,** 30–51.
3. del Campo, J. J., Opoku-Serebuoh, E., Isaacson, A. B., Scranton, V. L., Tucker, M., Han, M., and Mohler, W. A. (2005) Fusogenic activity of EFF-1 is regulated via dynamic localization in fusing somatic cells of *C. elegans. Curr Biol* **15,** 413–423.
4. Mohler, W. A., Shemer, G., del Campo, J., Valansi, C., Opoku-Serebuoh, E., Scranton, V., Assaf, N., White, J. G., and Podbilewicz, B. (2002) The type I membrane protein EFF-1 is essential for developmental cell fusion in *C. elegans. Dev. Cell* **2**, 355–362.
5. Mohler, W. A., Simske, J. S., Williams-Masson, E. M., Hardin, J. D., and White, J. G. (1998) Dynamics and ultrastructure of developmental cell fusions in the *Caenorhabditis elegans* hypodermis. *Curr. Biol.* **8**, 1087–1090.
6. Shemer, G., Suissa, M., Kolotuev, I., Nguyen, K. C. Q., Hall, D. H., and Podbilewicz, B. (2004) EFF-1 is sufficient to initiate and execute tissue-specific cell fusion in *C. elegans. Curr. Biol.* **14,** 1587–1591.
7. Gattegno, T., Mittal, A., Valansi, C., Nguyen, K. C. Q., Hall, D. H., Chernomordik, L. V., and Podbilewicz, B. (2007) Genetic control of fusion pore expansion in the epidermis of *Caenorhabditis elegans. Mol. Biol. Cell* **18,** 1153–1166.
8. Nguyen, C. Q., Hall, D. H., Yang, Y., and Fitch, D. H. A. (1999) Morphogenesis of the *Caenorhabditis elegans* male tail tip. *Dev. Biol.* **207,** 86–106.
9. Podbilewicz, B. and White, J. G. (1994) Cell fusions in the developing epithelia of *C. elegans. Dev. Biol.* **161,** 408–424.

10. Podbilewicz, B., Leikina, E., Sapir, A., Valansi, C., Suissa, M., Shemer, G., and Chernomordik, L. V. (2006) The *C. elegans* developmental fusogen EFF-1 mediates homotypic fusion in heterologous cells and in vivo. *Dev. Cell* **11,** 471–481.
11. Sapir, A., Choi, J., Leikina, E., Avinoam, O., Valansi, C., Chernomordik, L. V., Newman, A. P., and Podbilewicz, B. (2007) AFF-1, a FOS-1–regulated fusogen mediates fusion of the anchor-cell in *C. elegans. Dev. Cell* **12,** 683–698.
12. Mi, S., Lee, X., Li, X.-P., Veldman, G. M., Finnerty, H., Racie, L., LaVallie, E., Tang, X.-Y., Edouard, P., Howes, S., Keith, J. C., and McCoy, J. M. (2000) Syncytin is a captive retroviral envelope protein involved in human placental morphogenesis. *Nature* **403,** 785–789.
13. Greenstein, D. (2005) Control of oocyte meiotic maturation and fertilization, in *WormBook* (The *C. elegans* Research Community, ed.). Available at: http://www.wormbook.org.
14. L'Hernault, S. W. (2006) Spermatogenesis, in *WormBook* (The *C. elegans* Research Community, ed.). Available at: http://www.wormbook.org.
15. Yamamoto, I., Kosinski, M. E., and Greenstein, D. (2006) Start me up: cell signaling and the journey from oocyte to embryo in *C. elegans. Dev. Dyn.* **235,** 571–585.
16. Sulston, J. E., Schierenberg, E., White, J. G., and Thomson, J. N. (1983) The embryonic cell lineage of the nematode *Caenorhabditis elegans. Dev. Biol.* **100,** 64–119.
17. Yochem, J., Gu, T., and Han, M. (1998) A new marker for mosaic analysis in Caenorhabditis elegans indicates a fusion between hyp6 and hyp7, two major components of the hypodermis. *Genetics* **149,** 1323–1334.
18. Wood, W. B. (1988) *The Nematode Caenorhabditis elegans*. Cold Spring Harbor Laboratory, Cold Spring Harbor, NY.
19. Williams-Masson, E. M., Heid, P. J., Lavin, C. A., and Hardin, J. (1998) The cellular mechanism of epithelial rearrangement during morphogenesis of the *C. elegans* dorsal hypodermis. *Dev. Biol.* **204,** 263–276.
20. Sulston, J. E. and Horvitz, H. R. (1977) Postembryonic cell lineages of the nematode *Caenorhabditis elegans. Dev. Biol.* **56,** 110–156.
21. White, J. G. (1988) The Anatomy, in *The Nematode Caenorhabditis elegans* (Wood, W. B., ed.). Cold Spring Harbor Laboratory, Cold Spring Harbor, NY, pp. 81–122.
22. Sternberg, P. W. (2005) Vulval development, in *WormBook* (The *C. elegans* Research Community, ed.). Available at: http://www.wormbook.org.
23. Austin, J., and Kenyon, C. (1994) Cell contact regulates neuroblast formation in the *Caenorhabditis elegans* lateral epidermis. *Development* **120,** 313–324.
24. Priess, J. R. and Hirsh, D. I. (1986) *Caenorhabditis elegans* morphogenesis: the role of cytoskeleton in elongation of the embryo. *Dev. Biol.* **117,** 156–173.
25. Doberstein, S. K., Fetter, R. D., Mehta, A. Y., and Goodman, C. S. (1997) Genetic analysis of myoblast fusion: *blown fuse* is required for progression beyond the prefusion complex. *J. Cell Biol.* **136,** 1249–1261.
26. Podbilewicz, B. (2000) Membrane fusion as a morphogenetic force in nematode development. *Nematology* **2,** 99–111.

27. Shemer, G., Kishore, R., and Podbilewicz, B. (2000) Ring formation drives invagination of the vulva in *C. elegans:* Ras, cell fusion and cell migration determine structural fates. *Dev. Biol.* **221,** 233–248.
28. Alper, S. and Kenyon, C. (2002) The zinc finger protein REF-2 functions with the Hox genes to inhibit cell fusion in the ventral epidermis of *C. elegans. Development* **129,** 3335–3348.
29. Alper, S. and Kenyon, C. (2001) REF-1, a protein with two bHLH domains, alters the pattern of cell fusion in *C. elegans* by regulating Hox protein activity. *Development* **128,** 1793–1804.
30. Shemer, G. and Podbilewicz, B. (2002) LIN-39/Hox triggers cell division and represses EFF-1/fusogen-dependent vulval cell fusion. *Genes Dev.* **16,** 3136–3141.
31. Kontani, K., Moskowitz, I. P., and Rothman, J. H. (2005) Repression of cell–cell fusion by components of the *C. elegans* vacuolar ATPase complex. *Dev. Cell* **8,** 787–794.
32. Chen, Z. and Han, M. (2001) *C. elegans* Rb, NuRD, and Ras regulate *lin-39*–mediated cell fusion during vulval fate specification. *Curr. Biol.* **11,** 1874–1879.
33. Gattegno, T. (2003) *Isolation and Characterization of Cell Fusion Mutants in C. elegans.* PhD Thesis, Technion-Israel Institute of Technology, Haifa.
34. Koh, K., Peyrot, S. M., Wood, C. G., Wagmaister, J. A., Maduro, M. F., Eisenmann, D. M., and Rothman, J. H. (2002) Cell fates and fusion in the *C. elegans* vulval primordium are regulated by the EGL-18 and the ELT-6 GATA factors—apparent direct targets of the LIN-39 Hox protein. *Development* **129,** 5171–5180.
35. Koh, K. and Rothman, J. H. (2001) ELT-5 and ELT-6 are required continuously to regulate epidermal seam cell differentiation and cell fusion in *C. elegans. Development* **128,** 2867–2880.
36. Cassata, G., Shemer, G., Morandi, P., Donhauser, R., Podbilewicz, B., and Baumeister, R. (2005) *ceh-16/engrailed* patterns the embryonic epidermis of *Caenorhabditis elegans. Development* **132,** 739–749.
37. Cinar, H. N., Richards, K. L., Oommen, K. S., and Newman, A. P. (2003) The EGL-13 SOX domain transcription factor affects the uterine p cell lineages in *Caenorhabditis elegans. Genetics* **165,** 1623–1628.
38. Choi, J., Richards, K. L., Cinar, H. N., and Newman, A. P. (2006) N-ethylmaleimide sensitive factor is required for fusion of the *C. elegans* uterine anchor cell. *Dev. Biol.* **297,** 87–102.
39. Choi, J. (2006) *Genetic Study of Cell Fusion Factors in Caenorhabditis elegans.* PhD Thesis, Verna and Marrs McLean Department of Biochemistry and Molecular Biology, Baylor College of Medicine, Houston.
40. Hanna-Rose, W. and Han, M. (1999) COG-2, a Sox domain protein necessary for establishing a functional vulval–uterine connection in *Caenorhabditis elegans. Development* **126,** 169–179.
41. *WormBase* (2002) Available at: http://www.wormbase.org, release WS90.
42. Chernomordik, L. V. and Kozlov, M. M. (2003) Protein–lipid interplay in fusion and fission of biological membranes. *Annu. Rev. Biochem.* **72,** 175–207.
43. Chernomordik, L. V. and Kozlov, M. M. (2005) Membrane hemifusion: crossing a chasm in two leaps. *Cell* **123,** 375–382.

44. Clark, S. G., Chisholm, A. D., and Horvitz, H. R. (1993) Control of cell fates in the central body region of *C. elegans* by the homeobox gene *lin-39. Cell* **74,** 43–55.
45. Salser, S. J., Loer, C. M., and Kenyon, C. (1993) Multiple HOM-C gene interactions specify cell fates in the nematode central nervous system. *Genes Dev.* **7,** 1714–1724.
46. Wang, B. B., Muller-Immergluck, M. M., Austin, J., Robinson, N. T., Chisholm, A., and Kenyon, C. (1993) A homeotic gene cluster patterns the anteroposterior body axis of *C. elegans. Cell* **74,** 29–42.
47. Koh, K., Bernstein, Y., and Sundaram, M. V. (2004) The nT1 translocation separates vulval regulatory elements from the egl-18 and elt-6 GATA factor genes. *Dev. Biol.* **267,** 252–263.
48. Clandinin, T. R., Katz, W. S., and Sternberg, P. W. (1997) *Caenorhabditis elegans* Hom-C genes regulate the response of vulval precursor cells to inductive signal. *Dev. Biol.* **182,** 150–161.
49. Maloof, J. N. and Kenyon, C. (1998) The Hox gene *lin-39* is required during *C. elegans* vulval induction to select the outcome of Ras signaling. *Development* **125,** 181–190.
50. Ross, J. M., Kalis, A. K., Murphy, M. W., and Zarkower, D. (2005) The DM domain protein MAB-3 promotes sex-specific neurogenesis in *C. elegans* by regulating bHLH proteins. *Dev. Cell.* **8,** 881–892.
51. Ch'ng, Q. and Kenyon, C. (1999) *egl-27* generates anteroposterior patterns of cell fusion in *C. elegans* by regulating *Hox* gene expression and *Hox* protein function. *Development* **126,** 3303–3312.
52. Forrester, W. C., Dell, M., Perens, E., and Garriga, G. (1999) A *C. elegans* Ror receptor tyrosine kinase regulates cell motility and asymmetric cell division. *Nature* **400,** 881–885.
53. Herman, M. A., Vassilieva, L. L., Horvitz, H. R., Shaw, J. E., and Herman, R. K. (1995) The *C. elegans* gene lin-44, which controls the polarity of certain asymmetric cell divisions, encodes a Wnt protein and acts cell nonautonomously. *Cell* **83,** 101–110.
54. Whangbo, J., Harris, J. R., and Kenyon, C. (2000) Multiple levels of regulation specify the polarity of an asymmetric cell division in *C. elegans. Development* **127,** 4587–4598.
55. Yoda, A., Kouike, H., Okano, H., and Sawa, H. (2005) Components of the transcriptional Mediator complex are required for asymmetric cell division in *C. elegans. Development* **132,** 1885–1893.
56. Rougvie, A. E. and Ambros, V. (1995) The heterochronic gene *lin-29* encodes a zinc finger protein that controls a terminal differentiation event in *Caenorhabditis elegans. Development* **121,** 2491–2500.
57. Fielenbach, N., Guardavaccaro, D., Neubert, K., Chan, T., Li, D., Feng, Q., Hutter, H., Pagano, M., and Antebi, A. (2007) DRE-1: an evolutionarily conserved F box protein that regulates *C. elegans* developmental age. *Dev. Cell* **12,** 443–455.
58. Del Rio-Albrechtsen, T., Kiontke, K., Chiou, S. Y., and Fitch, D. H. (2006) Novel gain-of-function alleles demonstrate a role for the heterochronic gene *lin-41* in *C. elegans* male tail tip morphogenesis. *Dev. Biol.* **297,** 74–86.

59. Lee, R. C., Feinbaum, R. L., and Ambros, V. (1993) The *C. elegans* heterochronic gene *lin-4* encodes small RNAs with antisense complementarity to lin-14. *Cell* **75,** 843–854.
60. Pasquinelli, A. E. and Ruvkun, G. (2002) Control of developmental timing by micrornas and their targets. *Annu. Rev. Cell Dev. Biol.* **18,** 495–513.
61. Reinhart, B. J., Slack, F. J., Basson, M., Pasquinelli, A. E., Bettinger, J. C., Rougvie, A. E., Horvitz, H. R., and Ruvkun, G. (2000) The 21-nucleotide let-7 RNA regulates developmental timing in *Caenorhabditis elegans. Nature* **403,** 901–906.
62. Chen, E. H., Grote, E., Mohler, W., and Vignery, A. (2007) Cell–cell fusion. *FEBS Lett.* **581,** 2191–93.
63. Newman, A. P., White, J. G., and Sternberg, P. W. (1996) Morphogenesis of the *C. elegans* hermaphrodite uterus. *Development* **122,** 3617–3626.
64. Kimble, J. (1981) Alterations in cell lineage following laser ablation of cells in the somatic gonad of *Caenorhabditis elegans. Dev. Biol.* **87,** 286–300.
65. Newman, A. P., White, J. G., and Sternberg, P. W. (1995) The *Caenorhabditis elegans lin-12* gene mediates induction of ventral uterine specialization by the anchor cell. *Development* **121,** 263–271.
66. Sternberg, P. W. and Horvitz, H. R. (1986) Pattern formation during vulval development in *C. elegans. Cell* **14,** 761–772.
67. Sherwood, D. R. and Sternberg, P. W. (2003) Anchor cell invasion into the vulval epithelium in *C. elegans. Dev. Cell* **5,** 21–31.
68. Sherwood, D. R., Butler, J. A., Kramer, J. M., and Sternberg, P. W. (2005) FOS-1 promotes basement-membrane removal during anchor-cell invasion in *C. elegans. Cell* **121,** 951–962.
69. Broday, L., Kolotuev, I., Didier, C., Bhoumik, A., Gupta, B. J., Sternberg, P. W., Podbilewicz, B., and Ronai, Z. (2004) The small ubiquitin like modifier (SUMO) is required for gonadal and uterine–vulval morphogenesis in *C. elegans. Genes Dev.* **18,** 2380–2391.
70. Newman, A. P., Acton, G. Z., Hartwieg, E., Horvitz, H. R., and Sternberg, P. W. (1999) The *lin-11* LIM domain transcription factor is necessary for morphogenesis of *C. elegans* uterine cells. *Development* **126,** 5319–5326.
71. Shemer, G. and Podbilewicz, B. (2003) The story of cell fusion: big lessons from little worms. *BioEssays* **25,** 672–682.
72. Chitwood, B. G. and Chitwood, M. B. H. (1974) *Introduction to Nematology.* University Park Press, Baltimore.
73. Kenyon, C. (1986) A gene involved in the development of the posterior body region of *C. elegans. Cell* **46,** 477–487.

5

Myoblast Fusion in *Drosophila*

Susan M. Abmayr, Shufei Zhuang, and Erika R. Geisbrecht

Summary

Myogenic differentiation in *Drosophila melanogaster*, as in many other organisms, involves the generation of multinucleate muscle fibers through the fusion of myoblasts. Prior to fusion, the myoblasts become specified as one of two distinct cell types. They then become competent to fuse and express genes associated with cell recognition and adhesion. Initially, cell-type–specific adhesion molecules mediate recognition and fusion between these two distinct populations of myoblasts. Intracellular proteins that are essential for the fusion process are then recruited to points of cell–cell contact at the membrane, where the cell surface molecules have become localized. Many of these cytosolic proteins contribute to reorganization of the cytoskeleton through activation of small guanosine triphosphatases and recruitment of actin nucleating proteins. Following the initial fusion event, the ultimate size of the syncytia is achieved through multiple rounds of fusion between the developing syncytia and mononucleate myoblasts. Ultrastructural changes associated with cell fusion include recruitment of electron-dense vesicles to points of cell–cell contact, resolution of these vesicles into fusion plaques, fusion pore formation, and membrane vesiculation. This chapter reviews our current understanding of the genes, pathways, and ultrastructural events associated with fusion in the *Drosophila* embryo, giving rise to multinucleate syncytia that will be used throughout larval life.

Key Words: Myoblast fusion; adhesion; *Drosophila*; founder myoblast; fusion-competent myoblast; cytoskeleton.

1. Introduction

The larval musculature of *Drosophila* is an elaborate array of 30 segmentally repeated muscle fibers that are generated by the fusion of committed myoblasts. These muscles are, in contrast to vertebrate muscle, single myotubes. They develop over a period of approximately 12h during embryogenesis, beginning roughly 6–7h after fertilization. The muscles will be fully formed, attached to the overlying epidermis, and capable of contraction a few hours before hatching, after which they will be used for larval locomotion. The

From: *Methods in Molecular Biology, vol. 475: Cell Fusion: Overviews and Methods*
Edited by: E. H. Chen © Humana Press, Totowa, NJ

muscle fibers differ from each other in features that include location, pattern of innervation, site of attachment, and size. These features are controlled by information contained within a specific cell type termed the *founder* myoblast through expression of one or more "muscle identity" genes. The founder cell is responsible for patterning the musculature, and it appears that there is a single founder cell for each muscle fiber. The founder cell seeds the fusion process by recruiting fusion-competent myoblasts. As fusion occurs, the "naïve" fusion-competent myoblasts take on the identity and features of the original founder cell, and the protein product of the muscle identity gene is usually observed in all nuclei of the syncitia. In the first stage of myoblast fusion, a single founder cell fuses with two or three surrounding fusion-competent myoblasts to form a muscle precursor. The final size of the muscle fiber will be attained in subsequent rounds of fusion between the muscle precursor and additional fusion-competent myoblasts ***(1–3)***. Overall, the smallest muscles of the embryo will be formed by fusion of as few as 3–5 cells, whereas larger muscles appear to include 20–25 cells ***(4)***.

2. Founder and Fusion-Competent Myoblasts

As mentioned, myoblast fusion occurs between two types of muscle cells, founder myoblasts and fusion-competent myoblasts. In the absence of fusion, each of these cell types is committed to myogenesis and expresses genes associated with terminal differentiation of muscle. Both cell types are derived from the somatic mesoderm, which is set aside by high levels of the bHLH transcription factor Twist ***(5)*** and further subdivided into competence groups ***(6)***. Muscle progenitors, which undergo one cell division to generate muscle founder cells, segregate from these equivalence groups through lateral inhibition mediated by Notch ***(6–9)***. Each founder cell then begins to express a specific constellation of transcription factors termed the *muscle identity* genes that, at least in some cases, has been shown to be required for subsequent differentiation of a specific founder cell ***(10–19)***. This unique pattern of gene expression is one of the first molecular indications that the founder myoblasts possess distinct identities from each other.

The remaining cells in each equivalence group become fusion-competent myoblasts, which appear to be identical to each other. Their specification and differentiation are controlled by a single zinc finger containing the Gli superfamily member Lameduck/Myoblasts-Incompetent/Gleeful ***(20–22)***. Spatially, the muscle founder cells lie in the outer layer of the somatic mesoderm, in contact with the epidermis or nervous system, while fusion-competent myoblasts are present in both this layer and a more internal layer. Thus, the fusion-competent cells tend to migrate to more external positions as fusion proceeds, and unfused myoblasts are more commonly observed in the internal layer.

3. Myoblast Recognition, Migration, and Adhesion

Founder cells, which dictate the pattern of muscle fibers, prefigure characteristics of future muscles that include their unique position, orientation, size, attachment sites, and pattern of innervation. Each serves as an attractant for the fusion-competent cells and seeds the fusion process, initially fusing with one or two surrounding fusion-competent cells to form binucleated or trinucleated muscle precursors. Prior to their fusion, the fusion-competent myoblasts must identify, migrate to, and adhere to the founder cell or muscle precursor with which they will eventually fuse. This interaction is specific and asymmetric such that founder cells do not fuse with each other and fusion-competent cells do not fuse with other mononucleate fusion-competent cells. After the initial formation of muscle precursors, additional rounds of fusion between these precursors and neighboring fusion-competent cells continue until the multinucleated myotube has attained its proper size.

Genetic studies in *Drosophila* have identified three members of the immunoglobulin superfamily (IgSF) that are essential for recognition and adhesion between founder and fusion-competent myoblasts (**Table 1; Fig. 1**). These include sticks and stones (SNS) ***(23)***, dumfounded/kin of irreC (Duf/Kirre) ***(24)***, and irregular-chiasm-C/roughest (IrreC/Rst) ***(25)***. The *hibris* (*hbs*) gene, which encodes a fourth IgSF member in *Drosophila* myoblasts, appears to regulate the fusion process (*see* **Table 1** and **Fig. 1; refs.** ***26–28***). The expression patterns of these transmembrane proteins provide the fundamental asymmetry between the founder cells and fusion-competent myoblasts, and each is discussed in turn below.

3.1. Immunoglobulin Superfamily Members: Duf/Kirre and IrreC/Rst

Duf/Kirre, a single-pass transmembrane protein, is a member of the DM-GRASP/BEN/SC1 subfamily. The extracellular region includes five Ig-like domains. The cytoplasmic region, comprising 367 amino acids, includes a consensus autophosphorylation domain similar to that found in receptor tyrosine kinases, a candidate PDZ-binding domain, and a putative serine phosphorylation site. Early in development, expression of the *duf/kirre* transcript is observed at low levels in the developing mesoderm. This broad expression becomes restricted to a limited number of cells in the embryo from which the founder cells arise. The *duf/kirre* transcript remains detectable in muscle precursors as long as they are fusing, but its level drops quickly after fusion is completed. It is not expressed in the fusion-competent cells. The corresponding Duf protein is observed on the surface of the founder cells, where it becomes concentrated at points of cell–cell contact. Like its transcript, Duf protein also decreases following fusion ***(24)***. Mutant embryos lacking only *duf/kirre* have

Table 1
Proteins Implicated in *Drosophila* Myoblast Fusion

Protein	Structure	Localization	Vertebrate homolog	References
Transmembrane proteins				
Sticks and stones	Ig domains; fibronectin type III domain; kinase target sites	FCM	Nephrin	***23***
Dumfounded/Kin of irregular chiasm C	Ig domains; autophosphorylation domain; PDZ binding site	FO	Dm-GRASP/Ben/SC1	***24***
Irregular chiasm C/Roughest	Ig domains; autophosphorylation domain; PDZ binding site	FCM and FO	Dm-GRASP/Ben/SC1	***25***
Hibris	Ig domains; fibronectin type III domain; kinase target sites	FCM	Nephrin	***26, 27***
Intracellular proteins				
Antisocial/Rolling pebbles	Lipolytic enzyme signature sequence; ATP/GTP binding site; ankyrin repeats; TPR repeats; coiled-coil domain	FO	Mants1	***44–46***
Myoblast city	SH3 domain; DHR1 domain (PIP_3 binding); DHR2 (or Docker domain; Rac binding); Crk binding sites	Broad; enriched in mesoderm; FCM and FO	Dock180	***42, 43***
DCrk	SH2 and SH3 domains	Broad; enriched in mesoderm	Crk-II and CrkL	***60***
DRac1/DRac2	Monomeric GTPase	Ubiquitous	Rac	***47, 61***
D-Arf6	Monomeric GTPase	Ubiquitous	ARF6	***63***
D-titin	Ig domains; fibronectin type III domains; PEVK domains	FCM and FO	Titin	***68, 69***
Loner/Schizzo	IQ motif; Sec7 domain; PH domain; coiled-coil domain	FO	$ARF\text{-}GEP_1$	***63, 64***

Blown fuse	PH domain	FCM and FO	Unknown	***37, 66***
Kette	Six predicted hydrophobic domains	FCM and FO	HEM/NAP1	***66***
Parcas	Unconventional SH3 domain	enriched in FCM	Sab	***71***
Wasp	Wasp homology domain; PIP_2 binding sites; GTPase binding domain; proline-rich region; Verprolin homology-connecting acidic domain	Ubiquitous	Wasp	***48, 67***
Rolling stone	Transmembrane domains	FCM and FO	Unknown	***33, 34***
WIP/solitary	WH2 domain; proline-rich region; Wasp binding domain	FCM	Vrp/WIP	***41, 48***
Singles bar	Hydrophobic, transmembrane domains; MARVEL domain	FCM and FO	Unknown	***70***

Arf6, adenosine diphosphate ribosylation factor; ATP, adenosine triphosphate; Crk, CT10 regulator of kinase; CrkL, Crk-Like; Dock180, 180-kDa protein downstream of Crk; D prefix, homologous *Drosophila* protein; FCM, fusion competent myoblast; FO, founder cell; GEF, guanine nucleotide exchange protein; GTP, guanosine triphosphate; HEM/NAP1, hematopoetic protein/Nck-associated protein 1; IG, immunoglobulin; mants1, mouse antisocial (ants)1; PIP_2, phosphatidylinositol bisphosphate; PIP_3, phosphatidylinositol triphosphate; Rac, Rho GTPase; Sab, SH3-domain binding protein that preferentially associates with Bruton's tyrosine kinase (Btk); TRP, tetratricopeptide repeats; Vrp, verprolin; Wasp, Wiskott-Aldrich syndrome protein; WIP, Wasp-interacting protein.

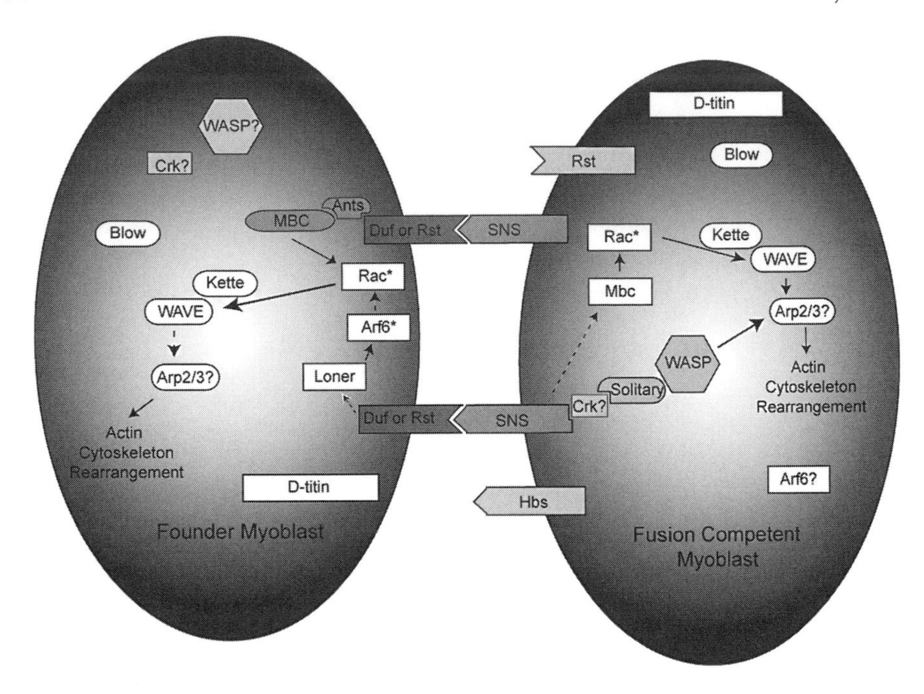

Fig. 1. A model based on known signal transduction pathways and proteins involved in myoblast fusion. Represented molecules include those for which a biochemical and/or genetic interaction has been demonstrated, or for which a function can be can be inferred from their structure. The bold arrows indicate established or likely associations, whereas the dashed arrows indicate hypothetical downstream targets of these cell surface receptors. Arf6, adenosine diphosphate ribosylation factor; Arp2/3: actin-related protein 2/3; Ants, Antisocial/Rolling Pebbles 7; Blow, Blown Fuse; Crk, CT10 regulator of kinase; Duf, Dumbfounded/Kin of IrreC; Hbs: Hibris; Rst, Irregular chiasm C/Roughest; Mbc, Myoblast city; Rac, Rho GTPase; SNS, Sticks and Stones; Wasp, Wiskott-Aldrich syndrome protein; WAVE, Wasp family Verprolin-homologous protein.

not been reported, primarily because the related gene *irreC/rst* can serve the same purpose.

The *irreC/rst* locus, identified on the basis of defects in axonal projections in the adult brain ***(29)***, is located 127 kilobases (kb) away from *duf/kirre*. It is the apparent result of a gene duplication event and encodes a protein with 45% similarity to Duf/Kirre. Like Duf/Kirre, IrreC/Rst has five extracellular Ig domains, a transmembrane region, and cytoplasmic tail ***(25,29,30)*** with both a PDZ binding site and potential phosphorylation sites. Notably, the IrreC/Rst pattern of expression is broader than that of Duf/Kirre and includes the founder cells as well as many of the fusion-competent myoblasts. No role has yet been identified for its expression in the latter cells. However, it serves a function redundant with that of Duf/Kirre in the founder cells. Specifically, embryos homozygous for a deletion that removes both the *duf/kirre* and *irreC/rst* genes exhibit severe muscle defects in which no myoblast fusion occurs. In addition, targeted mesodermal expression of either the IrreC/Rst or Duf/Kirre cDNA is sufficient to rescue myoblast fusion in the deficiency embryos. Notably, both Duf/Kirre and IrreC/Rst appear to act as attractants for fusion-competent cells, since expression of either protein in the embryonic ectoderm is sufficient to target migration of fusion-competent cells ***(24,25)***.

3.2. Immunoglobulin Superfamily Members: SNS and Hbs

Complimenting expression of Duf/Kirre or IrreC/Rst in the founder cells is the expression of two fusion-competent cell-specific IgSF members, SNS and Hbs, in the fusion-competent myoblasts. SNS has 8 extracellular Ig domains, a single fibronectin type-III domain, a membrane-spanning sequence, and 376 amino acid cytoplasmic regions ***(23)***. The cytoplasmic region has numerous sites for protein–protein interactions that commonly activate signal transduction cascades. These include sites for phosphorylation on serine and threonine, tyrosines, and conserved sites for interaction with PDZ domains. Both the *sns* transcript and protein are expressed exclusively in the fusion-competent cells of the somatic ***(23)*** and visceral ***(31,32)*** musculature, with no expression observed in the founder myoblasts. SNS appears on the cell surface just prior to fusion, often coincident with Duf/Kirre or IrreC/Rst at sites of cell–cell contact, and decreases rapidly as fusion is completed. Embryos lacking *sns* exhibit a complete absence of muscle fibers and a large number of unfused myosin-expressing myoblasts ***(23)***. The fusion-competent myoblasts of *sns* mutant embryos do not appear to migrate toward or associate with founder cells, remaining morphologically round instead of teardrop shaped with extended filopodia ***(1)***. In combination with interactions discussed below (*see* **Subheading 3.3**), these data suggest that SNS acts as a receptor on the surface of fusion-competent myoblasts through which these cells recognize and adhere to founder cells.

Of note, genetic and molecular studies revealed that ethyl methanesulfonate (EMS)-induced mutants in the *rolling stone (rost)* ***(33,34)*** locus are actually allelic to *sns* ***(23,35)***. This finding is particularly relevant because the EMS-induced *rost* ***(36)*** mutation, which was examined at the ultrastructural level by Doberstein et al. ***(37)***, actually represents a mutation in *sns* ***(23)***. Of particular note, electron-dense plaques accumulate in this allele, but the plasma membrane does not breakdown ***(37)***. However, no sequence lesion has been identified within the SNS coding sequence of *rost*^{15}, and myoblast fusion does not appear to be completely blocked ***(33)***. Thus, the accumulation of electron-dense plaques may reflect the presence of limited functional SNS in this particular allele.

Hbs is a paralogue of SNS, with 48% identity and 63% similarity overall. The Hbs extracellular region includes eight Ig-like domains, a "degenerate" Ig-like domain, and a single FN-III domain *(26–28)*. By comparison to this high degree of organizational and sequence homology with the SNS extracellular region, the Hbs cytodomain has less homology to that of SNS and is only 165 amino acids in length. It contains tyrosine residues and putative target sites for the cyclic adenosine monophosphate–dependent and cyclic guanosine monophosphate–dependent protein kinases C and CKII, some of which are conserved in SNS. It is expressed slightly earlier than SNS in the embryo and in a broader pattern that includes the trachea and malphigian tubules along with the visceral, somatic, and pharyngeal musculature. In the somatic musculature, Hbs is restricted to the fusion-competent myoblasts during fusion, where it declines slightly before SNS. In cells that express both proteins, SNS and Hbs colocalize at discrete points on the cell surface *(26)*. Embryos lacking *hbs* exhibit a modest increase in the number of unfused myoblasts, often with smaller or missing muscles. However, these defects are not sufficient to impair survival, and mutants survive to become semifertile adults.

3.3. Immunoglobulin Super Family Interactions

Mechanistically, recognition and adhesion between the founder and fusion-competent cells are mediated by the Duf/Kirre, IrreC/Rst, and SNS proteins on their surface. These interactions were most clearly demonstrated in aggregation assays using transfected *Drosophila* S2 cells in culture. S2 cells that express either Duf/Kirre or IrreC/Rst readily aggregate with SNS-expressing cells, and the proteins become enriched at points of cell–cell contact. SNS-expressing cells do not interact homotypically with other SNS-expressing cells. By contrast, Duf/Kirre and IrreC/Rst do mediate homotypic cell adhesion and are enriched at points of cell–cell contact. However, challenge experiments with SNS-expressing cells suggest that heterotypic interactions occur more rapidly and to a greater extent than homotypic interactions with other Duf/Kirre- or IrreC/Rst-expressing cells. Additionally, Duf/Kirre is detected upon immunoprecipitation of SNS from SNS:Duf aggregates, consistent with a physical interaction between Duf/Kirre and SNS that directs adhesion of founder cells and fusion-competent myoblasts *(38)*.

As in S2 cell interaction assays, Duf/Kirre and SNS are present at points of contact between founder and fusion-competent cells in the embryo *(38)*. One fundamental difference between the behavior of SNS and Duf/Kirre in the embryo and in S2 cells, however, is that the cytodomain of SNS is absolutely essential in the embryo. Whereas glycosylphosphatidylinositol (GPI)-anchored forms of SNS and Duf/Kirre are capable of directing aggregation of S2 cells, mesodermal expression of a GPI-anchored form of SNS that lacks both the transmembrane and cytoplasmic domains does not rescue muscle formation in *sns* mutant embryos. Recent examination of protein localization in the embryo has revealed that Duf/Kirre and SNS form a ring-shaped structure at the contact point of the precursor cell with a fusion-competent cell ***(39)***. These rings of Duf/Kirre or SNS assemble around an actin core or plug, referred to as a *fusion restricted myogenic-adhesive structure* (FuRMA) ***(39)***. Consistent with the expectation of critical protein–protein interactions with the cytodomains of the IgSF members, other fusion-associated proteins such as Ants/Rols (*see* **Subheading 5.1**) and blow (*see* **Subheading 5.5**) become enriched near these actin plugs ***(39)***.

In the past decade, novel cytoplasmic proteins as well as established components of known signaling pathways have been found to play roles in myoblast fusion. Some proteins may activate mechanisms downstream of SNS through which the fusion-competent cells actually migrate to the founder cells. Given the aforementioned sites in the SNS cytodomain, fusion-competent cell migration may be dependent on the binding of intracellular proteins to activate myoblast migration. Consistent with this model, fusion-competent myoblasts require SNS to migrate to sites of ectopic Duf/Kirre or IrreC. Moreover, the fusion-competent cells in Duf/Kirre–IrreC/Rst double-mutant embryos randomly extend filopodia but fail to attach to founder cells. By contrast, the fusion-competent cells of *sns* mutant embryos remain round and do not appear to extend filopodia. Together, these behaviors are consistent with a model in which the Duf/Kirre or IrreC/Rst attractants function as ligands for SNS and the idea that SNS-dependent filopodia direct migration of the fusion-competent cells upon detecting Duf/Kirre or IrreC/Rst on the founder cell.

In the S2 cell aggregation assay, Hbs-expressing cells do not interact with other cells that express Hbs, SNS, or IrreC/Rst but do interact with cells that express Duf/Kirre ***(27)***. The relevance of these differences to the developing musculature is not clear, because IrreC/Rst can clearly replace Duf/Kirre in the founder cells and Hbs is not essential for the fusion-competent cells to interact with these founder cells. As mentioned, embryos mutant for *hbs* have modest muscle defects and survive to become semifertile adults. SNS and Hbs colocalize at discrete points on the surface in fusion-competent cells in which both proteins are expressed. While this observation is consistent with direct interaction between SNS and Hbs, it could also reflect their independent localization to Duf-enriched sites of cell contact.

Of note, overexpression of Hbs in the somatic mesoderm partially disrupts myoblast fusion ***(26)***. Structure/function analysis has revealed the cytoplasmic domain to be responsible for this effect, because expression of the cytodomain alone mimics the Hbs overexpression phenotype and no phenotype is observed upon overexpression of either a secreted or membrane-bound extracellular domain. Loss of one copy of SNS enhances this myoblast fusion overexpression phenotype and dominantly suppresses the mild myoblast fusion defect observed in *hbs*-null embryos. Thus, these two proteins appear to antagonize each others' actions during mesoderm development, leading to the model that Hbs functions as a dose-dependent regulator of SNS ***(26)***. Mechanistically, Hbs could accomplish this goal by combining with SNS to form a "negative" receptor that responds differently to ligand than to the SNS receptor itself. Alternatively, Hbs and SNS may converge on an intracellular target such that Hbs sequesters this downstream component and it is not available for SNS. Both of these models accommodate Hbs as a nonessential regulator of SNS function. However, further study will be necessary to understand the exact role of Hbs in myoblast fusion.

4. Ultrastructural Events Associated With Myoblast Fusion

In a detailed morphological analysis of fusing myoblasts at the level of the electron microscope, Doberstein et al. ***(37)*** described a series of intracellular events that accompany fusion. The first obvious change is the accumulation of clusters of electron-dense vesicles on the cytoplasmic sides of opposed plasma membranes of two associated myoblasts. These vesicles align with one another across the intervening membranes to form paired vesicles, termed *prefusion complexes* ***(37)***. The vesicles then resolve into electron-dense plaques that are observed on the cytoplasmic sides of the corresponding plasma membranes. These plaques are reminiscent of structures identified in fusing vertebrate myoblasts ***(40)***. Little is known about the contents of these vesicles biochemically, although one recent study demonstrated that they become coated with actin prior to their membrane recruitment as an apparent consequence of migrating through an actin-rich field ***(41)***. As vesicles are being recruited and plaques form at sites of cell–cell contact, the myoblasts elongate to maximize contact points. Pores are then observed in the fusing membranes.

Similar studies have been carried out on mutant embryos in which myoblast fusion does not occur to determine the point at which fusion was arrested ***(37)***. The findings of these studies have been mentioned throughout the text as appropriate, but will also be summarized here. MBC protein is required at the first step of myoblast fusion, and *mbc* mutant embryos exhibit no multinucleate syncitia ***(42,43)***. At the ultrastructural level, the number of prefusion complexes seen in these embryos was significantly reduced. Ants/Rols7 is required for the second step of fusion, after formation of bi- or trinucleated precursor cells, but neither prefusion complexes nor electron-dense plaques are present in these multinucleate precursors in its absence ***(44–46)***. *Blow* mutant embryos are characterized by normal numbers of prefusion complexes and paired vesicles, with no apparent change in morphology ***(37)***. However, electron-dense plaques have not been observed. These plaques do accumulate in a nonnull allele of *sns*, but the plasma membrane does not breakdown. Although the ultrastructure of *sns* null alleles has not been analyzed, the first step of myoblast fusion does appear to occur. Expression of the constitutively active DRac1G12V GTPase is associated with a wild-type number of prefusion complexes and electron-dense membrane plaques. Moreover, the myoblasts appear to align normally. However, apposed plasma membranes between fusion partners have aberrant morphology, with few or no pores. The ultrastructure of loss-of-function alleles of *Drac1* and *Drac2*, in which myoblast fusion does not occur, has not been analyzed ***(47)***.

Two independent studies have implicated the actin cytoskeleton and Arp2/3 activation in myoblast fusion through identification of Solitary/D-Wasp-Interacting-Protein (*see* **Subheading 5.4**). Immunoelectron microscopy studies from Chen's laboratory have revealed actin-coated vesicles within pools of polymerized actin at sites of cell contact ***(41)***, perhaps recruited through an Sltr/D-WIP-mediated actin polymerization event. Using high-pressure freezing to preserve membrane ultrastructure prior to its visualization by electron microscopy and cytoplasmic green fluorescent protein (GFP) to monitor molecular transfer between cells, Kim and colleagues reported the absence of membrane pores in embryos mutant for *sltr/D-wip* ***(41)***. By contrast, Massarwa and colleagues used conventional electron microscopy to reveal defects in membrane vesiculation in these mutant embryos, arguing that actin polymerization via Sltr/D-WIP and Wasp is essential for expansion of the fusion pore but not for its formation ***(48)***. The latter group also observed the transfer of cytoplasmic GFP between founder and fusion-competent cells. Thus, the role of *sltr/D-wip* in pore formation and expansion awaits further analysis and clarification.

Finally, the exact roles of proteins in which the ultrastructure of mutant embryos was characterized prior to identification of the molecular lesion may require reevaluation with bona fide null alleles; others have yet to be evaluated in this manner. Nevertheless, it is apparent that ultrastructural analyses of the various mutants have revealed exciting new insights into the mechanism of myoblast fusion.

5. Cytoplasmic Proteins and Intracellular Signal Transduction Pathways

In the past decade, novel cytoplasmic proteins as well as established components of known signaling pathways have been found to play roles in myoblast fusion (*see* **Table 1** and **Fig. 1**). Some proteins may activate mechanisms downstream of SNS through which the fusion-competent cells actually migrate to the founder cells. Other proteins may be responsible for recruitment of the fusion machinery and the electron-dense vesicles to points of cell–cell contact. Still others may regulate cytoskeletal disassembly or reassembly that is coincident with membrane breakdown and formation of syncitia. While we are beginning to identify proteins that can interact with the cytoplasmic domains of the cell-type-specific adhesion molecules SNS, Duf/Kirre, and IrreC/Rst, remaining questions include whether these cytoplasmic interactions are triggered by cell recognition and adhesion, if they are sufficient to transmit the "need to fuse" signal, and how they regulate the extent of fusion. It also remains to be determined whether all of these events are downstream of SNS and Duf/Kirre or IrreC/Rst or if other cell surface triggers must exist. Finally, some of these proteins are expressed almost exclusively in myoblasts, as anticipated if their role is in fusion-specific events, while other molecules are broadly expressed and regulate pathways that function in many cell types.

5.1. Signaling Downstream of Immunoglobulin Super Family Receptors: Ants/Rols

The *antisocial/rolling pebbles (ants/rols)* locus was identified in three independent screens for mutations that disrupt embryonic myoblast fusion. It is required for the second step of fusion, after formation of bi- or trinucleated precursor cells, but neither prefusion complexes nor electron-dense plaques are present in these multinucleate precursors in its absence. The *rols* genomic locus gives rise to two transcripts, with *rols6* expression occurring predominantly in the early endoderm and *rols7* expression observed in the early mesoderm and later in the founder cells of the somatic muscle coincident with myoblast fusion. Moreover, Duf/Kirre and Ants/Rols7 are present on the myoblast membrane, and colocalize at points of cell–cell contact ***(45)***. The predicted Ants/Rols7 protein contains several domains with the potential to mediate protein–protein interaction, including a RING finger, nine ankyrin repeats, three tetratrico-peptide repeats (TPRs), and a coiled-coiled region ***(44–46)***. It binds to the cytoplasmic domain of Duf/Kirre through the ankyrin repeats in the C terminus and is preferentially translocated to the membrane of transfected S2 cells when Duf/Kirre is engaged in homotypic aggregates or heterotypic aggregates with SNS-expressing cells ***(44,49)***. This interaction, and the resulting Ants/Rols7 colocalization with Duf/Kirre at points of cell–cell contacts, is the basis for a positive feedback loop through which the amount of Duf/Kirre at the cell surface appears to be regulated ***(49)***. Altered amounts of Duf/Kirre at the cell surface, in turn, may regulate the extent of cell fusion. Additionally, Ants/Rols7 interacts with the N-terminal region of MBC in transfected S2 cells ***(44)***, which is discussed in more detail below (*see* **Subheading 5.2**). Finally, in embryos, D-titin is enriched at Duf-dependent foci when Ants/Rols7 is present ***(45)***. Thus, Ants/Rols7 may serve as a scaffold for the recruitment of cytoskeletal signaling proteins.

5.2. Signaling Downstream of Immunoglobulin Super Family Receptors: The CDM Pathway

As mentioned, Ants/Rols7 provides a critical link between the transmembrane protein Duf/Kirre and the downstream signaling protein MBC. Like Ants/Rols7, *mbc* mutant embryos are characterized by the absence of multinucleate muscle fibers and presence of large numbers of unfused myoblasts ***(43)***. The morphology of these mutant embryos differs from that seen in embryos lacking either *sns* or *duf/kirre* and *irreC/rst* in that the unfused fusion-competent cells migrate to, and cluster around, the founder cells ***(50)***. In contrast to Ants/Rols7, however, MBC is required at the first step of myoblast fusion, and mutant embryos exhibit no multinucleate syncytia. At the ultrastructural level, the number of prefusion complexes seen in these embryos was significantly reduced, suggesting that it might play a role in vesicle accumulation ***(37)***. MBC is expressed in a wide variety of tissues in the developing embryo ***(42)***. Consequently, the defects seen in *mbc* mutant embryos are not limited to myoblast fusion and include incomplete dorsal closure of the epidermis, abnormal fasciculation of the ventral nerve cord neurons, and severely impaired migration of border cells in the adult ovary ***(42,51,52)***. MBC is a founding member of the Ced-5, Dock 180, Myoblast city (CDM) superfamily of proteins that is conserved in *Caenorhabditis elegans* and mammalian organisms. In combination with Ced-12/Elmo, the CDM proteins function as unconventional bipartite guanine nucleotide exchange factors for the small GTPase Rac1 ***(53–56)***. Orthologs of *mbc* are involved in diverse processes that include cell engulfment, cell migration, epithelial morphogenesis, and oncogenic transformation ***(57,58)***.

Like other CDM proteins, MBC has an SH3 domain at the N terminus, a Dock homology region 1 (DHR1) domain, Dock homology region 2 (DHR2) or Docker domain, and proline-rich sites in the C terminus ***(42)***. The SH3 and Docker domains are required for interactions with Ced-12/ELMO and Rac1, respectively, in MBC orthologs, and transgenes with mutations in these regions of MBC fail to rescue myoblast fusion in *mbc* mutant embryos ***(59)***. Like its vertebrate orthologs, the DHR1 domain of MBC binds to phosphatidylinositol 3,4,5-triphosphate (PtdIns[3,4,5]P_3) and is also essential for myoblast fusion. Somewhat surprisingly, however, localization of MBC to the embryonic myoblast membrane occurs in the absence of this region ***(59)***.

Finally, the SH2–SH3 adaptor protein Crk is an essential component of the CDM pathway in *C. elegans* and mammals, and proline-rich sites in the C terminus of many CDM proteins mediate binding to this adaptor. In these systems, Crk targets the Dock180/ELMO complex to sites at the membrane and dramatically enhances activation of Rac1. Like MBC, *Drosophila* Crk is expressed quite broadly in tissues that include the embryonic musculature ***(60)***. Moreover, Crk binding to MBC is dependent on these proline-rich sites as expected. However, not only is Crk binding unnecessary for membrane localization of MBC in the musculature, but even MBC-mediated myoblast fusion occurs when these binding sites are deleted ***(59)***. Notably, a potential role for Crk in myoblast fusion that is independent of MBC is discussed below (*see* **Subheading 5.3**) Additional aspects of the CDM signal transduction pathway that are conserved in *Drosophila* myoblasts include activation of the downstream target GTPase DRac1. Specifically, both dominant-negative and constitutively active forms of DRac1 result in large numbers of unfused myoblasts ***(61)***. Moreover, loss-of-function mutations for all three *Drosophila Rac* genes confirmed that *DRac1* and *DRac2* are required, albeit redundantly, for proper fusion ***(62)***. Thus, Rac activation plays a critical role in myoblast fusion, and one pathway for this activation in myoblasts may involve MBC and conserved components of the CDM pathway.

5.3. Signaling Downstream of Immunoglobulin Super Family Receptors: Loner/Schizzo and Arf6

The fusion-competent myoblasts of embryos mutant for the *loner/schizzo* locus, which encodes a guanine nucleotide exchange factor in the Sec7 family ***(63,64)***, appear to recognize and extend filopodia toward the founder cells but do not fuse into syncytia ***(63)***. Of note, Loner/Schizzo gets recruited to the membrane at sites of cell contact in Duf-transfected S2 cells and remains diffuse within the cytoplasm of embryos lacking *duf/kirre* and *irreC/rst*. However, a direct interaction between Loner/Schizzo and Duf/Kirre could not be detected, suggesting that protein intermediaries are necessary.

In embryos, Loner/Schizzo protein can be seen in membrane puncta that also include Ants/Rols7, but this localization is not altered in an *ants/rols* mutant ***(63)***. The observation that Loner/Schizzo localization only partially overlaps that of Ants/Rols7 in both S2 cells and embryos suggests different protein interaction patterns and perhaps functions for the Ants/Rols7 and Loner/Schizzo proteins. Experiments rescuing the *loner* mutant phenotype have revealed that the GEF, Sec7, and PH domains are all required for proper muscle fusion. Because the Sec7 family of GEFs are known to regulate the adenosine diphosphate ribosylation factor (Arf) family of Ras GTPases, in vitro experiments were carried out to examine guanosine diphosphate/guanosine triphosphate (GDP/GTP) exchange of Arf6. Results demonstrated the ability of the Loner/Schizzo Sec7 domain to direct GDP/GTP exchange but not a mutant in which the GEF activity is abolished ***(63)***.

A role for this Loner/Arf6 interaction in *Drosophila* myoblasts is supported by the observation that mesodermal expression of a dominant-negative Arf6 results in unfused myoblasts. ARF6, in turn, is required for membrane localization of the Rac GTPase, and this localization is perturbed in *Drosophila* founder cells in the absence of a functional Loner/Arf6 complex ***(63)***. Arf6 also enhances Rac-mediated remodeling of the actin cytoskeleton in mammalian cells ***(65)***. Thus, both the CDM and Arf6 pathways may converge on Rac to influence critical, but separate, changes in the actin cytoskeleton.

5.4. Signaling Downstream of Immunoglobulin Super Family Receptors: Cytoskeletal Effector Proteins

The involvement of Rho family members in myoblast fusion, as downstream targets of both the CDM and the Loner/Arf6 pathways, has already implicated the actin cytoskeleton in this process. Studies now have identified proteins such as Kette that function as downstream effectors of Rho family GTPases to regulate the actin cytoskeleton in fusing myoblasts ***(66)***. Three additional studies have suggested that Wasp and Sltr/D-WIP act in a parallel pathway to regulate actin polymerization ***(41,48,67)***. In both cases, the Arp2/3 complex has been implicated as the ultimate target of these regulatory molecules.

Embryos bearing mutations in *kette* have defects in myoblast fusion in which it progresses to the second stage but no further ***(66)***. Later defects in muscle attachment are also observed in hypomorphic alleles of *kette*. Loss-of-function mutations in *kette* interact genetically with mutations in blown fuse (Blow) ***(37)***, a novel protein for which the function has not been fully elucidated ***(66)***. Interestingly, excess Kette can partially rescue the fusion defects seen in *blow* mutant embryos ***(66)***. The *kette* locus encodes a *Drosophila* ortholog of vertebrate Hem-2/Nap1, which regulates F-actin polymerization by sequestering the actin nucleation factor Wave (Scar in *Drosophila*) in a cytosolic complex that also contains Sra-1, Abi, and HSPC300 ***(66)***. Extrapolating from its known mechanism of action in other tissues, Wave is dissociated from the complex by either SH3 domain–containing proteins or by activated Rac and then becomes available to nucleate actin polymerization. Thus, Kette is proposed to play an inhibitory role in actin polymerization in myoblasts.

Mutations in *Drosophila wasp*, a second actin nucleation factor, also perturb myoblast fusion. Loss-of-function zygotic alleles of *wasp* apparently survive embryogenesis because of the persistence of a maternally provided gene product. By contrast, *wasp* mutants that lack the Arp2/3 binding domain stop after muscle precursor formation ***(67)***. This particular mutant phenotype is likely due to a dominant-negative effect of the truncated Wasp protein, because expression of a similar transgene in wild-type embryos interferes with myoblast fusion ***(67)***. Most notably, elimination of both maternal and zygotic *wasp* results in a severe embryonic muscle phenotype in which myoblasts cluster around elongated founder cells but fusion does not occur ***(41,48)***. Activation of Wasp is Cdc42 independent, as loss-of-functions mutants as well as dominant-negative and constitutively active mutants do not have severe myoblast fusion defects ***(67)***.

Genetic interaction experiments have suggested that Kette and Wasp function in separate pathways, because Kette is able to antagonize the myoblast mutant phenotype of *wasp* but does not appear to activate the Wasp protein ***(67)***.

The *Drosophila* ortholog of vertebrate WIP, Sltr/D-WIP, has two WH2 domains at its N terminus, a proline-rich region and a conserved Wasp-binding domain at its C terminus. It is expressed exclusively in the fusion-competent myoblasts and is later incorporated in the developing syncytia ***(41,48)***. Mutants in *sltr/D-wip* are lacking multinucleate syncytia, but the first step of myoblast fusion, giving rise to muscle precursors, still occurs. Moreover, as in *wasp* mutant embryos, the fusion-competent myoblasts cluster around these elongated muscle precursors ***(41,48)***. Sltr/D-WIP is recruited to points of contact between muscle precursors and fusion-competent myoblasts and, in turn, mediates localization of Wasp to these sites via its C-terminal region.

In embryos, the colocalization of D-WIP and Wasp coincides with the presence of the IgSF member Duf/Kirre at points of contact. Sltr/D-WIP expression is also coincident with the SNS IgSF member at points of cell–cell contact in transfected S2 cells and in embryos, and biochemical studies have suggested that this interaction is mediated by the SH2–SH3 adaptor protein Crk ***(41,48)***. Interestingly, a myristoylated form of Wasp that is targeted to the membrane in the absence of Sltr/D-WIP can partially rescue the *D-wip* mutant phenotype ***(48)***, consistent with the notion that the primary role of Sltr/D-Wip (and perhaps Crk) is to relay Wasp to sites of membrane fusion. While nucleation of actin polymerization at sites of cell–cell contact appears to have direct impact on the fusion machinery, the exact purpose of filamentous actin remains controversial. Moreover, the finding that Sltr/D-Wip also interacts biochemically with actin suggests that we do not yet fully comprehend all relevant interactions of these molecules ***(41)***. Nevertheless the observation that *sltr/D-wip* and *wasp* small inhibitory RNA disrupts myoblast fusion in mammalian C2C12 cells suggests that their role is conserved ***(41)***.

5.5. Other Intracellular Proteins Required for Myoblast Fusion

Several additional genes appear to be involved in myoblast fusion on the basis of a mutant or overexpression phenotype, although their specific biochemical roles remain unclear. For example, *blow* was identified in a screen for defective motoneuron axon guidance ***(37)***, but this mutant phenotype was found to be a secondary consequence of a major defect in myoblast fusion. *blow*, in which a PH domain is the only obvious structural signature, appears to be expressed in the cytoplasm of both founder cells and fusion-competent cells just prior to fusion ***(37)***. Ultrastructural studies of *blow* mutant embryos revealed the presence of prefusion vesicles but no electron-dense plaques. Placement at this point in the pathway is morphologically consistent with the observation that fusion-competent myoblasts lacking *blow* still migrate and associate tightly with elongated founder cells. Epistasis experiments put *blow* upstream of *kette* during myoblast fusion, and, consistent with this, overexpression of *kette* can rescue *blow* fusion defects and embryonic lethality ***(66)***.

Aside from its scaffolding role in muscle sarcomeres and chromosomes, the huge 1.8 MDa D-titin protein also plays a role in regulation of myoblast fusion ***(68,69)***. D-titin is composed of numerous Ig domains, fibronectin type-III domains, and PEVK domains, the latter of which mediates interaction with actin filaments. It is expressed in myoblasts prior to fusion and accumulates at sites of myotube–myoblast contact, often colocalizing with the Duf/Ants complex at these sites ***(49,68,69)***. Notably, D-titin becomes cytoplasmic in both *ants/rols7* and *duf/irreC* mutants ***(49)***.

Two final genes in which homozygous mutant embryos exhibit myoblast fusion defects are *rolling stone (rost) and singles bar*. Rost is predicted to be a membrane protein, and, consistent with this, it is enriched in embryonic membrane fractions ***(34)***. It is expressed in the mesoderm and is present throughout muscle development. Although the true *rost* loss-of-function phenotype remains unclear, embryos expressing antisense *rost* do exhibit a block in myoblast fusion ***(34)***.

The *singles bar* (*sing*) gene encodes a hydrophobic MARVEL domain-containing multipass transmembrane protein that is expressed in both founder and fusion-competent myoblasts and is essential for myoblast fusion ***(70)***. As in embryos mutant for *blow*, described earlier, the fusion-competent myoblasts of embryos mutant for *sing* migrate and adhere to the founder myoblasts ***(24)***, and the first stage of myoblast fusion can occur ***(70)***. However, ultrastructural analysis of myoblasts in *sing* mutant embryos revealed greater accumulation of electron-dense paired vesicles than observed in wild-type or *blow* mutant embryos ***(70)***. This mutant phenotype indicates that *sing* functions after cell migration and adhesion to allow progression past the prefusion complex. The presence of the MARVEL domain, which has been associated with vesicle trafficking in other systems, supports the intriguing possibility suggested by Estrada et al. ***(70)*** that Sing may function in membrane-associated events through which the vesicles fuse to the plasma membrane.

6. Conclusion

In the past decade, remarkable advances have been made in our understanding of *Drosophila* myogenesis in general and myoblast fusion in particular. A greater understanding of the molecules through which embryonic myoblasts migrate, recognize, and fuse to each other did not exist a decade ago. Most exciting is the possibility that evolutionarily conserved pathways, discussed herein, that are crucial for myoblast fusion in *Drosophila* will provide insights into vertebrate myoblast fusion. Some of these pathways, among them the CDM, Wasp, and Arp2/3 pathways, have in common the regulation of the actin cytoskeleton, which was implicated in vertebrate myoblast fusion long before today's sophisticated methods were available (*reviewed in* **ref. *50***).

Reciprocally, insights into these conserved pathways from their analysis in vertebrates have implications for *Drosophila* myoblast fusion. Beyond identification of the molecular components of fusion, technological advances that allow us to visualize myoblasts in living embryos or better preserve and examine ultrastructural details in mutant myoblasts are having a dramatic impact on our view of myoblast fusion.

Despite the major advances outlined herein, however, much remains to be understood. Genes known to play a critical role in myoblast fusion on the basis of their mutant phenotypes have yet to be understood mechanistically. Many enticing biochemical protein–protein interactions have yet to be validated by genetic interact studies or structure/function analysis in the embryo. Undoubtedly, genes that function in common pathways seem likely to be required at earlier stages of development, making their role in myoblast fusion more difficult to decipher. Thus, for individuals new to the field, as well as for old-timers who have seen the field blossom, the potential for many more exciting findings in the future is high.

References

1. Abmayr, S. M., Balagopalan, L., Galletta, B. J., Hong, S. J., Lawrence, I. G., Kostas, I., and Sarjeet, S. G. (2005) *Comprehensive Molecular Insect Science*. Elsevier, Amsterdam, pp. 1–43.
2. Chen, E. H. and Olson, E. N. (2004) Towards a molecular pathway for myoblast fusion in Drosophila. *Trends Cell Biol.* **14,** 452–460.
3. Horsley, V. and Pavlath, G. K. (2004) Forming a multinucleated cell: molecules that regulate myoblast fusion. *Cells Tissues Organs* **176,** 67–78.
4. Bate, M. (1990) The embryonic development of larval muscles in *Drosophila*. *Development* **110,** 791–804.
5. Baylies, M. K. and Bate, M. (1996) twist: a myogenic switch in *Drosophila*. *Science* **272,** 1481–1484.
6. Carmena, A., Bate, M., and Jimenez, F. (1995) *lethal of scute*, a proneural gene, participates in the specification of muscle progenitors during *Drosophila* embryogenesis. *Genes Dev.* **9,** 2373–2383.
7. Baker, R. and Schubiger, G. (1996) Autonomous and nonautonomous Notch functions for embryonic muscle and epidermis development in *Drosophila*. *Development* **122,** 617–626.
8. Bate, M. and Rushton, E. (1993) Myogenesis and muscle patterning in *Drosophila*. *C.R. Acad. Sci. Ser. III* **316,** 1047–1061.
9. Corbin, V., Michelson, A. M., Abmayr, S. M., Neel, V., Alcamo, E., Maniatis, T., and Young, M. W. (1991) A role for the *Drosophila* neurogenic genes in mesoderm differentiation. *Cell* **67,** 311–323.

10. Bate, M., Rushton, E., and Frasch, M. (1993) A dual requirement for neurogenic genes in *Drosophila* myogenesis. *Dev. Suppl.* 149–161.
11. Bourgouin, C., Lundgren, S. E., and Thomas, J. B. (1992) *apterous* is a *Drosophila* LIM domain gene required for the development of a subset of embryonic muscles. *Neuron* **9,** 549–561.
12. Dohrmann, C., Azpiazu, N., and Frasch, M. (1990) A new *Drosophila* homeobox gene is expressed in mesodermal precursor cells of distinct muscles during embryogenesis. *Genes Dev.* **4**, 2098–2111.
13. Frasch, M., Hoey, T., Rushlow, C., Doyle, H., and Levine, M. (1987) Characterization and localization of the even-skipped protein of *Drosophila. EMBO J.* **6,** 749–759.
14. Keller, C. A., Grill, M. A., and Abmayr, S. M. (1998) A role for *nautilus* in the differentiation of muscle precursors. *Dev. Biol.* **202,** 157–171.
15. Knirr, S., Azpiazu, N., and Frasch, M. (1999) The role of the NK-homeobox gene *slouch* (S59) in somatic muscle patterning. *Development* **126,** 4525–4535.
16. Nose, A., Isshiki, T., and Takeichi, M. (1998) Regional specification of muscle progenitors in *Drosophila*: the role of the *msh* homeobox gene. *Development* **125,** 215–223.
17. Nose, A., Mahajan, V. B., Goodman, C. S. (1992) Connectin: A homophilic cell adhesion molecule on a subset of muscles and the motoneurons that innervate them in *Drosophila. Cell* **70,** 553–567.
18. Ruiz-Gomez, M., Romani, S., Hartmann, C., Jackle, H., and Bate, M. (1997) Specific muscle identities are regulated by Kruppel during *Drosophila* embryogenesis. *Development* **124,** 3407–3414.
19. Jagla, T., Bellard, F., Lutz, Y., Dretzen, G., Bellard, M., and Jagla, K. (1998) *ladybird* determines cell fate decisions during diversification of *Drosophila* somatic muscles. *Development* **125,** 3699–3708.
20. Duan, H., Skeath, J. B., and Nguyen, H. T. (2001) *Drosophila* Lame duck, a novel member of the Gli superfamily, acts as a key regulator of myogenesis by controlling fusion-competent myoblast development. *Development* **128,** 4489–4500.
21. Furlong, E. E., Andersen, E. C., Null, B., White, K. P., and Scott, M. P. (2001) Patterns of gene expression during Drosophila mesoderm development. *Science* **293,** 1629–1633.
22. Ruiz-Gomez, M., Coutts, N., Suster, M. L., Landgraf, M., and Bate, M. (2002) myoblasts incompetent encodes a zinc finger transcription factor required to specify fusion-competent myoblasts in *Drosophila. Development* **129,** 133–141.
23. Bour, B. A., Chakravarti, M., West, J. M., and Abmayr, S. M. (2000) *Drosophila* SNS, a member of the immunoglobulin superfamily that is essential for myoblast fusion. *Genes Dev.* **14,** 1498–1511.
24. Ruiz-Gomez, M., Coutts, N., Price, A., Taylor, M. V., and Bate, M. (2000) *Drosophila* Dumbfounded: a myoblast attractant essential for fusion. *Cell* **102,** 189–198.
25. Strunkelnberg, M., Bonengel, B., Moda, L. M., Hertenstein, A., de Couet, H. G., Ramos, R. G., and Fischbach, K. F. (2001) rst and its paralogue kirre act redundantly during embryonic muscle development in *Drosophila. Development* **128,** 4229–4239.

26. Artero, R. D., Castanon, I., and Baylies, M. K. (2001) The immunoglobulin-like protein Hibris functions as a dose-dependent regulator of myoblast fusion and is differentially controlled by Ras and Notch signaling. *Development* **128,** 4251–4264.
27. Dworak, H. A., Charles, M. A., Pellerano, L. B., and Sink, H. (2001) Characterization of *Drosophila hibris*, a gene related to human nephrin. *Development* **128,** 4265–4276.
28. Dworak, H. A. and Sink, H. (2002) Myoblast fusion in *Drosophila. BioEssays* **24,** 591–601.
29. Ramos, R. G., Igloi, G. L., Lichte, B., Baumann, U., Maier, D., Schneider, T., Brandstatter, J. H., Frohlich, A., and Fischbach, K. F. (1993) The irregular chiasm C-roughest locus of *Drosophila*, which affects axonal projections and programmed cell death, encodes a novel immunoglobulin-like protein. *Genes Dev.* **7,** 2533–2547.
30. Strunkelnberg, M., de Couet, H. G., Hertenstein, A., and Fischbach, K. F. (2003) Interspecies comparison of a gene pair with partially redundant function: the *rst* and *kirre* genes in *D. virilis* and *D. melanogaster. J. Mol. Evol.* **56,** 187–197.
31. Klapper, R., Stute, C., Schomaker, O., Strasser, T., Janning, W., Renkawitz-Pohl, R., and Holz, A. (2002) The formation of syncytia within the visceral musculature of the *Drosophila* midgut is dependent on duf, sns and mbc. *Mech. Dev.* **110,** 85–96.
32. San Martin, B. and Bate, M. (2001) Hindgut visceral mesoderm requires an ectodermal template for normal development in *Drosophila. Development* **128,** 233–242.
33. Paululat, A., Burchard, S., and Renkawitz-Pohl, R. (1995) Fusion from myoblasts to myotubes is dependent on the *rolling stone* gene *(rost)* of *Drosophila. Development* **121,** 2611–2620.
34. Paululat, A., Goubeaud, A., Damm, C., Knirr, S., Burchard, S., and Renkawitz-Pohl, R. (1997) The mesodermal expression of *rolling stone (rost)* is essential for myoblast fusion in *Drosophila* and encodes a potential transmembrane protein. *J. Cell Biol.* **138,** 337–348.
35. Paululat, A., Holz, A., and Renkawitz-Pohl, R. (1999) Essential genes for myoblast fusion in *Drosophila* embryogenesis. *Mech. Dev.* **83,** 17–26.
36. Kestila, M., Lenkkeri, U., Mannikko, M., Lamerdin, J., McCready, P., Putaala, H., Ruotsalainen, V., Morita, T., Nissinen, M., Peltonen, L., Holmberg, C., Olsen, A., and Tryggvason, K. (1998) Positionally cloned gene for a novel glomerular protein—Nephrin—is mutated in congenital nephrotic syndrome. *Mol. Cell* **1,** 572–582.
37. Doberstein, S. K., Fetter, R. D., Mehta, A. Y., and Goodman, C. S. (1997) Genetic analysis of myoblast fusion: blown fuse is required for progression beyond the prefusion complex. *J. Cell Biol.* **136,** 1249–1261.
38. Galletta, B. J., Chakravarti, M., Banerjee, R., and Abmayr, S. M. (2004) SNS: adhesive properties, localization requirements and ectodomain dependence in S2 cells and embryonic myoblasts. *Mech. Dev.* **121,** 1455–1468.
39. Kesper, D. A., Stute, C., Buttgereit, D., Kreiskother, N., Vishnu, S., Fischbach, K. F., and Renkawitz-Pohl, R. (2007) Myoblast fusion in *Drosophila melanogaster* is

mediated through a fusion-restricted myogenic-adhesive structure (FuRMAS). *Dev. Dyn.* **236,** 404–415.

40. Rash, J. E. and Fambrough, D. (1973) Ultrastructural and electrophysiological correlates of cell coupling and cytoplasmic fusion during myogenesis in vitro. *Dev. Biol.* **30,** 166–186.
41. Kim, S., Shilagardi, K., Zhang, S., Hong, S. N., Sens, K. L., Bo, J., Gonzalez, G. A., and Chen, E. H. (2007) A critical function for the actin cytoskeleton in targeted exocytosis of prefusion vesicles during myoblast fusion. *Dev. Cell* **12,** 571–586.
42. Erickson, M. R. S., Galletta, B. J., and Abmayr, S. M. (1997) *Drosophila myoblast city* encodes a conserved protein that is essential for myoblast fusion, dorsal closure and cytoskeletal organization. *J. Cell Biol.* **138,** 589–603.
43. Rushton, E., Drysdale, R., Abmayr, S. M., Michelson, A. M., and Bate, M. (1995) Mutations in a novel gene, myoblast city, provide evidence in support of the founder cell hypothesis for *Drosophila* muscle development. *Development* **121,** 1979–1988.
44. Chen, E. H. and Olson, E. N. (2001) Antisocial, an intracellular adaptor protein, is required for myoblast fusion in *Drosophila. Dev. Cell* **1,** 705–715.
45. Menon, S. D. and Chia, W. (2001) *Drosophila* rolling pebbles: a multidomain protein required for myoblast fusion that recruits D-titin in response to the myoblast attractant Dumbfounded. *Dev. Cell* **1,** 691–703.
46. Rau, A., Buttgereit, D., Holz, A., Fetter, R., Doberstein, S. K., Paululat, A., Staudt, N., Skeath, J., Michelson, A. M., and Renkawitz-Pohl, R. (2001) *rolling pebbles (rols)* is required in *Drosophila* muscle precursors for recruitment of myoblasts for fusion. *Development* **128,** 5061–5073.
47. Hakeda-Suzuki, S., Ng, J., Tzu, J., Dietzl, G., Sun, Y., Harms, M., Nardine, T., Luo, L., and Dickson, B. J. (2002) Rac function and regulation during *Drosophila* development. *Nature* **416,** 438–442.
48. Massarwa, R., Carmon, S., Shilo, B. Z., and Schejter, E. D. (2007) WIP/WASp-based actin-polymerization machinery is essential for myoblast fusion in *Drosophila. Dev. Cell* **12,** 557–569.
49. Menon, S. D., Osman, Z., Chenchill, K., and Chia, W. (2005) A positive feedback loop between Dumbfounded and Rolling pebbles leads to myotube enlargement in *Drosophila. J. Cell Biol.* **169,** 909–920.
50. Abmayr, S. M., Balagopalan, L., Galletta, B. J., and Hong, S. J. (2003) Cell and molecular biology of myoblast fusion. *Int. Rev. Cytol.* **225,** 33–89.
51. Duchek, P., Somogyi, K., Jekely, G., Beccari, S., and Rorth, P. (2001) Guidance of cell migration by the Drosophila PDGF/VEGF receptor. *Cell* **107,** 17–26.
52. Nolan, K. M., Barrett, K., Lu, Y., Hu, K. Q., Vincent, S., Settleman, J. (1998) Myoblast city, the *Drosophila* homolog of DOCK180/CED-5, is required in a Rac signaling pathway utilized for multiple developmental processes. *Genes Dev.* **12,** 3337–3342.
53. Brugnera, E., Haney, L., Grimsley, C., Lu, M., Walk, S. F., Tosello-Trampont, A. C., Macara, I. G., Madhani, H., Fink, G. R., and Ravichandran, K. S. (2002) Unconventional Rac-GEF activity is mediated through the Dock180-ELMO complex. *Nat. Cell Biol.* **4,** 574–582.

54. Gumienny, T. L., Brugnera, E., Tosello-Trampont, A. C., Kinchen, J. M., Haney, L. B., Nishiwaki, K., Walk, S. F., Nemergut, M. E., Macara, I. G., Francis, R., Schedl, T., Qin, Y., Van Aelst, L., Hengartner, M. O., and Ravichandran, K. S. (2001) CED-12/ELMO, a novel member of the CrkII/Dock180/Rac pathway, is required for phagocytosis and cell migration. *Cell* **107,** 27–41.
55. Wu, Y. C., Tsai, M. C., Cheng, L. C., Chou, C. J., and Weng, N. Y. (2001) *C. elegans* CED-12 acts in the conserved crkII/DOCK180/Rac pathway to control cell migration and cell corpse engulfment. *Dev. Cell* **1,** 491–502.
56. Zhou, Z., Caron, E., Hartwieg, E., Hall, A., and Horvitz, H. R. (2001) The *C. elegans* PH domain protein CED-12 regulates cytoskeletal reorganization via a Rho/Rac GTPase signaling pathway. *Dev. Cell* **1,** 477–489.
57. Hasegawa, H., Kiyokawa, E., Tanaka, S., Nagashima, K., Gotoh, N., Shibuya, M., Kurata, T., Matsuda, M. (1996) DOCK180, a major CRK-binding protein, alters cell morphology upon translocation to the cell membrane. *Mol. Cell. Biol.* **16,** 1770–1776.
58. Wu, Y. C. and Horvitz, H. R. (1998) C. elegans phagocytosis and cell-migration protein CED-5 is similar to human DOCK180. *Nature* **392,** 501–504.
59. Balagopalan, L., Chen, M. H., Geisbrecht, E. R., and Abmayr, S. M. (2006) The CDM superfamily protein MBC directs myoblast fusion through a mechanism that requires phosphatidylinositol 3,4,5-triphosphate binding but is independent of direct interaction with DCrk. *Mol. Cell. Biol.* **26,** 9442–9455.
60. Galletta, B. J., Niu, X. P., Erickson, M. R., and Abmayr, S. M. (1999) Identification of a *Drosophila* homologue to vertebrate Crk by interaction with MBC. *Gene* **228,** 243–252.
61. Luo, L., Liao, Y. J., Jan, L. Y., and Jan, Y. N. (1994) Distinct morphogenetic functions of similar small GTPases: *Drosophila* Drac1 is involved in axonal outgrowth and myoblast fusion. *Genes Dev.* **8,** 1787–1802.
62. Ng, J., Nardine, T., Harms, M., Tzu, J., Goldstein, A., Sun, Y., Dietzl, G., Dickson, B. J., and Luo, L. (2002) Rac GTPases control axon growth, guidance and branching. *Nature* **416,** 442–447.
63. Chen, E. H., Pryce, B. A., Tzeng, J. A., Gonzalez, G. A., and Olson, E. N. (2003) Control of myoblast fusion by a guanine nucleotide exchange factor, loner, and its effector ARF6. *Cell* **114,** 751–762.
64. Onel, S., Bolke, L., and Klambt, C. (2004) The *Drosophila* ARF6-GEF Schizzo controls commissure formation by regulating Slit. *Development* **131,** 2587–2594.
65. Donaldson, J. G. (2003) Myoblasts fuse when loner meets ARF6. *Dev. Cell* **5,** 527–528.
66. Schroter, R. H., Lier, S., Holz, A., Bogdan, S., Klambt, C., Beck, L., and Renkawitz-Pohl, R. (2004) kette and blown fuse interact genetically during the second fusion step of myogenesis in *Drosophila. Development 131*, 4501–4509.
67. Schafer, G., Weber, S., Holz, A., Bogdan, S., Schumacher, S., Muller, A., Renkawitz-Pohl, R., and Onel, S. F. (2007) The Wiskott-Aldrich syndrome protein (WASP) is essential for myoblast fusion in *Drosophila. Dev. Biol.* **304,** 664–674.
68. Machado, C., Andrew, D. J. (2000) D-titin: a giant protein with dual roles in chromosomes and muscles. *J. Cell Biol.* **151,** 639–652.

69. Zhang, Y., Featherstone, D., Davis, W., Rushton, E., and Broadie, K. (2000) *Drosophila* D-titin is required for myoblast fusion and skeletal muscle striation. *J. Cell Sci.* **113,** 3103–3115.
70. Estrada, B., Maeland, A. D., Gisselbrecht, S. S., Bloor, J. W., Brown, N. H., and Michelson, A. M. (2007) The MARVEL domain protein, Singles Bar, is required for progression past the pre-fusion complex stage of myoblast fusion. *Dev. Biol.* **307,** 328–339.
71. Beckett K, B. M. (2006) Parcas, a regulator of non-receptor tyrosine kinase signaling, acts during anterior–posterior patterning and somatic muscle development in *Drosophila melanogaster. Dev. Biol.* **299,** 176–192.

6

Mammalian Fertilization Is Dependent on Multiple Membrane Fusion Events*

Paul M. Wassarman and Eveline S. Litscher

> The elements that unite are single cells, each on the point of death; but by their union a rejuvenated individual is formed, which constitutes a link in the eternal process of Life.
> *F. R. Lillie, Problems of Fertilization, University of Chicago Press, 1919*

Summary

Successful completion of fertilization in mammals is dependent on three membrane fusion events. These are *(1)* the acrosome reaction of sperm, *(2)* the fusion of sperm and egg plasma membranes to form a zygote, and *(3)* the cortical reaction of fertilized eggs. Extensive research into the molecular basis of each of these events has identified candidate proteins and factors involved in fusion of membranes during the mammalian fertilization process. Some of this information is provided here.

Key Words: Mammalian fertilization; sperm; eggs; acrosome reaction; sperm–egg fusion; cortical reaction.

1. Introduction

For two cellular membranes to fuse, lipid bilayers must be within ~15 Å of each other. Fusion of two membranes into a single continuous bilayer can occur between individual cells, between organelles and cellular membranes, and between membranes of organelles themselves ***(1–3)***. An example of each of these membrane fusion events is found during the process of fertilization in mammals, and they are indicated in **Figure 1**.

*This article is dedicated to the memory of David L. Garbers (1944–2006), whose research interests included mammalian fertilization.

From: *Methods in Molecular Biology, vol. 475: Cell Fusion: Overviews and Methods*
Edited by: E. H. Chen © Humana Press, Totowa, NJ

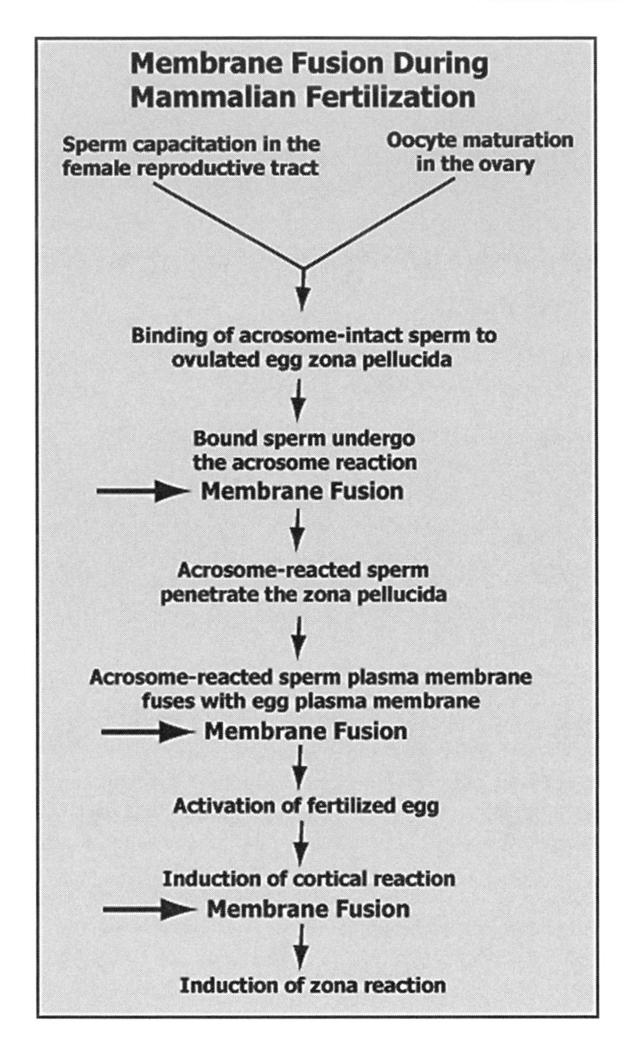

Fig. 1. Membrane fusion events during mammalian fertilization. Shown is an abbreviated pathway of fertilization in mammals, emphasizing the three membrane fusion events: acrosome reaction, sperm–egg fusion, and cortical reaction (indicated by arrows).

Among mammals, creation of a new individual of the species depends on fusion of sperm plasma membrane with unfertilized egg plasma membrane to form a one-cell embryo, or zygote. Clearly, membrane fusion is a critical step in the initiation of mammalian development. Membrane fusion plays a role as well during preparation of sperm for fusion with eggs ("acrosome reaction"; fusion of sperm outer acrosomal membrane and plasma membrane) and during the response of eggs to fusion with a single sperm ("cortical reaction"; fusion of egg cortical granule membrane and plasma membrane). These three membrane fusion events have been addressed in several recent reviews of mammalian fertilization ***(4–11)*** and are the subject of this chapter. Little attempt will be made here to review molecular details and models of membrane fusion in general (e.g., fusion of animal viruses with mammalian cells), but such information is readily available elsewhere ***(1–3,12–14)***.

2. Terms and Definitions

The following subsections briefly address the three instances of membrane fusion for mammalian gametes, eggs and sperm, during the fertilization process.

2.1. Acrosome Reaction

The acrosome, a membrane-enclosed vesicle that originates from the Golgi during spermiogenesis, is located just over the nucleus and under the plasma membrane in the apical region of the sperm head ***(15)***. The acrosome can be considered as a secretory granule that contains a variety of enzymes, including proteases (e.g., proacrosin), glycosidases, phosphatases, sulfatases, and phospholipases. The acrosome reaction is an exocytotic event that results in release/exposure of enzymes and other proteins that reside within the membrane-enclosed acrosomal vesicle. It takes place as a result of multiple fusions between the sperm's outer acrosomal membrane and plasma membrane, formation of hybrid vesicles, and exposure of the inner acrosomal membrane and acrosomal contents (**Fig. 2; refs.** ***6,16,17***). Only acrosome-reacted sperm, not acrosome-intact sperm, can penetrate the zona pellucida surrounding mammalian eggs and fuse with egg plasma membrane to form a zygote. That is, only acrosome-reacted sperm are "fusion competent."

2.2. Sperm–Egg Fusion

Zygote formation occurs as a result of fusion between plasma membrane in the equatorial region of the acrosome-reacted sperm head and plasma membrane of unfertilized eggs arrested at metaphase 2 of meiosis ***(6,17)***. Sperm–egg fusion occurs following ovulation and migration of eggs into the ampulla region of the oviduct and capacitation and migration of ejaculated sperm in the female reproductive tract. Fusion of sperm with eggs results in emission of a second polar body by fertilized eggs, restores a diploid chromosomal complement to eggs, and "activates" eggs (discussed in **Heading 5**) such that normal development ensues.

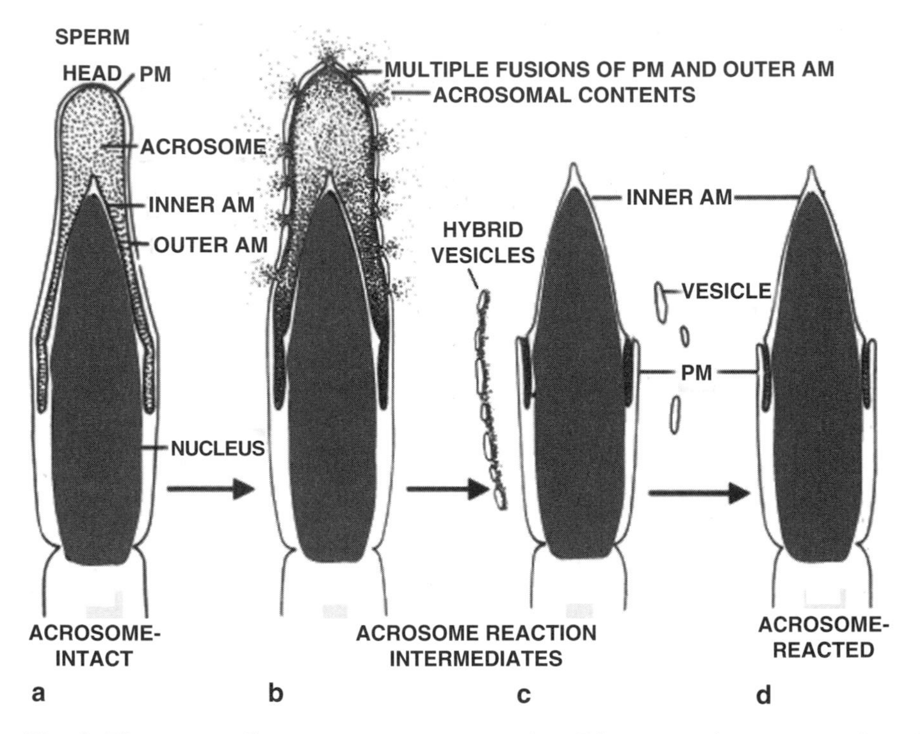

Fig. 2. The mammalian sperm acrosome reaction. Diagrammatic representation of mouse sperm undergoing the acrosome reaction **(A–D).** AM, acrosomal membrane; PM, plasma membrane. (Modified from **ref.** *17.)*

2.3. Cortical Reaction

Cortical granules, membrane-enclosed vesicles (0.1–1 µm in diameter) that originate from the Golgi during oogenesis, are found in the peripheral cortex (~2 µm thick) just beneath the egg plasma membrane ***(6,17,18)***. For example, each mouse egg has ~4,500 cortical granules, some of which appear to be attached to the cytoplasmic face of the egg plasma membrane. Cortical granules can be considered as secretory granules that contain a variety of enzymes, including proteases, ovoperoxidases, and glycosidases (e.g., N-acetylglucosaminidase). The cortical reaction is an exocytotic event that results in release of enzymes and other proteins that reside within cortical granules. It takes place as a result of fusion between cortical granule membrane and plasma membrane shortly after fertilization of eggs by a single sperm. Cortical granule contents are deposited into the perivitelline space, enter the zona pellucida, and modify zona pellucida glycoproteins, thereby establishing a so-called slow block to polyspermy *(6,17)*.

3. The Acrosome Reaction

Only sperm that have undergone the acrosome reaction can fuse with plasma membrane of ovulated eggs *(17,19)*. To enable sperm to undergo the acrosome reaction, they must first undergo a maturation process, called *capacitation*, in the female reproductive tract (in the isthmus of the oviduct) or in vitro under specific conditions. Uncapacitated sperm are unable to fertilize eggs. Capacitation is characterized in part by changes in cholesterol levels and phospholipid compositions of sperm membranes, changes in membrane fluidity, elevation of intracellular levels of Ca^{2+} and 3′-5′-cyclic adenosine monophosphate (cAMP), and changes in tyrosine phosphorylation of sperm proteins *(20)*. Capacitated sperm exhibit hyperactivated flagellar motility, perhaps necessary for penetration of sperm through the zona pellucida. Once capacitated, sperm are able to undergo the acrosome reaction, penetrate the zona pellucida, and fuse with egg plasma membrane.

For many mammalian species, including mice and human beings, only acrosome-intact sperm are reported to bind to the zona pellucida (**Fig. 3**). However, it is likely that during and after capacitation of sperm, transitional intermediates of membrane fusion are produced that take the form of intermediate membrane complexes or flickering fusion pores on capacitated sperm *(16,21,22)*. This may necessitate that the phrase "acrosome intact" be modified in describing the

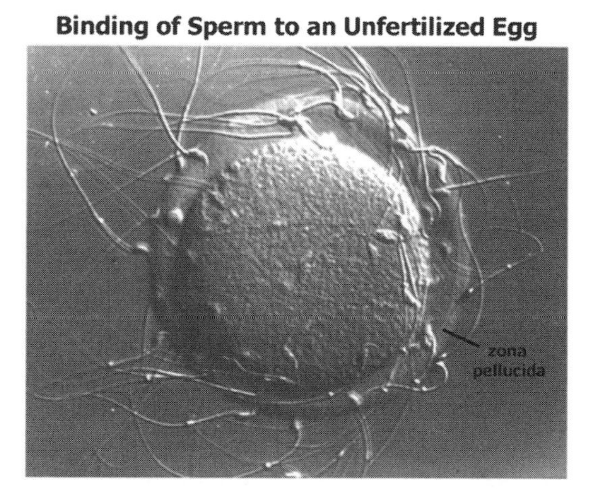

Fig. 3. Binding of sperm to the egg zona pellucida. Light micrograph (Nomarski differential interference contrast) of mouse sperm bound to the zona pellucida of an unfertilized mouse egg in vitro. (From **ref.** ***11.***)

binding of capacitated sperm to egg zona pellucida. Furthermore, a relatively low rate of "spontaneous" acrosome reactions occur in sperm populations, and, once again, this may reflect transitional intermediates of membrane fusion that are poised to complete the reaction in the absence of any additional stimulation.

It is widely accepted that capacitated, acrosome-intact sperm bind to the mouse egg zona pellucida and, subsequently, complete the acrosomal reaction (*see* **Fig. 2**). Consequently, it is not surprising that both solubilized egg zona pellucida and a purified egg zona pellucida glycoprotein, called ZP3, induce sperm to undergo the acrosome reaction in vitro ***(6,11,23–25)***. ZP3 stimulation depolarizes sperm membrane potential (~–25 mV), and this effect is not achieved with any other zona pellucida glycoprotein or with ZP3 from eggs of a different species ***(26)***.

Furthermore, several steps in ZP3 signal transduction during the acrosome reaction have been identified ***(4,6,23,27)***. These include activation of heterotrimeric G proteins, transient elevation of internal pH and level of cAMP, activation of phospholipase C, production of inositol 1,4,5-triphosphate (IP_3) and diacylglycerol from phosphatidylinositol 4,5-bisphosphate (PIP_2), transient elevation of Ca^{2+} due to release from internal stores (using IP_3 receptor channels), and Ca^{2+} entry by T-type Ca^{2+} channels (sustained Ca^{2+} levels). These steps result in a sustained elevation of intracellular Ca^{2+} to ~500 n*M*, carried out by transient receptor potential canonical (TRPC) Ca^{2+}-conducting cation channels, and is sufficient to induce completion of the acrosome reaction by sperm bound to the zona pellucida. Interestingly, TRPC2 colocalizes with ZP3 binding sites on sperm, and an antibody directed against the extracellular domain of TRPC2 prevents the sustained Ca^{2+} response to ZP3 ***(28)***. Additionally, there are indications that at least two components essential for intracellular membrane fusion in somatic cells, Rab3A GTPase and soluble N-ethylmaleimide–sensitive factor attachment protein receptors (SNAREs), are present in mammalian sperm and probably participate in the acrosome reaction ***(29–34)***. Clearly, the ZP3-activated acrosome reaction for mammalian sperm shares many features with receptor-regulated exocytosis for somatic cells ***(1,3)***.

Sperm that have undergone the acrosome reaction on the surface of the zona pellucida use proteolysis and motility to penetrate the zona pellucida and reach the plasma membrane. The former apparently is carried out by β-acrosin, a trypsinlike protease associated with the acrosomal matrix of acrosomal-intact sperm that is exposed and dispersed following the acrosome reaction ***(6,17)***; however, other acrosomal proteases as well may participate in this process. It is likely that acrosomal-reacted sperm remain bound to the zona pellucida because of interactions between sperm proacrosin, a zymogen, and sulfate groups on ZP2 ***(35)***. Penetrating sperm leave behind narrow slits in the zona pellucida. In mice, it takes ~5–20 min for acrosomal-reacted sperm to penetrate the egg zona pellucida in vitro.

4. Sperm–Egg Fusion

Once an acrosomal-reacted sperm reaches the perivitelline space of an ovulated egg (i.e., space between the plasma membrane and zona pellucida), plasma membrane at the equatorial segment of the sperm head, not the inner acrosomal membrane, fuses with plasma membrane present in microvilli over much of the egg surface (for many mammalian species the region overlying the metaphase II plate lacks microvilli) *(6)*. Fusion then extends to regions posterior to the head and the acrosomal region is engulfed by cortical cytoplasm of the egg; in many mammalian species, including mice, the entire sperm tail is incorporated into the egg cytoplasm (**Fig. 4**). For mice, sperm enter egg cytoplasm ~1 h after gametes are combined in vitro. Presumably, sperm–egg fusion leads to a fast (electrical) block to polyspermy at the level of the egg plasma membrane; however, the molecular basis of the block in mammals remains somewhat obscure.

Several egg and sperm proteins have been implicated in sperm–egg fusion (*see* **Fig. 4**). For nearly two decades, members of the ADAM (a disintegrin

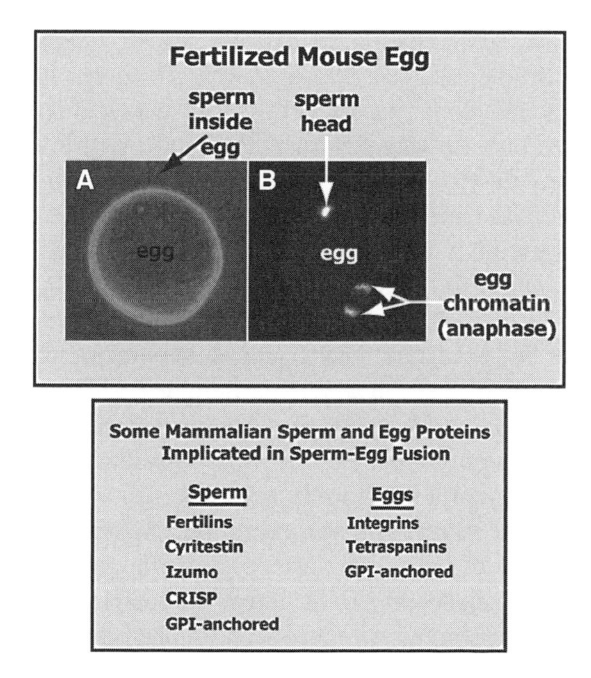

Fig. 4. A fertilized mouse egg. ***Top:*** An egg from which the zona pellucida was removed was inseminated for 45 min and then fixed and stained with diamidino-2-phenylindole (2 μg/mL). **(A)** Phase contrast image of the fertilized egg with the sperm tail visible outside the plane of focus (arrow). **(B)** The swollen DAPI-stained sperm nucleus and segregating anaphase chromosomes are visible within the egg cytoplasm (arrows). ***Bottom:*** List of some sperm and egg proteins implicated in the process of sperm–egg fusion. (B, modified from **ref.** ***70.)***

and metalloprotease) family on sperm and integrins on eggs have been considered as complementary partners in binding and fusion of sperm and eggs ***(4,6,7,9,11,36,37)***. The ADAM family proteins fertilin and cyritestin on sperm and integrin-α6β1 on eggs, in particular, have been strongly implicated in fertilization. Fertilin is a plasma membrane–anchored, heterodimeric complex composed of α- and β-subunits on the sperm surface. Fertilin-β (ADAM-2) is thought to support binding of sperm to eggs through its disintegrin domain and fertilin-α (ADAM-1) to induce fusion through its viruslike fusion peptide ***(7)***. However, despite extensive experimental evidence to the contrary, phenotypes of knockout mice cast suspicion on the roles of sperm fertilins and egg integrins in fertilization. Male mice in which fertilin-α and -β or cyritestin (ADAM-3) and female mice in which integrins (including integrin-α6β1) are deleted in homozygous nulls produce sperm and eggs, respectively, that can adhere to and fuse with each other normally ***(38–44)***. These findings suggest that other binding and fusion proteins may be present on sperm and eggs; other possibilities are also being considered ***(41,45)***.

Egg CD9 is likely to play an important role in sperm–egg fusion. CD9 is a member of the tetraspanin super family of integral plasma membrane proteins that associates with integrins ***(46,47)*** and facilitates fusion in cellular systems ***(48)***. Homozygous null females for CD9 exhibit severely reduced fertility due to greatly impaired sperm–egg fusion ***(49–51)***. Rescue experiments, using mRNAs encoding wild-type and mutated forms of CD9, suggest that its large extracellular loop participates in sperm–egg fusion ***(52)***. This is consistent with the report that an antibody directed against CD9 potently inhibits sperm–egg binding and fusion in vitro ***(42,53)***. It is possible that egg CD9 may serve as a receptor for sperm protein PSG17, a member of the immunoglobulin super family ***(54)***. It should be noted that other tetraspanins are present on mammalian eggs, and one or more of these may also participate in sperm–egg fusion ***(55)***.

The sperm protein Izumo (named after a Japanese shrine dedicated to marriage), a member of the immunoglobulin super family ***(56)***, and sperm CRISP (cysteine-rich secretory proteins) family proteins ***(57)*** may also play a role in sperm–egg fusion. The former is a transmembrane sperm protein, whereas the latter are associated with the sperm surface. Specific antibodies detect Izumo only on acrosome-reacted sperm, not on acrosome-intact sperm and prevent fertilization in vitro. Furthermore, sperm from homozygous null males for Izumo penetrate the zona pellucida and bind to egg plasma membrane but fail to fuse with the plasma membrane ***(56)***. On the other hand, injection of sperm from Izumo null males directly into eggs results in normal embryos that develop to term. It can be concluded that Izumo plays a critical role in sperm–egg fusion.

Finally, there is evidence to suggest that glycosylphosphatidylinositol (GPI)–anchored proteins may play a role in gamete fusion ***(9,45,58,59)***. For example, targeted mutagenesis of an enzyme essential in GPI-anchor biosynthesis results in female mice that are infertile because of a defect in fertilization *(58)*. Whether or not eggs from these mice lack one or more GPI-anchored proteins involved in sperm–egg fusion remains to be determined. In this context, recently it was suggested that release of GPI-anchored proteins from sperm by angiotensin-converting enzyme is essential for fertilization ***(60)***; however, it should be noted that different interpretations have been presented ***(61,62)***.

5. The Cortical Reaction

Fusion of sperm plasma membrane and egg plasma membrane leads rapidly to a so-called activation of the egg. Activation includes completion of meiosis with initiation of mitotic cell cycles, changes in membrane potential, release of Ca^{2+} from intracellular stores, and fusion of cortical granules with fertilized egg plasma membrane ***(63,64)***. Oocytes acquire the ability to undergo activation ("activation competence") close to the time of ovulation ***(65)***. The cortical reaction is followed by modifications of ZP2 and ZP3 by cortical granule components that establish a slow block to polyspermy ***(6,17,25,65–67)***. Modification of the zona pellucida by cortical granule components results in a hardening of the zona pellucida (i.e., a marked decrease in solubility) and in changes that prevent binding of free-swimming sperm to the zona pellucida. Collectively, these changes in the zona pellucida constitute what is termed the "zona reaction."

For the cortical reaction to occur, cortical granules must translocate as much as ~2 µm to the inner surface of egg plasma membrane (**Fig. 5**). To do so, cortical granules translocate in an actin microfilament-dependent manner utilizing actin-associated proteins, protein kinase C, and downstream proteins, such

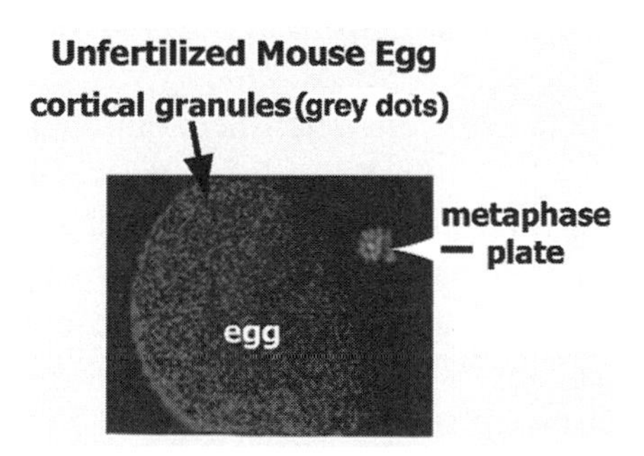

Fig. 5. Mammalian egg cortical granules. A mouse egg at 16 h after injection of human chorionic gonadotropin dually stained for cortical granules (gray dots stained with *Lens culinaris* agglutinin coupled with biotin and Texas red streptavidin) and chromatin (black stained with 4,6-diamidino-2-phenyl indole and Hoechst 33258). Note the cortical granule–free zone overlying the anaphase chromosomes (arrows). (Modified from **ref. *71.)***

as myristoylated alanine-rich C kinase substrate ***(66–68)***. The latter proteins crosslink actin filaments and anchor the actin network to the plasma membrane in somatic cells ***(69)***. Apparently, depolymerization or reorganization of cortical filamentous actin (F-actin) by activated protein kinase C and/or by other actin-associated proteins results in activated eggs undergoing the cortical reaction.

Like the sperm's acrosomal reaction, the cortical reaction requires an elevation of Ca^{2+} concentration in the egg, attributable in part to release of Ca^{2+} stores from the endoplasmic reticulum by IP_3 (generated together with diacylglycerol from PIP_2) and IP_3 receptor channels. For mouse eggs the Ca^{2+} concentration peaks at ~1 μM following egg activation. Inhibition of Ca^{2+} elevation prevents the cortical reaction, whereas artificial elevation of Ca^{2+} induces the cortical reaction in mammalian eggs. In this context, several Ca^{2+}-dependent proteins, including calmodulin, protein kinase C, Ca^{2+}/calmodulin-dependent protein kinase II, synaptotagmin, Rab3, and rabphilin-3A, have been considered as participants in the cortical reaction ***(66,67)***. Furthermore, SNARE proteins are also thought to be involved in mediating the cortical reaction. As is the case for the sperm's acrosome reaction, the cortical reaction of activated eggs shares many features with receptor-regulated exocytosis for somatic cells ***(1,3)***.

6. Conclusion

In this chapter, the three membrane fusion events that occur during the process of fertilization in mammals, the acrosome reaction, sperm–egg fusion, and cortical reaction, are described (*see* **Fig. 1**). The acrosome reaction of sperm and cortical reaction of eggs represent intracellular membrane fusion events, whereas sperm–egg fusion represents an intercellular membrane fusion event. Not surprisingly, many aspects of membrane fusion for mammalian gametes appear to mirror those described for somatic cells in that they share features with receptor-regulated exocytosis for somatic cells. Admittedly, some aspects of these events have not been described in sufficient detail here, and others have been omitted because of space limitations. Despite these deficiencies, it is hoped that the reader will be stimulated by the information that is provided and will consult some of the reference material for more details. The acrosome reaction, sperm–egg fusion, and cortical reaction are exciting areas of current research that will continue to provide important insights into the molecular basis of mammalian fertilization, in particular, and membrane fusion, in general. Such insights will undoubtedly suggest practical applications that bear on human fertility.

Acknowledgment

We are grateful to Elizabeth Chen for her kind invitation to write this chapter. Whenever possible, we have referred to articles and reviews published between 2000 and 2006. Our laboratory is supported in part by the National Institutes of Health (HD35105).

References

1. Alberts, B., Johnson, A., Lewis, J., Raff, M., Roberts, K., and Walter, P. (2002) *Molecular Biology of the Cell* (4th ed.). Garland Science, New York.
2. Jahn, R., Lang, T., and Sudhof, T. C. (2003) Membrane fusion. *Cell* **112,** 519–533.
3. Lodish, H., Berk, A., Matsudaira, P., Kaiser, C. A., Krieger, M., Scott, M. P., Zipursky, S. L., and Darnell, J. (2004) *Molecular Cell Biology* (5th ed.). Freeman, New York.
4. Evans, J. P. and Florman, H. M. (2002) The state of the union: the cell biology of fertilization. *Nat. Cell Biol.* **4 (Suppl. 1),** S57–S63.
5. Evans, J. P. (2002) The molecular basis of sperm–oocyte membrane interactions during mammalian fertilization. *Hum. Reprod. Update* **8,** 297–311.
6. Florman, H. M. and Ducibella, T. (2006) Fertilization in mammals, in *The Physiology of Reproduction* (Neill J.D., ed.), vol. 1. Elsevier, San Diego, pp. 55–112.
7. Primakoff, P. and Myles, D. G. (2002) Gamete fusion in mammals, in *Fertilization* (Hardy, D. M., ed.). Academic Press, San Diego, pp. 303–318.
8. Ramalho-Santos, J., Schatten, G., and Moreno, R. D. (2002) Control of membrane fusion during spermiogenesis and the acrosome reaction. *Biol. Reprod.* **69,** 254–260.
9. Stein, K. K., Primakoff, P., and Myles, D. (2004) Sperm–egg fusion: events at the plasma membrane. *J. Cell Sci.* **117,** 6269–6274.
10. Wassarman, P. M. (1999) Mammalian fertilization: molecular aspects of gamete adhesion, exocytosis, and fusion. *Cell* **96,** 175–183.
11. Wassarman, P. M., Jovine, L., and Litscher, E. S. (2001) A profile of fertilization in mammals. *Nat. Cell Biol.* **3,** E59–E64.
12. Chernomordik, L. V. and Kozlov, M. M. (2003) Protein–lipid interplay in fusion and fission of biological membranes. *Annu. Rev. Biochem.* **72,** 175–207.
13. Jahn, R. and Grubmuller, H. (2002) Membrane fusion. *Curr. Opin. Cell Biol.* **14,** 488–495.
14. Mayer, A. (2002) Membrane fusion in eukaryotic cells. *Annu. Rev. Cell Dev. Biol.* **18,** 289–314.
15. Eddy, E. M. (2006) The spermatozoon, in *The Physiology of Reproduction* (Neill J. D., ed.), vol. 1. Elsevier, San Diego, pp. 3–54.
16. Gerton, G. L. (2002) Function of the sperm acrosome, in *Fertilization* (Hardy, D. M., ed.). Academic Press, San Diego, pp. 265–302.
17. Yanagimachi, R. (1994) Mammalian fertilization, in *The Physiology of Reproduction* (Knobil E. and Neill J. D., eds.), vol. 1. Raven Press, New York, pp. 189–317.
18. Wassarman, P. M. and Albertini, D. F. (1994) Cellular and molecular biology of the mammalian ovum, in *The Physiology of Reproduction* (Knobil, E. Neill, J. D., eds.), vol. 1. Raven Press, NY, pp. 79–122.
19. Yanagimachi, R. (1981) Mechanisms of fertilization in mammals, in *Fertilization and Embryonic Development In Vitro* (Mastroianni L. and Biggers J. D., eds.). Plenum Press, New York, pp. 81–187.

20. Jaiswal, B. S. and Eisenbach, M. (2002) Capacitation, in *Fertilization* (Hardy, D. M., ed.). Academic Press, San Diego, pp. 57–117.
21. Kim, K.-S. and Gerton, G. L. (2003) Differential release of soluble and matrix components: evidence for intermediate states of secretion during spontaneous acrosomal exocytosis in mouse sperm. *Dev. Biol.* **264,** 141–152.
22. Kopf, G. S., Ning, X. P., Visconti, P. E., Purdon, M., Galantino-Homer, H., and Fornes, M. (1999) Signaling mechanisms controlling mammalian sperm fertilization competence and activation. In *The Male Gamete* (Gagnon C., ed.) 105–118. Cache River Press, Vienna, IL.
23. Florman, H. M., Kirkman-Brown, J., Brown, J. C., Jungnickel, M. K., and Sutton, K. A. (2003) The acrosome reaction: an example of egg-activated signal transduction in sperm. *ChemTracts Biochem. Mol. Biol.* **16,** 126–133.
24. Kopf, G. S. (2002) Signal transduction mechanisms regulating sperm acrosomal exocytosis, in *Fertilization* (Hardy, D. M., ed.). Academic Press, San Diego, pp. 181–223.
25. Wassarman, P. M., Bleil, J. D., Florman, H. M., Greve, J. M., Roller, R. J., Salzmann, G. S., and Samuels, F. G. (1985) The mouse egg's sperm receptor: what is it and how does it work? *Cold Spring Harb. Symp. Quant. Biol.* **50,** 11–19.
26. Arnoult, C., Zeng, Y., and Florman, H. M. (1996) ZP3-dependent activation of sperm cation channels regulates acrosomal secretion during mammalian fertilization. *J. Cell Biol.* **134,** 637–645.
27. Darszon, A., Acevedo, J. J., Galindo, B. E., Hernandez-Gonzalez, E. O., Nishigaki, T., Trevino, C. L., Wood, C., and Beltran, C. (2006) Sperm channel diversity and functional multiplicity. *Reproduction* **131,** 977–988.
28. Jungnickel, M. K., Marrero, H., Birnbaumer, L., Lemos, J. R., and Florman, H. M. (2001) Trp2 regulates entry of Ca^{2+} into mouse sperm triggered by egg ZP3. *Nat. Cell Biol.* **5,** 499–502.
29. De Blas, G. A., Roggero, C. M., Tomes, C. N., and Mayorga, L. S. (2005) Dynamics of SNARE assembly and disassembly during sperm acrosomal exocytosis. *PloS Biol.* **3,** e323.
30. Iida, H., Yoshinaga, Y., Tanaka, S., Toshimori, K., and Mori, T. (1999) Identification of Rab3A GTPase as an acrosome-associated small GTP-binding protein in rat sperm. *Dev. Biol.* **211,** 144–155.
31. Ramalho-Santos, J., Moreno, R. D., Sutovsky, P., Chan, A. W., Hewitson, L., Wessel, G. M., Simerly, C. R., and Schatten, G. (2000) SNAREs in mammalian sperm: possible implications for fertilization. *Dev. Biol.* **223,** 54–69.
32. Tomes, C. N., Michaut, M., De Blas, G. A., Visconti, P., Matti, U., and Mayorga, L. S. (2002) SNARE complex assembly is required for human sperm acrosome reaction. *Dev. Biol.* **243,** 326–338.
33. Tomes, C. N., De Blas, G. A., Michaut, M. A., Farre, E. V., Cherhitin, O., Visconti, P. E., and Mayorga, L. S. (2005) Alpha-SNAP and NSF are required in a priming step during the human sperm acrosome reaction. *Mol. Hum. Reprod.* **11,** 43–51.
34. Yunes, R., Michaut, M., Tomes, C., and Mayorga, L. S. (2000) Rab3A triggers the acrosome reaction in permeabilized human spermatozoa. *Biol. Reprod.* **62,** 1084–1089.

35. Howes, E. A. and Jones, R. (2003) Secondary binding of mammalian sperm to the zona pellucida. *ChemTracts Biochem. Mol. Biol.* **16,** 134–141.
36. Oh, E., Wortzman, G. B., Zhu, X., and Evans, J. P. (2003) Getting sperm and egg together: the molecules of gamete membrane interactions. *ChemTracts Biochem. Mol. Biol.* **16,** 142–157.
37. White, J. M. (2003) ADAMs: modulators of cell–cell and cell–matrix interactions. *Curr. Opin. Cell Biol.* **15,** 598–606.
38. Cho, C., Bunch, D. O., Faure, J.-E., Goulding, E. H., Eddy, E. M., Primakoff, P., and Myles, D. G. (1998) Fertilization defects in sperm from mice lacking fertilin β. *Science* **281,** 1857–1859.
39. Cho, C., Ge H., Branciforte, D., Primakoff, P., and Myles, D. G. (2000) Analysis of mouse fertilin in wild-type and fertilin $\beta^{-/-}$ sperm: evidence for C-terminal modification, α/β dimerization, and lack of essential role of fertilin b in sperm–egg fusion. *Dev. Biol.* **222,** 289–295.
40. He, Z. Y., Brakefusch, C., Fassler, R., Kreidberg, J. A., Primakoff, P., and Myles, D. G. (2003) None of the integrins known to be present on the mouse egg or to be ADAM receptors are essential for sperm–egg binding and fusion. *Dev. Biol.* **254,** 226–237.
41. Kim, E., Yamashita, M., Nakanishi, T., Park, K.-E., Kimura, M., Kashiwabara, S.-I., and Baba, T. (2006) Mouse sperm lacking ADAM1β/ADAM2 fertilin can fuse with the egg plasma membrane and effect fertilization. *J. Biol. Chem.* **281,** 5634–5639.
42. Miller, B. J., Georges-Labouesse, E., Primakoff, P., and Myles, D. G. (2000) Normal fertilization occurs with eggs lacking the integrin $\alpha6\beta1$ and is CD9-dependent. *J. Cell Biol.* **149,** 1289–1295.
43. Nishimura, H., Cho, C., Branciforte, D. R., Myles, D. G., and Primakoff, P. (2001) Analysis of loss of adhesive function in sperm lacking cyritestin or fertilin β. *Dev. Biol.* **233,** 204–213.
44. Shamsadin, R., Adham, I.M., Nayernia, K., Heinlein, U. A. O., Oberwinkler, H., and Engel, W. (1999) Male mice deficient for germ-cell cyritestin are infertile. *Biol. Reprod.* **61,** 1445–1451.
45. Stein, K. K., Go, J. C., Lane, W. S., Primakoff, P., and Myles, D. G. (2006) Proteomic analysis of sperm regions that mediate sperm–egg interactions. *Proteomics* **6,** 3533–3543.
46. Hemler, M. E. (1998) Integrin associated proteins. *Curr. Opin. Cell Biol.* **10,** 578–585.
47. Porter, J. C. and Hogg, N. (1998) Integrins take partners: cross-talk between integrins and other membrane receptors. *Trends Cell Biol.* **8,** 390–396.
48. Tachibana, I. and Hemler, M. E. (1999) Role of transmembrane 4 superfamily (TM4SF) proteins CD9 and CD81 in muscle cell fusion and myotube maintenance. *J. Cell Biol.* **146,** 893–904.
49. Kaji, K., Odu, S., Shikano, T., Ohnuki, T., Uematsu, Y., Sakagami, J., Tada, N., Miyazaki, S., and Kudo, A. (2000) The gamete fusion process is defective in eggs of CD9-deficient mice. *Nat. Genet.* **24,** 279–282.
50. Le Naour, F., Rubinstein, E., Jasmin, C., Prenant, M., and Boucheix, C. (2000) Severely reduced female fertility in CD9-deficient mice. *Science* **287,** 319–321.

51. Miyado, K., Yamada, G., Yamada, S., Hasuwa, H., Nakamura, Y., Ryu, F., Suzuki, K., Kosai, K., Inoue, K., Ogura, A., Okabe, M., and Mekada, E. (2000) Requirement of CD9 on the egg plasma membrane for fertilization. *Science* **287,** 321–324.
52. Zhu, G. Z., Miller, B. J., Boucheix, C., Rubinstein, E., Liu, C. C., Hynes, R. O., Myles, D. G., and Primakoff, P. (2002) Residues SFQ (173–175) in the large extracellular loop of CD9 are required for gamete fusion. *Development* **129,** 1995–2002.
53. Chen, M. S., Tung, K. S. K., Coonrod, S. A., Takahashi, Y., Bigler, D., Chang, A., Yamashita, Y., Kincade, P. W., Herr, J. C., and White, J. M. (1999) Role of integrin-associated protein CD9 in binding between sperm ADAM2 and the egg integrin $\alpha6\beta1$: implications for murine fertilization. *Proc. Natl. Acad. Sci., U.S.A.* **96,** 11830–11835.
54. Ellerman, D. A., Ha, C., Primakoff, P., Myles, D. G., and Dveksler, G. S. (2003) Direct binding of the ligand PSG17 to CD9 requires a CD9 site essential for sperm–egg fusion. *Mol. Biol. Cell* **14,** 5098–5103.
55. Takahashi, Y., Bigler, D., Ito, Y., and White, J. M. (2001) Sequence specific interaction between the disintegrin domain of mouse ADAM 3 and murine eggs: role of beta1 integrin-associated proteins CD9, CD81, and CD98. *Mol. Biol. Cell* **12,** 809–820.
56. Inoue, N., Ikawa, M., Isotani, A., and Okabe, M. (2005) The immunoglobin superfamily protein Izumo is required for sperm to fuse with eggs. *Nature* **434,** 234–238.
57. Ellerman, D. A., Da Ros, V. G., Cohen, D. J., Busso, D., Morgenfeld, M. M., and Cuasnicu, P. S. (2002) Expression and structure–function analysis of de, a sperm cysteine-rich secretory protein that mediates gamete fusion. *Biol. Reprod.* **67,** 1225–1231.
58. Alfieri, J. A., Martin, A. D., Takeda, J., Kondoh, G., Myles, D. G., and Primakoff, P. (2003) Infertility in female mice with an oocyte-specific knockout of GPI-anchored proteins. *J. Cell Sci.* **116,** 2149–2155.
59. Coonrod, S. A., Naaby-Hansen, S., Shetty, J., Shibahara, H., Chen, M., White, J. M., and Herr, J. C. (1999) Treatment of mouse oocytes with PI-PLC releases 70-kDa (pI 5) and 35- to 45-kDa (pI 5.5) protein clusters from the egg surface and inhibits sperm–oolemma binding and fusion. *Dev. Biol.* **207,** 334–349.
60. Kondoh, G., Tojo, H., Nakatani, Y., Komazawa, N., Murata, C., Yamagata, K., Maeda, Y., Kinoshita, T., Okabe, M., Taguchi, R., and Takeda, J. (2005) Angiotensin-converting enzyme is a GPI-anchored protein releasing factor crucial for fertilization. *Nat. Med.* **11,** 160–166.
61. Fuchs, S., Frenzel, K., Hubert, C., Lyng, R., Muller, L., Michaud, A., Xiao, H. D., Adams, J. W., Capecchi, M. R., Corvol, P., Shur, B. D., and Bernstein, K. E. (2005) Male fertility is dependent on dipeptidase activity of testis ACE. *Nat. Med.* **11,** 1139–1140.
62. Leisle, L., Parkin, E. T., Turner, A. J., Hooper, N. M. (2005) Angiotensin-converting enzyme as a GPIase: a critical reevaluation. *Nat. Med.* **11,** 1140–1142.

63. Runft, L. L., Jaffe, L. A., and Mehlmann, L. M. (2002) Egg activation at fertilization: where it all begins. *Dev. Biol.* **245,** 237–254.
64. Swann, K. and Jones, K. T. (2002) Membrane events of egg activation, in *Fertilization* (Hardy, D. M., ed.). Academic Press, San Diego, pp. 319–346.
65. Ducibella, T. (1996) The cortical reaction and development of activation competence in mammalian oocytes. *Hum. Reprod. Update* **2,** 29–42.
66. Abbott, A. L. and Ducibella, T. (2001) Calcium and the control of mammalian cortical granule exocytosis. *Front. Biosci.* **6,** d792–806.
67. Tsaadon, A., Eliyahu, E., Shtraizent, N., and Shalgi, R. (2006) When a sperm meets an egg: block to polyspermy. *Mol. Cell. Endocrinol.* **252,** 107–114.
68. Eliyahu, E., Tsaadon, A., Shtraizent, N., and Shalgi, R. (2005) The involvement of protein kinase C and actin filaments in cortical granule exocytosis in the rat. *Reproduction* **129,** 161–170.
69. Rossi, E. A., Li, Z., Feng, H., and Rubin, C. S. (1999) Characterization of the targeting, binding, and phosphorylation site domains of an A kinase anchor protein and a myristoylated alanine-rich C kinase substrate-like analog that are encoded by a single gene. *J. Biol. Chem.* **274,** 27201–27210.
70. Evans, J. P. (1999) Sperm disintegrins, egg integrins, and other cell adhesion molecules of mammalian gamete plasma membrane interactions. *Front. Biosci.* **4,** D114–D131.
71. Xu, Z., Abbott, A., Kopf, G. S., Schultz, R. M., and Ducibella, T. (1997) Spontaneous activation of ovulated mouse eggs: time-dependent effects on M-phase exit, cortical granule exocytosis, maternal messenger ribonucleic acid recruitment, and inositol 1,4,5-triphosphate sensitivity. *Biol. Reprod.* **57,** 743–750.

Additional References

Austin C. R. and Short R. V. (eds.) (1983) *Reproduction in Mammals,* Book 1, *Germ Cells and Fertilization*. Cambridge University Press, Cambridge, England.

Dunbar B. S. and O'Rand M.G. (eds.) (1991) *A Comparative Overview of Mammalian Fertilization*. Plenum Press, New York.

Gwatkin R. B. L. (1977) *Fertilization Mechanisms in Man and Mammals*. Plenum Press, New York.

Wassarman P. M. (ed.) (1991) *Elements of Mammalian Fertilization,* vols. 1 and 2. CRC Press, Boca Raton, FL.

7

Molecular Control of Mammalian Myoblast Fusion

Katie M. Jansen and Grace K. Pavlath

Summary

The fusion of postmitotic mononucleated myoblasts to form syncytial myofibers is a critical step in the formation of skeletal muscle. Myoblast fusion occurs both during development and throughout adulthood, as skeletal muscle growth and regeneration require the accumulation of additional nuclei within myofibers. Myoblasts must undergo a complex series of molecular and morphological changes prior to fusing with one another. Although many molecules regulating myoblast fusion have been identified, the precise mechanism by which these molecules act in concert to control fusion remains to be elucidated. A comprehensive understanding of how myoblast fusion is controlled may contribute to the treatment of various disorders associated with loss of muscle mass. In this chapter, we examine progress made toward elucidating the cellular and molecular pathways involved in mammalian myoblast fusion. Special emphasis is placed on the molecules that regulate myofiber formation without discernibly affecting biochemical differentiation.

Key Words: Myoblast fusion; myogenesis; myotube; skeletal muscle; myofiber; muscle growth; muscle regeneration.

1. Introduction

Skeletal muscle is composed of multinucleated myofibers that are formed via the fusion of mononucleated myoblasts during development. In the developing embryo, myoblasts derived from the somites must undergo a complex process of proliferation, myogenic differentiation, and migration. In adult muscle, myoblasts are derived from resident quiescent muscle precursor cells, called *satellite cells*. These myoblasts undergo fusion with one another and with existing myofibers in response to a growth stimulus or during regeneration from injury. Despite a long-held interest in understanding the process of skeletal muscle formation, regeneration, and growth, many aspects of mammalian myoblast fusion remain a mystery.

From: *Methods in Molecular Biology, vol. 475: Cell Fusion: Overviews and Methods*
Edited by: E. H. Chen © Humana Press, Totowa, NJ

The study of myogenesis has been greatly aided by the ability to isolate and grow myoblasts in vitro. Many of the early studies on myoblast fusion were performed using myoblasts isolated from chick and quail, but these findings have since been extrapolated to mammalian systems. Myoblasts remain in a proliferative state when cultured in high serum medium containing growth factors. Upon serum and mitogen withdrawal, myoblasts undergo an ordered process of myogenic commitment, cell cycle arrest, contractile protein expression, and finally cell fusion *(1)*. The confirmation that myofibers, referred to as *myotubes* in vitro, are formed through myoblast fusion occurred nearly 50 years ago. Prior to that time, alternative models of multinucleation of myofibers included amitosis or mitosis without cytokinesis. However, time lapse microscopic analysis of embryonic chick muscle explants demonstrated that new myofibers are formed through myoblast fusion *(2)*. Further evidence came from experiments showing that DNA synthesis is not required for myotube formation and that myotubes, unlike myoblasts, do not incorporate ^{3}H- thymidine in vitro *(3–5)*. Since that time, a tremendous amount of research has focused on uncovering the cellular and molecular processes involved in myoblast fusion.

Two molecularly distinct phases of cell fusion occur in mammalian muscle cells, a phenomenon that has also been observed in *Drosophila* myogenesis *(6,7)*. During the first phase of fusion, myoblast–myoblast fusion occurs to generate the initial multinucleated cell. This type of fusion occurs during embryonic development, regeneration, and hyperplasia. During the second stage of fusion, additional myoblasts fuse with the nascent myotube, leading to increased myonuclear number and cell size. By far this second phase of fusion is most prevalent, as it occurs not only in embryonic development, regeneration, and hyperplasia, but also in hypertrophy, growth after atrophy and maintenance of myofibers throughout the life of an individual.

Cell fusion occurs between muscle cells that have undergone early differentiation events such as expression of myogenin, cell cycle withdrawal, and expression of contractile proteins. Impairment or enhancement of these early differentiation events will affect myotube formation and growth. Therefore, altered myotube formation or growth may arise from changes in either early differentiation or cell fusion. For the purposes of this review and in the interest of space, we have limited our discussion to molecules that impact myotube formation or growth without detectably altering myogenin expression, cell cycle exit, or expression of creatine kinase or contractile proteins (**Table 1**). Some of the molecules that are commonly associated with myoblast fusion in the literature have either not been analyzed for, or are associated with, decreases in early differentiation and are not discussed here.

Table 1
Molecules That Regulate Myoblast Fusion in Mammals

Membrane-associated proteins	Intracellular molecules	Extracellular/secreted molecules
Acetylcholine receptor ***(43)***	Arf6 ***(101)***	Calcium ***(67,69)***
Integrin-α4 ***(28,29)***	cGMP ***(82,83)***	Follistatin ***(82,86)***
Integrin-β1 ***(26,27)***	c-src ***(97)***	Growth hormone ***(96)***
Caveolin-3 ***(41)***	Calpain 3 ***(64,65)***	Interleukin-4 ***(91,92)***
CD9 ***(30)***	FOXO1a ***(45,47,49)***	MT1-MMP ***(98)***
GRP94 ***(40)***	mTOR ***(51)***	Nitric oxide ***(82–84)***
Mannose receptor ***(94)***	NFATc2 ***(90)***	Prostaglandin E1 ***(99)***
M-cadherin ***(18–22)***	N-Wasp/WIP ***(102)***	Prostaglandin F2α ***(95)***
Myoferlin ***(42)***	Rho/ROCK ***(46)***	uPA ***(38,39)***
NCAM ***(23–25)***	SHP-2 ***(97)***	
uPAR ***(38,39)***	Smn ***(66)***	
	Trio ***(21)***	

Arf6, ADP-ribosylation factor 6; FOXO1a, forkhead box gene, group O 1a; GRP94, glucose regulated protein 94; MT1-MMP, membrane type 1-matrix metalloprotease; mTOR, mammalian target of rapamycin; NCAM, neural cell adhesion molecule; NFATc2, nuclear factor of activated T cells c2; SHP-2, SH2 domain-containing tyrosine phosphatase; Smn, survival motor neuron protein; uPA, urokinase plasminogen activator; uPAR, urokinase plasminogen activator receptor; Wasp, Wiskott-Aldrich syndrome protein; WIP, Wasp-interacting protein.

2. Morphological Studies

Myoblast fusion is the culmination of an ordered set of specific cellular events: recognition, adhesion, alignment, and membrane union. Time-lapse photography and electron microscopy of myoblast fusion in vitro have provided important clues about these events. These studies revealed that myoblasts were initially very motile, but cell locomotion slowed as differentiation proceeded ***(8)***. Interestingly, myoblasts preferentially moved into some fields and out of others as myotubes began to form ***(9)***, suggesting directed migration in response to chemotactic factors. Through a sequence of cellular interactions ***(10)***, myoblasts adhered to one another during myotube formation. End-to-end alignment of cells was commonly observed in time-lapse studies along with lateral to lateral alignment, but end-to-lateral alignment was the most infrequent ***(8)***. Following adhesion, electron microscopy demonstrated that alignment occurred through the parallel apposition of the membranes of elongated myoblasts with myotubes or other myoblasts ***(11)***. Cytoplasmic vesicles were observed in close proximity to the plasma membranes where membrane union occurred in small regions between the aligned plasma membranes ***(11)***. Membrane union ultimately resulted in the formation of a single multinucleated cell.

Electron microscopy studies in vivo have shed further light on the events occurring during fusion of myogenic cells. Unilamellar vesicles were also observed in close apposition to the fusing membranes of muscle cells during development *(12)* or muscle regeneration *(13)*. Furthermore, extensive cytoskeletal reorganization occurred before and after fusion *(14)*. In regenerating mouse muscle, cell fusion was most commonly observed between myogenic cells that lacked external lamina *(14)*. Presumptive fusion events in regeneration were observed between various types of myogenic cells: myoblast–myoblast, myoblast–myotube, myotube–myotube, and myotube–myofiber.

A question that arises when considering myoblast fusion with myotubes is whether fusion is directed to specific sites along the length of a myotube. Studies during rodent development have shed some light on this topic. Pulsed bromodeoxyuridine labeling was performed at specific times during myofiber formation in rats, and the position of labeled myonuclei were subsequently examined in myofibers right before birth using immunocytochemistry *(15)*. These results demonstrated that new nuclei tend to be added to the ends of growing myofibers. Similar results were obtained in perinatal muscle growth of mice, a period when myofibers are also rapidly elongating *(16,17)*. How such specificity in cell guidance and fusion occurs has not been studied. In adult muscles, growth also occurs by increasing in cross-sectional area. Whether cell fusion with myofibers occurs all along their length in this situation is unknown.

3. Membrane-Associated Proteins

Molecules on the surface of differentiating myoblasts have long been hypothesized to play an important role in myoblast fusion. Several classes of receptors have proposed functions in promoting myoblast fusion, including integrins, tetraspanins, and immunoglobulin super family cell adhesion molecules. Cell surface receptors could theoretically regulate myoblast migration, recognition, adhesion, or membrane breakdown/fusion, but the mechanism by which most known membrane-associated proteins regulate fusion is unclear.

3.1. Cadherins and Cell Adhesion Molecules

Prior to undergoing fusion, differentiating myoblasts recognize and adhere to one another and to nascent myotubes. Although several adhesion molecules have been suggested to function in this process, the precise relationship between these molecules remains elusive. In addition, results from in vitro and in vivo experiments examining the role of these molecules during myoblast fusion have been difficult to reconcile. For example, the calcium-dependent adhesion molecule M-cadherin has been suggested to function in myoblast fusion. Several studies in which the function of M-cadherin was blocked by peptides, antibodies, or mRNA expression knock-down support this hypothesis

(18–21). M-cadherin-mediated adhesion activated Rac1 and the Rho-GEF trio, an event that was a prerequisite for fusion of C2C12 myoblasts ***(21)***. However, M-cadherin null mice did not have defects in skeletal muscle formation or regeneration, showing that this adhesion molecule is not essential for myoblast fusion in vivo ***(22)***.

Other cell adhesion molecules may compensate for the lack of M-cadherin in vivo. For example, inhibition of neural cell adhesion molecule (NCAM), a member of the immunoglobulin super family of cell adhesion molecules, by antibodies or NCAM peptides blocked embryonic chick myoblast fusion in vitro ***(23)***. Conversely, ectopic expression of the differentiation-specific isoform of human NCAM in C2C12 myoblasts promoted fusion ***(24)***. However, myoblasts isolated from NCAM null mice fused to form myotubes in vitro in a manner indistinguishable from wild-type myoblasts ***(25)***. The recognition and adhesion of myoblasts prior to fusion likely involves several adhesion molecules, and the interplay between such molecules should be further investigated in the future.

3.2. Integrins

Integrins are a class of extracellular matrix receptors that consist of an α- and a β-subunit. A role for integrins in modulating myoblast fusion was first suggested following the observation that a monoclonal antibody, later demonstrated to recognize integrin-β1, could inhibit the differentiation and fusion of embryonic chick myoblasts in vitro ***(26)***. Subsequently, the essential role of integrin-β1 during myoblast fusion was confirmed in vivo ***(27)***. Mice lacking the integrin-β1 gene in skeletal muscle were generated using the CRE-Lox system. Although a few small myofibers were observed in these mice, β1 null myoblasts derived from these mice did not fuse in vitro despite being capable of undergoing biochemical differentiation. Observation of β1 null myoblasts by electron microscopy revealed that the inability of β1 null myoblasts to undergo fusion did not result from impaired cell adhesion but likely arose during membrane union ***(27)***. The identity of the α-subunit involved in integrin-dependent myoblast fusion remains elusive. Although antibodies recognizing integrin-α4 have been reported to block myoblast fusion, mouse muscle cells null for integrin-α4 readily contribute to skeletal muscle formation in vivo ***(28,29)***.

Cellular distribution of the tetraspanin CD9, a receptor known to interact with integrin-β1 in muscle cells ***(30)***, was disrupted in the absence of integrin-β1. This observation, coupled with a previous report that CD9 antibodies inhibited fusion of C2C12 myoblasts ***(30)***, prompted speculation that integrin-β1 and CD9 may act together to promote myoblast fusion. Tachibana and Helmer ***(30)*** also reported that the antibodies recognizing CD9 and CD81 (a closely related tetraspanin) inhibited myoblast fusion in an additive manner without disrupting differentiation. However, an essential role for CD9 or CD81 during myoblast fusion in vivo has not been reported, despite the generation of single and double CD9/CD81 null mice ***(31–33)***.

3.3. Urokinase Plasminogen Activator Receptor

The urokinase system is composed of the cell surface receptor urokinase plasminogen activator receptor (uPAR), the serine proteinase urokinase plasminogen activator (uPA), and the inhibitory molecules PAI-1 and PAI-2. Cell surface bound uPA can cleave plasminogen to plasmin, which degrades components of the extracellular membrane surrounding the cell, thereby aiding in cell migration. In addition, the uPA/uPAR system can act independently of plasmin to modulate cellular processes such as migration and adhesion ***(34,35)***. This later function appears to involve a tripartite complex composed of uPA, uPAR, and PAI-1. The uPA/uPAR is localized to the leading edge of migrating cells where it has been postulated to play a role in the cytoskeletal reorganizations necessary for migration by mediating mechanical force transfer across the plasma membrane ***(36)***. Components of the urokinase system are expressed by human muscle cells ***(37,38)***. Inhibition of any member of the tripartite complex, for example, by preventing uPA binding to uPAR or inhibiting uPA or PAI-1 activity by antibodies, leads to decreased fusion in human myoblasts in vitro ***(38)***. The decrease in myoblast fusion did not appear to be dependent on the proteolytic activity of plasmin. Similarly, inhibition of the tripartite complex did not affect differentiation. Subsequent work demonstrated that the uPA/uPAR/PAI-1 complex regulates migration of myoblasts prior to fusion ***(39)***. These experiments demonstrate that cell migration is critical for the development of cell–cell contacts necessary for myoblast fusion, a theme that is repeated with several other molecules to be discussed later.

3.4. Miscellaneous Membrane-Associated Molecules

Additional membrane-associated proteins are thought to play a role in regulating myoblast fusion, although the mechanism by which these proteins regulate fusion is still unclear. For example, glucose-regulated protein 94 (GRP94) is required for myoblast fusion. GRP94 is a muscle-specific protein and member of a class of resident endoplasmic reticulum proteins thought to function as molecular chaperones. Antisense knock-down of GRP94 in C2C12 myoblasts inhibited fusion, whereas GRP94 overexpression accelerated fusion ***(40)***. GRP94 is expressed on the surface of differentiating myoblasts, although the mechanism by which GRP94 regulates fusion is unknown.

Another membrane protein involved in myoblast fusion is the muscle-specific caveolin-3, as antisense knock-down of this transcript in C2C12 myoblasts inhibited fusion without affecting expression of markers of muscle differentiation ***(41)***. Recently, myoferlin, a Ca^{2+}-dependent phospholipid binding protein, was shown to play an important role in the second stage of myoblast fusion *(42)*. Primary myoblasts isolated from myoferlin null mice formed smaller myotubes in vitro and smaller myofibers during regeneration in vivo. Myoferlin is concentrated at sites of cell–cell contact during fusion and could hypothetically help regulate membrane fusion or breakdown at these sites, although the mechanism of myoferlin-dependent fusion has not been reported.

Nicotinic acetylcholine receptors (nAChRs) are a family of ligand-gated pentameric ion channels required for neurotransmission by acetylcholine at the neuromuscular junction. In addition, several studies have suggested a role for nAChR in myoblast fusion. Fusion, but not biochemical differentiation, of chick myoblasts was inhibited by α-bungarotoxin *(43)*. The role of nAChR in fusion seems to be conserved in other species, as activation or inhibition of nAChR also altered fusion of human myoblasts *(44)*. The mechanism of nAChR-mediated myoblast fusion is unknown but may be related to the influx of calcium triggered by this receptor.

4. Intracellular Pathways

4.1. FOXO1a and Rho/ROCK

The transcription factor FOXO1a (also called FKHR) is expressed in skeletal muscle and translocates to the nucleus at the onset of differentiation. Ectopic expression of a mutant form of FOXO1a lacking the transactivation domain severely inhibited fusion of primary mouse muscle cells without altering expression of early and late myogenic markers, whereas expression of a non-phosphorylatable FOXO1a enhanced the rate and extent of fusion *(45)*. The Rho/ROCK pathway likely acts upstream of FOXO1a, as ROCK can phosphorylate FOXO1a in vitro, and the addition of a ROCK inhibitor to differentiating C2C12 myoblasts led to nuclear accumulation of FOXO1a and accelerated myoblast fusion *(46)*. Interestingly, FOXO1a regulates transcription of cyclic GMP-dependent kinase 1 (cGK1), which in turn phosphorylates FOXO1a, abolishing its DNA binding activity *(47)*. This negative feedback loop may help control the rate of myoblast fusion. FOXO1a null mice are embryonic lethal; therefore, the requirement for this transcription factor in vivo has not been assessed *(48)*. However, transgenic mice overexpressing FOXO1a in skeletal muscle exhibited a marked decrease in muscle mass *(49)*. A more thorough investigation of the role FOXO1a plays in regulating the rate of myoblast fusion in vivo is needed in the future.

4.2. Mammalian Target of Rapamycin

The mammalian target of rapamycin (mTOR) is a serine/threonine protein kinase that is required for transmission of promyogenic signals during muscle differentiation *(50)*. In addition to its role in promoting expression of genes involved in myogenic differentiation, mTOR is required for myotube growth during the second stage of myoblast fusion *(51)*. While rapamycin inhibited differentiation of C2C12 myoblasts, ectopic expression of a rapamycin-resistant (RR) mTOR mutant allowed for differentiation in the presence of rapamycin. However, expression of a kinase-inactive (KI) RR mTOR in differentiating C2C12 myoblasts treated with rapamycin inhibited the second stage of myoblast fusion while allowing for myogenic differentiation. Interestingly, conditioned media from RR mTOR cells rescued the fusion defect of KI/RR mTOR-expressing cells, suggesting that the kinase activity of mTOR regulates a secreted factor important for fusion of myoblasts with growing myotubes. Identifying this fusion-promoting factor and understanding the mechanism by which this factor regulates fusion will provide valuable information regarding myotube growth.

4.3. Calpain

Calpains are calcium-activated intracellular cysteine proteases. Calpain activity is activated by a variety of factors, including calcium, phospholipids, and phosphorylation by the mitogen-activated protein kinase extracellular signal regulated kinase *(52)*. Conversely, calpastatin is a specific intracellular inhibitor of calpain. The calpain family is composed of two ubiquitously expressed members (μ-calpain and m-calpain). Skeletal muscle also expresses a muscle-specific calpain (calpain 3), which is the major isoform expressed in adult skeletal muscles. Alterations in calpain 3 cause limb-girdle muscular dystrophy type 2A *(53)*.

During in vitro muscle differentiation, the activity of calpains is increased as myotubes are formed *(54–56)*. A large body of literature implicates μ-calpain and m-calpain in myoblast fusion, but, in general, crucial tests distinguishing fusion effects from effects on biochemical differentiation were not performed. Several early papers demonstrated decreased creatine kinase activity, a marker of biochemical differentiation, with treatment of pharmacologic calpain inhibitors *(57,58)*. Experiments overexpressing the endogenous calpain inhibitor calpastatin differ in their effects on biochemical differentiation. Calpastatin overexpression in rat L8 myoblasts using an expression plasmid led to decreased expression of myogenin, a critical myogenic transcription factor for the initiation of differentiation *(59)*. In contrast, microinjection of purified calpastatin into primary adult rat myoblasts or C2C12 myoblasts decreased fusion without a change in the expression of myosin heavy chain, a differentiation-specific marker, although these later data were not quantified or shown *(60)*. Differences in the timing of calpastatin overexpression during myogenesis could contribute to these conflicting results. Future experiments should be directed toward addressing a role for the ubiquitous calpains in fusion independent of differentiation effects, as calpain activity is required for alteration of matrix proteins ***(60,61)***, reorganization of the cytoskeleton ***(54,62)***, and cell migration ***(63)***, any of which could contribute to cell fusion downstream of biochemical differentiation.

In contrast, a role for the muscle-specific calpain 3 in myoblast fusion is better established. Primary myoblasts from calpain 3 null mice gave rise to myotubes containing an increased number of myonuclei in vitro ***(64)***. Subsequent studies demonstrated that calpain 3 acts to control the levels of membrane-associated β-catenin and M-cadherin during myogenesis ***(65)***. Given the association of M-cadherin with fusion, increased levels of M-cadherin in calpain 3 null myoblasts may contribute to the enhanced fusion events.

4.4. Survival Motor Neuron (Smn) Protein

Mutations in the survival motor neuron (Smn) protein occur in patients suffering from spinal muscular atrophy (SMA). Although the precise function of Smn is unclear, it is a component of the spliceosome and may help regulate RNA splicing. A role for Smn in myoblast fusion was uncovered by knockdown of the Smn transcript in C2C12 myoblasts ***(66)***. While it is interesting to speculate that Smn may regulate splicing of RNA transcripts that promote myoblast fusion, this hypothesis has not been directly tested.

5. Extracellular and Secreted Factors

5.1. Calcium

Calcium was one of the earliest molecules identified to function in myogenesis and appears to play a role at multiple steps. Myogenesis is dependent on both extracellular ***(67–69)*** and intracellular ***(70–72)*** calcium. The requirement for extracellular calcium may be related to extracellular proteins that require calcium for normal activity during either myoblast adhesion or myoblast fusion ***(73,74)***. In addition, extracellular calcium may be important in the fusion of lipid bilayers ***(75)*** that occurs during myogenic cell fusion. Potentially many of the fusion effects ascribed to calcium may be related to recently demonstrated effects of calcium on early differentiation. Increases in intracellular calcium activate the phosphatase calcineurin, leading to downstream activation of myogenic regulatory factors such as myogenin and myocyte enhancer factor-2 ***(76–80)***. Thus, the precise role calcium plays in regulating cell fusion has been difficult to characterize.

5.2. Nitric Oxide and Follistatin

Transient increases in cyclic guanosine monophosphate (cGMP) at the onset of myoblast fusion occur in both primary chick myoblasts ***(81)*** and mouse myoblasts ***(82)***. This increase in cGMP is due to the production of nitric oxide (NO) by NO synthase ***(82,83)***. Pharmacological enhancement of NO or cGMP levels

(82–84) enhanced cell fusion, whereas decreased levels of these two molecules diminished cell fusion ***(81–83)***. Pisconti and coworkers demonstrated that pharmacological manipulation of NO levels was maximally effective only if performed at the beginning of the differentiation process ***(82)***. These authors performed careful analyses of cell differentiation and concluded that NO influenced myoblast fusion without influencing upstream differentiation events.

The effects of NO/cGMP on myoblast fusion were mostly mediated by follistatin, as the fusion index was decreased by antibodies against follistatin. Follistatin is a secreted protein that interacts with several transforming growth factor-β family members and regulates their biological activity. Importantly, follistatin inhibits the activity of myostatin, a negative regulator of skeletal muscle hypertrophy ***(85)***. Increases in cGMP were responsible for activating transcription of follistatin through the transcription factors myoD, cyclic adenosine monophosphate response element binding protein, and nuclear factor of activated T cells, known mediators of myogenesis. Follistatin has also been shown to regulate increases in myoblast fusion with myotubes in response to deacetylase inhibitors ***(86)***. Whether follistatin produced in response to NO/cGMP similarly acts only on the second phase of fusion is unknown.

5.3. Upstream Modulators and Downstream Targets of Nuclear Factor of Activated T Cells

The nuclear factor of activated T cells (NFAT) family of transcription factors consists of four calcium-activated members, NFATc1–c4, all of which are expressed in skeletal muscle ***(87,88)***. A central role for NFATc2 in orchestrating the second phase of myoblast fusion has emerged ***(89)***. Muscle regeneration in NFATc2$^{-/-}$ mice was characterized by normal formation of regenerating myofibers, but these myofibers were unable to grow at the same rate as wild type mice. NFATc2$^{-/-}$ myoblasts formed small multinucleated muscle cells in vitro because of a defect in the recruitment and/or fusion of myogenic cells with nascent myotubes ***(90)***. Subsequent experiments demonstrated that NFATc2 was required for the production of interleukin-4 (IL-4) during the second stage of myoblast fusion. Interleukin-4$^{-/-}$ myoblasts were also defective in the recruitment of myogenic cells with nascent myotubes ***(91)***. Furthermore, IL-4 was required for the fusion of mesenchymal stem cells with muscle ***(92)***. Interleukin-4 may promote cell fusion by enhancing myoblast migration ***(93)***.

Interleukin-4 likely promotes fusion in part by regulating expression of the mannose receptor, a cell surface endocytic C-type lectin that binds to sulfated glycoproteins or terminal mannose, fucose, or N-acetylglucosamine residues. Mannose receptor null myoblasts formed smaller myotubes in vitro and myofibers in vivo ***(92)***. Mannose receptor null myoblasts displayed a reduction in general motility as well as an impairment of directed migration to factors released by fusing muscle cells. The identity of these secreted factors remains to be determined.

Upstream activators of the NFATc2 pathway are also key players in regulating myonuclear addition to nascent myotubes. NFATc2 was required for the increase in myonuclear number due to prostaglandin $F_{2\alpha}$ ***(95)*** as well as growth hormone ***(96)***. Recent evidence suggests NFATc2 may be downstream of a c-src pathway in muscle ***(97)***. Fornaro et al. ***(97)*** demonstrated that the phosphatase SHP-2 stimulates c-src, leading to activation of NFAT and subsequent fusion of myoblasts with myotubes. Interleukin-4 was decreased in SHP-2 null muscle cells. These results support the idea that NFATc2 is a target for positive regulation by SHP-2 in skeletal muscle. However, whether the phenotype of SHP-2 null mice is a consequence of disrupting NFATc2 solely as opposed to other NFAT family members needs to be formally determined.

5.4. Miscellaneous Secreted Molecules

Rearrangement of the extracellular membrane by matrix metalloproteases (MMPs) is a key step during the formation of new myofibers ***(98)***. General inhibitors of MMPs have been shown to inhibit fusion of C2C12 myoblasts without affecting biochemical differentiation. The activity of membrane type 1-MMP (MT1-MMP) is important for fusion, as short hairpin RNA knock-down of MT1-MMP expression inhibits fusion and MT1-MMP null mice undergo impaired myofiber formation in vivo ***(98)***. Although additional metalloproteases have been speculated to regulate myoblast fusion, a more thorough investigation of the role these proteins play in regulating the myogenic differentiation program is needed.

Other prostaglandins in addition to prostaglandin F2α also play a role in myoblast fusion. When prostaglandin synthesis was inhibited either pharmacologically or genetically, cell fusion was inhibited. The inhibition of cell fusion of embryonic chick muscle cells by indomethacin was overcome by the addition of prostaglandin E1 but not E2 ***(99)***. Primary myoblasts from mice genetically deficient in cyclooxygenase 2 (COX2), an enzyme necessary for prostaglandin synthesis, formed nascent myotubes to the same degree as wild-type cells but were deficient in the second stage of myoblast fusion ***(100)***. Effects on biochemical differentiation were not specifically addressed in this study. Both prostaglandins E2 and F2α were able to significantly enhance fusion in the COX2 null myoblasts. The differences in the ability of prostaglandin E2 to rescue fusion in these two studies may be related to differences in the origins of the myoblast populations analyzed or to differential effects of pharmacological or genetic inhibition on prostaglandin synthesis.

6. Conclusion

Since the discovery nearly 50 years ago that multinucleated skeletal muscle cells arise from the fusion of mononucleated myoblasts, much progress has been made toward understanding this critical and complex process. The cellular events involved in myoblast fusion have been well characterized, but a thorough understanding of the molecules involved remains elusive. In recent years, mutational screens in *Drosophila melanogaster* have revealed a number of genes essential for myoblast fusion in fruit flies. A priority should be placed on determining if these genes have conserved roles during myoblast fusion in mammals.

Several proteins involved in cytoskeletal rearrangements such as the small guanosine triphosphatase adenosine diphosphate ribosylation factor 6, Wiskott-Aldrich syndrome protein (Wasp), and Wasp-interacting protein are required for myoblast fusion in both flies and mammalian cells ***(101,102)***. In addition, few studies have investigated the relationships among the molecules with known roles in mammalian myoblast fusion. Gaining a deeper understanding of these relationships will provide insight into the unique biological processes of cell–cell fusion. Finally, unbiased approaches, such as large-scale screens using small molecule libraries, may also identify new molecules with a role in myoblast fusion. Further research into the regulation of myotube formation and growth may provide insight into new or improved strategies for treatment of skeletal muscle diseases.

References

1. Andres, V. and Walsh, K. (1996) Myogenin expression, cell cycle withdrawal, and phenotypic differentiation are temporally separable events that precede cell fusion upon myogenesis. *J. Cell Biol.* **132(4),** 657–566.
2. Capers, C. R. (1960) Multinucleation of skeletal muscle in vitro. *J. Biophys. Biochem. Cytol.* **7,** 559–566.
3. Konigsberg, I. R. (1961) Some aspects of myogenesis in vitro. *Circulation* **24,** 447–457.
4. Konigsberg, I. R., McElvain, N., Tootle, M., and Herrmann, H. (1960) The dissociability of deoxyribonucleic acid synthesis from the development of multinuclearity of muscle cells in culture. *J. Biophys. Biochem. Cytol.* **8,** 333–343.
5. Stockdale, F. E. and Holtzer, H. (1961) DNA synthesis and myogenesis. *Exp. Cell Res.* **24,** 508–520.
6. Menon, S. D. and Chia, W. (2001) *Drosophila* rolling pebbles: a multidomain protein required for myoblast fusion that recruits D-titin in response to the myoblast attractant Dumbfounded. *Dev. Cell* **1(5),** 691–703.
7. Rau, A., Buttgereit, D., Holz, A., et al. (2001) rolling pebbles (rols) is required in Drosophila muscle precursors for recruitment of myoblasts for fusion. *Development* **128(24),** 5061–5073.

8. Powell, J. A. (1973) Development of normal and genetically dystrophic mouse muscle in tissue culture. I. Prefusion and fusion activities of muscle cells: phase contrast and time lapse study. *Exp. Cell Res.* **80(2),** 251–264.
9. Chazaud, B., Christov, C., Gherardi, R. K., and Barlovatz-Meimon, G. (1998) In vitro evaluation of human muscle satellite cell migration prior to fusion into myotubes. *J. Muscle Res. Cell Motil.* **19(8),** 931–936.
10. Knudsen, K. A. and Horwitz, A. F. (1977) Tandem events in myoblast fusion. *Dev. Biol.* **58(2),** 328–338.
11. Wakelam, M. J. (1985) The fusion of myoblasts. *Biochem. J.* **228(1),** 1–12.
12. Kalderon, N. and Gilula, N. B. (1979) Membrane events involved in myoblast fusion. *J. Cell Biol.* **81(2),** 411–425.
13. Robertson, T. A., Grounds, M. D., Mitchell, C. A., and Papadimitriou, J. M. (1990) Fusion between myogenic cells in vivo: an ultrastructural study in regenerating murine skeletal muscle. *J. Struct. Biol.* **105(1–3),** 170–182.
14. Fulton, A. B., Prives, J., Farmer, S. R., and Penman, S. (1981) Developmental reorganization of the skeletal framework and its surface lamina in fusing muscle cells. *J. Cell Biol.* **91(1),** 103–112.
15. Zhang, M. and McLennan, I. S. (1995) During secondary myotube formation, primary myotubes preferentially absorb new nuclei at their ends. *Dev. Dyn.* **204(2),** 168–177.
16. Kitiyakara, A. and Angevine, D. M. (1963) Further studies on regeneration and growth in length of striated voluntary muscle with isotopes P32 and thymidine-H3. *Nippon Byori Gakkai Kaishi* **52,** 180–183.
17. Aziz, U. and Goldspink, G. (1974) Distribution of mitotic nuclei in the biceps brachii of the mouse during post-natal growth. *Anat. Rec.* **179(1),** 115–118.
18. Zeschnigk, M., Kozian, D., Kuch, C., Schmoll, M., and Starzinski-Powitz, A. (1995) Involvement of M-cadherin in terminal differentiation of skeletal muscle cells. *J. Cell Sci.* **108(Pt 9),** 2973–2981.
19. Mege, R. M., Goudou, D., Diaz, C., et al. (1992) N-cadherin and N-CAM in myoblast fusion: compared localisation and effect of blockade by peptides and antibodies. *J. Cell Sci.* **103(Pt 4),** 897–906.
20. Charrasse, S., Comunale, F., Grumbach, Y., Poulat, F., Blangy, A., and Gauthier-Rouviere, C. (2006) RhoA GTPase regulates M-cadherin activity and myoblast fusion. *Mol. Biol. Cell* **17(2),** 749–759.
21. Charrasse, S., Comunale, F., Fortier, M., Portales-Casamar, E., Debant, A., and Gauthier-Rouviere. C. (2007) M-cadherin activates Rac1 GTPase through the Rho-GEF trio during myoblast fusion. *Mol. Biol. Cell* **18(5),** 1734–1743.
22. Hollnagel, A., Grund, C., Franke, W. W., and Arnold, H. H. (2002) The cell adhesion molecule M-cadherin is not essential for muscle development and regeneration. *Mol. Cell. Biol.* **22(13),** 4760–4770.
23. Knudsen, K. A., McElwee, S. A., and Myers, L. (1990) A role for the neural cell adhesion molecule, NCAM, in myoblast interaction during myogenesis. *Dev. Biol.* **138(1),** 159–168.
24. Dickson, G., Peck, D., Moore, S. E., Barton, C.H., and Walsh, F. S. (1990) Enhanced myogenesis in NCAM-transfected mouse myoblasts. *Nature* **344(6264),** 348–351.

25. Charlton, C. A., Mohler, W. A., and Blau, H. M. (2000) Neural cell adhesion molecule (NCAM) and myoblast fusion. *Dev. Biol.* **221(1),** 112–119.
26. Menko, A. S. and Boettiger, D. (1987) Occupation of the extracellular matrix receptor, integrin, is a control point for myogenic differentiation. *Cell* **51(1),** 51–57.
27. Schwander, M., Leu, M., and Stumm, M., et al. (2003) Beta1 integrins regulate myoblast fusion and sarcomere assembly. *Dev. Cell.* **4(5),** 673–685.
28. Rosen, G. D., Sanes, J. R., LaChance, R., Cunningham, J. M., Roman, J., and Dean, D. C. (1992) Roles for the integrin VLA-4 and its counter receptor VCAM-1 in myogenesis. *Cell* **69(7),** 1107–1119.
29. Yang, J. T., Rando, T. A., Mohler, W. A., Rayburn, H., Blau, H. M., and Hynes, R. O. (1996) Genetic analysis of alpha 4 integrin functions in the development of mouse skeletal muscle. *J. Cell Biol.* **135(3),** 829–835.
30. Tachibana, I. and Hemler, M. E. (1999) Role of transmembrane 4 superfamily (TM4SF) proteins CD9 and CD81 in muscle cell fusion and myotube maintenance. *J. Cell Biol.* **146(4),** 893–904.
31. Kaji, K., Oda, S., Miyazaki, S., and Kudo, A. (2002) Infertility of CD9-deficient mouse eggs is reversed by mouse CD9, human CD9, or mouse CD81; polyadenylated mRNA injection developed for molecular analysis of sperm–egg fusion. *Dev. Biol.* **247(2),** 327–334.
32. Takeda, Y., Tachibana, I., Miyado, K., et al. (2003) Tetraspanins CD9 and CD81 function to prevent the fusion of mononuclear phagocytes. *J. Cell Biol.* **161(5),** 945–956.
33. Rubinstein, E., Ziyyat, A., Prenant, M., et al. (2006) Reduced fertility of female mice lacking CD81. *Dev. Biol.* **290(2),** 351–358.
34. Seebacher, T., Manske, M., Zoller, J., Crabb, J., and Bade, E. G. (1992) The EGF-inducible protein EIP-1 of migrating normal and malignant rat liver epithelial cells is identical to plasminogen activator inhibitor 1 and is a component of the ECM migration tracks. *Exp. Cell Res.* **203(2),** 504–507.
35. Wei, Y., Waltz, D. A., Rao, N., Drummond, R. J., Rosenberg, S., and Chapman, H. A. (1994) Identification of the urokinase receptor as an adhesion receptor for vitronectin. *J. Biol. Chem.* **269(51),** 32380–32388.
36. Wang, N., Planus, E., Pouchelet, M., Fredberg, J. J., and Barlovatz-Meimon, G. (1995) Urokinase receptor mediates mechanical force transfer across the cell surface. *Am. J. Physiol.* **268(4 Pt 1),** C1062–C1066.
37. Quax, P. H., Frisdal, E., Pedersen, N., et al. (1992) Modulation of activities and RNA level of the components of the plasminogen activation system during fusion of human myogenic satellite cells in vitro. *Dev. Biol.* **151(1),** 166–175.
38. Bonavaud, S., Charriere-Bertrand, C., Rey, C., et al. (1997) Evidence of a non-conventional role for the urokinase tripartite complex (uPAR/uPA/PAI-1) in myogenic cell fusion. *J. Cell Sci.* **110(Pt 9),** 1083–1089.
39. Chazaud, B., Bonavaud, S., Plonquet, A., Pouchelet, M., Gherardi, R. K., and Barlovatz-Meimon, G. (2000) Involvement of the [uPAR:uPA:PAI-1:LRP] complex in human myogenic cell motility. *Exp. Cell Res.* **258(2),** 237–244.

40. Gorza, L. and Vitadello, M. (2000) Reduced amount of the glucose-regulated protein GRP94 in skeletal myoblasts results in loss of fusion competence. *FASEB J.* **14(3),** 461–475.
41. Galbiati, F., Volonte, D., Engelman, J. A., Scherer, P. E., and Lisanti, M. P. (1999) Targeted down-regulation of caveolin-3 is sufficient to inhibit myotube formation in differentiating C2C12 myoblasts. Transient activation of p38 mitogen-activated protein kinase is required for induction of caveolin-3 expression and subsequent myotube formation. *J. Biol. Chem.* **274(42),** 30315–30321.
42. Doherty, K. R., Cave, A., Davis, D. B., et al. (2005) Normal myoblast fusion requires myoferlin. *Development* **132(24),** 5565–5575.
43. Entwistle, A., Zalin, R. J., Warner, A. E., and Bevan, S. (1988) A role for acetylcholine receptors in the fusion of chick myoblasts. *J. Cell Biol.* **106(5),** 1703–1712.
44. Krause, R. M., Hamann, M., Bader, C. R., Liu, J. H., Baroffio, A., and Bernheim, L. (1995) Activation of nicotinic acetylcholine receptors increases the rate of fusion of cultured human myoblasts. *J. Physiol.* **489(Pt 3),** 779–790.
45. Bois, P. R. and Grosveld, G. C. (2003) FKHR (FOXO1a) is required for myotube fusion of primary mouse myoblasts. *EMBO J.* **22(5),** 1147–1157.
46. Nishiyama, T., Kii, I., and Kudo, A. (2004) Inactivation of Rho/ROCK signaling is crucial for the nuclear accumulation of FKHR and myoblast fusion. *J. Biol. Chem.* **279(45),** 47311–47319.
47. Bois, P. R., Brochard, V. F., Salin-Cantegrel, A. V., Cleveland, J. L., and Grosveld, G. C. (2005) FoxO1a-cyclic GMP-dependent kinase I interactions orchestrate myoblast fusion. *Mol. Cell. Biol.* **25(17),** 7645–7656.
48. Furuyama, T., Kitayama, K., Shimoda, Y., et al. (2004) Abnormal angiogenesis in Foxo1 (Fkhr)-deficient mice. *J. Biol. Chem.* **279(33),** 34741–34749.
49. Kamei, Y., Miura, S., Suzuki, M., et al. (2004) Skeletal muscle FOXO1 (FKHR) transgenic mice have less skeletal muscle mass, down-regulated Type I (slow twitch/red muscle) fiber genes, and impaired glycemic control. *J. Biol. Chem.* **279(39),** 41114–41123.
50. Cuenda, A. and Cohen, P. (1999) Stress-activated protein kinase-2/p38 and a rapamycin-sensitive pathway are required for C2C12 myogenesis. *J. Biol. Chem.* **274(7),** 4341–4346.
51. Park, I. H. and Chen, J. (2005) Mammalian target of rapamycin (mTOR) signaling is required for a late-stage fusion process during skeletal myotube maturation. *J. Biol. Chem.* **280(36),** 32009–32017.
52. Glading, A., Bodnar, R. J., Reynolds, I. J., et al. (2004) Epidermal growth factor activates m-calpain (calpain II), at least in part, by extracellular signal-regulated kinase-mediated phosphorylation. *Mol. Cell. Biol.* **24(6),** 2499–2512.
53. Richard, I., Broux, O., Allamand, V., et al. (1995) Mutations in the proteolytic enzyme calpain 3 cause limb-girdle muscular dystrophy type 2A. *Cell* **81(1),** 27–40.
54. Elamrani, N., Brustis, J. J., Dourdin, N., et al. (1995) Desmin degradation and Ca(2+)-dependent proteolysis during myoblast fusion. *Biol. Cell* **85(2–3),** 177–183.

55. Stockholm, D., Barbaud, C., Marchand, S., et al. (1999) Studies on calpain expression during differentiation of rat satellite cells in primary cultures in the presence of heparin or a mimic compound. *Exp. Cell Res.* **252(2),** 392–400.
56. Joffroy, S., Dourdin, N., Delage, J. P., Cottin, P., Koenig, J., and Brustis, J. J. (2000) M-calpain levels increase during fusion of myoblasts in the mutant muscular dysgenesis (mdg) mouse. *Int. J. Dev. Biol.* **44(4),** 421–428.
57. Kwak, K. B., Kambayashi, J., Kang, M. S., Ha, D. B., and Chung, C. H. (1993) Cell-penetrating inhibitors of calpain block both membrane fusion and filamin cleavage in chick embryonic myoblasts. *FEBS Lett.* **323(1–2),** 151–154.
58. Ueda, Y., Wang, M. C., Ou, B. R., et al. (1998) Evidence for the participation of the proteasome and calpain in early phases of muscle cell differentiation. *Int. J. Biochem. Cell Biol.* **30(6),** 679–694.
59. Barnoy, S., Maki, M., and Kosower, N. S. (2005) Overexpression of calpastatin inhibits L8 myoblast fusion. *Biochem. Biophys. Res. Commun.* **332(3),** 697–701.
60. Temm-Grove, C. J., Wert, D., Thompson, V. F., Allen, R. E., and Goll, D. E. (1999) Microinjection of calpastatin inhibits fusion in myoblasts. *Exp. Cell Res.* **247(1),** 293–303.
61. Dourdin, N., Brustis, J. J., Balcerzak, D., et al. (1997) Myoblast fusion requires fibronectin degradation by exteriorized M-calpain. *Exp. Cell Res.* **235(2),** 385–394.
62. Dourdin, N., Balcerzak, D., Brustis, J. J., Poussard, S., Cottin, P., and Ducastaing, A. (1999) Potential m-calpain substrates during myoblast fusion. *Exp. Cell Res.* **246(2),** 433–442.
63. Dedieu, S., Poussard, S., Mazeres, G., et al. (2004) Myoblast migration is regulated by calpain through its involvement in cell attachment and cytoskeletal organization. *Exp. Cell Res.* **292(1),** 187–200.
64. Kramerova, I., Kudryashova, E., Tidball, J. G., and Spencer, M. J. (2004) Null mutation of calpain 3 (p94) in mice causes abnormal sarcomere formation in vivo and in vitro. *Hum. Mol. Genet.* **13(13),** 1373–1388.
65. Kramerova, I., Kudryashova, E., Wu, B., and Spencer, M. J. (2006) Regulation of M-cadherin–{beta}-catenin complex by calpain 3 during terminal stages of myogenic differentiation. *Mol. Cell Biol.* **26(22),** 8437–8447.
66. Shafey, D., Cote, P. D., and Kothary, R. (2005) Hypomorphic Smn knockdown C2C12 myoblasts reveal intrinsic defects in myoblast fusion and myotube morphology. *Exp. Cell Res.* **311(1),** 49–61.
67. Shainberg, A., Yagil, G., and Yaffe, D. (1969) Control of myogenesis in vitro by Ca 2+ concentration in nutritional medium. *Exp. Cell Res.* **58(1),** 163–167.
68. Salzberg, S., Mandelboim, M., Zalcberg, M., Shainberg, A., and Mandelbaum, M. (1995) Interruption of myogenesis by transforming growth factor beta 1 or EGTA inhibits expression and activity of the myogenic-associated (2'-5') oligoadenylate synthetase and PKR. *Exp. Cell Res.* **219(1),** 223–232.
69. Przybylski, R. J., Szigeti, V., Davidheiser, S., and Kirby, A. C. (1994) Calcium regulation of skeletal myogenesis. II. Extracellular and cell surface effects. *Cell Calcium* **15(2),** 132–142.

70. Constantin, B., Cognard, C., and Raymond, G. (1996) Myoblast fusion requires cytosolic calcium elevation but not activation of voltage-dependent calcium channels. *Cell Calcium* **19(5),** 365–374.
71. David, J. D. and Higginbotham, C. A. (1981) Fusion of chick embryo skeletal myoblasts: interactions of prostaglandin E1, adenosine 3 :5 monophosphate, and calcium influx. *Dev. Biol.* **82(2),** 308–316.
72. Przybylski, R. J., MacBride, R. G., and Kirby, A. C. (1989) Calcium regulation of skeletal myogenesis. I. Cell content critical to myotube formation. *In Vitro Cell Dev. Biol.* **25(9),** 830–838.
73. Knudsen, K. A. (1985) The calcium-dependent myoblast adhesion that precedes cell fusion is mediated by glycoproteins. *J. Cell Biol.* **101(3),** 891–897.
74. Knudsen, K. A., Myers, L., and McElwee, S. A. (1990) A role for the Ca2(+) dependent adhesion molecule, N-cadherin, in myoblast interaction during myogenesis. *Exp. Cell Res.* **188(2),** 175–184.
75. Papahadjopoulos, D., Nir, S., and Duzgunes, N. (1990) Molecular mechanisms of calcium-induced membrane fusion. *J. Bioenerg. Biomembr.* **22(2),** 157–179.
76. Konig, S., Beguet, A., Bader, C. R., and Bernheim, L. (2006) The calcineurin pathway links hyperpolarization (Kir2.1)-induced Ca^{2+} signals to human myoblast differentiation and fusion. *Development* **133(16),** 3107–3114.
77. Konig, S., Hinard, V., Arnaudeau, S., et al. (2004) Membrane hyperpolarization triggers myogenin and myocyte enhancer factor-2 expression during human myoblast differentiation. *J. Biol. Chem.* **279(27),** 28187–28196.
78. Friday, B. B., Horsley, V., and Pavlath, G. K. (2000) Calcineurin activity is required for the initiation of skeletal muscle differentiation. *J. Cell Biol.* **149(3),** 657–666.
79. Friday, B. B., Mitchell, P. O., Kegley, K. M., and Pavlath, G. K. (2003) Calcineurin initiates skeletal muscle differentiation by activating MEF2 and MyoD. *Differentiation* **71(3),** 217–227.
80. Xu, Q., Yu, L., Liu, L., et al. (2002) p38 Mitogen-activated protein kinase-, calcium-calmodulin-dependent protein kinase-, and calcineurin-mediated signaling pathways transcriptionally regulate myogenin expression. *Mol. Biol. Cell* **13(6),** 1940–1952.
81. Choi, S. W., Baek, M. Y., and Kang, M. S. (1992) Involvement of cyclic GMP in the fusion of chick embryonic myoblasts in culture. *Exp. Cell Res.* **199(1),** 129–133.
82. Pisconti, A., Brunelli, S., Di Padova, M., et al. (2006) Follistatin induction by nitric oxide through cyclic GMP: a tightly regulated signaling pathway that controls myoblast fusion. *J. Cell Biol.* **172(2),** 233–244.
83. Lee, K. H., Baek, M. Y., Moon, K. Y., et al. (1994) Nitric oxide as a messenger molecule for myoblast fusion. *J. Biol. Chem.* **269(20),** 14371–14374.
84. Long, J. H., Lira, V. A., Soltow, Q. A., Betters, J. L., Sellman, J. E., and Criswell, D. S. (2006) Arginine supplementation induces myoblast fusion via augmentation of nitric oxide production. *J. Muscle Res. Cell Motil.* **27(8),** 577–584.
85. Lee, S. J. and McPherron, A. C. (2001) Regulation of myostatin activity and muscle growth. *Proc. Natl. Acad. Sci. U.S.A.* **98(16),** 9306–9311.

86. Iezzi, S., Di Padova, M., Serra, C., et al. (2004) Deacetylase inhibitors increase muscle cell size by promoting myoblast recruitment and fusion through induction of follistatin. *Dev. Cell* **6(5),** 673–684.
87. Abbott, K. L., Friday, B. B., Thaloor, D., Murphy, T.J., and Pavlath, G. K. (1998) Activation and cellular localization of the cyclosporine A-sensitive transcription factor NF-AT in skeletal muscle cells. *Mol. Biol. Cell* **9(10),** 2905–2916.
88. Jacquemin, V., Butler-Browne, G. S., Furling, D., and Mouly, V. (2007) IL-13 mediates the recruitment of reserve cells for fusion during IGF-1–induced hypertrophy of human myotubes. *J. Cell Sci.* **120(Pt 4),** 670–681.
89. Pavlath, G. K. and Horsley, V. (2003) Cell fusion in skeletal muscle—central role of NFATC2 in regulating muscle cell size. *Cell Cycle* **2(5),** 420–423.
90. Horsley, V., Friday, B. B., Matteson, S., Kegley, K. M., Gephart, J., and Pavlath, G. K. (2001) Regulation of the growth of multinucleated muscle cells by an NFATC2-dependent pathway. *J. Cell Biol.* **153(2),** 329–338.
91. Horsley, V., Jansen, K. M., Mills, S. T., and Pavlath, G. K. (2003) IL-4 acts as a myoblast recruitment factor during mammalian muscle growth. *Cell* **113(4),** 483–494.
92. Schulze, M., Belema-Bedada, F., Technau, A., and Braun, T. (2005) Mesenchymal stem cells are recruited to striated muscle by NFAT/IL-4–mediated cell fusion. *Genes Dev.* **19(15),** 1787-1798.
93. Lafreniere, J. F., Mills, P., Bouchentouf, M., and Tremblay, J. P. (2006) Interleukin-4 improves the migration of human myogenic precursor cells in vitro and in vivo. *Exp. Cell Res.* **312(7),** 1127–1141.
94. Jansen, K. M. and Pavlath, G. K. (2006) Mannose receptor regulates myoblast motility and muscle growth. *J. Cell Biol.* **174(3),** 403–413.
95. Horsley, V. and Pavlath, G. K. (2003) Prostaglandin F2(alpha) stimulates growth of skeletal muscle cells via an NFATC2-dependent pathway. *J. Cell Biol.* **161(1),** 111–118.
96. Sotiropoulos, A., Ohanna, M., Kedzia, C., et al. (2006) Growth hormone promotes skeletal muscle cell fusion independent of insulin-like growth factor 1 up-regulation. *Proc. Natl. Acad. Sci. U.S.A.* **103(19),** 7315–7320.
97. Fornaro, M., Burch, P. M., Yang, W., et al. (2006) SHP-2 activates signaling of the nuclear factor of activated T cells to promote skeletal muscle growth. *J. Cell Biol.* **175(1),** 87–97.
98. Ohtake, Y., Tojo, H., and Seiki, M. (2006) Multifunctional roles of MT1-MMP in myofiber formation and morphostatic maintenance of skeletal muscle. *J. Cell Sci.* **119(Pt 18),** 3822–3832.
99. Entwistle, A., Curtis, D. H., and Zalin, R. J. (1986) Myoblast fusion is regulated by a prostanoid of the one series independently of a rise in cyclic AMP. *J. Cell Biol.* **103(3),** 857–866.
100. Shen, W., Prisk, V., Li, Y., Foster, W., and Huard, J. (2006) Inhibited skeletal muscle healing in cyclooxygenase-2 gene-deficient mice: the role of PGE2 and PGF2alpha. *J. Appl. Physiol.* **101(4),** 1215–1221.

101. Chen, E. H., Pryce, B. A., Tzeng, J. A., Gonzalez, G. A., and Olson, E. N. (2003) Control of myoblast fusion by a guanine nucleotide exchange factor, loner, and its effector ARF6. *Cell* **114(6),** 751–762.
102. Kim, S., Shilagardi, K., Zhang, S., et al. (2007) A critical function for the actin cytoskeleton in targeted exocytosis of prefusion vesicles during myoblast fusion. *Dev. Cell* **12(4),** 571–586.

8

Placenta Trophoblast Fusion

Berthold Huppertz and Marcus Borges

Summary

It has been known for more than 150 years that syncytial fusion is a normal feature in biological systems. In humans there are two larger syncytial tissues: skeletal muscles fibers and placental syncytiotrophoblast. Other fusion events take place as well from fertilization of the oocyte to infection of human cells by enveloped viruses (however, the latter does not necessarily lead to syncytium formation).

Although knowledge of the fusion process is incomplete, it is clear that membranes do not fuse easily; specific proteins and other factors are required and are selectively activated. In this chapter, we describe the classic proteins, such as the syncytins, assumed to be involved in the fusion process. We also describe other factors that may play roles in the fusion process or in the preparation of the cells to fuse, such as charged phospholipids, divalent cations, and intracellular proteases. Finally, we speculate on why trophoblast cells fuse in vitro and deal with in vitro models of trophoblast fusion and how their fusion rates can be quantified.

Key Words: Trophoblast; fusion; phospholipid; cation; caspase; calpain; BeWo choriocarcinoma cells; forskolin.

1. Introduction

In certain tissues syncytial fusion of cells leads to the generation of syncytia such as in skeletal muscle fibers *(1)*. These structures are by definition no longer cells and may reach a size of several square meters. The decisive step in syncytialization is fusion of the two neighboring membranes, which takes place during many cellular processes, including membrane traffic, fertilization, and infection by enveloped viruses *(2,3)*. Fusion allows the exchange of contents between different membrane-encircled compartments. Membranes do not fuse easily under normal circumstances; this helps to maintain the individuality of the intracellular compartments and of the cell itself. Therefore, the fusion process requires specific proteins and is subject to selective control.

From: *Methods in Molecular Biology, vol. 475: Cell Fusion: Overviews and Methods*
Edited by: E. H. Chen © Humana Press, Totowa, NJ

The placental syncytiotrophoblast arises from and is maintained throughout pregnancy by fusion of villous cytotrophoblasts. In this chapter, we concentrate on fusion requirements that are known to play a role or that are speculated to be important for trophoblast fusion. We do not spend as much time on fusogenic proteins such as ADAM-12 (a disintegrin and metalloprotease) and syncytins, which have been addressed extensively elsewhere ***(4,5)***, but will extend our review to introduce other putative players that are involved in the preparation of fusion or the fusion process directly.

2. Human Placenta and Villous Trophoblast

During human development, fusion processes are fundamental for the beginning and maintenance of pregnancy. The major constituent of the human placenta is the trophoblast, the first cell lineage that develops before any embryonic tissue arises. At the stage between morula and blastocyst the trophoblast lineage differentiates from the embryoblast and develops a cover around the early embryo ***(6)***.

At the time of adhesion of the blastocyst, the first step of intercellular fusion of the neighboring trophoblast cells occurs at the embryonic pole of the blastocyst. They generate the early oligonucleated syncytiotrophoblast, which penetrates the uterine epithelium. Only with this first step of fusion can implantation of the blastocyst into uterine tissues be achieved. Soon after the blastocyst has implanted completely within the uterine tissues and has been covered by the uterine epithelium again, the syncytiotrophoblast changes its function and stops invasion. It now develops into the outermost layer of the placental villi and thus comes into direct contact with maternal blood later in pregnancy (**Fig. 1**).

In general, the human trophoblast differentiates into two major subtypes, both of which are critical for normal placental function ***(7)***. Those trophoblast cells that leave the placenta proper are termed *extravillous trophoblast* and invade maternal uterine tissues to adapt maternal blood flow toward the placenta for the maintenance of the fetus. The other subtype of trophoblast remains within the placenta proper and becomes the epithelial cover of the placental villi; it is thus termed *villous trophoblast*.

Villous trophoblast is composed of two layers: the syncytiotrophoblast and the cytotrophoblast. The cytotrophoblast displays high proliferative properties, whereas the differentiated syncytiotrophoblast has lost its generative capacity and is no longer able to proliferate. The syncytiotrophoblast is a multinucleated layer that forms the outer surface of placental villi and comes into direct contact with maternal blood ***(6,8,9)***. It is a single layer that is continuous and normally uninterrupted and that extends over the surface of all villous trees of a placenta ***(6)***. This syncytium is responsible for many of the functions performed by the

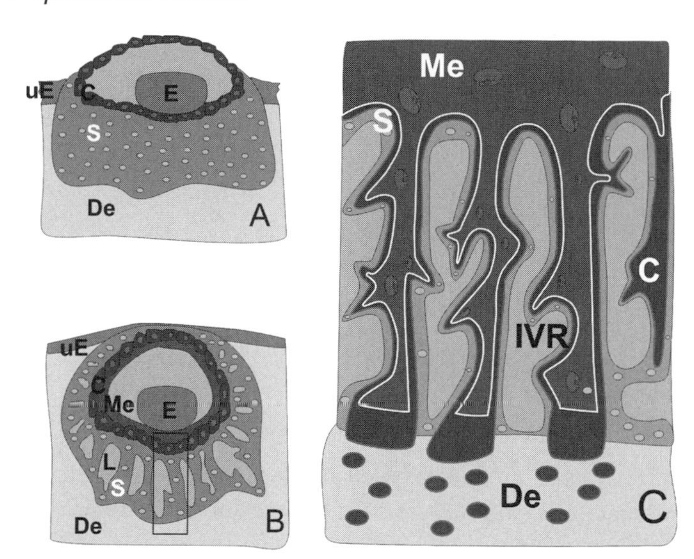

Fig. 1. Development of the syncytiotrophoblast. **(A)** Prelacunar stage. At days 7 to 8 postconception, the syncytiotrophoblast (S) has penetrated through the uterine epithelium (uE) into the decidua (De) and comes into direct contact with maternal cells. At the same time the cytotrophoblast (C) and the embryo (E) escape from contacting maternal tissues. **(B)** Lacunar stage. At days 8 to 9 postconception, fluid-filled spaces appear within the syncytiotrophoblast. These lacunae (L) grow and flow together to generate one blood-filled space, the intervillous space. Between embryo and cytotrophoblast, extraembryonic mesenchyme (Me) grows out from the embryo. **(C)** Villous stage. At about day 19 postconception, the first capillaries develop within the placental villi. The syncytiotrophoblast as the outer surface of the villi faces the intervillous space (IVR) and is no longer invasive as in the first stages of development. Rather, it has differentiated into an epithelial-like structure building the placental barrier between maternal blood and embryonic/fetal tissues.

placenta, such as transport of oxygen, nutrients, and waste products, hormone production required for fetal development, and immune tolerance *(6)*.

Compared with the first step of trophoblast fusion, the subsequent fusion events that take place throughout pregnancy differ in one important aspect. The initial fusion was a fusion event between two mononucleated trophoblast cells at the time of blastocyst adhesion. All following fusion events are between one mononucleated cell (the cytotrophoblast) and the multinucleated layer (the syncytiotrophoblast). Intercellular fusion between two mononucleated cells should no longer take place later in pregnancy, because this would diminish the pool of cytotrophoblast stem cells that maintain the syncytiotrophoblast, which would lead to necrotic degeneration of the layer within a few days *(10)*.

The syncytiotrophoblast needs to be maintained throughout pregnancy by continuous incorporation of underlying cytotrophoblast cells into this syncytial layer. In recent years it has become clear that coordination and appropriate control of trophoblast fusion are crucial to preserve a healthy pregnancy. Dysregulation and alterations of trophoblast fusion may be directly involved in pathological conditions such as preeclampsia ***(11,12)***.

3. Prerequisites for Fusion

There are several known or speculated constituents of a cell that are required for syncytial fusion with a neighboring cell.

3.1. Charged Phospholipids and Divalent Cations

In mammalian cells phosphatidylserine is normally restricted to the inner leaflet of the plasma membrane, pointing to the fact that these cells sustain an asymmetrical distribution of phospholipids ***(13)***. Under specific circumstances cells redistribute phosphatidylserine from the inner to the outer leaflet of the plasma membrane. This "flip" of phosphatidylserine is known to be a prerequisite for syncytial fusion ***(14)*** and is also known as a signal to eliminate cells during apoptosis ***(15)***.

There are two main multinucleated fusion systems in the human, skeletal muscle and placental trophoblast. In both systems it has become clear that the redistribution of phosphatidylserine is indeed a prerequisite for fusion ***(16,17)***. For myotube formation to generate skeletal muscles, a transient exposition of phosphatidylserine directly at cell–cell contact sites is required ***(17)***. In placental villi a few cytotrophoblast cells display the flip of phosphatidylserine without any signs of apoptosis ***(16)***. Furthermore, it has been demonstrated that the phosphatidylserine flip to the outer leaflet of the plasma membrane is also required for fusion of trophoblast-derived BeWo choriocarcinoma cells ***(14,18)***. Rote and coworkers ***(14,18)*** were able to inhibit fusion by applying a monoclonal antibody directed against phosphatidylserine.

It has recently become clear that fusogenic proteins need the redistribution of phosphatidylserine to enable fusion of cells. Martin et al. ***(19)*** demonstrated that the presence of phosphatidylserine is required for the insertion of fusogenic proteins into the lipid bilayer. These fusion proteins are organized as an α-helix and are oriented almost parallel to the lipid acyl chains. This orientation is known to be crucial for the mechanism of protein-induced membrane fusion ***(20)***.

Membrane fusion between phospholipid bilayers can be induced by a variety of chemicals, most simply by multivalent ions. The apparent role of multivalent ions is to bring two apposing lipid bilayers into contact. It appears that divalent cations such as calcium and magnesium play a major role in cell fusion. By examining the interactions of unilamellar vesicles containing phosphatidylcholine and phosphatidic acid in the presence of calcium and magnesium, Leventis et al. *(21)* were able to show the importance of these divalent cations for different fusion mechanisms. The importance of the combined action of specific phospholipid components and divalent cations has recently been reexamined. Faraudo and Travesset *(22)* describe phosphatidic acid as a key phospholipid that is involved in a wide range of biological processes, including membrane fusion. At the same time the presence of calcium is crucial for the biological functionality of phosphatidic acid, showing the interdependency of negatively charged phospholipids and divalent cations in supporting syncytial fusion.

3.2. Protease Activity

In humans, both the skeletal muscle and trophoblast fusion systems require the flip of phosphatidylserine for syncytial fusion. The same appears to be true for the activity of certain intracellular proteases.

Two families of intracellular nonlysosomal proteases seem to play crucial roles in fusion of muscle and trophoblast. These families of proteases mediate cleavage of specific target proteins in a large number of processes during differentiation, life, and death of cells *(23)*.

In the development of skeletal muscle from mononucleated myoblasts to multinucleated muscle tubes the activity of calpain is upregulated prior to fusion *(24)*. Calpains are Ca^{2+}-regulated cysteine proteases playing essential roles in a variety of processes *(23)*. Opposing this effect, the overexpression of the endogenous inhibitor of calpain, calpastatin, inhibits fusion of myoblasts *(25)*. The system of the villous trophoblast appears to make use of a similar system of proteases. Here it is not the family of calpain proteases but rather the family of caspases such as the initiator caspase 8 that is crucial for the fusion process *(26)*. Black et al. *(26)* exploited technologies to block caspase 8 expression and activity to block trophoblast fusion.

In both systems the cleavage of specific target proteins seems to be essential to initiate the fusion process, but the exact nature of the targets is still unknown. The flip of phosphatidylserine may be a first link between proteases and fusion. As has been described, the phosphatidylserine flip is known to occur early in apoptosis as well as prior to fusion. Because initiator caspases are known to induce this flip by cleavage of unknown proteins, a deeper knowledge of this link may help understanding trophoblast fusion *(27)*.

3.3. Proteins Involved in Cell Fusion

The members of the syncytin protein family, syncytin-1 and syncytin-2, appear to be some of the main players in the process of trophoblast fusion, although opinions differ as to their importance (for a more detailed analysis, see **refs.** ***4,5***). The balance of expression and exposition of syncytins and their receptors needs to be taken into account ***(4,28)***. It has become clear that more than one fusogenic protein is required to induce and orchestrate the complex mechanism of cell fusion. ADAM-12 plays a role in recruiting cells for muscle formation, and it may well be that ADAM-12 has a similar function for trophoblast. Other proteins, including connexin 43 ***(29,30)***, cadherin 11 ***(31)***, and CD98 ***(32)***, have been demonstrated to be involved in trophoblast fusion. In addition, proteases such as caspase 8 ***(26)*** and calpain ***(17)*** play roles in this process, as outlined earlier.

3.4. Syncytins and Trophoblast Fusion

About 8% of the human genome is of retroviral origin that was incorporated during evolution ***(33)***. Although most of these regions are not translated, some elements are actually translated. The most interesting elements expressed in the human placenta related to trophoblast fusion are the envelope genes (env regions) of ERV-3, HERV-W, and HERV-FRD ***(34)***. In 2000, the protein encoded by the HERV-W element was termed *syncytin* ***(35)***, which is now referred to as syncytin-1 (**Fig. 2**). The receptor for syncytin-1, RDR ***(36)***, is the D mammalian retrovirus receptor, known thus far as a neutral amino acid transporter system, ASCT2/ATB0/SLC1A5 ***(37)***.

There is evidence for a direct involvement of syncytin-1 in trophoblast fusion ***(38)***. Syncytin-1 mRNA and protein are restricted to the syncytiotrophoblast later in gestation and are present in both layers of villous trophoblast (syncytiotrophoblast and cytotrophoblast) early in gestation ***(35,36,39)***. Syncytin-1 protein expression is higher in first trimester villous trophoblast than in term tissues ***(40)***.

Only little is known about the syncytin-1 receptor(s) and their localization. Syncytin-1 has been found in the basal part of the syncytiotrophoblast ***(35)***, where also the ASC (a sodium-dependent transport system of neutral amino acids) receptor proteins have been detected ***(41)***. It is therefore tempting to speculate that both proteins are expressed in the basal membrane of the syncytiotrophoblast or on opposing plasma membranes of cytotrophoblast and syncytiotrophoblast.

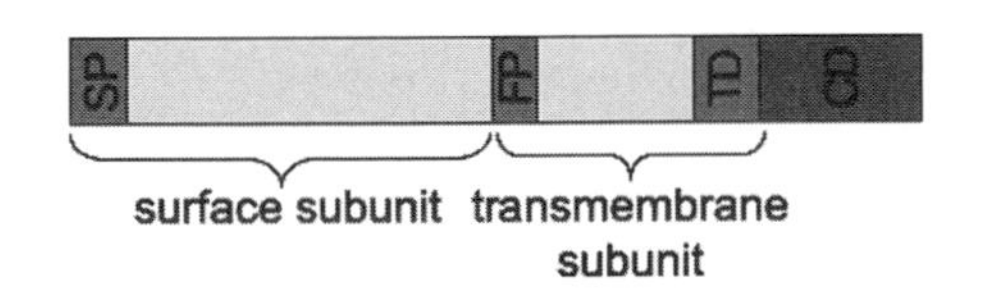

Fig. 2. Predicted domain structure of syncytin-1. SP, signal peptide; FP, fusion peptide; TD, transmembrane domain; CD, cytoplasmic domain. (Modified from **ref.** ***64.***)

Glial cell missing-1 (GCM1), a placenta-specific transcription factor, is a regulator of syncytin-1 expression. It regulates syncytin-1 mRNA expression via two GCM1 binding sites upstream the 5′-long terminal repeat of the syncytin-1 harboring HERV-W family members ***(39)***. GCM1 is localized in a subset of highly differentiated villous cytotrophoblasts ***(42)***. Because these cells are prone to fuse, GCM1 promotion of syncytin-1 expression could be crucial for syncytial fusion in villous trophoblast.

HERV-FRD is an endogenous retroviral sequence and its envelop gene, *syncytin-2*, is expressed in the placenta as well ***(43)***. Initiation of fusion in several cell lines after transient transfection with the HERV-FRD env protein points to its putative importance for cell fusion ***(43)***.

4. Trophoblast Fusion In Vitro

4.1. Considerations About Trophoblast Fusion In Vitro

One must bear in mind that all experiments performed with isolated primary trophoblast have a major drawback. As mentioned, cytotrophoblast cells isolated from placental villi should no longer fuse with each other but rather only with the syncytiotrophoblast. Why then do trophoblast cells fuse in vitro? Here are some hypotheses: 1. The isolation procedure forces at least some of the primary cells to restore the phenotype they had during implantation. During implantation the early trophoblast cells were able to fuse with each other as two mononucleated cells. 2. Primary villous trophoblast cells are able to change their phenotype and are known to become extravillous after isolation. This may also explain why isolated trophoblast cells no longer proliferate. During invasion some extravillous trophoblast cells fuse to generate the multinucleated giant cells found at the border between decidua and myometrium. Thus, the change of phenotype in isolated primary trophoblast cells from a villous phenotype to an extravillous phenotype may explain fusion of these cells in vitro. If this speculation is true, then we need to revisit the whole literature on trophoblast fusion in vitro. 3. The third hypothesis deals with alterations of trophoblast cells during the long isolation procedure. It has been shown that during isolation both mononucleated trophoblast cells and large numbers of syncytial fragments are isolated ***(44)***. The phenotype of these fragments ranges from anuclear to mononucleated or oligonucleated and sometimes even multinuclear. These fragments can be found in variable amounts in trophoblast isolates. After the first description of a very crude trophoblast isolation procedure 20 years ago ***(45)***, there have been intensive efforts to remove these fragments from the isolates of cytotrophoblast.

Finally, Guilbert et al. ***(46)*** showed how to generate cultures of highly pure mononucleated cytotrophoblast cells. But why do these cells still fuse in culture if they maintain a villous phenotype and if they are pure isolates of cytotrophoblast? Again we need to speculate: Isolation of villous trophoblast makes use of a trypsin digestion, which is known to disrupt membranes. Thus the possibility arises that hybrids of syncytiotrophoblast fragments and cytotrophoblast are generated during trypsinization. These mononucleated hybrids contain parts of the syncytial plasma membrane and thus are no longer "pure" cytotrophoblast cells, although they may look like such a cell. These hybrids may be the "seed crystal" for syncytial fusion in vitro since other pure cytotrophoblasts may recognize these hybrids as syncytiotrophoblast. As a result normal fusion between cytotrophoblast and syncytiotrophoblast occurs in vitro similar to the situation in vivo—but is identified as fusion between two cytotrophoblast cells.

4.2. Model Systems for Trophoblast Fusion In Vitro

Cytotrophoblast differentiation in vitro is stimulated by a number of factors, including growth factors and hormones. In vitro studies have established that soluble factors, such as epidermal growth factor and estriol ***(47–53)***, activate different intracellular signaling pathways to stimulate the differentiation of villous cytotrophoblast cells into syncytiotrophoblast. The role and direct involvement of cyclic adenosine monophosphate (cAMP)–dependent protein kinases has also been shown ***(54–56)***. However, because of their limited life span and the limitations in transfecting the primary cells, model systems have been established mimicking trophoblast fusion.

The human choriocarcinoma cell line BeWo as well as other immortalized lineages are frequently used as models to study trophoblast fusion ***(57)***. They can be manipulated and maintained in a nondifferentiated stage and undergo differentiation and concurrent intercellular fusion upon the addition of fusion promoters.

BeWo cells can be induced to undergo differentiation and fusion within 48 to 72 h by the addition of forskolin to the culture medium ***(14)***. Forskolin is a labdane diterpene that is produced by the plant *Plectranthus barbatus* and that is commonly used to raise levels of cAMP and to act via G-protein-coupled receptors ***(58)***. Data regarding cell physiology reveal that forskolin is capable of resensitizing cell receptors by activating the enzyme adenylyl cyclase and increasing the intracellular levels of cAMP.

4.3. Models to Quantify Trophoblast Fusion In Vitro

Borges et al. ***(59)*** have presented a model of a cell–cell fusion assay based on cell staining using green and red fluorescent cytoplasmic dyes, which become intracellularly mixed only after syncytial fusion, generating a yellowish cytoplasm. In this system cell–cell fusion was quantified by fluorescence microscopy in the choriocarcinoma cell lines BeWo, JAR, and JEG3 and in some nontrophoblastic cell lines. The authors found clear differences in fusion behavior between the cell lines, with clear cell–cell fusion only of BeWo cells. Another fusion model established by Kudo et al. ***(32,58)*** again applied two different fluorescent dyes. This time cell fusion was quantified by quantitative flow cytometry ***(32,58)***. The disadvantage of the second method is that aggregation of cells may be interpreted as fusion.

In addition to these two models that utilize fluorescent dyes, immunohistochemistry of a number of proteins has been applied to quantify trophoblast fusion including E-cadherin ***(31,60)*** and desmoplakin ***(60,61)***. Some of the hormones produced by the syncytiotrophoblast in vivo such as β-human chorionic gonadotropin (β-hCG) and human placental lactogen (hPL) can be used as markers for syncytium formation in vitro as well ***(62)***. Many authors use β-hCG as a marker to quantify syncytialization of primary trophoblast and trophoblast-related cells. Synthesis of β-hCG mRNA is a tool that has become nearly essential when studying syncytial formation, although choriocarcinoma cells such as BeWo cells may express it as mononucleated cells as well ***(63)***.

5. Conclusion

Despite the research herein outlined, the mechanisms, prerequisites, and regulatory steps of trophoblast fusion are still a mystery. We need to unravel the roles of numerous proteins involved in the process and sort out the array of spatially and temporally expressed genes (the proteome) involved in normal human placental development, growth, and pathology. Much more still has to be done, and the challenge is to apply specific methodological and technical advances for the characterization of trophoblast fusion.

References

1. Chen, E. H. and Olson, E. N. (2004) Towards a molecular pathway for myoblast fusion in *Drosophila. Trends Cell Biol.* **14,** 452–460.
2. Chen, E. H. and Olson, E. N. (2005) Unveiling the mechanisms of cell–cell fusion. *Science* **308,** 369–373.
3. Potgens, A. J., Schmitz, U., Bose, P., Versmold, A., Kaufmann, P., and Frank, H. G. (2002) Mechanisms of syncytial fusion: a review. *Placenta* **23(Suppl. A),** S107–S113.
4. Potgens, A. J., Drewlo, S., Kokozidou, M., and Kaufmann, P. (2004) Syncytin: the major regulator of trophoblast fusion? Recent developments and hypotheses on its action. *Hum. Reprod. Update* **10,** 487–496.
5. Huppertz, B., Bartz, C., and Kokozidou, M. (2006) Trophoblast fusion: fusogenic proteins, syncytins and ADAMs, and other prerequisites for syncytial fusion. *Micron* **37,** 509–517.
6. Benirschke, K., Kaufmann, P., and Baergen, R. (2006). *Pathology of the Human Placenta*, 5th ed. Springer, New York.

7. Kuzmin, P. I., Zimmerberg, J., Chizmadzhev, Y. A., and Cohen, F. S. (2001) A quantitative model for membrane fusion based on low-energy intermediates. *Proc. Natl. Acad. Sci. U.S.A.* **98,** 7235–7240.
8. Midgley, A., Pierce, G., Denau, G., and Gosling, J. (1963) Morphogenesis of syncytiotrophoblast in vivo: an autoradiographic demonstration. *Science* **141,** 350–351.
9. Panigel, M. (1993) The origin and structure of extraembryonic tissues, in *The Human Placenta* (C. Redman, I. Sargent, and P. Starkey, eds.). Blackwell Scientific Publications, London, pp. 3–32.
10. Castellucci, M., Kaufmann, P., and Bischof, P. (1990) Extracellular matrix influences hormone and protein production by human chorionic villi. *Cell Tissue Res.* **262,** 135–142.
11. Huppertz, B., Kaufmann, P., and Kingdom, J. C. P. (2002) Trophoblast turnover in health and disease. *Fetal Maternal Med. Rev.* **13,** 17–32.
12. Huppertz, B. and Kingdom, J. C. (2004) Apoptosis in the trophoblast—role of apoptosis in placental morphogenesis. *J. Soc. Gynecol. Invest.* **11,** 353–362.
13. Bevers, E. M., Comfurius, P., and Zwaal, R. F. (1996) Regulatory mechanisms in maintenance and modulation of transmembrane lipid asymmetry: pathophysiological implications. *Lupus* **5,** 480–487.
14. Lyden, T. W., Ng, A. K., and Rote, N. (1993) Modulation of phosphatidylserine epitope expression by BeWo cells during forskolin treatment. *Placenta* **14,** 177–186.
15. Savill, J. (1998) Apoptosis. Phagocytic docking without shocking. *Nature* **392,** 442–443.
16. Huppertz, B., Frank, H. G., Kingdom, J. C. P., Reister, F., and Kaufmann, P. (1998) Villous cytotrophoblast regulation of the syncytial apoptotic cascade in the human placenta. *Histochem. Cell Biol.* **110,** 495–508.
17. van den Eijnde, S. M., van den Hoff, M. J., Reutelingsperger, C. P., van Heerde, W. L., Henfling, M. E., Vermeij-Keers, C., Schutte, B., Borgers, M., and Ramaekers, F. C. (2001) Transient expression of phosphatidylserine at cell-cell contact areas is required for myotube formation. *J. Cell Sci.* **114,** 3631–3642.
18. Adler, R. R., Ng, A. K., and Rote, N. (1995) Monoclonal antiphosphatidylserine antibody inhibits intercellular fusion of the choriocarcinoma line JAR. *Biol. Reprod.* **53,** 905–910.
19. Martin, I., Pecheur, E. I., Ruysschaert, J. M., and Hoekstra, D. (1999) Membrane fusion induced by a short fusogenic peptide is assessed by its insertion and orientation into target bilayers. *Biochemistry* **38,** 9337–9347.
20. Decout, A., Labeur, C., Goethals, M., Brasseur, R., Vandekerckhove, J., and Rosseneu, M. (1998) Enhanced efficiency of a targeted fusogenic peptide. *Biochim. Biophys. Acta* **1372,** 102–116.
21. Leventis, R., Gagne, J., Fuller, N., Rand, R. P., and Silvius, J. R. (1986) Divalent cation induced fusion and lipid lateral segregation in phosphatidylcholine-phosphatidic acid vesicles. *Biochemistry* **25,** 6978–6987.
22. Faraudo, J. and Travesset, A. (2007) Phosphatidic acid domains in membranes: effect of divalent counterions. *Biophys. J.* **92,** 2806–2818.

23. Bartoli, M. and Richard, I. (2005) Calpains in muscle wasting. *Int. J. Biochem. Cell Biol.* **37,** 2115–2133.
24. Barnoy, S., Glaser, T. and Kosower, N.S. (1998) The calpain–calpastatin system and protein degradation in fusing myoblasts. *Biochim. Biophys. Acta* **1402,** 52–60.
25. Barnoy, S., Maki, M., and Kosower, N.S. (2005) Overexpression of calpastatin inhibits L8 myoblast fusion. *Biochem. Biophys. Res. Commun.* **332,** 697–701.
26. Black, S., Kadyrov, M., Kaufmann, P., Ugele, B., Emans, N., and Huppertz, B. (2004) Syncytial fusion of human trophoblast depends on caspase 8. *Cell Death Differ.* **11,** 90–98.
27. Das, M., Xu, B., Lin, L., Chakrabarti, S., Shivaswamy, V., and Rote, N. S. (2004) Phosphatidylserine efflux and intercellular fusion in a BeWo model of human villous cytotrophoblast. *Placenta* **25,** 396–407.
28. Drewlo, S., Leyting, S., Kokozidou, M., Mallet, F. and Potgens, A.J. (2006) C-Terminal truncations of syncytin-1 (ERVWE1 envelope) that increase its fusogenicity. *Biol. Chem.* **387,** 1113–20.
29. Frendo, J.L., Cronier, L., Bertin, G., Guibourdenche, J., Vidaud, M., Evain- Brion, D., Malassine, A. (2003) Involvement of connexin 43 in human trophoblast cell fusion and differentiation. *J. Cell Sci.* **116,** 3413–3421.
30. Cronier, L., Defamie, N., Dupays, L., Theveniau-Ruissy, M., Goffin, F., Pointis, G., and Malassine, A. (2002) Connexin expression and gap junctional communication in human first trimester trophoblast. *Mol. Hum. Reprod.* **8,** 1005–1013.
31. Getsios, S. and MacCalman, C. D. (2003) Cadherin-11 modulates the terminal differentiation and fusion of human trophoblastic cells in vitro. *Dev. Biol.* **257,** 41–54.
32. Kudo, Y. and Boyd, C. A. (2004) RNA interference-induced reduction in CD98 expression suppresses cell fusion during syncytialization of human placental BeWo cells. *FEBS Lett.* **577,** 473–477.
33. de Parseval, N., Lazar, V., Casella, J. F., Benit, L., and Heidmann, T. (2003) Survey of human genes of retroviral origin: identification and transcriptome of the genes with coding capacity for complete envelope proteins. *J. Virol.* **77,** 10414–10422.
34. Rote, N. S., Chakrabarti, S., and Stetzer, B. P. (2004) The role of human endogenous retroviruses in trophoblast differentiation and placental development. *Placenta* **25,** 673–683.
35. Mi, S., Lee, X., Li, X.-P., Veldman, G. M., Finnerty, H., Racie, L., LaVallie, E., Tang, X.-Y., Edouard, P., Howes, S., Keth, J. C., and McCoy, J. M. (2000) Syncytin is a captive retroviral envelope protein involved in human placental morphogenesis. *Nature* **403,** 785–789.
36. Blond, J. L., Lavillete, D., Cheynet, V., Bouton, O., Oriol, G., Chapel-Fernandes, S., Mandrand, B., Mallet, F., and Cosset, F. L. (2000) An envelope glycoprotein of the human endogenous retrovirus HERV-W is expressed in the human placenta and fuses cells expressing the type D mammalian retrovirus receptor. *J. Virol.* **74,** 3321–3329.
37. Lavillette, D., Marin, M., Ruggieri, A., Mallet, F., Cosset, F. L., and Kabat, D. (2002) The envelope glycoprotein of human endogenous retrovirus type W uses

a divergent family of amino acid transporters/cell surface receptors. *J. Virol.* **76,** 6442–6452.

38. Frendo, J. L., Olivier, D., Cheynet, V., Blond, J. L., Bouton, O., Vidaud, M., Rabreau, M., Evain-Brion, D., and Mallet, F. (2003) Direct involvement of HERV-W Env glycoprotein in human trophoblast cell fusion and differentiation. *Mol. Cell Biol.* **23,** 3566–3574.
39. Yu, C., Shen, K., Lin, M., Chen, P., Lin, C., Chang, G. D., and Chen, H. (2002) GCMa regulates the syncytin-mediated trophoblast fusion. *J. Biol. Chem.* **277,** 50062–50068.
40. Smallwood, A., Papageorgiou, A., Nicolaides, K., Alley, M. K, Alice, J., Nargund, G., Ojha, K., Campbell, S., and Banerjee, S. (2003) Temporal regulation of syncytin (HERV-W), maternally imprinted PEG10, and SGCE in human placenta. *Biol. Reprod.* **69,** 286–293.
41. Cariappa, R., Heath-Monnig, E., and Smith, C. H. (2003) Isoforms of amino acid transporters in placental syncytiotrophoblast: plasma membrane localization and potential role in maternal/fetal transport. *Placenta* **24,** 713–726.
42. Baczyk, D., Satkunaratnam, A., Nait-Oumesmar, B., Huppertz, B., Cross, J. C., and Kingdom, J. C. P. (2004) Complex patterns of GCM1 mRNA and protein in villous and extravillous trophoblast cells of the human placenta. *Placenta* **25,** 553–559.
43. Blaise, S., de Parseval, N., Bénit, L., and Heidmann, T. (2003) Genomewide screening for fusogenic human endogenous envelopes identifies syncytin 2, a gene conserved on primate evolution. *Proc. Natl. Acad. Sci. U.S.A.* **100,** 13013–13018.
44. Huppertz, B., Frank, H. G., Reister, F., Kingdom, J., Korr, H., and Kaufmann P. (1999) Apoptosis cascade progresses during turnover of human trophoblast: Analysis of human cytotrophoblast and syncytial fragments in vitro. *Lab. Invest.* **12,** 1–16.
45. Kliman, H. J., Nestler, J. E., Sermasi, E., Sanger, J. M., and Strauss, J. F. 3rd. (1986) Purification, characterization, and in vitro differentiation of cytotrophoblasts from human term placentae. *Endocrinology* **118,** 1567–1582.
46. Guilbert, L. J., Winkler-Lowen, B., Sherburne, R., Rote, N. S., Li, H., and Morrish, D. W. (2002) Preparation and functional characterization of villous cytotrophoblasts free of syncytial fragments. *Placenta* **23,** 175–183.
47. Cronier, L., Guibourdenche, J., Niger, C., and Malassine, A. (1999) Oestradiol stimulates morphological and functional differentiation of human villous cytotrophoblast. *Placenta* **20,** 669–676.
48. Alsat, E., Haziza, J., and Evain-Brion, D. (1993) Increase in epidermal growth factor receptor and its messenger ribonucleic acid levels with differentiation of human trophoblast cells in culture. *J. Cell Physiol.* **154,** 122–128.
49. Morrish, D., Linetsky, E., Bhardwaj, D., Li, H., Dakour, J., Marsh, R., Paterson, M., and Godbout, R. (1996) Identification by subtractive hybridization of a spectrum of novel and unexpected genes associated with in vitro differentiation of human cytotrophoblast cells. *Placenta* **17,** 431–441.
50. Garcia-Lloret, M., Morrish, D., Wegmann, T., Honore, L., Turner, A., and Guilbert, L. (1994) Demonstration of functional cytokine-placental interactions: CSF-1 and

GM-CSF stimulate human cytotrophoblast differentiation and peptide hormone secretion. *Exp. Cell Res.* **214,** 46–54.
51. Cronier, L., Alsat, E., Hervé, J.C., Delèze, J., and Malassiné, A. (1998) Dexamethasone stimulates Gap junctional communication peptide hormone production and differentiation in human term trophoblast. *Trophoblast Res.* **11,** 35–49.
52. Cronier, L., Bastide, B., Hervé, J. C., Delèze, J., and Malassiné, A. (1994) Gap junctional communication during human trophoblast differentiation: influence of human chorionic gonadotropin. *Endocrinology* **135,** 402–408.
53. Shi, Q., Lei, Z., Rao, C., and Lin, J. (1993) Novel role of human chorionic gonadotropin in differentiation of human cytotrophoblasts. *Endocrinology* **132,** 1387–1395.
54. Keryer, G., Alsat, E., Tasken, K., and Evain-Brion, D. (1998) Cyclic AMP–dependent protein kinases and human trophoblast cell differentiation in vitro. *J. Cell Sci.* **111,** 995–1004.
55. Frendo, J.L., Thérond, P., Bird, T., Massin, N., Muller, F., Guibourdenche, J., Luton, D., Vidaud, M., Anderson, W., and Evain-Brion, D. (2001) Overexpression of copper zinc superoxide dismutase impairs human trophoblast cell fusion and differentiation. *Endocrinology* **142,** 3638–3648.
56. Frendo, J. L., Thérond, P., Guibourdenche, J., Bidart, J. M., Vidaud, M., and Evain-Brion, D. (2000) Modulation of copper/zinc superoxide dismutase expression and activity with in vitro differentiation of human villous cytotrophoblast. *Placenta* **21,** 773–781.
57. Insel, P. A. and Ostrom, R. S. (2003) Forskolin as a tool for cxamining adenylyl cyclase expression, regulation, and G protein signaling. *Cell Mol. Neurobiol.* **23,** 305–14.
58. Kudo, Y., Boyd, C. A., Kimura, H., Cook, P. R., Redman, C. W., and Sargent, I. L. (2003) Quantifying the syncytialisation of human placental trophoblast BeWo cells grown in vitro. *Biochim. Biophys. Acta* **1640,** 25–31.
59. Borges, M., Bose, P., Frank, H. G., Kaufmann, P., and Pötgens, A. J. G. (2003) A two colour fluorescence assay for the measurement of syncytial fusion between trophoblast-derived cell lines. *Placenta* **24,** 959–964.
60. Alsat, E., Wyplosz, P., Malassine, A., Guibourdenche, J., Porquet, D., Nessmann, C., and Evain-Brion, D. (1996) Hypoxia impairs cell fusion and differentiation process in human cytotrophoblast, in vitro. *J. Cell Physiol.* **168,** 346–353.
61. Saleh, L., Prast, J., Haslinger, P., Husslein, P., Helmer, H., and Knofler, M. (2007) Effects of different human chorionic gonadotrophin preparations on trophoblast differentiation. *Placenta* **28,** 199–203.
62. Handwerger, S. (1991) The physiology of placental lactogen in human pregnancy. *Endocrinology* **12,** 329–336.
63. Gauster, M., Siwetz, M., and Huppertz, B. (2007) Is upregulation of ßhCG expression a marker of syncytialization of BeWo cells? *Placenta* **28,** A71.
64. Chang, C., Chen, P. T., Chang, G. D., Huang, C. J., and Chen, H. (2004) Functional characterization of the placental fusogenic membrane protein syncytin. *Biol. Reprod.* **71,** 1956–1962.

9

Macrophage Fusion

Molecular Mechanisms

Agnès Vignery

Summary

Macrophages are the most versatile, plastic, and mobile cells in the animal kingdom. They are present in all tissues and might even define a true "body-wide" network that maintains health and ensures the repair of tissues and organs. In specific and rare instances, macrophages fuse to form multinucleate osteoclasts and giant cells in bone and in chronic inflammatory reactions, respectively. While macrophages lose most of their plasticity and mobility after they become multinucleate, at the same time they acquire the capacity to resorb calcified tissues, such as bone, and foreign bodies, such as pathogens and implants, and they mediate the replacement of the resorbed tissue by new tissue. There is evidence to suggest that macrophages might also fuse with somatic cells to repair tissues and with tumor cells to trigger the metastatic process. The molecular machinery of macrophage fusion remains poorly characterized, but it is likely to be shared by all fusing macrophages.

Key Words: Macrophage; fusion; osteoclast; giant cell; plasma membrane.

1. Introduction

Macrophages are mononucleate cells that are present in all tissues but that are highly mobile. They have the unique ability to fuse with themselves, in specific and rare instances, to form osteoclasts in bone and giant cells at sites of chronic inflammatory reactions. There is evidence to suggest that macrophages might also fuse with somatic and tumor cells to repair tissues and to trigger the metastatic process, respectively. Fusion of macrophages allows them to more efficiently perform fewer, but more specific, tasks such as resorption of bone and foreign bodies, repair of tissues, and promotion of cancer. The molecular mechanisms that mediate the fusion of macrophages remain poorly understood. Here, I discuss the functional versatility of macrophages and the ways in which multinucleation

From: *Methods in Molecular Biology, vol. 475: Cell Fusion: Overviews and Methods*
Edited by: E. H. Chen © Humana Press, Totowa, NJ

affects their function. I also discuss our current understanding of the mechanism whereby macrophages fuse with each other and, possibly, with other cells.

2. Macrophages

Macrophages originate from the bone marrow, in which their precursors, the monoblasts, proliferate and differentiate into premonocytes prior to their migration, as monocytes, to other tissues via the bloodstream. For monocytes to leave the blood compartment and gain access to tissues, they must become adherent and, therefore, differentiate into macrophages. Macrophages perform numerous functions in tissues, and their roles in many processes remain to be fully characterized *(1–3)*. Macrophages have gained some notoriety recently with the discovery of the family of Toll-like receptors (TLRs; **ref.** *4*) that recognize pathogen-associated molecular patterns. The identification of TLRs has resulted in the recognition of the fact that macrophages do, indeed, represent the first line of defense against invasive pathogens and of their role as key participants in innate immunity.

2.1. Macrophage Heterogeneity

Macrophages have the ability to adjust to their environment and to most, if not all, physiological and pathological conditions. Thus, not surprisingly, macrophages exhibit a high level of heterogeneity *(2)*. In addition, their precursors, the blood monocytes, also exhibit considerable heterogeneity *(2)*. Attempts to categorize macrophages abound, and yet, because of the extreme versatility of macrophages, their phenotypes remain difficult to pin down. One of the salient features of macrophages is their ability to become activated or "angry" upon exposure to soluble factors or foreign bodies, such as implants and pathogens. Macrophages respond to this type of activation by expressing specific sets of surface determinants that are dependent on the signals that they receive: expression of some determinants is suppressed while that of others is enhanced; some determinants are newly expressed while others disappear. Such changes have led to the classification of macrophages in terms of the levels of expression of a repertoire of surface determinants *(2)*. In addition to their extraordinary functional plasticity and adaptability, as represented by their changing molecular attributes, it seems likely that macrophages might also have the unique ability to dedifferentiate, possibly returning to a younger type of precursor, and become less mature, a phenomenon that might allow them to differentiate along a new pathway. This intriguing possibility remains to be confirmed.

2.2. Macrophages Perform Many Tasks

Prior to the recent surge of interest in macrophages, it was generally accepted that their functions included recycling of apoptotic cells and nonfunctional extracellular components; secretion of cytokines and growth factors; presentation antigens to T lymphocytes in the context of major histocompatibility class I and class II complexes; and phagocytosis of foreign bodies, such as pathogens and implants. Macrophages are professional "dispatchers" that direct the immune response according to the nature of the internalized molecule or foreign body. In addition, macrophages command cells to turn over, resorbing tissues and then directing repair via the secretion of growth factors. Thus, in addition to their role as dispatchers, macrophages can be viewed as the "health care coordinators" of tissues and organs, and, because of their high mobility and their widespread distribution, they can be considered as "cells without borders" ***(5)***. Indeed, it seems very plausible that macrophages might control tissue and organ homeostasis by defining a network in which they are connected to one another directly via cell–cell contact and also indirectly via soluble factors, such as cytokines. If this hypothesis is correct, macrophages might save time and energy by informing each other about tissue health without having to travel far. Although this new hypothetical and metaphor-laden concept is supported by the observation that macrophages are present in tissues at a relatively constant density, it remains to be validated.

In addition to defining a putative network, macrophages can undergo clonal expansion in situ such that they are able to rapidly increase in number and, thus, to accelerate their response to a specific exacting situation. Instead of multiple macrophages being recruited from the surrounding tissues and from blood monocytes to one specific anatomical site that is in need of help, clonal expansion provides a short cut that allows macrophages to intervene quickly and in large numbers to protect the host ***(2)***. Indeed, clonal expansion of macrophages might precede fusion, with daughter cells becoming able to recognize one another prior to merging their contents ***(6,7)***.

The question of whether or not macrophages are able to recirculate has not yet been addressed. If macrophages are highly mobile and do circulate from tissue to tissue and from organ to organ, their recirculation, via lymphatics or blood vessels, could have several major functional consequences. For example, recirculation would facilitate the integration of information that macrophages might carry around as possible "memory cells." It would also accelerate the speed at which information is delivered from one cell to another, from one tissue to another, and from one organ to another. It has been long known that there is a difference, in terms of mechanisms of repair, between injuries that occur simultaneously and those that occur sequentially. The hypothetical concept of interaction between tissues might be easier to validate if macrophages do indeed recirculate and travel via lymphatics or even perhaps via blood vessels. If such is the case, the mechanisms that allow macrophages to recirculate should be investigated in detail. If macrophages do, in fact, use blood vessels to travel around the body, such exploitation of the bloodstream might explain in part the heterogeneity that monocytes display because they come from many sources. The question then becomes "do macrophages dedifferentiate and disguise themselves as monocytes or stem cells so that they can travel blood vessels? Does the blood itself induce the expression of specific phenotypic markers on macrophages? Blood is a liquid connective tissue that remains poorly characterized and understood. Although the concept that macrophages might recirculate, via lymphatics, arteries, or veins, is far-fetched, we have to remain open minded when we think about macrophages.

In addition to their extraordinary plasticity, macrophages have the ability to expand clonally in situ, to define a potential network, and, possibly to recirculate. However, their versatility undergoes a drastic metamorphosis when macrophages merge to form multinucleate cells.

3. Macrophage Multinucleation

Multinucleation has a dramatic effect on macrophage function. In contrast to mononucleate macrophages that are long lived (they live for months), heterogeneous, versatile, mobile, and able to adapt to new environments and to perform a multiplicity of functions, multinucleate macrophages are short lived (with a half-life of approximately 3 days); they have a well-defined molecular phenotype and are relatively homogeneous; they are highly focused and perform a limited number of tasks, for example, resorption of the substrate on which they have formed and induction of the repair of the resorbed tissue; and their mobility is limited to their birthplace. Thus, multinucleation endows macrophages with powerful destructive ability, allowing them to resorb substrates as hard and solid as bone. Because they are so large, multinucleate macrophages can resorb large components that cannot be internalized by a single mononucleate cell. Indeed, the number of nuclei packed within a multinucleate macrophage appears to be proportional to the size of the target such cells are able to resorb. The molecular mechanism by which multinucleate macrophages, such as osteoclasts, resorb bone, was reviewed recently *(8)*. In brief, osteoclasts attach firmly to bone, their target, via a sealing zone, which defines a ring that seals off an extracellular compartment. The contents of the compartment are acidic, and the low pH facilitates the dissolution of the target (e.g., bone) or the killing of pathogens (in the case of giant cells), as well as the activation of lysosomal enzymes. Hence, the compartment is considered to be an "extracellular lysosome."

Multinucleate macrophages are more than the sum of their parts because multinucleation is an essential step in the differentiation of osteoclasts. Mononucleate macrophages cannot resorb bone efficiently, as illustrated in diseases in which macrophages cannot fuse, such as some forms of osteopetrosis, which are associated with bones that are both thick and abnormally brittle.

After osteoclasts have completed their task, their fate remains ill-defined. It is likely that osteoclasts undergo apoptosis, although the mechanisms that lead to the death of osteoclasts are poorly understood. By contrast to osteoclasts, giant cells appear to stay around for extended periods of time within granulomas. But the fate of giant cells, like that of osteoclasts, remains to be determined.

3.1. Osteoclasts and Giant Cells: One Type of Cell?

Osteoclasts and giant cells are multinucleate cells that result from the fusion of macrophages. However, whereas osteoclasts differentiate on bone, giant cells form at sites of chronic inflammatory reactions in response to foreign bodies. Osteoclasts and giant cells resorb bone and foreign bodies, respectively, the substrates on which they have formed by fusion and on which they reside. They promote the differentiation and the activation of osteoblasts, which form new bone, or fibroblasts, which encapsulate the foreign body and form a granuloma. One feature that distinguishes osteoclasts from giant cells, in addition to the fact that giant cells differentiate only at sites of chronic inflammatory reactions, is that osteoclasts ensure that the amount of bone they resorb is replaced by a similar amount of bone, formed by osteoblasts. It is the coordinated and coupled activities of osteoclasts and osteoblasts that maintain a relatively constant bone mass. By contrast, giant cells mediate the physical isolation of the foreign body to which they have adhered from the rest of the tissue by stimulating the deposition by fibroblasts of a thick layer of collagen, which enfolds the foreign body to form a granuloma. This collagenous envelope limits the growth and the spread of pathogens and hence contains the infection. Thus, osteoclasts and giant cells, in their own ways, protect the body by mediating its repair and by isolating pathogens, respectively.

3.2. Osteoclasts and Giant Cells Differentiate Via Different Signaling Pathways

It is now well accepted that osteoclasts form from bone marrow–derived macrophages that are stimulated with macrophage colony-stimulating factor (M-CSF) and receptor activator of nuclear factor kappa B (RANKL; **ref.** ***9***). Mice that lack M-CSF or its receptor, M-CSFR (also called c-fms) lack macrophages, and hence lack osteoclasts. As a result, they develop osteopetrosis, a condition as noted above, characterized by bones that are thick and brittle because they cannot undergo remodeling. Similarly, mice that lack RANKL or its receptor, RANK, lack osteoclasts and develop osteopetrosis (*for reviews, see* **refs.** ***10–12***). It is also well accepted that giant cells form from bone marrow–derived macrophages that are stimulated with M-CSF or granulocyte-macrophage CSF (GM-CSF) and with differentiation-inducing cytokines such as and interleukin (IL)-3, IL-4, and IL-13 ***(13,14)***.

It is, however, unclear whether these cytokines are required for the formation of giant cells in vivo. Recently, Helming and Gordon ***(15)*** reported that peritoneal macrophages, activated by thioglycollate in vivo fuse in vitro in response to IL-4. This is an interesting observation in the light of results from our own laboratory (Vignery, unpublished data), which indicate that in contrast with tissue macrophages, such as those from rodent lungs and peritoneum, which achieve 99% and 50% fusion in vitro, respectively, when cultured in medium supplemented with 5% to 10% fetal calf serum only ***(16,17)***, thioglycollate-activated peritoneal macrophages no longer fuse spontaneously in vitro. Indeed, rodent alveolar and peritoneal macrophages are a useful source of fusing cells that have allowed the cloning of cDNAs that code for molecules that participate in fusion as well as the identification of novel putative fusion molecules ***(6)***. It is of interest to note that IL-4, which promotes giant cell formation, inhibits formation of murine osteoclast in vitro ***(18)***. In addition, the chemokine monocyte chemotactic protein-1, which promotes the migration of macrophages, has been shown to stimulate the formation of both mouse giant cells in vivo and human osteoclasts in vitro ***(19, 20)***.

Thus the commonalities and the differences between osteoclasts and giant cells remain ill-defined. However, it appears that these cells share a common fusion machinery ***(21)***. Bone marrow macrophages undergo the first major step in their differentiation when they enter the bloodstream where they become monocytes. They undergo a second major step in their differentiation when they leave the blood for specific tissues, in which they differentiate into macrophages in response to cytokines such as RANKL and tumor necrosis factor-α. Nevertheless, because it is likely that macrophages need to recognize each other as "self" in order to fuse, it is possible that they do so as a result of clonal expansion.

4. Do Somatic and Tumor Cells Fuse With Macrophages?

It has become increasingly clear that cells that belong to the myeloid lineage have the ability to fuse with cells from other lineages for the repair of organs, for example, with liver cells ***(22)***. However, it is unclear whether such myeloid cells are macrophages. Because macrophages remain the only type of myeloid cell that is known to have the ability to fuse, it is possible that they are the fusing myeloid cells. In addition, it has been suggested that macrophages might fuse with tumor cells to create diversity, thereby inducing metastasis and chromosomal aberrations and mediating epigenic regulation ***(23)***. The mechanisms exploited by macrophages and other myeloid cells for recognition and fusion with somatic and tumor cells remain uncharacterized.

5. The Mechanics of Macrophage Fusion

Despite the well-recognized fact that cells of the monocyte–macrophage lineage fuse with one another and, possibly, with somatic and tumor cells, the molecular fusion machinery remains poorly characterized. It has been proposed that fusion of macrophages to generate osteoclasts and giant cells employs a similar machinery. It has also been proposed that the expression of the components of the putative fusion machinery is induced only transiently at the onset of fusion because macrophages are, for the most part, mononucleate and rarely fuse.

Attempts to identify components of the fusion machinery have led to the cloning of macrophage fusion receptor (MFR), now called signal-regulatory protein-α (SIRPα; **ref.** ***17***). Macrophage fusion receptor/SIRPα was recognized by monoclonal antibodies that blocked the fusion of rat alveolar macrophages ***(16)***. It was then discovered that MFR/SIRPα, like CD4, which is the receptor for human immunodeficiency virus (HIV), is a transmembrane protein that belongs to the superfamily of immunoglobulins (IgSF). However, in contrast to CD4, which is expressed by myeloid cells and T lymphocytes, MFR/SIRPα is expressed predominantly by myeloid cells. Subsequently, CD47 was identified as the ligand for MFR/SIRPα. CD47 is homologous to proteins expressed by Vaccinia and Variola viruses ***(24)***.

The viral protein A38L is not known as an actual fusion protein, as is CD47. However, A38L promotes the entry of Ca^{2+} ions into cells possibly via formation of pores ***(25)***. Because pore formation is a tactic used by parasites to enter host cells, a mechanism exploited by yeast cells for fusion during mating ***(26)***, and a mechanism exploited by nematode cells when they fuse during the differentiation of various organs ***(27)***, it might also be used by macrophages to fuse with one another. The overexpression of CD47 or A38L leads to cell death ***(28)***, raising the possibility that, once membranes from adjacent cells are closely apposed and stable, CD47 might form pores that trigger cell–cell fusion. Although this possibility is extremely speculative, it suggests interesting avenues of research.

Sterling et al. ***(29)*** proposed that CD44 might be a component of the putative macrophage fusion machinery, and Cui et al. ***(30)*** reported subsequently that both the extra- and the intracellular domains of CD44 are cleaved sequentially during macrophage fusion. They also reported that the intracellular domain of CD44 translocates to the nucleus where it promotes the activation of nuclear factor kappa B (NF-κB), a transcription factor that is required for the differentiation of osteoclasts, including fusion ***(31,32)***. The extracellular domain of MFR/SIRP*α* is also shed from the plasma membrane of macrophages during fusion (Cui and Vignery, unpublished observation), suggesting the possibility that the intracellular domain of MFR/SIRPα also moves to the nucleus to alter gene expression.

Most recently, Cui et al. ***(33)*** generated evidence that CD200 and its receptor CD200R play a role in macrophage fusion/multinucleation. Although CD200 is not expressed in macrophages, its experssion is highly induced at the onset of fusion. These authors showed that mice that lack CD200 have a higher bone mass than wild types in part because they have fewer osteoclasts. It is interesting that CD200 and CD200R, like MRF/SIRPα and CD47, belong to the super family of immunoglobulins. It appears that CD200R plays a role in activation of the NF-κB and mitogen-activated protein kinase signaling pathways downstream of RANK, a receptor that plays a central role in the differentiation of macrophages leading to the formation of osteoclasts. The authors concluded that CD200 engagement of the CD200R at the initiation of macrophage fusion regulates further differentiation to osteoclasts. Whether CD200 and its receptor mediate fusion remains a possibility.

Additional cell surface molecules that might also play a role in the attachment of macrophages to a substrate prior to fusion include CD9 and CD81, which, like SPE-38 from *Caenorhabditis elegans*, are tetraspan membrane proteins ***(34,35)***. Cadherin and the purigenic receptor P2X7 appear to facilitate the fusion of macrophages to yield osteoclasts and giant cells, respectively, in vitro ***(36,37)***. However, mice that lack the P2X7 receptor have normal osteoclasts ***(38)***. Inhibitors of expression of mannose receptors prevent macrophage fusion in vitro ***(39)***, and integrin-β1 and -β2 mediate the adhesion of macrophages at the onset of fusion ***(40)***. Although each of these macromolecules might participate at some level in cell–cell recognition and/or attachment, none is absolutely required for fusion.

Most fusion of macrophages to yield osteoclasts is impaired in mice that lack the Vo subunit d2 of vacuolar-type adenosine triphosphatase (V-ATPase; **ref.** ***41)***, and these mice develop a mild form of osteopetrosis. Therefore, the V-ATPase is probably involved in the fusion of macrophages.

Recently, dendritic cell-specific transmembrane protein (DC-STAMP) has been shown to be a prerequisite for the fusion of macrophages ***(42)***. Mice that lack DC-STAMP lack multinucleate osteoclasts and giant cells, and they develop a mild form of osteopetrosis. Because DC-STAMP is a receptor with seven-transmembrane domains, it is reminiscent of CXCR4, the coreceptor for HIV that is required for the fusion of HIV with host cells, and of yeast Ste2 and Ste3, two G-protein-coupled receptors that are responsible for the initiation of fusion in mating yeast cells ***(43,44)***.

Ligation of DC-STAMP to an as yet unidentified ligand, might regulate rather than mediate fusion. While we still have many questions about the role of DC-STAMP, one question of immediate interest is whether DC-STAMP interacts with or regulates the expression of MFR/SIRPα, CD47, and CD44.

The possibility that macrophages fuse with somatic cells as part of the repair process, and with cancer cells as part of the metastatic process, has been discussed and remains open ***(5,7)***. Indeed, unlike viruses, which often have a single type of coat protein, plasma membranes are associated with a larger variety of proteins, both integral and membrane associated, which are themselves modified posttranslationally by the addition of lipids and sugar moieties. The complexity of plasma membrane proteins, as well as their intracellular domains, which transduce signals downstream, suggests that the cell fusion machinery is more complicated than originally anticipated and that its components

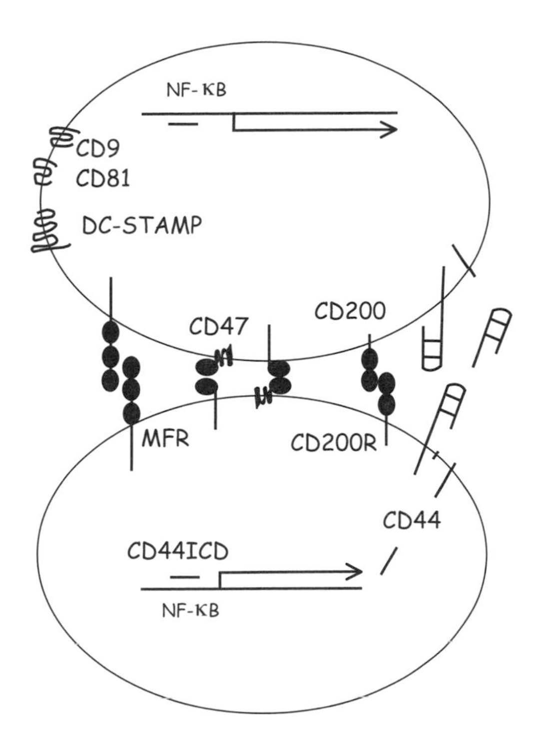

Fig. 1. Macrophages recognize each other when macrophage fusion receptor (MFR)/ signal-regulatory protein-α (SIRPα) binds to CD47. The stepwise association of the long form of MFR and then the short form of MFR (MFR-s) with CD47 brings plasma membranes close to one another. The shedding of the extracellular domain of MFR might facilitate this association (Cui and Vignery, unpublished observation). The distance between macrophage plasma membranes could be reduced to 5–10 nm if MFR-s and CD47 bend upon binding. The shedding of the extracellular domain of CD44 further facilitates plasma membranes from opposite cells to get closer and fuse. When the intracellular domain of CD44 is cleaved by a γ-secretase complex, it translocates to the nucleus to promote the activation of nuclear factor kappa B (NF-κB), which is a transcription factor indispensable for osteoclastogenesis. Ligation of CD200 with its receptor CD200R promotes multinucleation, while activation of DC-STAMP by its unidentified ligand is required for fusion. DC-STAMP, dendritic cell-specific transmembrane protein.

might act in a sequential manner to mediate fusion. The requirement for cell–cell recognition, attachment, and, finally, fusion, in addition to regulatory mechanisms such as those that involve DC-STAMP, suggest that we are only just beginning to understand the mechanics of macrophage fusion.

6. Conclusion

Macrophage fusion has only recently captured the attention of cell biologists, who are now exploiting molecular tools and mouse models in attempts to dissect this important process. It has become clear that the fusion of viruses with host cells and the fusion of intracellular membranes during trafficking are mediated by specific sets of proteins that define specific fusion machineries, and many of these proteins have been identified. The molecular mechanisms that mediate cell–cell fusion remain, by contrast, largely uncharacterized. Nonetheless, it appears that the fusion of macrophages, and possibly that of other types of cell, follows a well-ordered program.

First, macrophages receive a signal or are activated by a ligand that is secreted by either one of the fusing partner cells or, perhaps, by other cells. Such a ligand might be, for example, the still unidentified macromolecule that binds to DC-STAMP. Expression of DC-STAMP by one of a pair of fusing macrophage partners is sufficient to initiate a cascade of still-unidentified molecular events that lead to attraction, attachment, and fusion. As in the case of other fusing cells, such as mating yeast cells, receptor-mediated attraction triggers activation of the expression of fusion-specific genes, which leads to cell–cell attachment that is mediated by members of the immunoglobulin super family and also, perhaps, by tetraspan proteins, such as CD9 and CD81. Fusion ensues, and new functional multinucleate cells, such as osteoclasts and giant cells, are generated. As new components of the macrophage fusion machinery are identified and fusion mechanisms are characterized, our understanding of fusion will increase and facilitate the development of new therapies to prevent and treat diseases, such as some forms of osteopetrosis, in which the normal fusion of macrophages is impaired.

References

1. Greaves, D. R. and Gordon, S. (2002) Macrophage-specific gene expression: current paradigms and future challenges. *Int. J. Hematol.* **76,** 6–15.
2. Taylor, P. R., Martinez-Pomares, L., Stacey, M., Lin, H. H., Brown, G. D., Gordon, S. (2005) Macrophage receptors and immune recognition. *Annu. Rev. Immunol.* **23,** 901–944.
3. Hume, D. A. (2006) The mononuclear phagocyte system. *Curr. Opin. Immunol.* **18,** 49–53.
4. Medzhitov, R. and Janeway, C. A. Jr. (2002) Decoding the patterns of self and nonself by the innate immune system. *Science* **296,** 298–300.

5. Vignery, A. (2005) Macrophage fusion: are somatic and cancer cells possible partners? *Trends Cell Biol.* **15,** 188–193.
6. Vignery, A. (2005) Osteoclasts and giant cells: macrophage–macrophage fusion mechanism. *Int. J. Exp. Pathol.* **81,** 291–304.
7. Vignery, A. (2000) Macrophage fusion: the making of osteoclasts and giant cells. *J. Exp. Med.* **202,** 337–340.
8. Bruzzaniti, A. and Baron, R. (2006) Molecular regulation of osteoclast activity. *Rev. Endocr. Metab. Disord.* **7,** 123–139.
9. Boyle, W. J., Simonet, W. S., and Lacey, D. L. (2003) Osteoclast differentiation and activation. *Nature* **423,** 337–342.
10. Teitelbaum, S. L. and Ross, F. P. (2003) Genetic regulation of osteoclast development and function. *Nat. Rev. Genet.* **8,** 638–649.
11. Sharma, S. M., Hu, R., Bronisz, A., Meadows, N., Lusby, T., Fletcher, B., Hume, D. A., Cassady, A. I., and Ostrowski, M. C. (2006) Genetics and genomics of osteoclast differentiation: integrating cell signaling pathways and gene networks. *Crit. Rev. Eukaryot. Gene Expr.* **16,** 253–277.
12. Asagiri M. and Takayanagi, H. (2007) The molecular understanding of osteoclast differentiation. *Bone* **40,** 251–264.
13. McNally, A. K. and Anderson, J. M. (1995) Interleukin-4 induces foreign body giant cells from human monocytes/macrophages. Differential lymphokine regulation of macrophage fusion leads to morphological variants of multinucleated giant cells. *Am. J. Pathol.* **147,** 1487–1499.
14. DeFife, K. M., Jenney, C. R., McNally, A. K., Colton, E., and Anderson, J. M. (1997) Interleukin-13 induces human monocyte/macrophage fusion and macrophage mannose receptor expression. *J. Immunol.* **158,** 3385–3390.
15. Helming, L. and Gordon, S. (2007) Macrophage fusion induced by IL-4 alternative activation is a multistage process involving multiple target molecules. *Eur. J. Immunol.* **37,** 33–42.
16. Saginario, C., Qian, H.-Y., and Vignery, A. (1995) Identification of an inducible surface molecule specific to fusing macrophages. *Proc. Natl. Acad. Sci. U.S.A.* **92,** 12210–12214.
17. Saginario, C., Sterling, H., Beckers, C., Kobayashi, R.-J., Solimena, M., Ullu, E., and Vignery, A. (1998) MFR, a putative receptor mediating the fusion of macrophages. *Mol. Cell. Biol.* **18,** 6213–6223.
18. Shioi, A., Teitelbaum, S. L., Ross, F. P., Welgus, H. G., Suzuki, H., Ohara, J., and Lacey, D. L. (1991) Interleukin 4 inhibits murine osteoclast formation in vitro. *J. Cell Biochem.* **47,** 272–277.
19. Kyriakides, T. R., Foster, M, J., Keeney, G. E., Tsai, A., Giachelli, C. M., Clark-Lewis, I., Rollins, B. J., and Bornstein, P. (2004) The CC chemokine ligand, CCL2/MCP1, participates in macrophage fusion and foreign body giant cell formation. *Am. J. Pathol.* **165,** 2157–2166.
20. Kim, M. S., Day, C. J., Selinger, C. I., Magno, C. L., Stephens, S. R., and Morrison, N. A. (2006) MCP-1–induced human osteoclast-like cells are tartrate-resistant acid phosphatase, NFATc1, and calcitonin receptor–positive but require receptor activator of NFkappaB ligand for bone resorption. *J. Biol. Chem.* **281,** 1274–1285.

21. Chen, E., Grote, E., Mohler, W., and Vignery, A. (2007) Membrane exchange special issue: cell–cell fusion. *FEBS Lett.* **581(11),** 2181–2193, 2007.
22. Willenbring, H., Bailey, A. S., Foster, M., Akkari, Y., Dorrell, C., Olson, S., Finegold, M., Fleming, W. H., and Grompe, M. (2004) Myelomonocytic cells are sufficient for therapeutic cell fusion in liver. *Nat. Med.* **10,** 744–748.
23. Duelli, D. and Lazebnik, Y. (2003) Cell fusion: a hidden enemy? *Cancer Cell* **3,** 445–448.
24. Parkinson, J. E., Sanderson, C. M., and Smith, G. L. (1995) The Vaccinia virus A38L gene product is a 33-kDa integral membrane glycoprotein. *Virology* **214,** 177–188.
25. Sanderson, C. M., Parkinson, J. E., Hollinshead, M., and Smith, G. L. (1996) Overexpression of the *Vaccinia* virus A38L integral membrane protein promotes Ca^{2+} influx into infected cells. *J. Virol.* **70,** 905–914.
26. Nolan, S., Cowan, A. E., Koppel, D. E., Jin, H., and Grote, E. (2006) FUS1 Regulates the Opening and Expansion of Fusion Pores between Mating Yeast. *Mol. Biol. Cell* **17,** 2439–2450.
27. Mohler, W. A., Simske, J. S., Williams-Masson, E. M., Hardin, J. D., and White, J. G. (1998) Dynamics and ultrastructure of developmental cell fusions in the *Caenorhabditis elegans* hypodermis. *Curr. Biol.* **8,** 1087–1090.
28. Nishiyama, Y., Tanaka, T., Naitoh, H., Mori, C., Fukumoto, M., Hiai, H., and Toyokuni, S. (1997) Overexpression of integrin-associated protein (CD47) in rat kidney treated with a renal carcinogen, ferric nitrilotriacetate. *Jpn. J. Cancer Res.* **88,** 120–128.
29. Sterling, H., Saginario, C., and Vignery, A. (1998) CD44 occupancy prevents macrophage multinucleation. *J. Cell Biol.* **843,** 837–847.
30. Cui, W., Ke, J. Z., Zhang, Q., Ke, H. Z., Chalouni, C., and Vignery, A. (2006) The intracellular domain of CD44 promotes the fusion of macrophages. *Blood* **107,** 796–805.
31. Franzoso, G., Carlson, L., Xing, L., Poljak, L., Shores, E. W., Brown, K. D., Leonardi, A., Tran, T., Boyce, B. F., and Siebenlist, U. (1997) Requirement for NF-kappaB in osteoclast and B-cell development. *Genes Dev.* **11,** 3482–3496.
32. Iotsova, V., Caamano, J., Loy, J., Yang, Y., Lewin, A., Bravo, R. (1997) Osteopetrosis in mice lacking NF-kappaB1 and NF-kappaB2. *Nat. Med.* **3,** 1285–1289.
33. Cui, W., Cuartas, E., Ke, J., Zhang, Q., Einarsson, H. B., Sedgwick, J. D., Li, J., and Vignery, A. CD200 and its receptor, CD200R, modulate bone mass via the differentiation of osteoclasts. *Proc. Natl. Acad. Sci. U.S.A.* Epub 2007 August 28.
34. Takeda, Y., Tachibana, I., Miyado, K., Kobayashi, M., Miyazaki, T., Funakoshi, T., Kimura, H., Yamane, H., Saito, Y., Goto, H., Yoneda, T., Yoshida, M., Kumagai, T., Osaki, T., Hayashi, S., Kawase, I., and Mekada, E. (2003) Tetraspanins CD9 and CD81 function to prevent the fusion of mononuclear phagocytes. *J. Cell Biol.* **161,** 945–956.
35. Chatterjee, I., Richmond, A., Putiri, E., Shakes, D. C., and Singson, A. (2005) The Caenorhabditis elegans spe-38 gene encodes a novel four-pass integral membrane protein required for sperm function at fertilization. *Development* **132,** 2795–2808.

36. Mbalaviele, G., Chen, H., Boyce, B. F., Mundy, G. R., and Yoneda T. (1995) The role of cadherin in the generation of multinucleated osteoclasts from mononuclear precursors in murine marrow. *J. Clin. Invest.* **95,** 2757–2765.
37. Lemaire, I., Falzoni, S., Leduc, N., Zhang, B., Pellegatti, P., Adinolfi, E., Chiozzi, P., and Di Virgilio, F. (2006) Involvement of the purinergic P2X7 receptor in the formation of multinucleated giant cells. *J. Immunol.* **177,** 7257–7265.
38. Ke, H. Z., Qi, H., Weidema, A. F., Zhang, Q., Panupinthu, N., Crawford, D. T., Grasser, W. A., Paralkar, V. M., Li, M., Audoly, L. P., Gabel, C. A., Jee, W. S., Dixon, S. J., Sims, S. M., and Thompson, D. D. (2003) Deletion of the P2X7 nucleotide receptor reveals its regulatory roles in bone formation and resorption. *Mol. Endocrinol.* **17,** 1356–1367.
39. McNally, A. K., DeFife, K. M., and Anderson, J. M. (1996) Interleukin-4–induced macrophage fusion is prevented by inhibitors of mannose receptor activity. *Am. J. Pathol.* **149,** 975–985.
40. McNally, A. K. and Anderson, J. M. (2002) Beta1 and beta2 integrins mediate adhesion during macrophage fusion and multinucleated foreign body giant cell formation. *Am. J. Pathol.* **160,** 621–630.
41. Lee, S. H., Rho, J., Jeong, D., Sul, J. Y., Kim, T., Kim, N., Kang, J. S., Miyamoto, T., Suda, T., Lee, S. K., Pignolo, R. J., Koczon-Jaremko, B., Lorenzo, J., and Choi, Y. (2006) v-ATPase V0 subunit d2–deficient mice exhibit impaired osteoclast fusion and increased bone formation. *Nat. Med.* **12,** 1403–1409.
42. Yagi, M., Miyamoto, T., Sawatani, Y., Iwamoto, K., Hosogane, N., Fujita, N., Morita, K., Ninomiya, K., Suzuki, T., Miyamoto, K., Oike, Y., Takeya, M., Toyama, Y., and Suda, T. (2005) DC-STAMP is essential for cell–cell fusion in osteoclasts and foreign body giant cells. *J. Exp. Med.* **202,** 345–351.
43. Bardwell, L. A. (2005) Walk-through of the yeast mating pheromone response pathway. *Peptides* **26,** 339–350.
44. Elion, E. A. (2000) Pheromone response, mating and cell biology. *Curr. Opin. Microbiol.* **3,** 573–581.

II

Cell Fusion Assays

10

Cell Fusion Assays for Yeast Mating Pairs

Eric Grote

Summary

Yeast mating provides an accessible genetic system for the discovery of fundamental mechanisms in eukaryotic cell fusion. Although aspects of yeast mating related to pheromone signaling and polarized growth have been intensively investigated, fusion itself is poorly understood. This chapter describes methods for measuring the overall efficiency of yeast cell fusion and for monitoring various stages of the fusion process including cell wall remodeling, plasma membrane fusion, and nuclear fusion.

Key Words: Cell fusion; mating; membrane fusion; cell wall remodeling; karyogamy.

1. Introduction

Saccharomyces cerevisiae typically grows as a diploid in its natural environment. Meiosis is induced under nutrient limited conditions to produce 4 haploid spores, which are retained within an ascus. Each ascus contains two mating type (MAT) *MAT***a** spores and two *MAT*α spores, whose mating type is determined by genes that shuttle into the *MAT* genetic locus *(1)*. When nutrient availability improves, the spores germinate and resume growth by forming buds that mature into daughter cells. The mother cell can then switch mating types and fuse with its daughter or another neighboring cell to reform a diploid *(2)*. Laboratory yeast strains have a mutation in the *HO* endonuclease gene to prevent mating type switching. This allows the propagation of haploid clones and greatly facilitates genetic analysis.

An overview of the mating pathway is shown in **Figure 1**. Mating initiates with an exchange of peptide pheromones *(3,4)*. MAT**a** cells secrete **a**-factor, which binds to a receptor expressed exclusively on *MAT*α cells, and vice versa. The two pheromones activate a common G protein/mitogen-activated protein (MAP) kinase signaling pathway that results in cell cycle arrest prior to DNA

From: *Methods in Molecular Biology, vol. 475: Cell Fusion: Overviews and Methods*
Edited by: E. H. Chen © Humana Press, Totowa, NJ

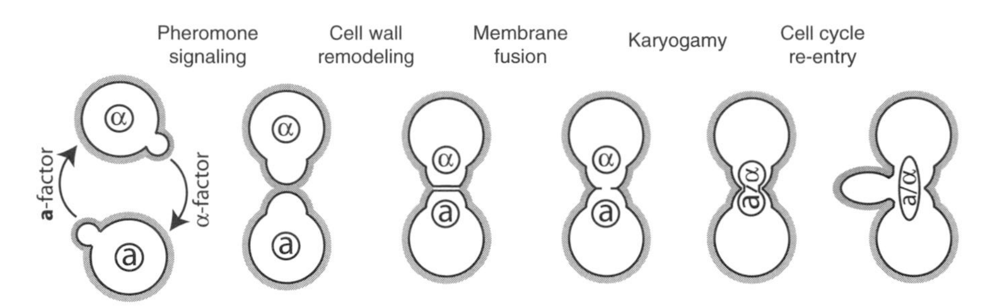

Fig. 1. Stages of the yeast mating pathway.

replication, polarized growth toward a potential mating partner, and transcriptional activation of mating-specific genes. Two cells of opposite mating type bind to each other and generate a unified cell wall surrounding the mating pair. The cell wall at the contact site can then be degraded without compromising the osmotic stability of the mating pair. The plasma membranes of the two cells then come into contact and ultimately fuse *(5,6)*. After plasma membrane fusion, the haploid nuclei from the two parent cells are transported into contact along microtubules and then fuse *(7)*. The fused zygote then reenters the cell cycle, and a diploid daughter cell buds from the neck connecting the remnants of its two parent cells. The last organelles to fuse are vacuoles *(8)*. Vacuoles remain segregated in their respective halves of the zygote until after the zygotic bud emerges. They then abruptly move along actin cables into the emerging daughter cell, fuse, and quickly redistribute their contents into the three lobes of the budded zygote.

There are different classes of mating mutants that arrest at distinct stages in the mating process. Sterile mutants have pheromone signaling defects and fail to form mating pairs. Cell polarity and cell wall remodeling mutants arrest with pairs of cells that have bound to each other but failed to fuse. Plasma membrane fusion mutants complete cell wall remodeling but either fail to achieve cytoplasmic continuity or do so with delayed kinetics. Karyogamy mutants complete plasma membrane fusion but retain two distinct nuclei. This chapter describes methods to monitor each step in the cell fusion process and to distinguish between these different classes of mating mutants.

A general feature of the following assays is that mating is initiated on the surface of a filter layered over a nutrient agar plate (**Subheading 3.2**). *Saccharomyces cerevisiae* mates approximately twofold more efficiently on filters than directly on the surface of an agar plate and mates 50-fold more efficiently on filters than in an aerated liquid culture, where sheer forces oppose mating pair assembly. Filter mating also bypasses the usual requirement for agglutinin proteins associated with the cell wall *(9)*.

Genetic complementation is the traditional assay used to measure the efficiency of the entire mating process. The complementation assay described below takes advantage of the recessive *met15* and *lys2* mutations present in BY4741 and BY4742, the strains used to construct the yeast knockout collection. In this assay, the *MAT***a** met15 strain fails to grow on medium lacking methionine and the *MAT*α *lys2* strain fails to grow on medium lacking lysine. After mating, the resulting *MAT***a**/MATα MET15/met15 LYS2/lys2 diploid can grow on medium lacking both methionine and lysine.

Microscopic assays are used to identify which steps in the mating process are inhibited by a particular mutation or condition. Microscopic assays can also reveal mating defects that are too subtle to be detected by genetic complementation. The simplest assay to start with is to observe mating reactions by differential interference contrast (DIC) microscopy (**Fig. 2**). This assay reveals

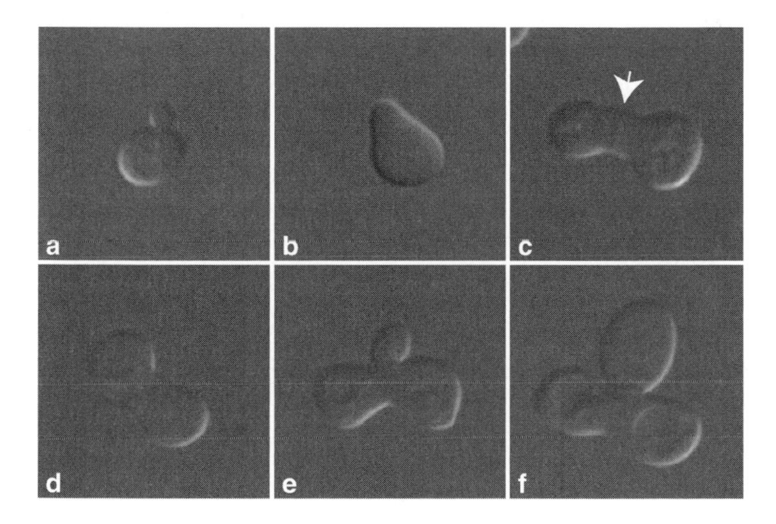

Fig. 2. Differential interference contrast (100×) images of mating yeast. **(A)** A *ste5*^{ts} cell that failed to respond to mating pheromone. The normal pheromone response includes a G_1 cell-cycle arrest. Bud formation and mitotic growth continue when pheromone signaling is disrupted. **(B)** A *MAT***a** cell treated with synthetic α-factor has extended a mating projection in the absence of a potential mate. Cells with this appearance are commonly referred to as *shmoos* because of their resemblance to a 1950s cartoon character. **(C)** A mating pair with a central septum (arrow). The contact zone has expanded, as typically occurs in cell fusion mutants with cell wall remodeling defects. However, plasma membrane fusion can occur in some mating pairs with a central septum. **(D)** A mating pair that appears to have recently fused. **(E)** A presumably fused zygote with a small bud growing from the neck connecting the two parent cells. **(F)** A presumably fused zygote with a large bud that will soon separate from the zygote to form a diploid daughter cell.

whether cells of the two strains under examination have formed mating pairs and can provide guidance regarding which additional assays will provide further information regarding the specific defect. Mating pairs are peanut shaped, with two lobes connected by a wide neck, whereas individual cells are oval shaped and often have a daughter cell budding from one side. After fusion, the zygote matures to form a three-lobed structure with a diploid daughter cell budding from the neck connecting her two parents. However, after a failed fusion attempt, one of the cells in the prezygote can reorient its axis of polarity and either mate with a different partner or bud off a haploid daughter cell ***(10)***. Thus, the appearance of a three-lobed structure does not provide definitive evidence for cell fusion. Prior to the completion of cell wall remodeling, a septum is often visible at the interface between the two parent cells. The contact zone and septum are often extended in mutants with cell polarity defects. In contrast, some *fig1* mating pairs arrest prior to cell wall remodeling with long, narrow necks ***(11)***. Importantly, a visible septum does not provide conclusive evidence for cell fusion failure, because fusion often occurs across a septum. By analogy, it is impossible to see the hole in a doughnut when looking from one edge. When the cells in a mating reaction fail to form mating pairs, cell morphology can provide additional information regarding the mating defect. Cells responding to pheromone will form mating projections, which have a wide base, whereas mitotic growth by budding continues in the absence of a pheromone response.

Whenever two mutant strains have a cell fusion defect when mated to each other, the individual strains should be tested for mating to wild-type mating partners. Most cell fusion mutants have a bilateral defect, meaning that they mate poorly to a mutant strain of the opposite mating type but have only minor defects when mated to a wild-type partner ***(12)***. In the case of cell wall remodeling mutants, this is easily explained by the possibility to degrade a cell wall from either side. Unilateral mating defects are often mating type (**a** or α) specific and are found in mutants with defects in pheromone biosynthesis, pheromone release, and pheromone receptors. Test crosses with *MAT***a** and *MAT*α partners will also reveal if one strain was mislabeled with an incorrect mating type.

Strains of opposite mating type that fail to form mating pairs are likely to have pheromone signaling defects. Halo assays can be used to test for the ability to release and respond to pheromones. To test for pheromone release, cells are patched onto a low-density lawn of supersensitive tester cells and then incubated for 2 days. If the cells in the patch release pheromone, there will be a halo surrounding the patch where the supersensitive tester cells failed to grow because they responded to the pheromone by arresting their cell cycle. To test for pheromone sensitivity, a filter paper disk soaked with mating pheromone is placed over a low-density lawn of the unknown strain. After incubation, a halo of no cell growth surrounding the disk indicates that the unknown strain has undergone a cell-cycle arrest in response to the pheromone. Another convenient pheromone response assay is to measure transcriptional activation of a *FUS1-lacZ* reporter gene in response to mating pheromone. Detailed methods for these assays are described elsewhere ***(13)***.

The mating defects in strains with impaired cell fusion can be further defined using fluorescent markers. FM4-64 is a styryl dye that fluoresces bright red when incorporated into membranes. Although FM4-64 is often used as a tracer of endocytic membrane traffic in yeast ***(14)***, it is restricted to the plasma membrane of cells maintained at 4°C. Within a membrane bilayer, FM4-64 diffuses rapidly. Thus, the double layer of plasma membranes at the junction between the two cells of an unfused mating pair will stain brightly (**Fig. 3**). Indeed, the bright fluorescence of FM4-64 is far more obvious than the cell wall septum found in DIC images. However, as with the septum, a small fusion pore connecting the two FM4-64-stained membranes within the intercellular junction can be impossible to detect. The shape of the boundary separating two FM4-64-labeled cells indicates whether or not cell wall remodeling is complete. Intact cell walls provide structural support to the membranes to maintain a flat boundary. Once the membranes come into contact, they can flex into one of the parent cells, forming a finger or bubble of membrane-bound cytoplasm ***(8,15)***. Mating pairs with intact cell walls and a flat interface are defined as early prezygotes, whereas mating pairs with plasma membrane contact and cytoplasmic fingers are referred to as late prezygotes.

Green fluorescent protein (GFP) is an ideal marker for cytoplasmic mixing ***(15)***. As a globular protein with dimensions of 2.3 × 2.3 × 2.7 nm, GFP can diffuse through a fusion pore linking the plasma membranes of two cells that is 100-fold too small to be convincingly detected by DIC microscopy or FM4-64 staining (*see* **Fig. 3**). In an unfused prezygote, the intercellular junction can be clearly delineated when the two cells express fluorescent proteins with different colors (**Fig. 4**). This technique facilitates detection of the smallest cytoplasmic fingers, which project less than 0.2 mm into the cytoplasm of the mating partner. An even higher resolution view of the interface between cells can be obtained from electron microscopic images as described in Chapter 11. Cytoplasmic GFP is also a useful marker for the integrity of the plasma membrane. In *prm1* mutant mating pairs, defects in plasma membrane fusion often lead to lysis, resulting in a loss of cytoplasmic GFP fluorescence ***(15)***.

The rate at which cytoplasmic GFP diffuses from one cell into its mating partner is proportional to the size of the fusion pore ***(16)***. This relationship is described by Fick's law, which can be used to calculate pore permeance. Although the actual dimensions of a pore can only be estimated from its

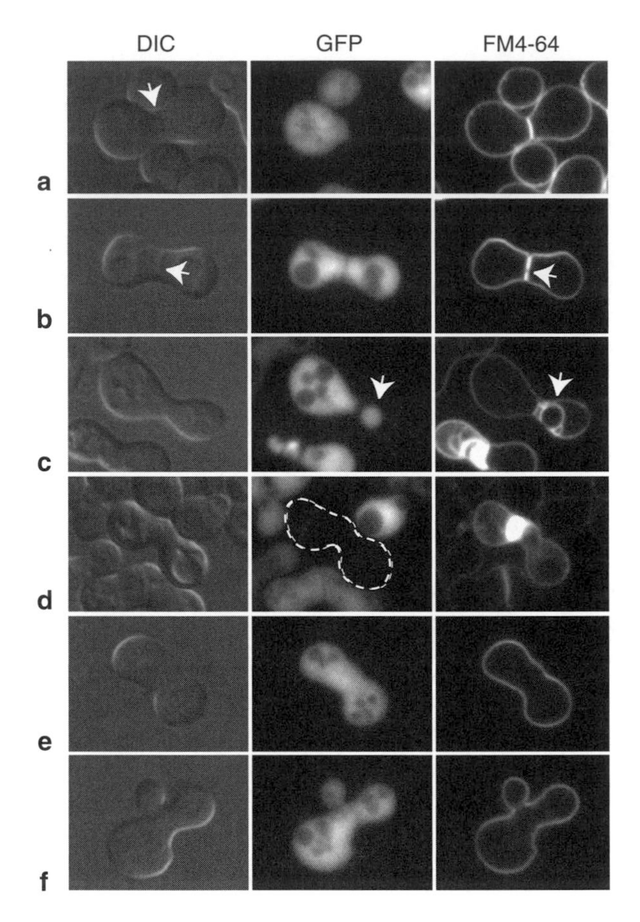

Fig. 3. Mating pairs stained with FM4-64, with cytoplasmic green fluorescent protein (GFP) expressed by the *MAT***a** partner. **(A)** An early prezygote. Green fluorescent protein is retained in the *MAT***a** cell, indicating that the plasma membranes have not fused. The intervening cell wall is visible as a central septum in the differential interference contrast (DIC) image (arrow) and has FM4-64-stained plasma membrane on both sides. **(B)** A fused zygote with a cell wall remnant. Green fluorescent protein has diffused from the *MAT***a** cell into the *MAT*α cell, indicating that the plasma membranes have fused. The cell wall remnant appears as a central septum in the DIC image and has FM4-64-stained plasma membrane on either side. In this particular zygote, it is possible to see a discontinuity in the cell wall (arrow). Green fluorescent protein is excluded from the large vacuoles in each half of the zygote. The vacuole membrane does not stain with FM4-64 because growth and endocytosis were arrested by collecting the mating pairs in azide buffer at 4°C. **(C)** A late prezygote with a finger (bubble) of membrane-bound cytoplasm extending from the *MAT***a** GFP cell into its *MAT*α partner (arrows). **(D)** A lysed mating pair. Green fluorescent protein has diffused out of the cytoplasm into the surrounding medium. **(E)** A fused zygote. Fusion is indicated by both GFP transfer and the absence of FM4-64-stained membrane between the parental cells. **(F)** A fused zygote with a bud that will mature into a diploid daughter cell.

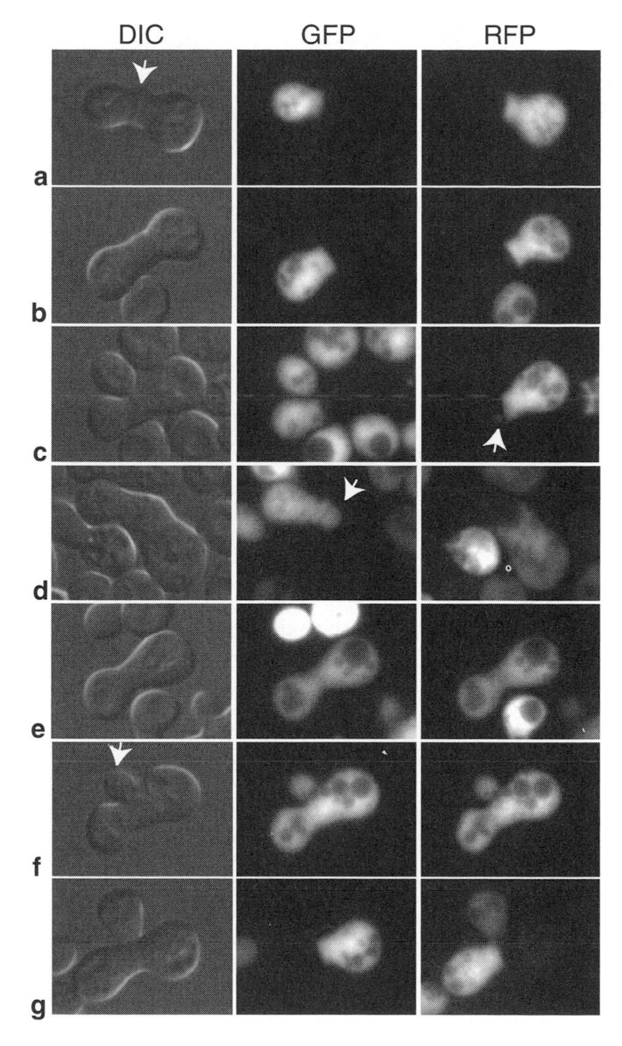

Fig. 4. Mating pairs between *MAT***a** GFP and *MATα RFP* cells. **(A)** An early prezygote. Green fluorescent protein and red fluorescent protein (RFP) remain within their parental cells, indicating that the plasma membranes have not fused. Cell wall, visible as a central septum in the differential interference contrast (DIC) image (arrow), prevents contact and fusion between the two plasma membranes and supports the plasma membranes to maintain a flat interface between the two cells. **(B)** An early prezygote with a cell wall remnant that is not visible in the DIC image. **(C)** A late prezygote with a small cytoplasmic finger extending from the *MATα RFP* cell (arrow). **(D)** A late prezygote with a large cytoplasmic finger extending from the *MAT***a** GFP cell (arrow). **(E)** A fused zygote. **(F)** A fused zygote with a small bud (arrow). **(G)** A *MATα RFP* cell starting to mate with the *MAT***a** GFP cell of an arrested mating pair.

permeance, a typical fusion pore between mating yeast cells opens abruptly to a diameter of approximately 30 nm and then gradually expands at a constant rate. In a few exceptional mating pairs, the rate of GFP transfer suddenly increases, suggesting the opening of a second fusion pore, or stops, suggesting that the fusion pore has closed. *fus1* mutant mating pairs have small, slowly expanding pores, suggesting that Fus1 regulates the fusion machinery ***(16)***.

Nuclear fusion (karyogamy) typically occurs approximately 15 min after plasma membrane fusion ***(17)***. Nuclear fusion can be monitored with fluorescent markers for chromosomal DNA and the nuclear envelope (**Fig. 5**). Karyogamy mutants complete cell wall remodeling and plasma membrane fusion, but not nuclear fusion, and accumulate a mating intermediate called a *cytoductant*, which has unified cytoplasm but two distinct nuclei. Once a cytoductant reenters the mitotic cell cycle, the resulting daughter cells have haploid nuclei but mitochondria from both parents, providing the basis for a genetic test described elsewhere ***(18)***. If the two nuclei of a cytoductant are in contact, the karyogamy defect lies in nuclear envelope fusion. However, if the two nuclei remain within their respective halves of the cytoductant, either microtubule-dependent transport or fusion pore expansion might be inhibited.

Vacuole fusion occurs very late in the mating process in the BY4741 genetic background. Vacuoles from the two parent cells are restricted to their respective halves of the zygote for over 1 h after plasma membrane fusion. Once the cell cycle resumes and a diploid bud starts to grow, the vacuoles suddenly and simultaneously stream into the bud, where they fuse and then distribute throughout the zygote. The signal to initiate movement may arise from binding of newly synthesized Vac17 to Vac8 on the vacuole membrane ***(19)***. Vacuole fusions can be monitored by labeling vacuoles in the two haploid cultures with different markers before they are mated. Vacuoles in the *MAT***a** GFP cells are labeled by endocytosis of FM4-64, and *MATα* vacuoles are labeled with CellTracker Blue ***(14,20)***. Fused vacuoles have FM4-64-stained membranes surrounding a CellTracker Blue-labeled core (**Fig. 6**).

2. Materials

2.1. Growing Yeast Cultures

2.1.1. Equipment

1. Waterbath set at 55°C to cool agar before poring plates.
2. Shaking incubators set at 250 RPM: 25°C, 30°C.
3. Culture tubes: 25 × 150 mm glass tubes with KAP-UTS (Bellco) lids. Autoclave.
4. Spectrophotometer (Visible).
5. Disposable microcuvettes.
6. Hemacytometer to calibrate the spectrophotometer.
7. Sterile pipettes (glass or plastic): 25 mL, 10 mL, 5 mL, 1 mL

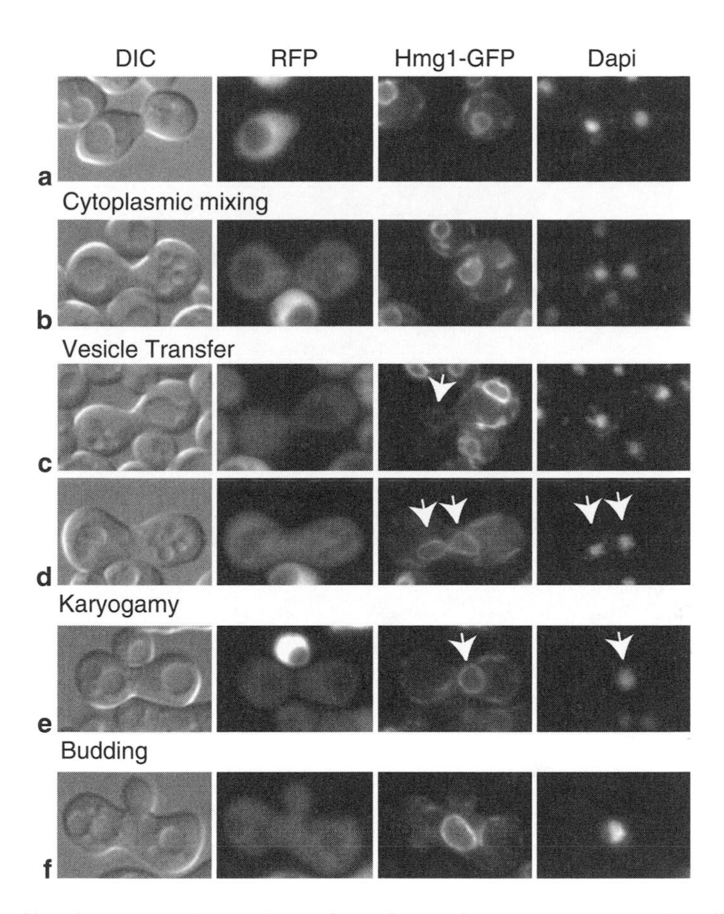

Fig. 5. Simultaneous observation of mating pair assembly, cytoplasmic mixing, small vesicle transfer, and karyogamy. *MAT***a** cells expressing Hmg1-GFP as a marker for the nuclear envelope and endoplasmic reticulum were mated to *MAT*α cells expressing cytoplasmic red fluorescent protein (RFP). The mating mix was stained with DAPI immediately before imaging. Differential interference contrast (DIC), RFP (red channel), Hmg1-GFP (green channel), and DAPI (blue channel) images are shown for each mating pair. **(A)** A prezygote, or unfused mating pair. Cytoplasmic RFP is restricted to the *MAT*α mating partner, and Hmg1-GFP is restricted to the *MAT***a** mating partner. **(B)** A fused mating pair with RFP filling the unified cytoplasm. Nuclei are aligned across the fusion site, but Hmg1-GFP remains restricted to the *MAT***a** cell. **(C)** A fused mating pair with aligned nuclei. A limited amount of Hmg1-GFP is visible in the nuclear envelope from the *MAT*α cell (arrow). Hmg1-GFP is likely to have transferred between cells in the membrane of small vesicles that budded off from the endoplasmic reticulum or nuclear envelope of the *MAT***a** cell and then fused with the nuclear envelope of the *MAT*α nucleus. **(D)** A fused mating pair where cytoplasmic RFP and nuclear envelope localized Hmg1-GFP have both diffused across the fusion pore. The two DAPI-stained nuclei are aligned but have not yet fused (arrows). **(E)** A fused zygote with a single fused nucleus (arrows). **(F)** A fused zygote that has completed karyogamy, reentered the cell cycle, and initiated bud formation. (Adapted from **ref. *16*.**)

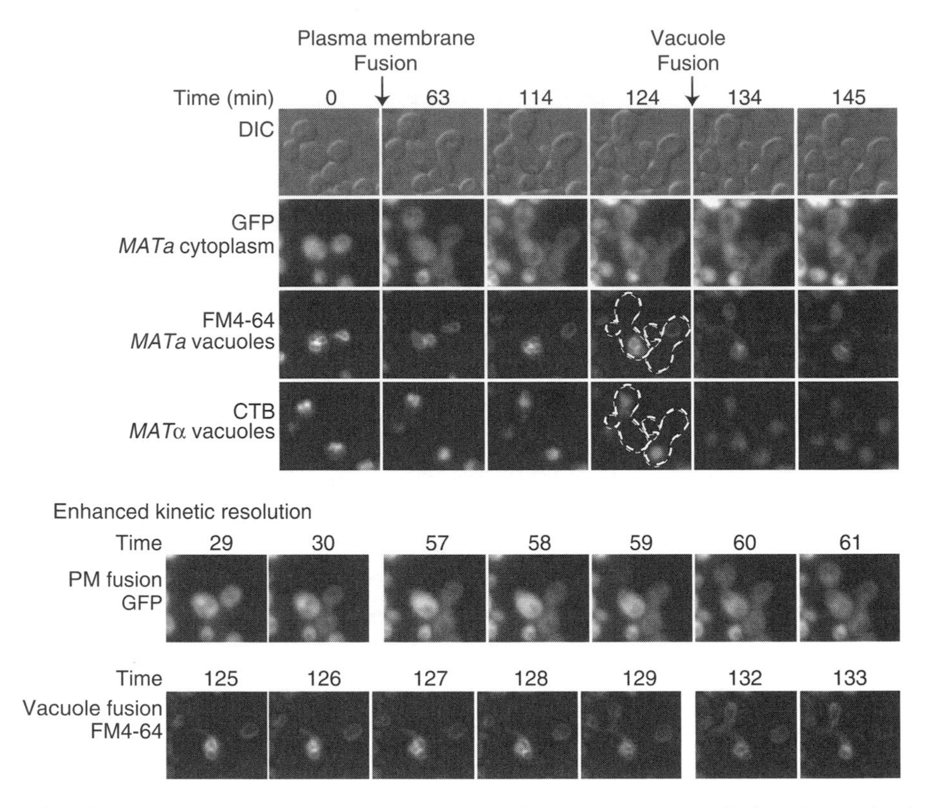

Fig. 6. Time-lapse imaging of cytoplasmic mixing and vacuole fusion. Vacuoles in *MAT***a** GFP cells were labeled by FM4-64 uptake. *MAT*α vacuoles were labeled with CellTracker Blue CMAC. The two populations of cells were mixed, collected on a filter, and incubated on an SC plate for 45 min. The mating mix was then transferred to a microscope slide with an agar pad for time-lapse imaging. Time 0 represents the start of image collection at approximately 60 min after the initiation of mating. Two adjacent mating pairs are shown in each image. The top panel presents differential interference contrast (DIC), green fluorescent protein (GFP), FM4-64, and CTB images collected at ~10-min intervals. The lower panels present GFP and FM4-64 images collected at 1-min time intervals to reveal the rate of GFP transfer between cells and the kinetics of vacuole migration into the bud and fusion. In the mating pair on the right, GFP transfer is completed within the 1-min interval from 29 to 30 min, and vacuole transport and fusion occur within the 1-min interval form 132 to 133 min. In the mating pair on the left, GFP requires 5 min to transfer between cells, indicating a small fusion pore, and vacuole fusion does not occur until 4 min after the FM4-64-labeled vacuoles from the *MAT***a** cell are transported toward the bud, suggesting that contact between vacuoles from the two halves of the zygote is still restricted by a small fusion pore.

8. Pipet-Aid or other mechanical pipetting device.
9. Pipetmen (or equivalent): 1000 mL, 200 mL, 20 mL.
10. Sterile pipette tips.
11. Refrigerator, 4°C.
12. Freezer, −80°C
13. Sterile 50% glycerol.
14. 2 mL cryovials.

2.1.2. Yeast Strains

1. BY4741—*MAT***a** *Δura3 Δleu2 Δhis3 Δmet15* (Open Biosystems).
2. BY4742—*MATα Δura3 Δleu2 Δhis3 Δlys2* (Open Biosystems).

2.1.3. Growth Media

1. YP: 10 g/L yeast extract, 20 g/L peptone. Autoclave.
2. 40% dextrose (glucose): Dissolve in H_2O. Autoclave.
3. YPD media: Add 5% glucose to YP immediately before use.
4. Petri dishes, 100 mm.
5. YPD plates: Mix 10 g yeast extract, 20 g peptone, 20 g agar in a 2-L flask containing 950 mL of H_2O and a large stir bar. Autoclave for 45 min with slow exhaust (liquid cycle). Cool to 55°C in a waterbath. Carefully add 50 mL of 40% dextrose, avoiding bubbles by pouring down the side of the flask. Mix slowly on a magnetic stir plate. Pour 30 mL into each 100-mm Petri dish.
6. Amino acid powder stock without methionine, lysine, and uracil (for 5 L): 0.1 g adenine, 0.1 g L-arginine HCl, 0.5 g L-aspartic acid, 0.5 g L-glutamic acid, 0.1 g L-histidine, 0.15 g L-isoleucine, 0.5 g L-leucine, 0.25 g L-phenylalanine, 2 g L-serine, 1 g L-threonine, 0.1 g L-tryptophan, 0.15 g L-tyrosine, 0.75 g L-valine. Store at room temperature with desiccant.
7. 100× amino acid stock solutions: 2 mg/mL L-methionine, 3 mg/mL L-lysine, 2 mg/mL uracil.
8. SC-MK plates: Autoclave 20 g of agar in a 2-L flask containing 800 mL of H_2O and a large stir bar. Cool to 55°C in a waterbath. Prepare a 5× SC-MK stock by mixing 6.7 g yeast nitrogen base (Difco), 20 g glucose, 0.63 g amino acid powder stock, and 10 mL 100× uracil into 200 mL H_2O. Filter sterilize. Warm to 55°C in a waterbath. Gently mix the 5× SC-MK stock into the agar. Pour 30 mL into each Petri dish.
9. SC-K plates: Same as above except add 10 mL 100× methionine and 10 mL of 100× uracil to the 5× SC-MKU stock.
10. SC-U plates: Same as above except add 10 mL 100× methionine and 10 mL 100× lysine to the 5× SC-MKU stock.

2.2. Filter Mating

1. Filters, 25 mm diameter, 0.45 mm pore, mixed cellulose esters (Millipore HAWP025000).
2. Filter forceps.

3. 1225 Sapling Vacuum Manifold (Millipore).
4. #6 rubber stoppers (×II).
5. Vacuum line and tubing.
6. Mixing/collection tubes: 17 × 100 mm polypropylene or polystyrene.
7. TAF buffer: 20 mm Tris-HCl (pH 7.4), 20 m*M* NaN_3, 20 m*M* NaF, store at 4°C.
8. 1.5-mL microcentrifuge tubes.
9. Microcentrifuge.

2.3. Genetic Complementation

1. Glass beads, 3 mm, autoclaved.
2. Hand tally counter.

2.4. Imaging

1. Microscope slides, standard 1 × 3 inch.
2. Coverslips, #1 thickness:

   a. 22 × 22 mm for standard images.
   b. 18 × 18 mm for time-lapse imaging.

3. Immersion oil.
4. Axioplan2 imaging microscope (Zeiss, Thornwood, NY).
5. HBO100 mercury arc lamp (Zeiss).
6. ebq isolated lamp power supply (Zeiss).
7. Plan-Apochromat 100× 1.4 Oil DIC objective lens (Zeiss).
8. Plan-Apochromat 63× 1.4 Oil DIC objective lens (Zeiss).
9. Tempcontrol-37 analog temperature controllers for stage and objective lenses (Zeiss).
10. Orca-ER camera and camera controller (Hamamatsu, Bridgewater, NJ).
11. Openlab with Automator microscope control software (Improvision, Waltham, MA).
12. Neutral density filter (UVND 1.3) for time-lapse imaging.
13. Filter sets for epifluorescence (*see* **Note 1**):

Green (GFP)	Excitation filter:	HQ470/40
(Chroma Endow GFP)	Dichroic mirror:	Q495 LP
	Emission filter:	HQ 525/50
Red (RFP, FM4-64)	Excitation filter:	HQ545/30
(Chroma TRITC)	Dichroic mirror:	Q570 LP
	Emission filter:	HQ620/60M
UV (DAPI, CellTracker)	Excitation filter:	G365
(Zeiss, set 49)	Dichroic mirror:	FT 395
	Emission filter:	BP 445/50

2.5. FM4-64 Staining

1. FM4-64 stock (Molecular Probes/Invitrogen, Carlsbad, CA): 8 m*M* (5 µg/µL, 100×). Store at –20°C.
2. 0.5-mL tubes or 96-well plate, round-bottom.

2.6. Green Fluorescent Protein × Red Fluorescent Protein Matings: Yeast Transformation

The yeast integrating plasmids listed below were constructed in a common vector. The vector contains a *GPD* promoter for strong constitutive expression and a *URA3* selectable marker. An integrating vector was chosen to ensure uniform staining within a clonal population of cells. Because BY4741 has a complete deletion of its *URA3* gene, an extra segment of chromosomal DNA from the *SSO1* gene was inserted into the vector to target integration to the *SSO1* locus. This segment includes 840 bp of coding sequence from the 3′ end of the *SSO1* gene and 453 bp of 3′ untranslated DNA and includes unique restriction enzyme recognition sequences for *Bgl*II and *Afl*II. After transformation and homologous recombination, functional and nonfunctional copies of *SSO1* surround the integrated DNA.

1. eGFP pEG220 ***(15)***.
2. DsRed pEG223 ***(15)***.
3. mCherry* pEG463 ***(16)***.
4. Gag-GFP pEG440 ***(16)***.
5. Hmg1-GFP pEG218 ***(16)***.
6. *Bgl*II restriction enzyme (NEB).
7. Restriction digest reaction buffer 3 (NEB).
8. 15-mL conical centrifuge tubes, sterile.
9. 1.5-mL microcentrifuge tubes, sterile.
10. Heating block set at 42°C.
11. Tabletop agitator (Vortex Genie2).
12. H_2O, sterile.
13. Salmon sperm DNA (10 mg/mL).
14. LTE: 100 m*M* lithium acetate, 10 m*M* Tris-HCl (pH 8.0), 1 m*M* EDTA. Sterile.
15. PEG/LTE: 35% (w/v) polyethylene glycol (3300 g/mol) in LTE, sterile.

2.7. Time-Lapse Imaging

1. Agarose (molecular biology grade).
2. VALAP: Add 10 g vaseline, 10 g lanolin, and 10 g paraffin to a 50-mL conical tube. Heat in a boiling water bath to dissolve. Transfer 10 mL to a 25-mL Pyrex beaker and reserve the rest. Place the end of a cotton-tipped applicator (Q-tip) into the VALAP and let the VALAP wet the applicator as it cools and solidifies. Solid VALAP can be stored at room temperature.

2.8. Permeance Calculations

1. Openlab (Improvision).
2. Excel (Microsoft).

---

* Use of pEG463 requires a material transfer agreement from R. Tsien, UCSD, for mCherry ***(21)***.

2.9. Karyogamy

1. DAPI stock: 1 mg/mL (400×).

2.10. Vacuole Fusion

1. CellTracker Blue CMAC (Molecular Probes/Invitrogen) 10 m*M* (10,000×) stock in DMSO.
2. Hepes/glucose buffer: 20 m*M* Hepes (pH 7.4), 1% glucose.

3. Methods

3.1. Growing Yeast Cultures

Yeast is grown in either rich medium (YPD) or synthetic medium (SC). Rich medium is for routine growth. Synthetic medium can be deficient in defined nutrients and is therefore useful to test for auxotrophies and as a selective medium for the genetic markers often found on plasmids. Yeast grown in synthetic medium also has less autofluorescence. Yeast grow well either on agar plates or in aerated liquid cultures. Growth of wild-type yeast is optimal at 30°C, but we routinely grow liquid cultures overnight at 25°C, because we occasionally use temperature-sensitive mutants.

The accepted practice among yeast researchers is to begin an experiment with a culture grown overnight to early log phase. Mating efficiency and kinetics are particularly sensitive to the growth state of the culture, so it is important to ensure that cultures are neither undergrown nor overgrown before they are mated. The density of a yeast culture is typically measured using a spectrophotometer. As a rule of thumb, an OD_{600} reading of 1 corresponds to 10^7 cells per mL (*see* **Note 2**). Early log phase YPD culture densities range from OD_{600} 0.1 to 0.8. Log phase cultures grown in synthetic media have an OD_{600} ranging from OD_{600} 0.1 to 0.4.

Yeast strains can be stored as colonies on a plate or as saturated (stationary phase) cultures. Yeast will survive in YPD media for several months at 4°C, but viability declines with a half-life of less than ~2 weeks in synthetic media at 4°C. For long-term storage, mix 1 mL of a saturated culture with 500 µL of 45% (w/v) glycerol (final 15% glycerol), and place in a −80°C freezer. Strains recovered from −80°C stocks should be struck to single colonies on a nutrient agar plate before use.

Although every strain has its own growth rate, a useful rule of thumb for setting up a log phase culture of a wild-type strain is to dilute cells from a saturated YPD culture 1:1000 into fresh YPD media and grow for 16 h at 25°C in a shaking incubator. A less dilute inoculation is required for slow-growing strains or when the starting culture was grown in synthetic medium. It is often advantageous to set up several overnight cultures with different initial cell densities so that at least one of the cultures will be in log phase growth the following morning. Undergrown cultures can be transferred to a 30°C shaker for more rapid growth or concentrated by centrifugation. Overgrown cultures with an OD_{600} < 2.0 in YPD or <1.0 in SC can be diluted 1:6 and grown for 3 h before use, although this practice is not optimal. As a control for sterility of the growth media, it is useful to prepare an extra culture tube with media but no yeast inoculation.

3.2. Filter Mating

This procedure has been optimized for maximum fusion efficiency. An equal number of cells of each mating type are collected onto a filter at approximately 25% confluence. The filter is then placed cell side up onto a nutrient agar plate.

3.2.1. Day 1

Inoculate individual *MAT***a** and *MATα* cultures in 5 mL of medium. Incubate overnight in a rotating shaker at 25°C. Rich YPD medium is preferred unless dropout medium is required to maintain an episomal plasmid.

3.2.2. Day 2

1. Check the density of the cultures by eye. Cultures that will overgrow in the next hour should be diluted with growth medium. Undergrown cultures can be transferred to a 30°C shaker for faster growth.
2. Prepare the filtration apparatus with one filter for each mating reaction. Filters are best handled with a Millipore filter forceps. Make certain that the O-rings are in the proper location to form an airtight seal. Cover unused filtration places with a #6 rubber stopper to prevent leakage of the vacuum.
3. Prepare a mixing tube with 5 mL of growth medium for each mating reaction.
4. Label filter locations on the back of a YPD plate (*see* **Note 3**). Up to eight filters will fit in a standard 10-cm Petri dish.
5. Measure the density of the cultures. Optimal mating requires early log phase cultures.
6. Calculate the volume equivalent to 10^6 cells for each culture.
7. Mix 10^6 *MAT***a** and 10^6 *MATα* cells in the appropriate tube for each mating reaction.
8. Transfer the mating mixtures to the filtration apparatus.
9. Apply a vacuum to collect the cells on the filters. Release the vacuum by removing one rubber stopper approximately 5 s after the medium has been aspirated through the filters (*see* **Note 4**). Remove the top plate from the filtration manifold.
10. Transfer each filter cell side up to a nutrient agar plate.
11. Incubate the mating reactions at 30°C (*see* **Note 5**). The incubation time depends on the goal of the experiment (*see* **Note 6**). For time-lapse experiments (*see* **Subheading 3.9**), cells are transferred to microscope slides after 45 min. At this time, most cells have assembled into mating pairs, but few have fused. Genetic complementation assays (*see* **Subheading 3.3**) are incubated for 90 min to allow mating pairs

to form and initiate fusion before being transferred to selective medium. Endpoint assays are typically stopped after 2.5 h to allow most mating pairs to fuse but minimize the number of diploid cells that have budded off from the fused zygotes.

12. During the 30°C incubation, prepare labeled collection tubes to recover the mating cells from the filters for the next step in the experiment. For genetic complementation assays and time-lapse experiments, the collection tubes will contain 1 mL of selective medium at room temperature. For endpoint experiments, cells are collected in 1 mL of TAF buffer at 4°C.
13. Stop the mating reactions by picking up each filter with a forceps and transferring it to a collection tube. The side of the filter with the mating pairs should face inward. Vortexing to wet the sides of the tube will make it easier to slide the filter down. The filter should rest flat on the side of the tube and contact the liquid at the bottom. Vortex vigorously to wash the mating pairs off of the filter.
14. For endpoint assays, transfer the TAF buffer with suspended yeast cells to a 1.5-mL snap cap tube. Mating reactions can now be stored on ice for up to 3 days before imaging. A limited amount of lysis occurs over time during storage.

3.3. Genetic Complementation Assay

1. Prepare a *sterile* filter mating reaction between *MAT***a** *met15* and *MAT*α *lys2* strains (*see* **Note 7**). A critical feature of this assay is that the mating products will be incubated for 3 days on plates with selective growth medium instead of being collected in ice cold TAF buffer. Thus, aseptic technique is essential to guard against contamination. Follow the method described in **Subheading 3.2.** with the following modifications:
   a. Use sterile culture tubes to mix the *MAT***a** and *MAT*α cells before filtering.
   b. Use autoclaved pipettes and pipette tips. Flame pipettes (if glass) before use.
   c. Wash the filtration apparatus with water and rinse with 70% ethanol before use. Allow to air dry before loading the filters.
   d. Sterilize the forceps by immersing it in 95% ethanol and flaming before each use.
   e. Incubate the mating reaction for 90 min at 30°C on an SC plate.
2. During the incubation, prepare sterile 15-mL tubes with 1 mL of SC medium lacking methionine (M) and lysine (K) to collect mated cells from the filters.
3. Prepare sterile 1.5-mL microcentrifuge tubes with 900 µL of SC-MK medium to dilute the cells to an appropriate concentration for mating. Then apply 10^6 cells of each mating type to the filter. It is convenient to spread 100 µL on a plate and to count 10^2 (100) cells. Thus, three 1:10 serial dilutions will be required for each mating reaction.
4. Collect the mated cells from the filters into 1 mL of SC-MK medium. *Do not use TAF buffer*, as this will kill the cells.
5. Serially dilute each sample three times by transferring 100 µL to the next 1.5-mL tube and mixing to yield 1:10, 1:100, and 1:1000 dilutions.
6. Plate 100 µL of the 1:1000 diluted cells on an SC-MK plate to select for diploids and on an SC-K plate to select for both MAT**a** haploids and mated diploids. If a mating frequency of less than 10% is anticipated, plate 100 µL of the 1:100

dilution on a second SC-MK plate. The 1:10 dilution and neet (undiluted) cells can also be plated on SC-MK if a mating frequency of <1% or <0.1% is anticipated. The diluted mating mixes must be spread evenly on the plates to have well-distributed colonies for counting. This can be done using autoclaved 3-mm glass beads. Pour approximately 10 beads on each plate, add the diluted mating mixture, and then roll the beads across and around the surface of the plate to spread out the cells. Once the liquid has been absorbed into the plates, the glass beads can be poured off and recycled.

7. Incubate the plates for 3 days at 30°C.
8. Count the number of colonies on each plate using a hand tally counter (*see* **Note 8**). Calculated the mating efficiency by dividing the number of colonies on an SC-MK plate by the number on the SC-M plate and correcting for the use of a less diluted sample if required.

3.4. Collecting Still Images of Yeast Mating Pairs

Mating reactions should remain in TAF buffer on ice until they are stained or transferred to a microscope slide.

1. Concentrate the mating reactions. Centrifuge for 10 s at low speed (2000 *g*) and then aspirate all but approximately 20 µL of the TAF buffer. Resuspend each mating mixture with a pipette tip or by vortexing.
2. Stain an aliquot of each mating reaction with FM4-64 (*see* **Subheading 3.5**) or DAPI (*see* **Subheading 3.7**) immediately before imaging.
3. Transfer 1.8 µL to a microscope slide. Gently place a coverslip on top (*see* **Note 9**).
4. Identify random fields with an appropriate cell density by DIC microscopy. Image each field in the DIC and fluorescent channels, as appropriate. We preset the imaging parameters (filter sets, exposure time, etc.) and then use an automated procedure to collect image sets and prepare composite images. Multiple fields should be imaged from each slide to ensure that there are a sufficient number of mating pairs for quantitative analysis.

3.5. Staining Yeast Mating Pairs With FM4-64

1. Prepare a 2× working solution of FM6-64 by diluting the 8 mg/mL stock 1:50 in TAF buffer.
2. Prepare 1.8 µL aliquots of the FM4-64 working solution in 0.5 mL tubes or in a round bottom 96-well plate. Store on ice.
3. Add 1.8 µL of a concentrated mating reaction, mix with the pipette tip, and immediately transfer 1.8 µL to a microscope slide for imaging.
4. Image with a red filter set. See **Figure 3** for examples of FM4-64-labeled mating pairs at various stages of mating (*see also* **Note 10**).

3.6. Green Fluorescent Protein × Red Fluorescent Protein Mating Assay

We have designed GFP and RFP yeast expression vectors that provide bright uniform cytoplasmic staining (*see* **Note 11**). We conventionally express GFP in *MAT***a** cells and RFP in *MAT*α cells. To score mating efficiency and identify the distribution of arrested mating intermediates, *MAT***a** *GFP* and *MAT*α *RFP* strains are mated on filters as described in **Subheading 3.2** and collected in TAF buffer. Strains expressing genetically encoded fluorescent proteins are also used in the time-lapse mating assays described in **Subheading 3.9**.

3.6.1. Yeast Transformation

The following procedure for expressing eGFP in *MAT***a** *ura3* yeast cells is based on the standard yeast transformation procedure developed by Geitz ***(22)***. Higher efficiency methods are also available ***(23)***.

3.6.1.1. Day 1

1. Prepare an *Afl*II or *Bgl*II digest of the desired plasmid. To express cytoplasmic eGFP, use pEG220.
   a. Mix 5 μL pEG220 (0.5–1.0 mg/mL), 1.1 μL 10× NEB Rxn buffer 3, 0.5 μL *Bgl*II, and H_2O to 11 μL.
   b. Incubate at 37°C for 1.5 h.
   c. Run 1 μL of the digest and 0.5 μL of uncut pEG220 on an 0.8% Agarose gel to confirm that a unique site was cut to linearize the DNA.
   d. Store the digest at −20°C.
2. Inoculate a 10-mL culture of BY4741 yeast in 10 mL of sterile YPD. Incubate for ~16 h at 250 RPM, 25°C, in a shaking incubator.

3.6.1.2. Day 2

1. Place the salmon sperm DNA in a boiling waterbath for 20 min to denature. Transfer to an ice water bath for 10 min. Store on ice.
2. Measure the cell density with a spectrophotometer. Cells in late log phase, $OD_{600\,nm} = 0.9 - 1.2$ (~10^7cells/mL) are best for transformation (*see* **Note 2**).
3. Transfer the cells to a 15-mL conical centrifuge tube.
4. Note the cell density and volume. Calculate the total number of cells and the volume required for a suspension of 5×10^8 cells/mL; 200 μl is appropriate for a 10-mL culture at $OD_{600\,nm} = 1.0$.
5. Collect the cells by centrifugation for 5 min at 2500 *g*. Aspirate the supernatant with a sterile pipette tip.
6. Wash the cell pellet in 10 mL sterile H_2O. Collect the cells by centrifugation for 5 min at 2500 *g*. Aspirate the supernatant with a sterile pipette tip.
7. Resuspend the washed cells in 600 μL LTE. Transfer to a 1.5-mL microfuge tube. Collect the cells by centrifugation for 10 s in a microcentrifuge. Aspirate the supernatant with a sterile pipette tip.
8. Resuspend the competent cells in LTE at a density of 5×10^8 cells/mL. Competent cells can be used immediately or stored on ice for up to 12 h.
9. Prepare a transformation reaction containing

a. 100 μL competent cells.
b. 10 μL boiled salmon sperm DNA.
c. 10 μL *Bgl*II digested eGFP DNA.

10. Mix and then add 700 μL PEG/LTE.
11. Incubate at 30°C for 45 min. Invert the tube after 20 min to mix.
12. Transfer to the 42°C heating block. Incubate for 8 min.
13. Collect the cells by centrifugation for 15 s in a microfuge. Aspirate the supernatant with a sterile pipette tip.
14. Wash the cell pellet with 800 μL sterile H_2O. It is not necessary resuspend the pellet. Aspirate the supernatant with a sterile pipette tip.
15. Resuspend the pellet in 100 μL of sterile H_2O.
16. Spread the cells on an SC-uracil plate.
17. Incubate for 2 days at 30°C.

3.6.1.3. Days 4–5

1. Several medium-sized colonies should be visible on the SC-uracil plate.
2. Choose three colonies and streak them on a second SC-uracil plate to purify the clones (*see* **Note 12**).

3.6.1.4. Days 6–8

1. Transfer one isolated colony from each uracil-positive clone to a culture tube containing 5 mL of sterile YPD.
2. Inoculate a control 5-mL YPD culture with untransformed cells.
3. Grow overnight in a shaking incubator at 25°C.

3.6.1.5. One Day Later: Identify and Store GFP-Positive Clones

1. Transfer 100 μL from each culture to a 1.5-mL microfuge tube. Centrifuge at 2500 *g* for 10 s to collect the cells. Aspirate all but 5 μL of the supernatant. Vortex to resuspend the pellet.
2. Transfer 1.8 μL to a microscope slide. Gently overlay a coverslip.
3. Test for GFP expression by viewing the cells by epifluorescence microscopy. Select a clone in which all cells in the population have bright cytoplasmic fluorescence (*see* **Note 13**).
4. Continue incubating the YPD culture at 25°C until it grows to saturation at the end of the day. Stationary YPD cultures are stable at 4°C for 2 months.
5. For longer term archival storage, mix 1 mL of stationary culture with 500 μL of 50% glycerol in a 2-mL cryovial. Store at −80°C.

3.6.2. Scoring the Products of a MAT **a** GFP x MATα RFP Mating

MAT **a** *GFP* cells are mated to *MATα RFP* cells, and microscopic images are collected as described in **Subheadings 3.2** and **3.3**. For an unbiased analysis, image fields must be selected randomly, and all mating pairs within a field should be scored. Analysis of 200–400 mating pairs is usually sufficient to yield a statistically significant result comparing the percentages of fused pairs, early prezygotes, late prezygotes, and lysed pairs. Examples of fused and lysed mating pairs as well as early and late prezygotes are shown in **Figure 4**. It saves time to count the fused mating pairs first, as they are easily identified. Lysed mating pairs are identified by a combination of factors, including shriveled vacuoles in the DIC image and loss of cytoplasmic GFP (*see* **Note 14**). Careful attention should be paid to the distinction between early and late prezygotes. Cytoplasmic fingers indicating plasma membrane contact can be quite small and are best identified by examination of the individual GFP and RFP grayscale images. The DIC image is best for determining whether two adjacent cells with different mating types (colors) are part of a mating pair or if they simply happen to be close to each other on the slide.

3.7. Karyogamy

For this assay, *MAT***a** cells transformed with *HMG1-GFP* are mated to *MATα RFP* cells (*see* **Fig. 6**). Hmg1-GFP marks the endoplasmic reticulum, which is contiguous with the nuclear envelope. RFP transfer marks the time of plasma membrane fusion. Nuclear DNA is stained with DAPI. The completion of karyogamy is scored by the presence of a single DAPI-stained nucleus surrounded by an Hmg1-GFP-labeled nuclear envelope. A fraction of the Hmg1-GFP escapes the endoplasmic reticulum in COPII vesicles and is then recycled back. This vesicle transport pathway allows Hmg1-GFP to transfer between the two nuclei of a fused mating pair prior to karyogamy.

1. Prepare a filter mating reaction between *MAT***a** HMG1-GFP cells and *MATα RFP* cells as described in **Subheading 3.2**. Collect the mating reaction in TAF buffer and concentrate to 20 μL.
2. Prepare 200 mL of a 2× DAPI working stock in TAF buffer.
3. Mix 1.8 μL of the mated cell suspension with 1.8 μL DAPI working stock (*see* **Note 15**). Transfer 1.8 μL to a microscope slide and apply a coverslip.
4. Collect images in the following order: DIC, DAPI, RFP, GFP (*see* **Note 16**).
5. Prepare GFP + RFP + DAPI composite images.
6. Score for nuclear congression and fusion within the population of fused mating pairs identified by cytoplasmic RFP transfer as illustrated in **Figure 5**.

3.8. Vacuole Fusion

To follow vacuole movement and fusion, vacuoles in *MAT***a** GFP cells and standard *MATα* cells are separately labeled before mating with FM4-64 and CellTracker Blue (*see* **Fig. 6**). Vacuole fusion can be monitored with still images collected at different time points or by time-lapse imaging as described in **Subheading 3.9**. The number of CellTracker Blue images that can be collected in a time-lapse series is limited because ultraviolet excitation is detrimental to cell fusion.

3.8.1. Labeling Vacuoles With FM4-64

1. Prepare a log phase culture of *MAT***a** GFP cells grown in YPD media (*see* **Note 17**).
2. Prepare an FM4-64–labeling mix by diluting 0.8 μL 100× FM4-64 stock into 800 μL of YPD medium.
3. Pellet 3 × 10^6 cells in a 15 mL round bottom tube by centrifugation at 3000 *g* for 3 min.
4. Aspirate the YPD and resuspend in 600 μL of YPD/FM4-64.
5. Incubate at 30°C in a shaking incubator for 30 min. The *MATα* cells can be labeled with CellTracker Blue during this incubation.
6. Transfer the suspension of *MAT***a** GFP cells to a 1.5-mL tube.
7. Wash the cells with YPD medium. Pellet the cells in a personal microfuge for 10 s. Aspirate the supernatant. Resuspend the cell pellet in 700 μL of room temperature YPD. Repeat three times (*see* **Note 18**).

3.8.2. Labeling Vacuoles With CellTracker Blue

1. Prepare a log phase culture of *MATα* cells grown in YPD media.
2. Prepare a 1 μ*M* solution of CellTracker Blue CMAC in Hepes/glucose buffer.
3. Transfer 3 × 10^6 cells to a 1.5-mL tube.
4. Pellet the cells by centrifugation for 10 s in a personal microfuge and aspirate off the growth medium.
5. Wash the pellet with Hepes/glucose buffer. Pellet the cells. Aspirate off the wash buffer.
6. Resuspend the pellet in 700 μL CellTracker Blue solution.
7. Incubate for 5 min at room temperature.
8. Wash once with 700 μl of YPD
9. Resuspend the pellet in 700 μL YPD.

3.8.3. Mating

1. Transfer the two populations of labeled cells to disposable microcuvettes. Measure cell density in a spectrophotometer.
2. Mix 1 × 10^6 cells from each microcuvette into 5 mL of YPD. Prepare a filter mating as described in **Subheading 3.2**.
3. For still images, collect the mating reactions in TAF buffer.
4. Acquire images in the following order: DIC, CellTracker Blue, FM4-64, GFP (*see* **Note 16**).
5. Prepare GFP + CellTracker Blue + FM4-64 composite images.

3.9. Time-Lapse Imaging of Yeast Mating Pairs

Time-lapse imaging has been used for a variety of purposes in yeast cell fusion research. Time-lapse microscopy played an integral role in establishing the relative timing of cytoplasmic mixing and vacuole fusion (*see* **Fig. 6**). In the analysis of plasma membrane fusion in wild-type and *prm1* mutant mating pairs, time-lapse microscopy was used to distinguish cell fusion intermediates from off-pathway end products. For example, small cytoplasmic fingers were observed before fusion in 10% of wild-type mating pairs and before virtually all contact-dependent lysis events in the *prm1* mutant. In contrast, fusion and lysis rarely occurred after a cytoplasmic finger grew to more than 1 μm ***(15)***. In the analysis of fusion pore opening and expansion, time-lapse microscopy provided high-resolution kinetic data for measurements of GFP transfer between cells ***(16)***.

Yeast mating is not a synchronous process, and there are currently no controlled methods to reversibly accumulate mating intermediates. Thus, the best way to capture fusions is to choose an appropriate time interval and select a microscopic field with a large number of potential mating pairs. The time interval from when *MAT***a** and *MAT*α cells are first mixed until the initiation of plasma membrane fusion ranges from 1 to 2.5 h and is influenced by asynchronous growth in the starting cultures (with respect to the cell cycle) and by the local cell density of potential mating partners and competitors. In addition, mutations such as *fig1* and *fus1* can delay fusion ***(16,24)***.

3.9.1. Sample Preparation

1. Turn on the mercury arc lamp and microscope at least 1 h before starting the mating reaction. Set the stage and objective temperature controllers to 30°C.
2. Prepare a filter mating reaction as described in **Subheading 3.2** (*see* **Note 19**).
3. Incubate for 45 min at 30°C.
4. During the incubation period, melt the VALAP on a hotplate set to low heat (#2; *see* **Note 20**).
5. Prepare a slide with a 1-mm-thick SC agarose pad as illustrated in **Figure 7**.
   a. Place approximately six layers of labeling tape (1-mm thick) on the surface of a glass microscope slide. Repeat with a second slide. Arrange the two taped slides on either side of a third slide on a flat surface.
   b. Mix 0.15 g of agarose into 5 mL SC medium to prepare a 3% solution. Place the mixture in a boiling waterbath to dissolve the agarose (*see* **Note 21**).
   c. Transfer 600 μL of the SC-agarose solution to the center of the middle glass slide. Use a 1-mL pipette tip with its end cut off to transfer this viscous solution. Immediately but carefully place a fourth slide across the other three. The two ends of the fourth slide should rest on the taped surface, and the SC-agarose should be sandwiched between the two glass surfaces.
6. Collect the mating cells. Transfer the filter with the mating reaction into a collection tube containing 1 mL of SC medium. Mix vigorously to wash the cells off of the filter.
7. Concentrate the cells. Transfer the 1-mL cell suspension to a 1.5-mL microcentrifuge tube. Pellet the cells by low speed centrifugation (2000 *g*) for 10 s. Remove all but approximately 10 μL of the SC medium. Aggressively resuspend the pellet with a pipette tip.

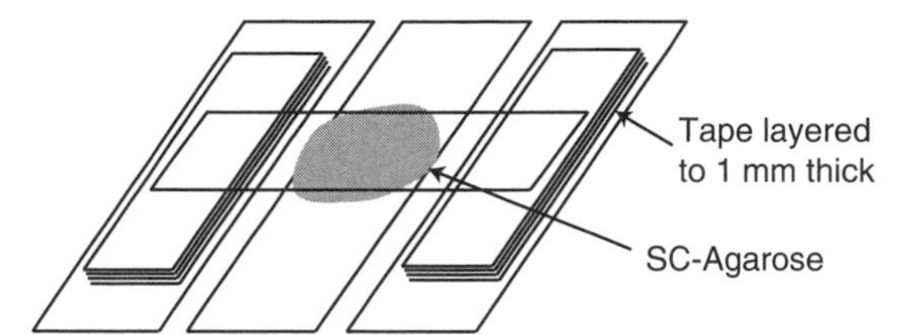

Fig. 7. Illustration of the method for producing a microscope slide with a 1-mm-thick SC-agarose pad with a flat surface.

8. Gently separate the two glass slides of the SC agarose sandwich. The SC-agarose pad will adhere to one of the slides. There should not be any air bubbles between the SC-agarose pad and the slide.
9. Transfer 1.5 µL of the cell suspension to the top of the SC-agarose pad.
10. Gently apply an 18-mm^2 coverslip. The cells should spread out to an even layer.
11. Trim away the excess SC-agarose with a razor blade.
12. Seal the slide by coating the edges of the SC-agarose pad and coverslip with VALAP using the cotton-tipped applicator.

The slide is ready for imaging once the VALAP solidifies.

3.9.2. Imaging

The quality of a series of time-lapse images is limited by the brightness and photostability of the fluorophores and by the potential of intense illumination to inhibit biological processes (*see* **Note 22**). Among the fluorescent markers described above, cytoplasmic GFP provides the best combination of brightness and photostability. Cytoplasmic DsRed is also reasonably bright and photostable, but its utility is limited by the large variability in fluorescence intensity among cells in a clonal population, which is related to the slow folding and maturation rate of the fluorophore. At an intermediate exposure time, some cells are too dim while others are too bright, which reduces the number of mating pairs in a field that produce useful data. mCherry has uniform and reasonably bright fluorescence, but it fades faster than GFP. FM4-64 has extremely bright fluorescence, but it also fades faster than GFP. The blue-emitting fluorophores DAPI and CTB are the least useful in time-lapse imaging because the ultraviolet excitation light causes photodamage to the cells that inhibits cell fusion.

Given the limitations of available probes, careful choices must be made to define imaging parameters that are appropriate for the experimental goals. The parameters for four different imaging protocols are presented in **Table 1** (*see* **Note 23**).

For the first protocol, the goal is to detect the cytoplasmic fingers that appear before plasma membrane fusion or lysis. Because cytoplasmic fingers can

Table 1
Parameters of Four Imaging Protocols

	Fusion/Lysis	Permeance	Permeance	Vacuole fusion
	a *prm1* GFP α *prm1* mCherry	**a** GFP α	Extended kinetics a GAG-GFP α mCherry	**a** GFP + FM4-64 α + CTB
Lens/ binning	100×/1×	63×/2×	63×/2×	63×/2×
Total time	5 min	15 min	1 h	2 h
DIC interval	2.5 s	30 s	30 s	1 min
GFP interval	2.5 s	2 s	15 s	1 min
Red interval	2.5 s	n/a	5 s for 15 min	1 min
Blue interval	n/a	n/a	n/a	10 min

extend from either cell in a mating pair, GFP images of the *MAT***a** cytoplasm and mCherry images of the *MAT*α cytoplasm are both required. The DIC images are also collected to monitor contact-dependent lysis. Because cytoplasmic fingers can be transient structures that extend less than 0.5 μm past the original boundary between cells, a 100× objective lens is used to maximize spatial resolution, and the interval between successive images is as short as possible (*see* **Note 24**). The drawback of this high spatial and temporal resolution is that the GFP and mCherry signals fade quickly. A partial solution to this problem is to reduce the initial exposure time so that the brightest pixels in the GFP and mCherry images are at 50% of saturation and to gradually increase the exposure times for later images in each series as the fluorescence emission declines. Although the total imaging time is limited to 5 min, multiple fields can be imaged on each slide during the period when the cells are actively mating.

In the second protocol, the goal is to measure the total amount of GFP in each cell of a mating pair in order to calculate pore permeance. Spatial resolution is not critical, so a 63× objective lens was selected and 2 × 2 binning is used to combine the light from four pixels on the CCD chip into one pixel in the image in order to reduce the required exposure time. These compromises in spatial resolution allow 450 GFP images to be collected at 2-s intervals over a 15-min period, thereby providing *(1)* sufficient kinetic resolution to calculate the permeance of a large pore and *(2)* a relatively long time interval for GFP to diffuse to equilibrium through small, late opening pores, which is useful to minimize the number of pores that cannot be analyzed because of incomplete GFP transfer.

In the third protocol, the goal is to combine precise measurements of the pore opening time, initial permeance, and permeance expansion rate using mCherry, with observations of the later stages of pore expansion using Gag-GFP. mCherry fades faster than GFP, so images are collected less frequently and for a shorter time. Because Gag-GFP diffuses more slowly through fusion pores, Gag-GFP images are collected less frequently over a longer time interval.

In the final protocol, the goal is to measure the relative times of plasma membrane fusion and vacuole fusion. There is a >1-h lag between these events, so images are collected at relatively long intervals. A major limitation is that the ultraviolet light used to excite CTB inhibits cell fusion. Therefore, the frequency of CTB images is limited to once per 10 min.

Focal drift caused by thermal fluctuations and shrinkage of the agarose pad over time is a serious problem that can impact image quality and the resulting permeance calculations. Thermal fluctuations can be minimized by allowing the imaging system to warm up and equilibrate for at least 2 h before use. In addition, images should be monitored as they are collected and minor focal adjustments made when necessary. The required focal adjustments are often predictable and can usually be made without interrupting the data collection.

3.10. Fusion Pore Permeance Measurements

The permeance calculation is based on Fick's law of diffusion, which states that the rate at which GFP moves through a pore depends on the permeance of the pore and the difference between the GFP concentrations of the two cells (*see* **Note 25**). Permeance equals the area divided by the length of the pore. Because GFP concentrations cannot be conveniently measured, Fick's equation was solved by integration to express permeance as a function of the intensity of GFP fluorescent over time and the cytoplasmic volumes of the two mating cells ***(16)***.

$$P(t) = -\frac{V_D V_R}{V_D + V_R}\frac{d}{dt}\ln\left[I_D(t) - I_R(t)\,{}^{V_D}\!/_{V_R}\right]$$

where P is permeance, t is time, V is cytoplasmic volume, I is GFP fluorescent intensity, D is GFP donor (*MAT***a**) cell, and R is GFP recipient (*MAT*α) cell.

Cytoplasmic volumes are measured in two separate ways. The $\frac{V_D V_R}{V_D + V_R}$ factor is calculated from microscopic measurements of cellular dimensions. The V_D/V_R factor is calculated from postfusion GFP intensities, because the concentration of cytoplasmic GFP in the two cells is equal at equilibrium.

1. Scan through a time-lapse series of images to identify all mating pairs that fuse (*see* **Note 26**). Exclude pairs if GFP starts to diffuse before the first image, if GFP has not diffused to equilibrium by the last image, or if there is a major change in the focal plane during the time interval when GFP is transferring among cells. Label the remaining mating pairs that fuse.

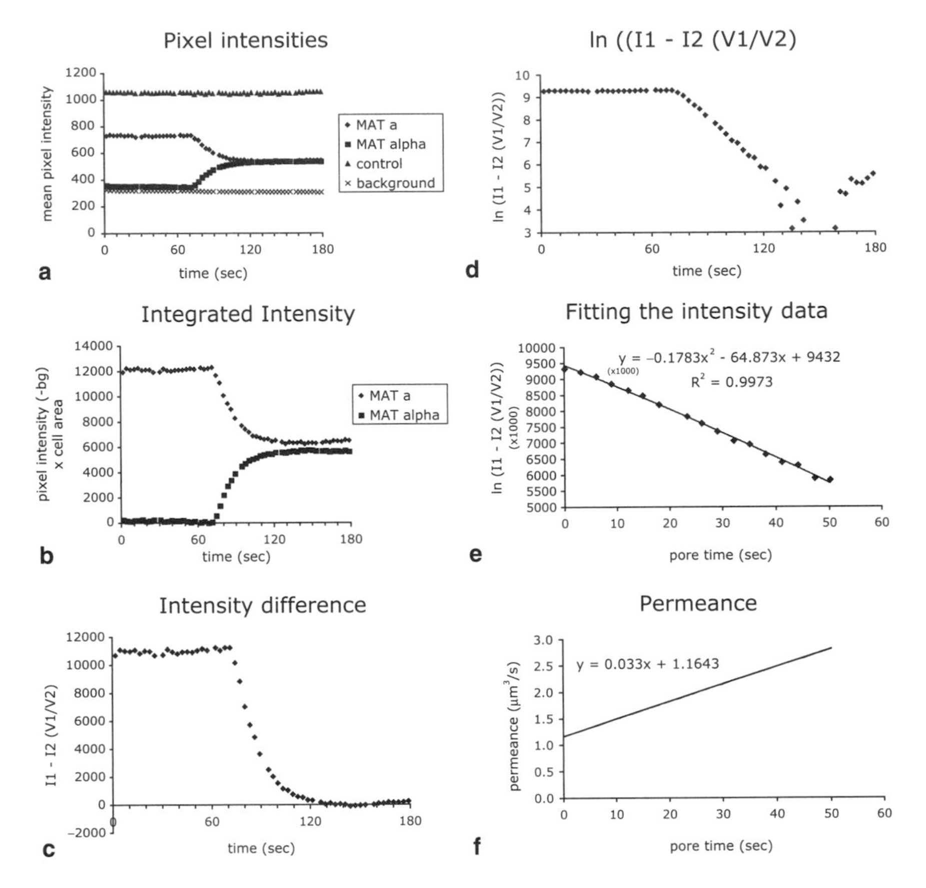

Fig. 8. Permeance calculation for a typical fusion pore. The permeance equation $P(t) = -\left(\frac{V_D V_R}{V_D + V_R}\right) d\,\mathrm{In}\left[I_D(t) - I_R(t)\left(\frac{V_D}{V_R}\right)\right] \Big/ dt$ was derived from Fick's law of diffusion. **(A)** Fluorescence intensity measurements. A series of fluorescent images of a field of mating *MAT***a** green fluorescent protein (GFP) and *MAT*α cells was collected at 2.5-s intervals. Boundaries were drawn around the two cells of a mating pair, and the mean fluorescence intensity in each cell was measured for each image. To control for photobleaching and other extrinsic factors, fluorescence intensity was also measured for a set of adjacent *MAT***a** GFP cells that did not fuse and for a background region. **(B)** The raw fluorescence intensity data were corrected for background fluorescence and photobleaching and then multiplied by the area of the cell. **(C)** The volume adjusted intensity difference between the cells approaches 0 as GFP diffuses to equilibrium. **(D)** The natural log of the volume adjusted intensity difference was calculated to fit the form of the permeance equation shown above. **(E)** A second order polynomial equation was fit to the logarithmic data for the interval corresponding to the start of GFP movement until GFP approached equilibrium. **(F)** The first differential of the fitted curve was multiplied by a volume constant to yield permeance in $\mu m^3/s$. (Adapted from **ref. *16*.**)

2. For each pair, note the start time when GFP is first detected in the *MAT*α cell and the end time when GFP appears to have diffused to equilibrium. Select a time interval for analysis that includes at least 5 images before the start time and at least 10 images after the end time.
3. At the start time, measure the length and width of the two cells in micrometers. Scroll through images from the next ~20 time points and estimate whether the vacuoles in each cell, which exclude GFP, occupy 15%, 25%, or 35% of the total cellular volume (*see* **Note 27**). Use these measurements to calculate the cytoplasmic volumes (in cubic micrometers) of the GFP donor (*MAT***a**) and acceptor (*MAT*α) cells according to the following formula.

$$V = \frac{\Pi}{6} 1ww \left(\frac{100\% - Vac\%}{100\%} \right)$$

4. Draw boundaries surrounding the two cells in the mating pair. These boundaries can include unoccupied space outside of the cell pair but should exclude adjacent cells with GFP fluorescence. Verify for each image during the GFP-transfer interval that the two cells remain within their boundaries and that adjacent GFP cells do not enter the region of interest. Measure the area, in micrometers squared, defined by each boundary.
5. Draw two additional boundaries, one surrounding a region with no GFP cells and a second surrounding adjacent GFP cells that do not fuse during the GFP diffusion interval. Measurements from the empty region are used for background subtraction, and measurements of the stable GFP cells are used to correct for photobleaching and changes in exposure times and lamp intensity. Measure the area in micrometers squared defined by each boundary.
6. Measure the mean GFP fluorescence within the four boundaries at each time point (**Fig. 8A**). We use an Openlab™ Automation for this task. Transfer the data to a Microsoft™ Excel spreadsheet for subsequent calculations. The time measurements should be converted into seconds and hundredths of seconds.
7. Subtract the mean background fluorescence from the mean fluorescence of the GFP donor, GFP recipient, and stable GFP regions (*see* **Note 28**).
8. Calculate an intensity correction ratio for each time point using data from the stable GFP region. Divide the background-subtracted mean fluorescence at the start time by the background-subtracted mean fluorescence at each time point.
9. At each time point, multiply the background-subtracted mean GFP intensities of the GFP donor and GFP recipient regions by the corresponding intensity correction ratio to yield normalized mean GFP intensities.
10. Multiply the normalized mean GFP intensities of the GFP donor and GFP recipient cells at each time point by the corresponding area of the measured region to yield the integrated GFP intensity, which represents the total GFP present in each cell. Add the GFP donor and GFP recipient intensities to calculate the total GFP intensity in the cell pair.
11. Plot the donor, recipient, and total integrated GFP intensities as a function of time (**Fig. 8B**). Confirm that GFP has diffused to equilibrium during the measured time

interval and that the total GFP remains constant before, during, and after fusion (*see* **Note 29**).

12. Calculate V_D/V_R from the I_D/I_R ratio at equilibrium. At equilibrium, the GFP concentration (I/V) is equal in the two cells. If $I_D/V_D = I_R/V_R$, then $V_D/V_R = I_D/I_R$.
13. Plot $I_D(t) - I_R(t)\ V_D/V_R$ versus time (**Fig. 8C**). This plot should approach 0 at equilibrium. Select a time interval for the permeance calculation ranging from the start time until GFP has diffused to within 5% of equilibrium. At later time points, the data represent a small difference between two large numbers and are too noisy to be useful.
14. Plot $\ln\left[I_D(t) = I_R(t)\ V_D/V_R\right]$ versus time (**Fig. 8D**).
15. Define the start time as t = 0. Fit the $\ln\left[I_D(t) = I_R(t)\ V_D/V_R\right]$ versus time plot with a second order polynomial equation $\langle y(t) = at^2 + bt + c\rangle$ over the time interval selected above (*see* **Notes 30** and **31** and **Fig. 5E**).
16. Calculate the first differential of the resulting equation $\langle y(t) = 2at + b\rangle$. Multiply by $\frac{V_D V_R}{V_D + V_R}$ to produce $P(t)$. $P(t)$ will be a first order equation with an initial permeance $\left(-b\frac{V_D V_R}{V_D + V_R}\right)$ and a linear rate $\left(-2a\frac{V_D V_R}{V_D + V_R}\right)$ of permeance increase (**Fig. 8F**).
17. Repeat steps 2 through 16 for all mating pairs in the field. Calculate the mean initial permeance and the mean rate of permeance increase.

4. Notes

1. Bandpass filters are used for fluorescent imaging to avoid crosstalk and minimize photodamage.
2. Because spectrophotometers are calibrated to measure absorbance rather than light scattering, the actual conversion from OD_{600} to cell density depends on the light path of the spectrophotometer and should be calibrated using a hemacytometer.
3. The nutrient composition of the media will influence the efficiency of mating. We typically use YPD plates. Although yeast are modestly autofluorescent when grown in YPD, this does not interfere with the ability to detect the bright fluorescence of GFP, DsRed, mCherry, FM4-64, or DAPI. Synthetic media can also be used. The plasma membrane fusion defect of the *erg6* mutant is enhanced on SC plates.
4. It is useful to have a light vacuum remaining when disassembling the filtration apparatus to ensure that the filters remain seated on the support and do not stick to the O-rings on the lid.
5. Increasing the temperature from 25°C to 30°C increases the rates of growth and mating. Mating is inefficient at temperatures above 34°C because of instability of the **a**-factor mating pheromone.

6. In a wild-type mating reaction at 30°C, cytoplasmic mixing is first detected after a lag of 45 min. Eighty percent of the cells in a wild-type mating mixture will fuse in 2.5 h. At this time, many of the fused zygotes will have a large bud, but few buds have separated from the mother cell. The unfused cells probably did not have an available mating partner.
7. Other common genetic markers such as the *ura3*, *leu2*, and *his3* auxotrophies and the G418-resistance cassette KanMX can also be used to select for diploids. One consideration regarding the choice of genetic markers is that yeast cells can often continue mating for some time after being transferred to a plate lacking an essential nutrient or supplemented with a drug. For example, BY4741 cells grown in rich YPD medium store sufficient methionine to continue growing for up to three cell divisions after transfer to SC-methionine plates. Less growth on methionine-free medium is found if the cells are precultured in medium with limited (3 μg/mL) methionine. Transfer from YPD to media lacking uracil, leucine, histidine, or lysine results in a cessation of growth after less than one to two cell divisions for the *ura3*, *leu2*, *his3,* and *lys2* mutants, respectively. Continued growth after transfer to selective media is a critical parameter to consider for kinetic assays. Of course, the considerable dilution that occurs before plating on selective media ensures that only mating pairs assembled before dilution have an opportunity to complete mating.
8. To avoid counting the same colony twice, mark its location on the bottom of the plate with a washable marker while counting.
9. The cells should spread out and form a monolayer between the slide and coverslip. The cell density often varies across the slide, so it is useful to screen for a field of appropriate density by DIC microscopy. In the microscope, cells often appear to float or stream within the slide. This will stop within 1 min if the cell suspension has spread out properly. If not, another slide should be prepared. It is not necessary to seal the slide. However, slides should be imaged within 10 min after preparation (before they dry out) and then discarded.
10. FM4-64–stained cells are particularly sensitive to drying out or being smashed between the slide and coverslip. Damaged cells have a punctuate appearance. It is usually possible to find a large region of the slide with rim-stained cells.
11. The GFP plasmid pEG220 contains yeast optimized enhanced GFP ***(25)***. Two RFP plasmids were constructed. One contains mCherry, a rapidly folding monomeric red fluorescent protein ***(21)***. The second contains DsRed. DsRed is a tetramer, so it diffuses more slowly than GFP and mCherry ***(16)***. DsRed fluorescence is less uniform because the half-time for in vitro maturation of the fluorophore is 24 h. GAG-GFP encodes a fluorescent subunit of the L-a virus core, which assembles into a 39 nm icosahedral viral particle ***(16,26)***.
12. Once a uracil-positive clone has been isolated, it is no longer necessary to grow in SC-uracil medium to maintain the transformed plasmid because the GFP gene has integrated into the genome.
13. It is not uncommon for all three clones to be identical.
14. Protonation of the GFP fluorophore contributes to the rapid loss of GFP fluo-

rescence in time-lapse images ***(15)***. In the neutral pH TAF buffer used to arrest mating reactions, residual GFP can occasionally be found in lysed cells, but it has a characteristically punctate appearance easily distinguishable from the even distribution of live cells. An alternative method to score for lysis is to stain the mating reaction with 0.02% methylene blue before imaging ***(24)***.

15. DAPI will stain RNA and mitochondrial DNA at high concentrations or after long incubation periods. Nuclear DNA is easily identified because it is surrounded by an Hmg1-GFP-labeled nuclear envelope.
16. The GFP image should be collected last when mating pairs are stored in TAF buffer. Under hypoxic conditions, a small fraction of the GFP excited with 470 nm light will transition to a red fluorescent state.
17. FM4-64 uptake is significantly reduced when cells are grown in SC media. If dropout medium is required for plasmid maintenance, cells from an overnight culture grown in selective medium can be transferred to YPD medium and grown for 30 min before labeling.
18. Care should be taken to avoid aspirating cells from the pellet or sides of the tube when washing after FM4-64 and CellTracker Blue labeling. Some cell loss is inevitable, but 1×10^6 cells of each mating type are required per mating reaction.
19. A higher percentage of the cells form mating pairs when mating is initiated on a filter, but this step is not essential.
20. Do not let the VALAP boil or it will make an awful mess. VALAP can be kept in a liquid state for hours but should not be left on the hotplate overnight or it will degrade and turn dark brown.
21. YPD medium and crude agar cause autofluorescence and poor optical clarity and are thus unacceptable alternatives.
22. Sample illumination can be reduced with a neutral density filter. In addition, the mercury arc lamp shutter should be opened immediately before collecting a fluorescent image and closed immediately afterward to minimize the time that the cells are exposed to intense light.
23. Openlab Automations developed for data collection and analysis are available upon request.
24. Filter cubes for DIC, GFP, and RFP imaging are placed in adjacent positions on a rotating filter turret to minimize the time required to switch filters.
25. Permeance is routinely measured using only GFP transfer data. mCherry transfers between cells at the same rate as GFP but fades faster. DsRed is a tetramer that diffuses more slowly than GFP. Small pores can allow GFP, but not DsRed, to pass ***(16)***. DsRed also allows expansion to be monitored for a longer time. Permeance is impossible to measure with Gag-GFP, because a variable fraction of Gag-GFP forms immobile aggregates.
26. In wild-type mating pairs, no correlation was found between the time of fusion and the pore size. The time-lapse series should be extended for as long as practical in order to avoid excluding small slowly expanding pores, which require a long time for GFP to diffuse to equilibrium, from the statistical analysis.
27. Approximately 24% ± 11% (n = 40) of the total cellular volume in wild-type mating pairs is occupied by vacuoles, as determined by confocal microscopy ***(16)***.

Mutants with vacuolar transport defects may have smaller vacuoles. Vacuolar volume is also regulated in response to the nutritional environment.

28. After background subtraction, the mean intensity of the GFP recipient cell before fusion should be 0, but the actual measurement is often slightly higher because of out of focus light from adjacent GFP cells.
29. Changes in sum of the integrated GFP intensities indicate either that GFP has leaked out of the cell pair during fusion or that the GFP intensity measurements are not linear.
30. In Microsoft Excel, the fit is more accurate if the $I_D(t)$, and $I_R(t)$ values are multiplied by 1000 before calculating the fit to avoid truncation of significant figures. The coefficients in the fit curve must then be divided by 1000.
31. A second order polynomial equation will provide a poor fit to the data if the fusion pore closes before GFP has diffused to equilibrium or if a second fusion pore opens between the same pair of cells ***(16)***. In this case, ensure that the V1/V2 ratio is measured at a time when GFP has truly diffused to equilibrium, and then fit the polynomial equation to the data from the start time until the discontinuity in the curve.

Acknowledgment

This work was supported by a Research Scholar Award from the American Cancer Society. Scott Nolan participated in the development of the karyogamy, vacuole fusion, and permeance assays.

References

1. Haber, J. E. (1998) Mating-type gene switching in *Saccharomyces cerevisiae. Annu. Rev. Genet.* **32,** 561–599.
2. Cosma, M. P. (2004) Daughter-specific repression of Saccharomyces cerevisiae HO: Ash1 is the commander. *EMBO Rep.* **5(10),** 953–957.
3. Elion, E. A. (2000) Pheromone response, mating and cell biology. *Curr. Opin. Microbiol.* **3(6),** 573–581.
4. Bardwell, L. (2005) A walk-through of the yeast mating pheromone response pathway. *Peptides* **26(2),** 339–350.
5. White, J. M. and Rose, M. D. (2001) Yeast mating: getting close to membrane merger. *Curr. Biol.* **11(1),** R16–R20.
6. Chen, E. H., Grote, E., Mohler, W., and Vignery, A. (2007) Cell–cell fusion. *FEBS Lett.* **581(11),** 2181–2193.
7. Rose MD. (1996) Nuclear fusion in the yeast *Saccharomyces cerevisiae. Annu. Rev. Cell Dev. Biol.* **12,** 663–695.
8. Weisman, L. S. and Wickner, W. (1988) Intervacuole exchange in the yeast zygote: a new pathway in organelle communication. *Science* **241(4865),** 589–591.
9. Roy, A., Lu, C. F., Marykwas, D. L., Lipke, P. N., and Kurjan, J. (1991) The AGA1 product is involved in cell surface attachment of the *Saccharomyces cerevisiae* cell adhesion glycoprotein a-agglutinin. *Mol. Cell. Biol.* **11(8),** 4196–4206.
10. Heiman, M. G. and Walter, P. (2000) Prm1p, a pheromone-regulated multispanning membrane protein, facilitates plasma membrane fusion during yeast mating. *J. Cell Biol.* **151(3),** 719–730.

11. Erdman, S., Lin, L., Malczynski, M., and Snyder, M. (1998) Pheromone-regulated genes required for yeast mating differentiation. *J. Cell Biol.* **140(3),** 461–483.
12. Trueheart, J., Boeke, J. D., and Fink, G. R. (1987) Two genes required for cell fusion during yeast conjugation: evidence for a pheromone-induced surface protein. *Mol. Cell. Biol.* **7(7),** 2316–2328.
13. Sprague, G. F., Jr. (1991) Assay of yeast mating reaction. *Methods Enzymol.* **194,** 77–93.
14. Vida, T. A. and Emr, S. D. (1995) A new vital stain for visualizing vacuolar membrane dynamics and endocytosis in yeast. *J. Cell Biol.* **128(5),** 779–792.
15. Jin, H., Carlile, C., Nolan, S., and Grote, E. (2004) Prm1 prevents contact-dependent lysis of yeast mating pairs. *Eukaryot. Cell* **3(6),** 1664–1673.
16. Nolan, S., Cowan, A. E., Koppel, D. E., Jin, H., and Grote, E. (2006) FUS1 Regulates the opening and expansion of fusion pores between mating yeast. *Mol. Biol. Cell* **17(5),** 2439–2450.
17. Maddox, P., Chin, E., Mallavarapu, A., Yeh, E., Salmon, E.D., and Bloom, K. (1999) Microtubule dynamics from mating through the first zygotic division in the budding yeast *Saccharomyces cerevisiae. J. Cell Biol.* **144(5),** 977–987.
18. Berlin, V., Brill, J. A., Trueheart, J., Boeke, J. D., and Fink, G. R. (1991) Genetic screens and selections for cell and nuclear fusion mutants. *Methods Enzymol.* **194,** 774–792.
19. Tang, F., Kauffman, E. J., Novak, J. L., Nau, J. J., Catlett, N. L., and Weisman, L. S. (2003) Regulated degradation of a class V myosin receptor directs movement of the yeast vacuole. *Nature* **422(6927),** 87–92.
20. Stefan, C. J. and Blumer, K. J. (1999) A syntaxin homolog encoded by VAM3 mediates down-regulation of a yeast G protein–coupled receptor. *J. Biol. Chem.* **274(3),** 1835–1841.
21. Shaner, N. C., Campbell, R. E., Steinbach, P. A., Giepmans, B. N., Palmer, A. E., and Tsien, R. Y. (2004) Improved monomeric red, orange and yellow fluorescent proteins derived from *Discosoma* sp. red fluorescent protein. *Nat. Biotechnol.* **22(12),** 1567–1572.
22. Schiestl, R. H. and Gietz, R. D. (1989) High efficiency transformation of intact yeast cells using single stranded nucleic acids as a carrier. *Curr. Genet.* **16(5-6),** 339–346.
23. Gietz RD, Woods RA. (2006) Yeast transformation by the LiAc/SS carrier DNA/PEG method. *Methods Mol. Biol.* **313,** 107–120.
24. Aguilar, P. S., Engel, A., and Walter, P. (2006) The plasma membrane proteins Prm1 and Fig1 ascertain fidelity of membrane fusion during yeast mating. *Mol. Biol. Cell* **18(2),** 547–556.
25. Cormack, B. P., Bertram, G., Egerton, M., Gow, N. A., Falkow, S., and Brown, A. J. (1997) Yeast-enhanced green fluorescent protein (yEGFP)a reporter of gene expression in *Candida albicans. Microbiology* **143(Pt 2),** 303–311.
26. Ribas, J. C. and Wickner, R. B. (1998) The Gag domain of the Gag-Pol fusion protein directs incorporation into the L-A double-stranded RNA viral particles in *Saccharomyces cerevisiae. J Biol. Chem.* **273(15),** 9306–9311.

Color Plates

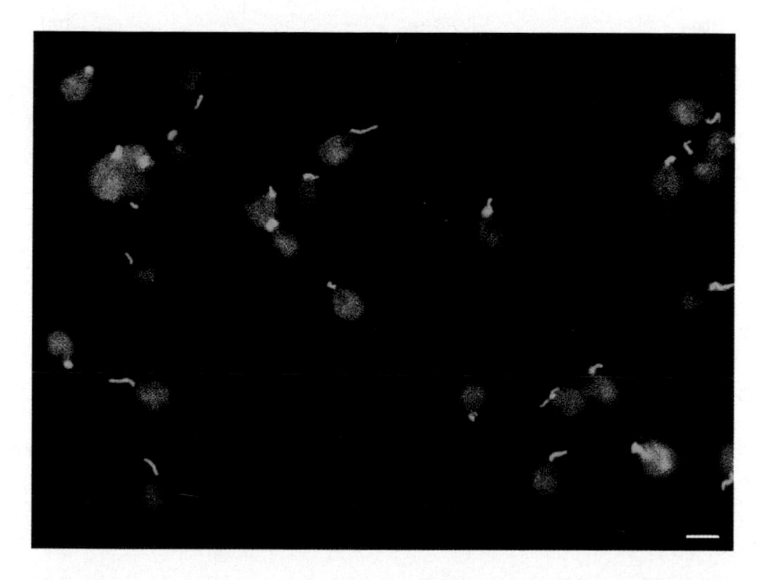

Plate 1. The mt(+) fertilization tubule. mt(+) gametes were activated by incubation with dibutrylyl cAMP and papaverine for 20 min. Fertilization tubules (green) were visualized by fluorescent microscopy after staining actin filaments with BODIPY phallacidin. Following capture of the image in the fluorescein channel, the residual chlorophyll present in cell bodies (red) was imaged by examination in the rhodamine channel. Bar = 5 μm (see Figure 2, Chapter 3; page 44).

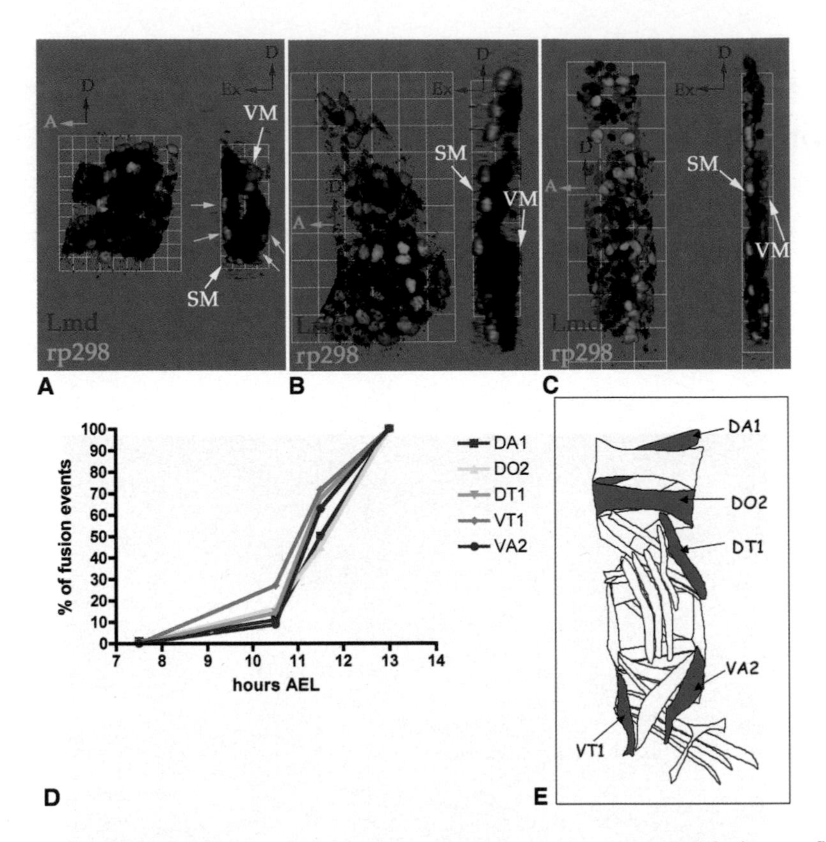

Plate 2. Founder cell and fusion-competent myoblast arrangements and fusion profiles of individual muscles (see Figure 1, Chapter 15; page 265).

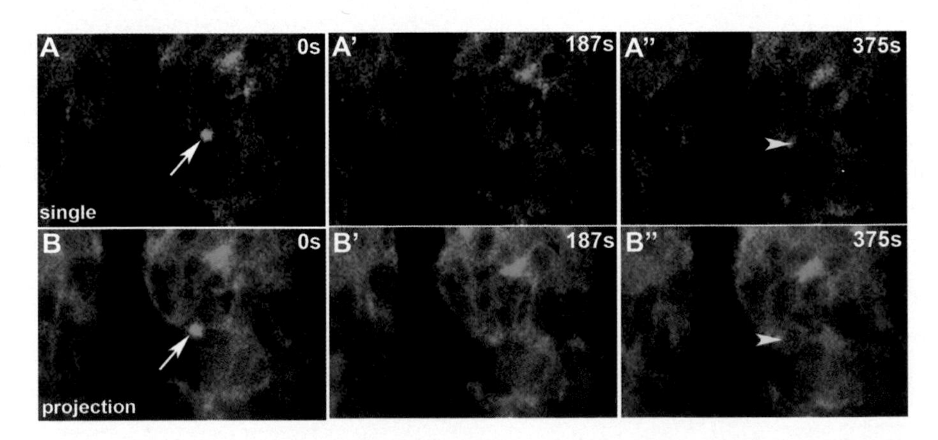

Plate 3. Single myoblast fusion event. Lateral view of live *twist* promoter *–GFP::actin, apME580-NLS::dsRed.T4* embryo (see Figure 4, Chapter 15; page 269).

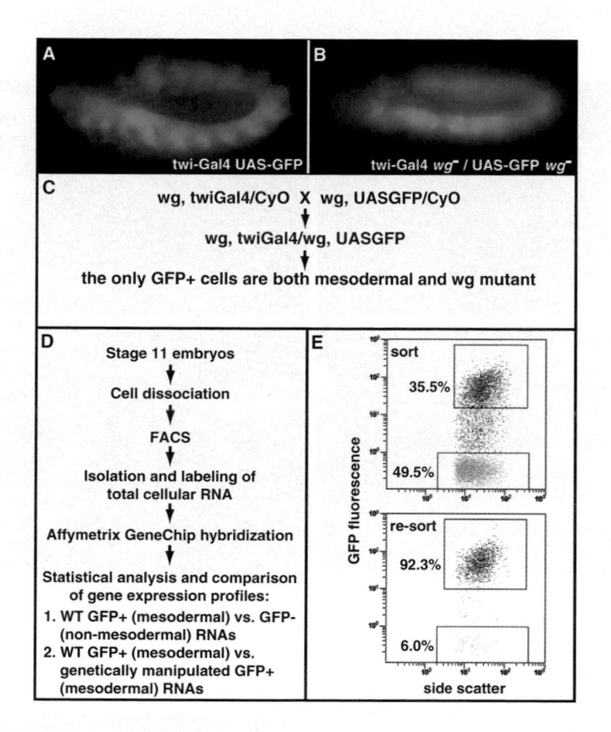

Plate 4. Experimental strategy to obtain gene expression signatures of purified *Drosophila* embryonic mesodermal cells (see Figure 1, Chapter 17; page 301).

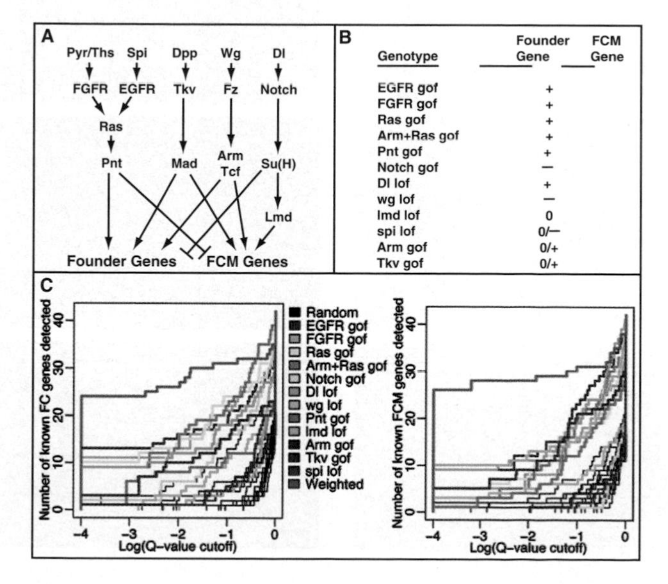

Plate 5. Statistical metaanalysis of an expression profiling compendium to predict myoblast-specific gene signatures (see Figure 2, Chapter 17; page 302).

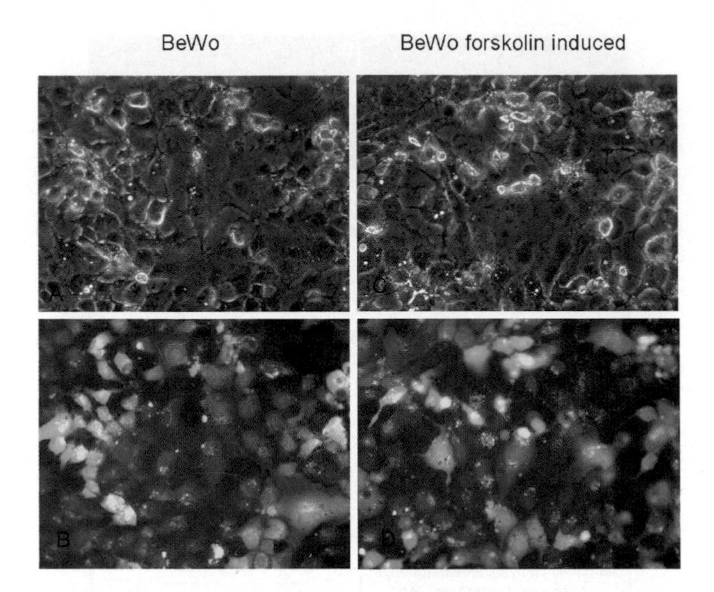

Plate 6. (**I**) BeWo cells color labeled with green and red fluorescence dye mixed together (see Figure 4, Chapter 21; page 374).

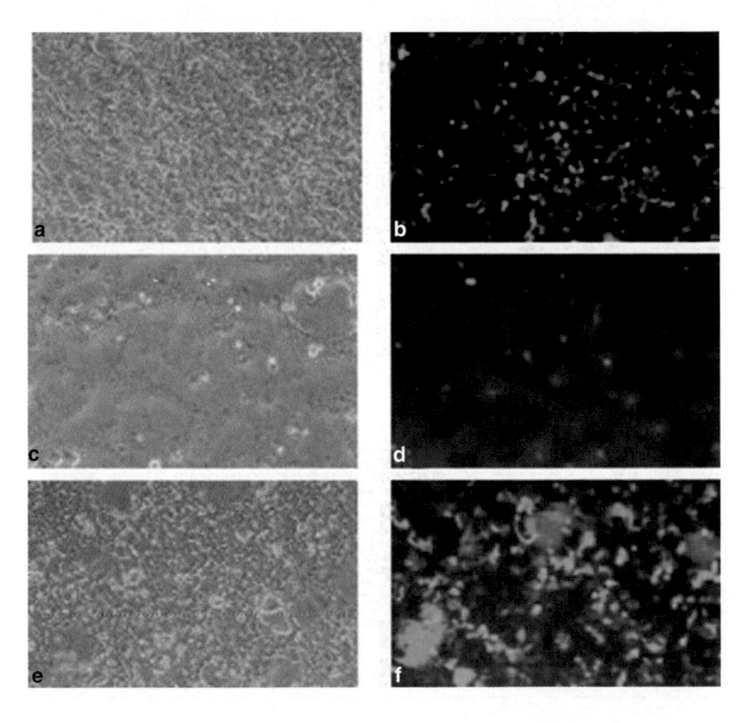

Plate 7. **(II)** HEK239 cells were transfected with plasmids encoding syncytin-1 483 + EGFP (1:1) or syncytin-1 antisense + EGFP (1:1) (see Figure 4, Chapter 21; page 375).

11

Ultrastructural Analysis of Cell Fusion in Yeast

Alison E. Gammie

Summary

The process of creating a single cell from two progenitor cells requires molecular precision to coordinate the events leading to cytoplasmic continuity while preventing lethal cell lysis. Cell fusion characteristically involves the mobilization of fundamental processes, including signaling, polarization, adhesion, and membrane fusion. The yeast *Saccharomyces cerevisiae* is an ideal model system for examining the events of this critical and well-conserved process. Researchers employ yeast cells because they are rapidly growing, easy to manipulate, amenable to long-term storage, genetically tractable, readily transformed, and nonhazardous. The genetic and morphological characterizations of cell fusion in wild-type and fusion mutants have helped define the mechanism and temporal regulation required for efficient cell fusion. Ultrastructural studies, in particular, have contributed to the characterization of and revealed striking similarities within cell fusion events in higher organisms. This chapter details two yeast cell fusion ultrastructural methods. The first utilizes an ambient temperature chemical fixation, and the second employs a combination of high-pressure freezing and freeze substitution.

Key Words: Ultrastructure; cell fusion; yeast; mating; electron microscopy; freeze substitution; high-pressure freezing.

1. Introduction

1.1. Preparation of Cell Fusion Intermediates: Mating Parameters

A diploid yeast cell forms upon cell fusion of two haploid cells of opposite mating types (*MAT***a** and *MAT*α). Mating haploid cells poised to fuse constitute a prezygote, and the initial diploid cell created upon fusion is termed a *zygote (for recent reviews, see* **refs. *1,2***). Yeast grow efficiently in haploid or diploid states; therefore, cell fusion is not essential for yeast survival. For this reason, cell fusion mutants are easy to identify using genetic methods *(3)*. The analysis of mutants allows researchers to examine prezygotes arrested at a particular step in cell fusion *(4)*. The mutants are particularly useful for investigating short-lived steps in the process.

From: *Methods in Molecular Biology, vol. 475: Cell Fusion: Overviews and Methods*
Edited by: E. H. Chen © Humana Press, Totowa, NJ

Before embarking on the lengthy procedures required for visualization of cell fusion at the ultrastructural level, researchers should optimize the mating conditions. Even under ideal conditions, the percentage of zygotes in a mating mixture rarely exceeds 40%. Furthermore, in wild-type matings prezygotes are particularly elusive, comprising no more than 1%–2% of zygotes. In mating mixtures with cell fusion defects, a higher percentage of pairs are blocked as prezygotes (up to 90%); however, other parameters sometimes interfere with characterization. For example, in most cell fusion mutants the pathway is partially functional; thus, given excessive time to mate, the defect might be missed. Moreover, many mutants show reduced rates of prezygote formation in addition to the specific arrest point in cell fusion; therefore, in these cases it proves difficult to identify significant numbers of prezygotes. All of these factors underscore the importance of optimizing the mating reaction.

Four experimental parameters for efficient mating require particular attention: (1) growth phase, (2) cell ratio/density, (3) temperature, and (4) time. Efficient matings are derived from freshly grown cells in the early logarithmic phase of growth combined at a 1:1 ratio of mating partners, at a density of ~10^5 cells/mm^2, and incubated at 23°–30 °C for a minimum of a doubling time. The overall efficiency of zygote formation is diminished when cells are at stationary phase, at an uneven mating partner ratio, too sparse or too dense, at an elevated temperature (e.g., greater than 35 °C), or not in contact for a doubling time.

The first procedure detailed in this chapter is designed to optimize the mating reaction. The second allows for the assessment of the efficiency of zygote/prezygote formation using conventional light microscopy. Successful microscopic analysis of cell fusion depends on the accurate identification of a zygote. To the untrained eye, a zygote is often confused with a large-budded cell or with two closely associated cells (**Fig. 1A,B**). Zygotes may be distinguished from mitotic cells by their distinct cell–cell contact region. Prezygotes (*see* **Fig. 1C**) and zygotes (*see* **Fig. 1D**) have a smooth, curved junction between the parent cells, whereas mitotic cells have a sharp constriction at the mother/bud neck (*see* **Fig. 1A,B**). Wild-type prezygotes have slightly elongated cell bodies with a small region of contact (*see* **Fig. 1B**), whereas mitotic cells are spherical (*see* **Fig. 1A,B**). Cell fusion defective early (*see* **Fig. 1D**) and late (*see* **Fig. 1E**) prezygotes have a pronounced cell wall barrier between the mating cells. After gaining proficiency in preparing mating mixtures with a high percentage of zygotes/prezygotes, researchers may continue with one of the two ultrastructural procedures detailed in this chapter.

1.2. Ultrastructural Analysis of Yeast Cell Fusion

Published techniques for transmission electron microscopy of yeast cells often result in the destruction of the ultrastructure of the cell wall to improve

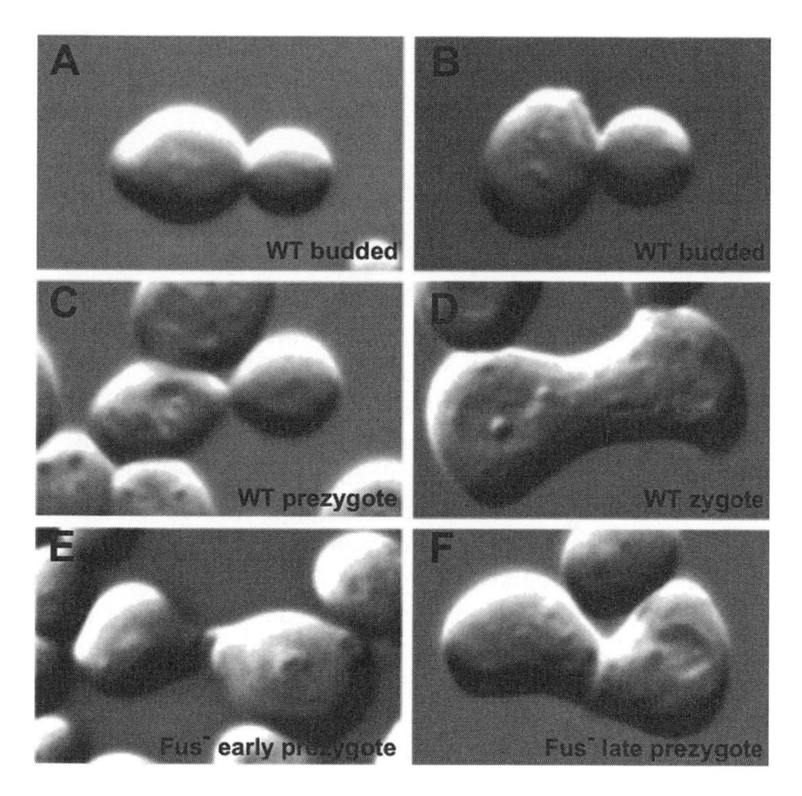

Fig. 1. Phenotypes of cell and nuclear fusion mutant zygotes detected by conventional microscopy. Cells were examined for prezygote/zygote morphology with differential interference contrast optics. **(A)** Wild-type (WT) budded cells. **(B)** Wild-type *MAT***a** × *MAT*α prezygote. **(C)** Wild-type zygote from mating in B. **(D)** Early cell fusion defective (Fus⁻) prezygote from a *MAT***a** *fus2* × *MAT*α *fus1 fus2* mating. **(E)** Late Fus⁻ prezygote/zygote from the mating in D.

resin infiltration (*reviewed in* **ref. 5**). Investigation of cell fusion requires examination of zygotes with well-preserved cell wall and membrane structures. I outline two effective methods for visualizing yeast cell fusion. The first utilizes an ambient temperature chemical fixation *(4,6)*, and the second employs high-pressure freezing and freeze substitution *(7,8)*. The first method depends on a combined fixation with glutaraldehyde for an array of cellular structures and with potassium permanganate for membranes that are refractory to aldehyde fixation. The fixation is followed by periodate oxidation of the cell wall to improve resin infiltration with minimal perturbations of the cell wall ultrastructure. The second method employs the rapid immobilization of cellular structures using high-pressure freezing followed by the gradual fixation of frozen specimens with an osmium tetroxide, uranyl acetate, methanol fixative.

Following embedment, sectioning, and optional poststaining, both methods allow for visualization of the ultrastructure of key components, including membrane structures, during yeast cell fusion.

1.2.1. Ambient Temperature Chemical Fixation

The ambient temperature chemical fixation method in this chapter has the advantage of being feasible in standard transmission electron microscopy laboratories. Yeast cell fusion intermediates in mating mixtures are initially immobilized using a buffered, osmotically balanced solution with glutaraldehyde as the primary fixative. Glutaraldehyde is a bifunctional cross-linker that typically reacts with α-amines of lysines to form linkages between proteins ***(9–11)***. The treatment creates an irreversibly cross-linked network throughout the cell in the range of seconds to minutes. Glutaraldehyde reacts to a certain extent with other molecules, such as nucleic acids and carbohydrates; however, lipids evade fixation ***(9–11)***. After fixation, the glutaraldehyde is washed out with a buffered solution, and the cells are exposed to potassium permanganate. Incubation of aldehyde-fixed samples with potassium permanganate allows for the visualization of membrane structures that would otherwise be extracted in subsequent steps of the procedure ***(5)***. Permanganate is a strong oxidizer presumably reduced by the hydrophilic head groups on membranes. Upon reduction, a precipitate of MnO_2 forms over lipid-rich areas ***(9–11)***. Hence, permanganate may function primarily as a stain rather than a fixative ***(5)***.

After treatment with permanganate, the cells are washed and treated with sodium periodate. Periodate treatment oxidizes vicinal diols, cleaving the carbon–carbon bond and replacing the hydroxyl groups with aldehyde groups (*reviewed in* **ref. *12***). Many sugar residues, including those in the yeast cell wall, contain vicinal diols. The treatment converts carbohydrates from a cyclic to a linear form; however, the carbohydrates are not destroyed or extracted, and therefore the cell wall ultrastructure is retained. After oxidation, cells are washed with a buffered solution and treated with ammonium chloride to quench free aldehydes ***(13)***.

After the periodate treatment, the cells are negatively stained with uranyl acetate. Deposition of the heavy metal uranium on cellular structures allows for increased contrast during transmission electron microscopy ***(10)***. In addition, uranyl acetate treatment helps to preserve nucleic acid–containing structures, including the nucleoplasm ***(5)***. Finally, the treatment eliminates poststaining of thin sections with uranyl acetate.

Following the uranyl acetate treatment the cells are dehydrated to replace water with a solvent that is miscible with the embedding resin. For the first ultrastructural procedure outlined in this chapter, ethanol is the dehydrating solvent. The dehydration is accomplished with short incubations each with increasing percentages of ethanol, ending in a treatment with absolute ethanol. After

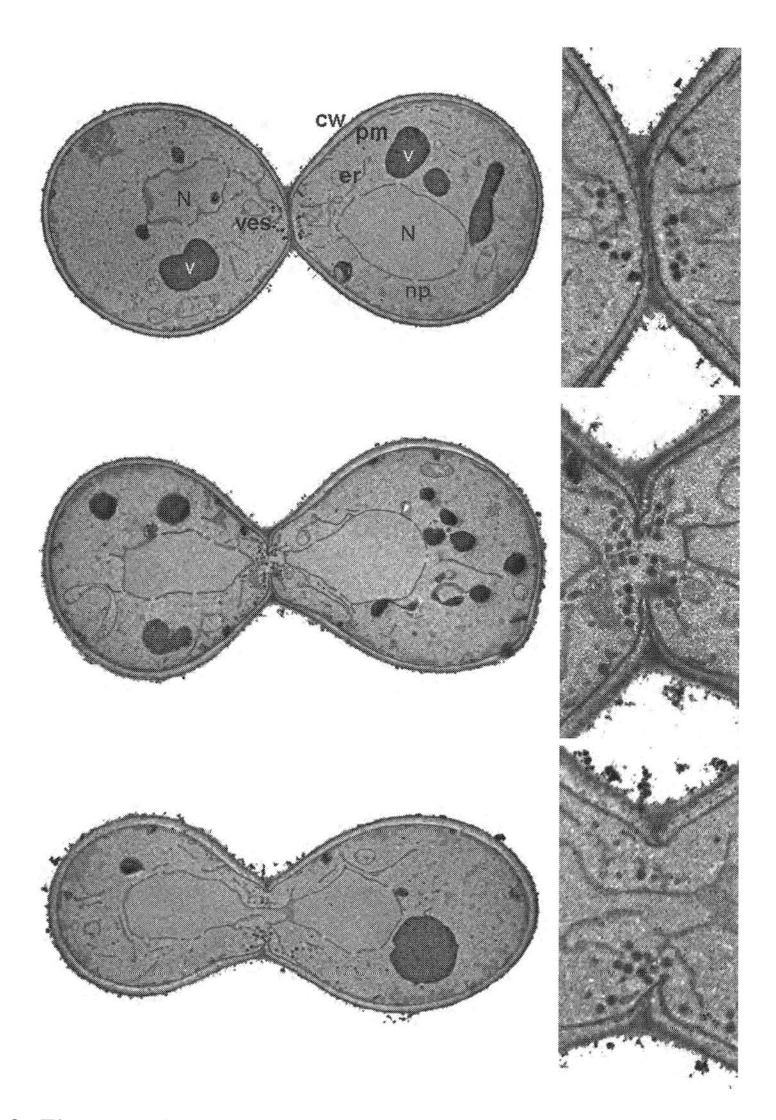

Fig. 2. Electron micrograph of a wild-type prezygote using ambient temperature chemical fixation. The zygote was fixed and stained with glutaraldehyde and permanganate followed by sodium metaperiodate oxidation of the cell wall. The micrograph is of a wild-type prezygote prior to cell fusion (top), a zygote just after cell fusion but before nuclear fusion (middle), and a zygote after cell and nuclear fusion (bottom). The left panels show the entire prezygote, and the right panels are magnifications of the zone of cell fusion. The nuclei (N) are outlined by the darkly staining nuclear envelope nuclei, with nuclear pores (np) appearing as light gaps. The structures with dark interiors are vacuoles (v). The darkly staining membranes on the cell perimeter are the plasma membranes (pm). The cell wall (cw) surrounds the plasma membrane and displays thinning in the zone of cell fusion in the prezygote. The long membranous structures are peripheral endoplasmic reticulum (er). The presence of numerous darkly stained vesicles (ves) at the zone of cell fusion (fusion zone) is characteristic of prezygotes. Also present in this micrograph are mitochondria (not labeled). An image of the middle panel zygote has been published by the author previously ***(4)***.

dehydration, the sample is infiltrated with a blend of low-viscosity hydrophilic and acrylic monomers that rapidly penetrate the cells. The resin is solidified in a heated vacuum oven overnight. Sections of 70–90 nm may be poststained with a second negative stain, lead citrate, if necessary and visualized using transmission electron microscopy. With this protocol, the cell wall is well preserved, membranes appear dark, and proteinaceous structures are light (**Fig. 2**).

1.2.2. High-Pressure Freezing and Freeze Substitution

The second ultrastructural method described in this chapter is a combination of high-pressure freezing of the mating mixture followed by fixation and dehydration at low temperatures *(7,14)*. The major advantage is that immobilized cellular structures are cross-linked and the specimen dehydrated while still frozen. The major disadvantage is that specialized, expensive equipment is required for the high-pressure freezing and freeze substitution.

Yeast cell fusion intermediates in mating mixtures are subjected to high-pressure freezing, allowing for rapid immobilization of components and the conversion of water to vitreous ice *(8,14)*. Thus, during high-pressure freezing cellular structures are instantly immobilized with a minimum of damaging or obscuring ice crystals. The frozen mating mixture is transferred to a vial containing an osmium, glutaraldehyde, uranyl acetate acetone fixative at −90 °C. The temperature of the specimen is gradually raised over several days so that fixation of immobilized cellular structures occurs at extremely low temperatures (e.g., starting at −50 °C for glutaraldehyde and −30 °C for osmium tetroxide), and thus chemical fixation occurs with minimal artifacts *(8)*. In addition, during the procedure the cellular water is replaced with a resin miscible solvent, acetone. Recall that glutaraldehyde is an excellent cross-linking agent for proteins but not for lipids. In the freeze-substitution fixative the osmium increases the retention and contrast of lipids in the cells. Osmium reacts with membrane lipids and to varying degrees with proteins *(10)*. The deposition of osmium on the hydrophilic regions of lipids allows for visualization of the lipid bilayer. The uranyl acetate in the freeze substitution functions similarly as described above to provide contrast and retain certain cellular structures *(5)*. One important feature of the fixative used in the freeze substitution is the inclusion of 1%–5% water. Previous analysis proved that the yeast membrane ultrastructure was preserved in the presences of up to 5% water in the freeze-substitution solvent *(15)*.

The temperature of the specimen is gradually increased to room temperature and rinsed with anhydrous acetone to remove the free osmium and glutaraldehyde. The process of embedment is accomplished by sequential exposure to increasing percentages of resin to acetone mixtures, culminating in an incubation in 100% resin. The resin is solidified at high temperature to complete the embedment. Sections (30–70 nm) are placed on grids and poststained with uranyl acetate and aqueous lead citrate if more contrast is required. Samples are visualized using conventional transmission electron microscopy. Using this protocol, membranes appear dark, and a range of cellular structures such as the microtubule-organizing center are preserved and visible *(7)*.

2. Materials

2.1. Preparing Mating Mixtures From Liquid Cultures

1. Yeast cells *MAT*α and *MAT***a** strains.
2. Sterile liquid culture medium, YEPD: 1% yeast extract, 2% peptone, 2% glucose. Combine 10 g yeast extract, 20 g peptone, and 1 L of distilled water. Cover with loosened cap. Autoclave using the liquid setting. Allow the liquid to cool. Add 50 mL filter-sterilized 40% glucose stock. Secure cap and mix well. Store at room temperature for months.
3. Sterile culture tubes.
4. Incubator and Roller Drum (New Brunswick Scientific, Inc., Edison, NJ).
5. YEPD nutrient agar plates: 1% yeast extract, 2% peptone, 2% glucose, 2% agar. Combine 10 g yeast extract, 20 g peptone, 20 g agar, 1 L of distilled water, and a magnetic stir bar in a 2-L flask. Cover with foil. Autoclave using the liquid setting. Place in a 55°–60 °C water bath. Add 50 mLfilter-sterilized 40% glucose stock upon cooling to 55°–60 °C while stirring on magnetic stir plate. Pour approximately 25 mL per 100-mm-diameter plate such that the dish is approximately half-filled. This recipe yields approximately 40 plates (2 sleeves). Allow for solidification and drying. Store in plastic plate sleeves at 4 °C for months.
6. Filtration device and 0.45 µm pore size filters (Millipore Co., Billerica, MA).

2.2. Confirmation of Mating Efficiency

1. PBS, phosphate buffered saline: 137 m*M* NaCl, 2.7 m*M* KCl, 1.4 m*M* KH_2PO_4, 4.3 m*M* Na_2HPO_4, pH 7.3. 10× stock solution (1 L): 80 g NaCl, 2 g KCl, 11.5 g $Na_2HPO_4 \times 7H_2O$, 2 g KH_2PO_4. Concentrated or diluted solutions may be stored at room temperature for months.
2. Methanol: Acetic acid (3:1) fixative, freshly prepared: prepare 1 mL (750 µL MeOH, 250 µL glacial acetic acid) per mating mixture. Chill on ice before use.
3. Bath sonicator.
4. Glass microscope slides: 3" × 1" (76.2 × 25.4 mm); thickness: 0.038" to 0.043" (0.96 to 1.10 mm).
5. Cover slips: 0.17 mm thickness.
6. Transmitted light microscope equipped with a 100× objective lens and differential interference contrast optics.

2.3. Ambient Temperature Fixation

1. Materials from **Subheading 2.1**.
2. 0.5 *M* potassium phosphate, pH 7.4: Mix 19 mL of 1 *M* KH_2PO_4 (136.1 g KH_2PO_4

per liter), 81 mL of 1 *M* K_2HPO_4 (174.2 g K_2HPO_4 per liter), and 100 mL distilled H_2O. Check pH of diluted solution. Store at room temperature for months.

3. 0.1 *M* $CaCl_2$: Dissolve 1.47 g calcium chloride dihydrate, F.W. 147.02, in distilled water to a final volume of 100 mL. Store at room temperature for months.
4. 0.1 *M* $MgCl_2$: Dissolve 2.03 g magnesium chloride, hexahydrate, F.W. 203.30, in distilled water to a final volume of 100 mL. Store at room temperature for months.
5. 1 *M* D-sorbitol: Dissolve 18.2 g D-sorbitol per 100 mL of distilled water. Filter sterilize or autoclave on liquid setting. Store at room temperature for months.
6. FIX: 40 m*M* potassium phosphate, pH 7.4; 1 m*M* $CaCl_2$; 1 m*M* $MgCl_2$;; 0.2 *M* D-sorbitol; 2% glutaraldehyde. Prepare 1.5–2 mL per mating mixture immediately before use. For 10 mL, combine: 800 µL of 0.5 *M* potassium phosphate, pH 7.4; 100 µL of 0.1 *M* $CaCl_2$; 100 µL of 0.1 *M* $MgCl_2$; 2 mL of 1 *M* D-sorbitol; 2.5 mL of 8% gluteraldehyde EM Grade in vacuum-sealed ampoules (PolySciences, Inc., Warrington, PA); and 4.5 mL of distilled water.
7. 4% potassium permanganate (0.04 g/mL of distilled water): Prepare at least 1 mL per mating mixture. The solubility of potassium permanganate in water is such that the solution should be started approximately 30 min to 1 h before use with constant stirring or agitation during that time interval. Shortly before use, filter the solution through a Whatman #1 filter to remove undissolved crystals. The quality of the potassium permanganate crystals is important for ultrastructural work. A low-mercury stock from Sigma-Aldrich (St. Louis, MO) has been reported to give superior results ***(5)***.
8. 1% sodium metaperiodate (0.01 g/mL of distilled water). Prepare 2 mL per mating mixture immediately before use.
9. 50 m*M* potassium phosphate, pH 7.4. Combine 10 mL of 0.5 *M* potassium phosphate, pH 7.4 (see above), and 90 mL distilled water. Store at room temperature for months.
10. 50 m*M* ammonium chloride: Dissolve 0.267 g ammonium chloride, F.W. 53.49, in 100 mL distilled water. Store at room temperature for months.
11. 2% uranyl acetate in water: Wear protective clothing and work in a ventilated hood if possible. Combine 0.1 g of uranyl acetate and 5 mL distilled water in a screw cap tube. Secure the cap and cover with foil. Agitate to dissolve (using a roller drum or rocking platform). Filter through a 0.45-µm syringe filter before use. Make immediately before use. Contact your local environmental health and safety office for appropriate disposal of the hazardous uranyl acetate waste.
12. Absolute ethanol.
13. LR White resin (PolySciences, Inc.). Work with resins using protective clothing in a ventilated fume hood. Liquid resins must be disposed of in an appropriate chemical waste container.
14. BEEM embedding molds (BEEM, Inc., West Chester, PA) prebaked overnight at 60°C.
15. Roller drum (New Brunswick Scientific) or rocking platform (Bio-Rad Laboratories, Inc., Hercules, CA).
16. Vacuum oven (with nitrogen flushing capabilities).

17. Lead citrate staining solution: Combine 1.33 g lead nitrate, 1.76 g sodium citrate, and 30 mL distilled water. Mix well for 1 min, and allow the solution to stand for 30 min with occasional shaking. Add 8 mL of 1 N NaOH and mix. Dilute to 50 mL with distilled water. The final pH should be pH 12. Store up to 6 months. Lead citrate is also commercially available (Ted Pella, Inc., Redding, CA).
18. Microtome with diamond knife.
19. Electron microscope specimen grids, 200 mesh.
20. Transmission electron microscope.

2.4. Low-Temperature Fixation

2.4.1. High-Pressure Freezing

1. Materials from **Subheading 2.1**.
2. Brass specimen carrier holder (hats or planchettes).
3. Flat toothpick, sharpened wooden stick, or spatula.
4. Whatman 542 filter paper.
5. Bal-Tech RMC (Brookline, NH) or Leica Microsystems (Bannockburn, IL) high-pressure freezing machine.
6. 2% uranyl acetate in water (*see* **Subheading 2.3**).
7. 1% osmium tetroxide plus 0.1% uranyl acetate in 95% acetone: Osmium tetroxide is toxic. Inhalation, ingestion, or skin contact with material may cause severe injury or death. Use protective clothing and caution when weighing dry material. Add 1 g OsO_4 crystals to 90 mL of acetone in a screw-cap bottle. Secure the cap and mix until completely dissolved. Add 5 mL 2% uranyl acetate in water (*see* **Subheading 2.3**) and bring the volume to 50 mL with acetone. Mix well. Label ~ 70 Nalgene 2.0 mL cryovials with pencil to indicate the fixative. Place vials in rack with caps removed. Dispense 1.5 mL into each cryovial. Cap cryovials immediately. Carefully lower rack into liquid nitrogen so that the cryovials remain upright and submerged. When thoroughly frozen, transfer vials to a liquid nitrogen storage device until ready to use. After use, dispose of dry and liquid waste in the appropriate hazardous waste containers.

2.4.2. Freeze Substitution

1. Bal-Tech RMC or Leica Microsystems freeze substitution machine.
2. 1% osmium tetroxide plus 0.1% uranyl acetate in 95% acetone (see **Subheading 2.4.1** for preparation and hazards).
3. Epon-Araldite: Combine 6.2 g Epon 812, 4.4 g Araldite 502, and 12.2 g dodecenylsuccinic anhydride (DDSA); mix well on a rocking platform for 10–20 min. For the final embedment, also add 0.8 mL benzyl dimethyl amine (BDMA). Make the resin immediately before use. Work with resins wearing protective clothing in a ventilated fume hood. Liquid resins must be disposed of in an appropriate chemical waste container.
4. BEEM capsules (*see* **Subheading 2.3**).
5. Vacuum oven (*see* **Subheading 2.3**).
6. Microtome.

7. Electron microscope specimen grids (200 mesh).
8. 2% uranyl acetate in 70% methanol; Wear protective clothing and work in a ventilated hood if possible. Combine 0.1 g of uranyl acetate and 5 mL of 70% methanol distilled water in a screw-cap tube. Secure the cap and cover with foil. Agitate to dissolve (using a roller drum or rocking platform). Filter through a 0.45- μμ syringe filter before use. Make immediately before use. Contact the local environmental health and safety office for appropriate disposal of the hazardous uranyl acetate waste.
9. Lead citrate (*see* **Subheading 2.3**).
10. Transmission electron microscope.

3. Methods

3.1. Preparing Mating Mixtures From Liquid Cultures

1. Microbial sterile technique must be maintained until the fixation step.
2. Day 1: Inoculate separate culture tubes each containing 5 mL of sterile YEPD medium with the *MAT*α and *MAT***a** cells to be analyzed. Grow overnight with aeration at 30 °C (*see* **Note 1**).
3. Day 2: Approximately 16 h prior to the mating experiment, dilute each of the saturated overnight cultures into 20 mL fresh medium using a range of dilutions (e.g., for YEPD, 5×10^{-4}, 10^{-4}, and 5×10^{-3}; for synthetic medium: 5×10^{-3}, 10^{-3}, and 5×10^{-2}). If the strains have a growth defect, the proper dilution must be determined empirically. Grow the cultures overnight at 30 °C with aeration (*see* **Note 2**).
4. Day 3: Measure the optical density at 600 nm (OD_{600}) of the samples. The cultures should be in early exponential growth (between 0.2 and 0.5 at OD_{600}) for efficient mating. If the cultures have overgrown, dilute the cultures with fresh medium to an OD_{600} of 0.1 and allow the cells to reach exponential phase.
5. When the strains are at early exponential phase, proceed with the mating. Mix together 20 mL each of the *MAT***a** and *MAT*α strains in a 50-mL tube and centrifuge to pellet the cells. Pour off most of the liquid, leaving about 1–5 mL. Resuspend the mating mixture in the residual liquid.
6. Concentrate the mating mixture on 0.45-μm pore size nitrocellulose filter discs using a sterilized vacuum-filtration system. If the efficiency of zygote formation is not affected in the mutant mating, 3–5 filters may be used for the mating. If the mating involves mutants that display a reduced efficiency of zygote formation, it is advised to spread the mating mix over 5–10 filters. To concentrate the mating mixture, the support base for the filter is inserted into the mouth of a side-arm flask attached to a vacuum line. The base and funnel may be sterilized with 70% ethanol. Using sterilized forceps, place a filter disc on the porous base. Concentrate the mating mixture onto the filter by applying the cells with the vacuum turned on (*see* **Note 3**).
7. After dispensing the mating mixtures over the appropriate number of 0.45-μm filters, place the filters cell-side up on prewarmed YEPD plates. Incubate the strains at 30 °C for 1.5–3 h (*see* **Note 4**).

3.2. Confirmation of Mating Efficiency

1. Prepare mating mixtures (*see* **Subheading 3.1**). Lift a filter from the mating plate. Curl and place one edge within a microcentrifuge tube with the mating mixture on the inside of the curled filter. Slowly rinse the mating mixtures from the filter with 1 mL of ice-cold PBS into the microcentrifuge tube. Keep the sample on ice except when manipulating.
2. Pellet briefly (30 s) in a microcentrifuge, and resuspend the mating mixture in 1 mL cold PBS. Repeat the wash twice.
3. Pellet the mating mixtures, and resuspend in 1 mL cold methanol:acetic acid (3:1). Place the cells on ice for 60 min. Pellet and remove as much of the liquid as possible with a pipette. Resuspend the cells in 1 mL PBS, and incubate on ice for at least 30 min to allow the cells to rehydrate. Pellet and wash the cells twice in 1 mL PBS.
4. Resuspend the final pellet in 100 µL of PBS. The mating mixtures may be kept at 4 °C for up to 1 week before examination. However, the cellular morphology will deteriorate over time.
5. Sonicate the samples in a bath sonicator at low power for approximately 3 minutes to break up mating aggregates.
6. Prepare single, wet-mount slides by placing 3 µL of the cell suspension on a standard glass microscope slide. Place the edge of a coverslip into the sample, and slowly lower the coverslip at an angle such that air bubbles are expelled at the far edge. Excess liquid should be removed by patting the slide with a lint-free absorbent tissue. If the cells and zygotes are not well separated, dilute the sample, repeat the sonication, and prepare a fresh slide.
7. Examine the slides by light microscopy to assess mating efficiency. Microscopic analysis of cell fusion is best determined using differential interference contrast (DIC) to assess the zygote/prezygote formation efficiency (*see* **Note 5**).

3.3. Ultrastructural Analysis of Cell Fusion

3.3.1. Ambient Temperature Chemical Fixation

1. Follow the protocol for preparing mating mixtures from liquid cultures (*see* **Subheading 3.1**). Lift a filter from the mating plate. Curl and place one edge within a microcentrifuge tube with the mating mixture on the inside of the curled filter. Wash the mating mixture into a microcentrifuge tube with 1 mL of FIX (40 m*M* potassium phosphate, pH 7.4; 1 m*M* $CaCl_2$; 1 m*M* $MgCl_2$; 0.2 M sorbitol; 2% glutaraldehyde). Repeat the process with all of the filters until a single mating mixture is within one tube (*see* **Note 6**).
2. Centrifuge cells and resuspend in 500 µL of fresh FIX. The total length of time in FIX should be 30 min (including the initial washing into the tubes) at room temperature. After the fixation, wash three times with 50 m*M* potassium phosphate, pH 7.4 (*see* **Note 7**).
3. Resuspend cells in 1 mL of fresh 4% potassium permanganate (*see* **Note 8**).
4. Mix the cells in the permanganate solution at 4 °C for 2–6 h with gentle rotation or rocking. Wash the cells at least four times with dH_2O until the supernatant is clear (i.e., no longer purple).

5. Resuspend the cells in 1 mL of 1% sodium metaperiodate. Centrifuge the cells and resuspend in 1 mL of 1% sodium metaperiodate. Incubate for a total of 15 min. Centrifuge and wash once with 50 m*M* potassium phosphate, pH 7.4 (*see* **Note 9**).
6. Resuspend the pellet in 1 mL of 50 m*M* ammonium chloride. Centrifuge and resuspend in 1 mL of 50 m*M* ammonium chloride. Incubate for a total of 15 min. Wash twice with dH_2O.
7. Resuspend the pellet in 2% uranyl acetate. Incubate at 4 °C overnight with mixing. Contact the local environmental health and safety office for appropriate disposal of the uranyl acetate waste.
8. Centrifuge the cells and perform the following dehydration washes with ethanol solutions: 50% ethanol, 5 min, two times; 70% ethanol, 5 min, two times; 95% ethanol, 5 min, one time; 100% ethanol, 5 min, three times. Use a freshly opened bottle of absolute ethanol for the final dehydration step (*see* **Note 10**).
9. Resuspend the final pellet in a 50:50 of 100% ethanol:LR white resin. Work in a ventilated hood and wear gloves when using unpolymerized resin. Incubate several hours with gentle rotation. Centrifuge and resuspend in 100% LR white. Incubate overnight with gentle rotation (*see* **Note 11**).
10. Centrifuge and resuspend in 100% LR white resin. Centrifuge and resuspend in ~200–500 µL of 100% LR white and transfer to prebaked (overnight at 60 °C) BEEM embedding molds. Check that there are no tiny bubbles in the bottom of the tube.
11. Prepare labels and place the labels in the tubes. Fill the tubes to the top with fresh resin. Place the samples under vacuum for about 30 min to remove air bubbles (*see* **Note 12**).
12. Incubate in a 60 °C vacuum oven (preflushed with nitrogen) for about 24 h (*see* **Note 13**).
13. Cut 70–90 nm sections. Place sections on grids.
14. Stain the sections for 5 min with lead citrate if more contrast is needed. Rinse the grids with distilled water (*see* **Note 14**).
15. Air dry and examine by transmission electron microscopy.

3.3.2. Low-Temperature Fixation

3.3.2.1. High-Pressure Freezing Procedure

1. Prepare matings (*see* **Subheading 3.1**). After the mating is completed, scrape cells off the filters using a flat toothpick, a sharpened wooden stick, or spatula (see **Note 15**).
2. Place the specimen carrier well in the loading apparatus (*see* **Note 16**).
3. Transfer cells from the tip of the stick to the well of a specimen carrier. The cells should completely fill the well. Carefully wipe off excess cells, cover, and secure the specimen carrier. Proceed to HPF immediately. The elapsed time of the above steps from placing the filtered cells on the agar plate to freezing should take less that 1 minute.

4. Perform the high pressure freezing immediately. Bal-Tech RMC and Leica Microsystems manufacture HPF machines. Please refer to the manufacturer's specifications for the use and maintenance of each machine.
5. After the high-pressure freezing step, transfer the carrier to a storage vial or to a vial of pre-frozen fixative. If the fixation parameters are known, storage in the fixative is advised. For visualization of yeast membranes use 1% osmium tetroxide plus 0.1% uranyl acetate in 95% acetone fixative.

3.3.2.2. Freeze Substitution Procedure

1. Bal-Tech RMC and Leica Microsystems manufacture freeze substitution machines. Follow the manufacturer's specifications for each machine. Using the 1% osmium tetroxide plus 0.1% uranyl acetate in 95% acetone fixative, begin the freeze substitution at −90 °C for at least 24 h.
2. Increase the temperature to −25 °C at a rate of 5 °C per hour (~15 h).
3. Hold at −25 °C for 12 h.
4. Increase the temperature to room temperature at a rate of 5 °C per hour (~9 h).
5. Rinse twice with anhydrous acetone for 15 min and embed with Epon-Araldite as detailed below (*see* **Note 17**).
6. Begin embedding with a solution of 10% resin/90% acetone overnight followed by incubations with 25%, 50%, and 75% resin in acetone each for 4–16 h.
7. Add 100% resin and incubate overnight.
8. Mix 100% resin and accelerator and incubate for 2 h.
9. Transfer samples to BEEM capsules or flat-ended molds for polymerization at 60 °C.
10. Cut sections (70 nm). Place sections on grids.
11. Poststain with 2% uranyl acetate in 70% methanol for 4 min and aqueous lead citrate for 2 min if more contrast is required. Rinse well with water and air dry if postsectioning staining is performed (*see* **Note 14**).
12. View samples with a transmission electron microscope.

4. Notes

1. Use rich YEPD medium unless selecting for plasmids. Best growth is achieved at 30 °C; however, modulation of the temperature may be necessary if working with a temperature-sensitive strain.
2. A roller drum in an incubator allows for adequate aeration.
3. Disposable sterile filtration units are also available from Millipore Co.
4. Many conventional laboratory strains are inherently temperature sensitive for mating, and the frequency of zygotes will be significantly reduced at 35 °C and above. For most matings, a doubling time (typically 90 min) is sufficient.
5. Differential interference contrast optical methods are preferable, as the bright halo surrounding cells in phase contrast optics may obscure details of the cell fusion zone. For detailed visual observation of the zygote morphology, 100× objective lenses are required. **Figure 1** provides examples of prezytoges/zygotes to aid in the determination of mating efficiency.

6. This method works well for S288C-derived strains; however, it may not work for all yeast strains.
7. The cross-linking reaction causes a significant release of protons, resulting in a drop in pH; thus, buffering is important. If the concentration of glutaraldehyde is too high, the formation of rapid cross-links is inhibited.
8. Permanganate concentrations for yeast range from 0.5% to 6%. It has been reported that permanganate is incompatible with many buffers (**5**). For this reason, the permanganate should be dissolved in water rather than buffered solutions.
9. Treatment of the cells with 0.5%–1.0% periodate does not alter cell wall appearance but significantly improves infiltration of the resin.
10. The incubations with ethanol should not exceed 5 min. Prolonged exposure to ethanol may lead to extraction of lipids and other molecules. To achieve proper infiltration of the resin, the final rinses must be with anhydrous ethanol (i.e., use a freshly opened bottle of absolute ethanol).
11. A slow roller drum provides gentle agitation.
12. Labels should be printed in small font with a laser printer or written in pencil. Do not use pen or ink-jet printed labels, because the ink rapidly diffuses in the resin.
13. Oxygen inhibits LR White polymerization; thus, removal of air bubbles in the sample and flushing the oven are important.
14. Older transmission electron microscopes may require more contrast for visualization of the ultrastructure. When using newer transmission electron microscopes, the poststaining step may be dispensed with.
15. The amount of moisture in the concentrated mating mixture is important for the high-pressure freezing step. To become proficient at obtaining the proper "texture," observe the concentrated cells through a dissecting scope. Use a flattened stick or spatula to scrape cells on the surface. If the cells appear fluid, place a piece of dry Whatman 542 filter paper under the filter to wick off excess moisture. If too dry, apply pressure to the filter on the agar plate while scraping to moisten the cells. The proper texture has been described as having an applesauce-like appearance.
16. These wells are also referred to as *hats* or *planchettes*. For yeast, brass specimen carriers work well.
17. Epon or Epon-Araldite mixtures are commonly used resins for electron microscopy because of superior cutting properties and stability under the electron beam. Epon is currently sold under the name EMbed 812. Do not to add accelerator until the final step of embedment.

Acknowledgments

I thank Kent McDonald, Max Heiman, and Peter Walter for sharing valuable information about the high-pressure freezing and freeze substitution procedures and Margaret Bisher for helping to implement the procedure at Princeton University.

References

1. Elion, E. A. (2000) Pheromone response, mating and cell biology. *Curr. Opin. Microbiol.* **3,** 573–581.
2. White, J. M. and Rose, M. D. (2001) Yeast mating: getting close to membrane merger. *Curr. Biol.* **11,** R16–R20.
3. Marsh, L. and Rose, M. D. (1997) The pathway of cell and nuclear fusion during mating in *S. cerevisiae*, in *The Molecular and Cellular Biology of the Yeast Saccharomyces: Cell Cycle and Cell Biology* (J. R. Pringle, J. R. Broach, and E. W. Jones. eds.), vol. 3. Cold Spring Harbor Laboratory Press, Cold Spring Harbor, NY, pp. 827–888.
4. Gammie, A. E., Brizzio, V., and Rose, M. D. (1998) Distinct morphological phenotypes of cell fusion mutants. *Mol. Biol. Cell.* **9,** 1395–1410.
5. Wright, R. (2000) Transmission electron microscopy of yeast. *Microsc. Res. Tech.* **51,** 496–510.
6. Gammie, A. E. and Rose, M. D. (2002) Assays of cell and nuclear fusion. *Methods Enzymol.* **351,** 477–498.
7. Heiman, M. G., Engel, A., and Walter, P. (2007) The Golgi-resident protease Kex2 acts in conjunction with Prm1 to facilitate cell fusion during yeast mating. *J. Cell Biol.* **176,** 209–222.
8. McDonald, K. L. and Auer, M. (2006) High-pressure freezing, cellular tomography, and structural cell biology. *Biotechniques* **41,** 137, 139, 141 *passim.*
9. Dawes, C. J. (1979) *Biological Techniques for Transmission and Scanning Electron Microscopy*. Ladd Research Industries, Burlington, VT.
10. Hayat, M. A. (2000) *Principles and Techniques of Electron Microscopy: Biological Applications*, 4th ed. Cambridge University Press, Cambridge, England.
11. Robinson, D. G. (1987) *Methods of Preparation for Electron Microscopy: An Introduction for the Biomedical Sciences*. Springer-Verlag, Berlin.
12. Perlin, A. S. (2006) Glycol-cleavage oxidation. *Adv. Carbohydr. Chem. Biochem.* **60,** 183–250.
13. van Tuinen, E. and Riezman, H. (1987) Immunolocalization of glyceraldehyde-3-phosphate dehydrogenase, hexokinase, and carboxypeptidase Y in yeast cells at the ultrastructural level. *J. Histochem. Cytochem.* **35,** 327–333.
14. McDonald, K. (1999) High-pressure freezing for preservation of high resolution fine structure and antigenicity for immunolabeling. *Methods Mol. Biol.* **117,** 77–97.
15. Walther, P. and Ziegler, A. (2002) Freeze substitution of high-pressure frozen samples: the visibility of biological membranes is improved when the substitution medium contains water. *J. Microsc.* **208,** 3–10.

12

Isolation and In Vitro Binding of Mating Type Plus Fertilization Tubules From *Chlamydomonas*

Nedra F. Wilson

Summary

During fertilization in *Chlamydomonas*, adhesion and fusion of gametes occur at the tip of specialized regions of the plasma membrane, known as *mating structures* ***(1,2)***. The mating type minus (mt[–]) structure is a slightly raised dome-shaped region located at the apical end of the cell body. In contrast, the activated mating type plus (mt[+]) structure is an actin-filled, microvillouslike organelle. Interestingly, a similar type of "fusion organelle" is conserved across diverse groups ***(3)***. *Chlamydomonas* provides an ideal model system for studying the process of gametic cell fusion in that it is amenable to genetic manipulations as well as cell and molecular biological approaches. Moreover, the ease of culturing *Chlamydomonas* combined with the ability to isolate the mt(+) fertilization tubule and the development of in vitro assays for adhesion makes it an ideal system for biochemical studies focused on dissecting the molecular mechanisms that underlie the complex process of gametic cell fusion ***(4)***.

Key Words: *Chlamydomonas*; gamete; cell fusion; fertilization tubule; flagella; signal transduction.

1. Introduction

Gametes of the haploid eukaryote *Chlamydomonas* are generated by differentiation of mating type plus (mt+) and/or mating type minus (mt–) vegetative cells ***(5,6)***. Similar to fertilization in multicellular organisms, an initial adhesion between mt(+) and mt(–) gametes activates a signal transduction pathway that generates gametes competent to undergo cell adhesion and fusion ***(7)***. This signal transduction pathway can be induced in gametes of a single mating type by incubation with flagella isolated from the opposite mating type or by incubation with dibutyryl cyclic adenosine monophosphate (cAMP). The increase in intracellular cAMP levels induces loss of cell walls and activation of mating structures, which are the sites of cell body adhesion and fusion ***(1,2,7)***. The adhesion between gametes mediated by flagella orients

From: *Methods in Molecular Biology, vol. 475: Cell Fusion: Overviews and Methods*
Edited by: E. H. Chen © Humana Press, Totowa, NJ

gametes to optimize contact between the apical regions of their mating structures. The microvillouslike nature of the mt(+) mating structure, the fertilization tubule, has allowed the development of methods for its isolation and use in in vitro binding assays described below ***(4)***. Here I describe conditions for culturing *Chlamydomonas* vegetatively, inducing gametogenesis, and activating gametes by incubation with either flagella isolated from gametes of the opposite mating type or with dibutyryl cAMP. Finally, I describe the isolation of fertilization tubules and an in vitro binding assay using isolated fertilization tubules and activated mt(–) gametes.

2. Materials

2.1. Chlamydomonas *Strains and Media*

1. *Chlamydomonas reinhardtii* strains 21gr (mt+) [66-1690] and 6145c (mt–) [cc-1691] (available from the *Chlamydomonas* Genetics Center, Duke University).
2. Stock solutions for media: 10% Na Citrate–$2H_2O$ (10 g/100 mL dH_2O); 1% $FeCl_3$–$6H_2O$ (1 g/100 mL dH_2O); 5.3% $CaCl_2$–$2H_2O$ (5.3 g/100 mL dH_2O); 10% $MgSO_4$–$7H_2O$ (10 g/100 mL dH_2O); 10% KH_2PO_4 (10 g/100 mL dH_2O); 2.2 *M* Na acetate. All stock solutions are made with ddH_2O and stored at 4 °C ***(8)***.
3. 10× trace metals stock solution: 16.2 m*M* H_3BO_3, 3.48 mM $ZnSO_4$, 1.79 m*M* $MnSO_4$, 0.84 m*M* $CoCl_2$, 0.83 m*M* Na_2MoO_4, 0.25 m*M* $CuSO_4$. Trace metal stock solution is made with ddH_2O and stored at 4 °C ***(8)***.
4. ½ R medium: For 1 L of medium, add the following to ddH_2O (*see* **Note 1**): 1.0 mL 10× trace metal stock, 5.0 mL 10% Na Citrate–$2H_2O$, 1.0 mL 1% $FeCl_3$–$6H_2O$, 3.0 mL 10% $MgSO_4$–$7H_2O$, 3.0 mL 10% NH_4NO_3, 2.1 mL 10% KH_2PO_4, 3.6 mL 10% K_2HPO_4–$3H_2O$, 5 mL 2.2 *M* Na acetate, pH to 6.8. Adjust pH down with 10% KH_2PO_4 or up with 10% K_2HPO_4–$3H_2O$. Autoclave media to sterilize and remove immediately from autoclave (*see* **Note 2; ref. *8***).
5. M-N medium: For 1 L of medium, add the following to ddH_2O: 1 mL of 10× trace metal stock, 5 mL 10% Na citrate–$2H_2O$, 1 mL 1% $FeCl_3$–$6H_2O$, 1 mL 5.3% $CaCl_2$–$2H_2O$, 3 mL 10% $MgSO_4$–$7H_2O$, 2.4 mL 10% K_2HPO_4–$3H_2O$. Adjust pH to 7.0 as described for ½ R medium. Autoclave media to sterilize and remove immediately from autoclave (*see* **Note 2; ref. *8***).

2.2. Activation of mt(+) Fertilization Tubules

1. Dibutyryl cAMP solution: 0.3 mM in M-N medium ***(4)***.
2. 100× papaverine: 15 m*M* papaverine prepared in fresh 100% DMSO. It is important to use a freshly opened vial of DMSO to prepare the papaverine stock (*see* **Note 3; ref. *4***).
3. Cell wall loss buffer: 0.075% Triton X-100, 0.5 m*M* EDTA ***(4)***.

2.3. Phallacidin Staining of Activated mt(+) Fertilization Tubules

1. FIX solution: 10 m*M* Hepes-OH, pH 7.2; 4% paraformaldehyde. Make fresh.
2. 0.2 *M* Sorensen's phosphate buffer, pH 7.2: Mix together 36 mL of 0.2 *M* Na_2HPO_4 and 14 mL of 0.2 *M* NaH_2PO_4.

3. 80% acetone solution: 2 m*M* Sorensen's phosphate buffer, pH 7.2; 30 m*M* NaCl, 80% acetone. Make fresh and bring to −20 °C before use.
4. 100% acetone solution: 100% acetone at −20 °C.
5. PBS: 150 m*M* NaCl, 10 m*M* Sorensen's phosphate buffer, pH 7.2.
6. BODIPY phallacidin staining solution: Pipette one aliquot of 200 U/mL stock solution of BODIPY phallacidin in 100% methyl alcohol into 1.5 mL microcentrifuge tube and dry down in the dark. Resuspend dried sample in PBS to yield a 50-U/mL solution (i.e., four volumes of PBS).
7. Mounting medium: Fluoromount-G mounting medium containing 2.5% of the quenching agent 1,4-diazabicyclo-[2.2.2]octane (Sigma-Aldrich Chemical Co, St. Louis, MO).

2.4. Isolation of mt(+) Fertilization Tubules

1. FTSB: 40 m*M* KCl, 50 m*M* EGTA, 0.5 m*M* EDTA, 1 m*M* AEBSF, 1 m*M* leupeptin (100× stock in dH_2O), 1 m*M* pepstatin A (100× stock in 100% methyl alcohol), 20 m*M* chymostatin (100× stock in 100% DMSO), 10 m*M* imidazole, pH 7.2.
2. 60% sucrose: 30 g sucrose/50 mL FTSB.
3. 50% sucrose: 25 g sucrose/50 mL FTSB.
4. 40% sucrose: 20 g sucrose/50 mL FTSB.
5. 20% sucrose: 10 g sucrose/50 mL FTSB.
6. 15% sucrose: 7.5 g sucrose/50 mL FTSB.
7. FTSB-sucrose: 150 m*M* NaCl, 8% sucrose, FTSB.
8. 30% Percoll: 1.8 mL 100% Percoll, 0.8 mL 60% sucrose, and 2.8 mL FTSB.

2.5. Isolation of g-Lysin

1. Hepes-Ca^{2+}: 1 m*M* $CaCl_2$, 10 m*M* Hepes.

2.6. Isolation of Flagella

1. 7% sucrose solution: 7% sucrose, 10 m*M* TrisCl, pH 7.2.
2. 25% sucrose solution: 25% sucrose, 10 m*M* TrisCl, pH 7.2.

2.7. Binding of Isolated mt(+) Fertilization Tubules

1. 1% glycine: 0.5 g glycine in 50 mL M-N medium.

3. Methods

3.1. Cell Culture

1. Vegetative cells are grown in 6-L flasks in ½ R medium with aeration at 22 °C on 13/11 hour light/dark cycle.
2. Allow cells to accumulate in bottom of 6-L flasks by negative phototaxis for 30–60 min under illumination by a bank of fluorescent lights (*see* **Note 4; ref. *4***).
3. Remove supernatant by aspiration with a water pump (Model #1P579E; Teel Water Systems, Dayton Electric Mfg., Co., Chicago, IL).

4. Sediment cells by centrifugation in 1-L polycarbonate centrifuge bottles at 4600 *g* for 20 min using a Dupont Sorvall H6000A rotor in a Dupont Sorvall RC-3B centrifuge.
5. Resuspend cells to 2×10^6 cells/mL in M-N medium and transfer to 6-L flasks.
6. Gametogenesis is induced by incubation in M-N medium overnight with aeration and under constant light ***(4–6)***.

3.2. Isolation of g-Lysin

1. Obtain flagellated, highly motile mt(+) and mt(–) gametes from 12-L cultures by negative phototaxis and centrifugation at 4600 *g* for 20 min (*see* **Subheading 3.1** and **Note 4**).
2. Resuspend mt(+) and mt(–) gametes separately in Hepes-Ca^{2+} at 3×10^8 cells/mL.
3. Mix equal numbers of mt(+) and mt(–) gametes together and incubate with gentle aeration for 30 min (*see* **Note 5**).
4. Remove cells by centrifugation at 4600 *g* for 20 min in 1-L polycarbonate centrifuge bottles in a Dupont Sorvall H6000A rotor in a Dupont Sorvall RC-3C centrifuge.
5. Transfer supernatant, which contains g-lysin, into 50-mL polycarbonate centrifuge tubes and clear by centrifugation at 21,000 *g* for 30 min in Dupont Sorvall SA-600 rotor in a Dupont Sorvall RC-5C centrifuge.
6. Filter-sterilize g-lysin and store in small aliquots at –80 °C.

3.3. Isolation of Flagella

1. Highly motile cells are obtained by negative phototaxis (*see* **Subheading 3.1** and **Note 4; refs. *4,9***).
2. Harvest cells by centrifugation in 1-L polycarbonate centrifuge bottles at 4600 *g* for 20 min at 4 °C.
3. Discard supernatant and resuspend cell pellet in ice cold 7% sucrose solution. ***All remaining steps on ice or at 4 °C.***
4. Transfer cells to chilled beaker containing stir bar.
5. While gently stirring cells, pH shock cells by dropwise addition of 1 N HAc to yield a final pH of 4.4. Examine cells by phase contrast microscopy to confirm loss of flagella.
6. After 1 min, increase pH to 7.2 by dropwise addition of 1 N KOH.
7. Transfer 25 mL of cells to 50-mL conical polycarbonate centrifuge tubes and underlay with 12 mL of ice cold 25% sucrose solution.
8. Centrifuge cells at 2600 *g* for 10 min.
9. Transfer upper layers of step gradient, including material at interface, to new centrifuge tube and underlay again with 25% sucrose solution.
10. Repeat centrifugation at 2600 *g* for 10 min.
11. Harvest flagella from upper layer of step gradient by centrifugation at 37,000 *g* for 20 min.
12. Resuspend flagella in M-N medium, flash freeze, and store in liquid N_2 until used.

3.4. Activation of Gametes

1. Highly motile gametic cells are obtained by negative phototaxis (*see* **Subheading 3.1** and **Note 4**).
2. Gametes are concentrated approximately 300-fold over their concentration by resuspension in ~40 mL of M-N medium (i.e., 15 mL of M-N medium per 6-L starting cell volume; final cell concentration ~3 × 10^9 cells/mL) and incubated in the light with vigorous aeration (*see* **Note 6; ref. *4***).
3. To cells, add dibutyryl cAMP to a final concentration of 15 m*M* and 100× papaverine to a final concentration of 0.15 m*M* and continue incubation in light with vigorous aeration for 25–60 min (*see* **Note 3**).
4. Check for activation of gametes by cell wall loss assay (*see* **Subheading 3.5**).

3.5. Cell Wall Loss Assay

1. Add 30 µL of cell suspensions (1–3 × 10^6 cells/mL) to 1 mL of cell wall loss buffer in a 1.5-mL microfuge tube.
2. Vortex briefly, and centrifuge for 20 s at 14,000 rpm at room temperature.
3. Check for absence of chlorophyll in sedimented material compared with nondetergent-treated control (*see* **Note 7; ref. *10***).

3.6. Isolation of mt(+) Fertilization Tubules

1. Obtain flagellated, highly motile cells from 12 L of mt(+) gametes by negative phototaxis and centrifugation at 4600 *g* for 20 min (*see* **Subheading 3.1; ref. *4***).
2. Concentrate sedimented cells approximately 300-fold over starting concentration by resuspension in 15 mL M-N medium/6 L original culture volume.
3. Activate mt(+) gametes by incubation with 15 m*M* dibutyryl cAMP and 0.15 m*M* papaverine (*see* **Subheading 3.2**).
4. Confirm activation of mt(+) gametes by cell wall loss and the appearance of fertilization tubules as determined by BODIPY phallacidin staining (*see* **Subheading 3.7**). ***All remaining steps on ice or at 4 °C.***
5. Wash activated gametes twice with ice cold FTSB by centrifugation at 2800 *g* for 4 min (*see* **Note 8**).
6. Resuspend activated gametes in ~40 mL per 6 L original culture volume of FTSB and homogenize on ice with an Omni 5000 homogenizer and 35-mm generator probe (Omni International, Gainsville, VA) until 80%–90% of cells are disrupted as determined by phase contrast microscopy.
7. Remove unbroken cells, cell fragments, and larger organelles by centrifugation at 2600 *g* for 4 min in 50-mL conical polycarbonate centrifuge tubes.
8. Transfer supernatant to 50-mL round-bottom polycarbonate centrifuge tubes and clear again by centrifugation at 6000 *g* for 6 min in a Dupont Sorvall SA-600 rotor.
9. Harvest fertilization tubules from supernatant by centrifugation at 37,000 *g* for 20 min in 50 mL polycarbonate round-bottom centrifuge tubes (*see* **Note 9**).
10. Resuspend fertilization tubules in 2.4 mL of 60% sucrose (*see* **Note 10; ref. *4***).
11. Prepare four sucrose step gradients by pipetting into four 6-mL conical centrifuge

tubes, 1 mL of each sucrose solution in the following order: 60%, 50%, and 40%.

12. Overlay the 40% sucrose step with 0.6 mL of the fertilization tubules resuspended in 60% sucrose from **step 10** (*see* **Note 10**).
13. Pipette on top of the fertilization tubule sample 1 mL of 20% sucrose followed by 1 mL of 15% sucrose.
14. Spin the sucrose step gradient at 14,000 *g* for 25 min in a Dupont Sorvall HB-4 rotor.
15. Collect fractions (0.5 mL) from top of each gradient, and identify fractions containing fertilization tubules by staining with BODIPY phallacidin (*see* **Subheading 3.7**).
16. Fractions enriched in fertilization tubules are pooled and fertilization tubules harvested by centrifugation at 100,000 *g* for 30 min in a Beckman TLA100.3 rotor.
17. Discard supernatant and resuspend pellet containing fertilization tubules in 1.2 mL of 8% sucrose and shear by passage through 25-gauge needle seven times.
18. Prepare two gradients of 30% Percoll. To do this, in 6-mL conical centrifuge tube add 1.8 mL 100% Percoll, 0.6 mL fertilization tubule sample, 0.8 mL 60% sucrose, and 2.8 mL FTSB. Mix solution by inversion (*see* **Note 11**).
19. Percoll gradient is formed in situ by centrifugation at 37,000 *g* for 22 min in a Dupont Sorvall SA-600 rotor (*see* **Note 11**).
20. Collect fractions (0.5 mL) from top of each gradient from above and assay for presence of fertilization tubules by staining with BODIPY phallacidin (*see* **Subheading 3.7**).
21. To harvest fertilization tubules and remove Percoll, pool fractions enriched in fertilization tubules, dilute with FTSB, and centrifuge for 1 h at 100,000 *g* in a Beckman TLA100.3 rotor.
22. Resuspend fertilization tubules, which form a fluffy, white sediment on top of the clear Percoll pellet, in a small volume of FTSB, flash freeze, and store in liquid N_2 until used.

3.7. BODIPY Phallacidin Staining of Fertilization Tubules

1. Precoat wells of eight-well glass slide (Cel-Line Associates, Inc., Newfield, NJ) with 10 µL of an aqueous solution of 0.1% polyethylenimine for 2 min (***4,11***).
2. Excess polyethylenimine is removed from the well by blotting with a Kim-Wipe from the side of the well. Allow the slide to air dry.
3. Samples of cells and cell fractions are fixed by mixing with an equal volume of freshly prepared FIX buffer (see **Subheading 2.3**) and 5 µL portions applied to wells of the precoated glass slide.
4. Samples are dried down onto slide.
5. Immerse slides for 6 min in 80% acetone at −20 °C (*see* **Note 12**).
6. Transfer slides to 100% acetone at −20 °C for 6 min.
7. Air dry slides.
8. Incubate slides with 20 µL/well of BODIPY phallacidin staining solution for 25 min at 37 °C in the dark (*see* **Note 13**).
9. Wash slides by serial immersion in PBS for 6 min for a total of three changes.

10. Apply a drop of Fluoromount-G mounting medium containing a quenching agent to each well of slide and cover with No. 0 coverslip.
11. Seal coverslip with clear fingernail polish.
12. Store slides in dark until examination.

3.8. Quantification of Fertilization Tubule Enrichment

To quantify enrichment of fertilization tubules, samples of homogenized cells, the pellet of harvested fertilization tubules from the sucrose gradient, and the pellet of harvested fertilization tubules from the Percoll gradient are serially diluted and stained with BODIPY phallacidin, and the number of fertilization tubules is counted using a fluorescent microscope *(4)*.

1. Apply 5 μL of fertilization tubules from the various isolation steps that have been serially diluted with FTSB to 0.1% polyethylenimine coated eight-well slides (*see* **Subheading 3.7**).
2. The following dilutions for fractions are typically used: homogenized cells, 1:4, 1:8, 1:16, 1:32; pellets from sucrose gradient and Percoll gradient fractions, 1:10, 1:20, 1:40, 1:80, 1:160, 1:320, 1:640, 1:1280. Dilute samples in FTSB.
3. Fertilization tubules are stained with BODIPY phallacidin (*see* **Subheading 3.7**).
4. Count the number of fertilization tubules within the area defined by the photographic field in the microscope.
5. At least 20 random fields per sample and two to three different dilutions per sample are counted in duplicate (*see* **Note 14**).
6. The total number of fertilization tubules in the sample is determined by calculating the average number of fertilization tubules per field based on the dilution and the calculated number of photographic fields per well.

3.9. Binding of Isolated Fertilization Tubules to Activated mt(–) Gametes

1. Activate mt(–) gametes by incubation with flagella isolated from mt(+) gametes at a ratio of 10 flagella/cell twice for 10 min each (*see* **Note 15; refs.** ***2,12***).
2. Confirm mt(–) gametic activation with the cell wall loss assay (*see* **Subheading 3.5**).
3. Fix activated mt(–) gametes by incubation for 6 min at room temperature with an equal volume of freshly prepared fix solution (*see* **Subheading 3.5** and **Note 16**).
4. Wash fixed cells three times with an excess of 1% glycine (*see* **Subheading 2.7**).
5. Resuspend fixed mt(–) cells at 2×10^7 cells/mL in M-N medium.
6. In a 1.5-mL microcentrifuge tube, place 100 μL of mt(–) cells and add 1 μL of freshly isolated fertilization tubules to yield ~10 fertilization tubules/mt(–) cell.
7. Incubate 15 min at room temperature on a reciprocal shaking platform.
8. Remove unbound fertilization tubules by adding an excess (1.4 mL) of M-N medium, and centrifuge at 14,000 rpm for 20 s.

9. Discard supernatant, and resuspend sedimented cells in 1.5 mL M-N medium and centrifuge at 14,000 rpm for 20 s.
10. Discard supernatant and resuspend cells in M-N medium.
11. To visualize bound fertilization tubules, cells are fixed by mixing with an equal volume of FIX solution (*see* **Subheading 2.3**) and stained with BODIPY phallacidin (*see* **Subheading 3.7**).

4. Notes

1. It is important to add medium components in the order listed to prevent precipitation of reagents with one other.
2. Medium should be immediately removed from the autoclave upon completion of the liquid cycle to prevent precipitation of components. If precipitation does occur, the precipitated components will occasionally return to solution as the medium cools. Do not use medium that remains precipitated.
3. The use of old DMSO to make up the papaverine stock results in poor activation of gametes. For best activation of gametes, use a container of DMSO that has not been opened for longer than 1 week. In addition, do not use a papaverine stock solution that is older than 5 days.
4. I observed that preselecting for cells (as both vegetative cells and gametes) that were able to undergo rapid negative phototaxis resulted in gametes that were optimal for activation by dibutyryl cAMP and papaverine. Cells unable to undergo negative phototaxis most likely reflect failure at signal transduction and are therefore removed by aspiration with a water pump (*4*). Moreover, I have found that when all of the cells in a culture are unable to undergo negative phototaxis, I obtained very poor gametic activation as assayed by either staining with BODIPY phallacidin to detect fertilization tubules or using the cell wall loss assay.
5. The 30-min incubation time reflects the average amount of time required for ~100% of the cells to shed their walls as assayed by the cell wall loss assay ***(4)***.
6. To achieve maximum activation of gametes and generation of fertilization tubules, it is necessary to vigorously aerate the activating gametes. To do this, concentrated mt(+) gametes are placed in a glass beaker. A pipette attached to an air supply is placed in the center of the beaker, and the flow of air is adjusted such that cells are almost bubbled out of the beaker ***(4)***.
7. To detect loss of cell walls, an aliquot of cells is placed into a 1.5-mL microcentrifuge tube containing an excess of the cell wall loss buffer. After brief vortexing, the cells are pelleted in the microcentrifuge. If walls have been lost, the detergent present in the cell wall loss buffer will solubilize the plasma and chloroplast membrane, releasing chlorophyll into the supernatant. As a result, the supernatant will appear green and the cell pellet will be white. For comparison purposes, an equal volume of cells in the absence of cell wall loss buffer can be pelleted. In the absence of detergent, chlorophyll is not extracted from the cells, and therefore the cell pellet remains green ***(10)***.
8. This wash step is to remove the cell walls that are released upon activation of gametes. The protocol described here suggests two washes; however, samples should

be microscopically examined for the presence of cell walls to determine the exact number of washes needed.

9. This centrifugation step will pellet both the isolated fertilization tubules as well as flagella.
10. To generate a solution whose density is equivalent to 30% sucrose, it is necessary to resuspend the isolated fertilization tubules in FTSB containing 60% sucrose ***(4)***.
11. The Percoll gradient is formed *in situ* during the centrifugation. Samples are subsequently diluted with FTSB and mixed with Percoll before the centrifugation step.
12. It is important to make sure that the 80% and 100% acetone solutions are at approximately −16 °C to −20 °C before using them to permeabilize and extract cells prior to phallacidin staining.
13. A plastic container with a tight-fitting lid is used as a staining chamber. The addition of moist paper towels to the bottom of the container keeps the chamber humidified. Once BODIPY phallacidin is added to the slides, place slides on the wet paper towels, add the container lid, and wrap entire container with aluminum foil.
14. All of the samples from a dilution series are stained in duplicate with BODIPY phallacidin and examined by fluorescence microscopy. Fertilization tubules are counted only in dilutions in which clearly isolated fertilization tubules could be observed. If the numbers of fertilization tubules are different in the duplicate samples, new dilutions of those samples should be stained and fertilization tubules counted again.
15. Although I describe adding flagella isolated from mt(+) gametes twice for 10 min each to mt(−) gametes, flagella should be added until the mt(−) gametes have undergo activation as evidenced by the loss of their cell walls in the cell wall loss assay described in **Subheading 3.5.**
16. The mt(−) gametes were fixed with paraformaldehyde following activation with either dibutyryl cAMP or flagella isolated from mt(+) gametes. This allowed visualization of binding but not fusion of the isolated fertilization tubules with the activated mt(−) mating structure. For studies examining fusion of these isolated organelles with mt(−) gametes, do not fix the mt(−) gametes after activation.

References

1. Friedmann, I., Colwin, A. L., and Colwin, L. H. (1968) Fine-structural aspects of fertilization in *Chlamydomonas reinhardi. J. Cell Sci.* **3,** 115–128.
2. Goodenough, U. W. and Weiss, R. L. (1975) Gametic differentiation in *Chlamydomonas reinhardtii* III. Cell wall lysis and microfilament-associated mating structure activation in wild-type and mutant strains. *J. Cell Biol.* **67,** 623–637.
3. Wilson, N. F. and Snell, W. J. (1998) Microvilli and cell–cell fusion during fertilization. *Trends Cell Biol.* **8,** 93–96.
4. Wilson, N. F., Foglesong, M. J., and Snell, W. J. (1997) The *Chlamydomonas* mating type plus fertilization tubule, a prototypic cell fusion organelle: isolation, characterization, and in vitro adhesion to mating type minus gametes. *J. Cell Biol.* **30,** 1537–1553.

5. Sager, R. and Granick, S. (1954) Nutritional control of sexuality in *Chlamydomonas reinhardi*. *J. Gen. Physiol.* **37,** 729–742.
6. Pan, J.-M, Haring, M. A., and Beck, C. F. (1996) Dissection of the blue-light dependent signal-transduction pathway involved in gametic differentiation of *Chlamydomonas reinhardtii*. *Plant Physiol.* **112,** 303–309.
7. Pasquale, S. M. and Goodenough, U. W. (1987) Cyclic AMP functions as a primary sexual signal in gametes of *Chlamydomonas reinhardtii*. *J. Cell Biol.* **105,** 2279–2292.
8. Sager, R. and Granick, S. (1954) Nutritional control of sexuality in *Chlamydomonas reinhardtii*. *J. Gen. Physiol.* **37,** 729–742.
9. Witman, G. B., Carlson, K., Berliner, J., and Rosenbaum, J. L. (1972) *Chlamydomonas* flagella. I. Isolation and electrophoretic analysis of microtubules, matrix, membranes, and mastigonemes. *J. Cell Biol.* **54,** 507–539.
10. Buchanan, M. J., Imam, S. H., Eskue, W. A., and Snell, W. J. (1989) Activation of the cell wall degrading protease, lysin, during sexual signalling in *Chlamydomonas*. The enzyme is stored as an inactive, higher relative molecular mass precursor in the periplasm. *J. Cell Biol.* **108,** 199–207.
11. Detmers, P. A., Carboni, J. M., and Condeelis, J. (1985) Localization of actin in *Chlamydomonas* using antiactin and NBD-phallacidin. *Cell Motil.* **5,** 415–430.
12. Misamore, M. J., Gupta, S., and Snell, W. J. (2003) The *Chlamydomonas* Fus1 protein is present on the mating type plus fusion organelle and required for a critical membrane adhesion event during fusion with minus gametes. *Mol. Biol. Cell.* **14,** 2530–2542.

13

Optical Imaging of Cell Fusion and Fusion Proteins in *Caenorhabditis elegans*

Star Ems and William A. Mohler

Summary

Cell fusion is a very dynamic process in which the entire membrane and cellular contents of two or more cells merge into one. Strategies developed to understand the component processes that make up a full fusion event require imaging to be performed over a range of space and time scales. These strategies must cover detection of nanometer-sized pores, monitoring cytoplasmic diffusion and the dynamic localization of proteins that induce fusion competence, and three-dimensional reconstruction of multinucleated cells. *Caenorhabditis elegans*' small size, predictable development, and transparent body make this organism optimal for microscopic investigations. In this chapter, focus is placed on light microscopy techniques that have been used thus far to study developmental fusion events in *C. elegans* and the insights that have been gained from them. There is also a general overview of the developmental timing of the cell fusion events. Additionally, several protocols are described for preparing both fixed and live specimens at various developmental stages of *C. elegans* for examination via optical microscopy.

Key Words: Cell fusion; microscopy; Nomarksi; laser scanning confocal microscopy; multiphoton microscopy; *C. elegans*; antibody staining.

1. Introduction

1.1. Cell Fusion in Caenorhabditis elegans

Syncytium formation in *C. elegans* occurs by a conserved sequence of fusion events during establishment of the body plan *(1,2)*. Cell fusion is part of the normal developmental program that governs the formation of the hypodermis, uterus, vulva, male tail, and pharynx in *C. elegans (2,3)*. The largest syncytium (hypodermal cell 7[hyp7]) in the adult hermaphrodite contains 138 nuclei (144 nuclei in males); other syncytia contain 15, 12, 9, 8, 6, and 4 nuclei, with

From: *Methods in Molecular Biology, vol. 475: Cell Fusion: Overviews and Methods*
Edited by: E. H. Chen © Humana Press, Totowa, NJ

numerous trinucleate and binucleate cells distributed throughout other fusing tissues *(4)*. Next is a general overview of the developmental timing and control of cell fusion events in *C. elegans* that can serve as a guide for studying specific cell–cell fusion events.

1.1.2. Embryonic Fusion

The first fusion event (after fusion between the sperm and egg) in embryonic development occurs just prior to the onset of elongation, during hypodermal cell migration and ventral enclosure *(1)*. Because the embryo is nonmotile at this stage, these early hypodermal fusions have yielded the best time-lapse images of cell–cell fusion. Cell fusions start in the anterior region of the *C. elegans* embryo and progress toward the posterior region *(1)*. The majority of embryonic hypodermal fusions occur as the embryo elongates (~3 h after they are born) and are completed between the 1.5-fold (~430 min after first cleavage) and 2-fold elongation stages *(1)*, giving rise to eight distinct syncytia (hyp1–7 and hyp10) *(4)*. Little is known about the precise timing of fusions in the pharynx *(4)*.

1.1.3. Larval Fusion

The uterus, vulva, male tail, and hypodermis all undergo additional cell fusions during larval development. In these organs, specification of cell fusion fate (a developmentally regulated alternative to other mononucleated cell fates) is often revealed by analysis of genetic mutants.

The anchor cell controls cell fate specification in the uterus and vulva, which begins at mid larval stage 2 (L2; **ref. 5**). There are several types of component parts of the uterus (all syncytial), including the du cell, toroidal cells, and utse and uv cells. Uterine syncytia are formed by cell fusion among descendents of the dorsal and ventral uterine precursor cells during mid L4 *(5)*. Fusion between a subset of π cell progeny forms the utse, which later engulfs and fuses with the anchor cell during early to mid L4 *(5)*.

Vulva development is closely linked to uterus development and the continued development of the hypodermis. Vulva development begins at the first larval stage *(3)*. There are six vulval precursor cells (P[3–8].p), two of which produce daughter cells that fuse to hyp7 (P4.p and P8.p) during mid L3 *(6)*. The cells of the vulva primordium undergo longitudinal and transverse migration and subsequent fusion events in five (vulA, vulC, vulD, vulE, and vulF) of the seven stacked toroids surrounding the vulval lumen *(7)*. The resultant vulva structure is composed of six syncytia: vulA, vulC, vulD, vulE, vulF, and the utse/anchor cell *(7)*.

Within the epidermis, a series of daughters of the seam hypodermal cells fuse in four waves with hyp7 as the larva elongates and grows during L1–L4 *(1)*. In addition, specialized fusion shapes the male tail. Cell fusions in the tail start anteriorly and progress posteriorly during mid to late L4 ***(8)***. Male-specific cell fusions occur among the tail–tip cells hyp8–hyp11 and hyp13; onset is marked by hyp8 fusion to hyp11, which subsequently fuses to hyp9 and hyp10 ***(8)***. Sex-specific fusion is also observed in the hyp13 cell, which fuses to hyp7 during L2 in hermaphrodites but remains an isolated binucleate cell in males until fusing with hyp7 in the L4 stage ***(8)***. The final morphology of the mature tail is involves a combination of additional cell migrations and cell fusions ***(8)***.

1.1.4. Fusogen Proteins

The *eff-1* gene encodes an integral membrane protein with a large N-terminal extracellular domain containing a hydrophobic peptide region and a short cytoplasmic C-terminal tail ***(9)***. This protein has been shown to be required for most fusion events in the hypodermis, pharynx, and vulva of *C. elegans* ***(9,10)***, whereas sperm–egg fusion involves a different (unknown) fusogen ***(11)***. EFF-1 overexpression in nonfusing *C. elegans* cell types is capable of inducing membrane permeance with similar kinetic characteristics to that of wild-type fusing cells (*see* **Fig. 3,** later; **refs. *10,11***) and can induce fusion in insect cells ***(12)***. Thus, EFF1 is both necessary and sufficient for cell fusion in *C. elegans* ***(9–11)***. Recently, AFF-1, a paralog of EFF-1, has also been found to be both necessary and sufficient for fusion of a distinct set of cell types, including the anchor cell and the adult seam ***(13)***.

1.2. Imaging

Several microscopic techniques that have been used to study cell fusion in *C. elegans* are discussed. The imaging techniques are presented in stages of increasing complexity, with the least complex optics (Nomarski) presented first.

1.2.1. Nomarski Differential Interference Contrast Microscopy

Imaging cells or cellular structures in *C. elegans* has predominantly utilized differential interference contrast (Nomarski or DIC) microscopy. Nomarski microscopes convert the refractive index differences within a transparent specimen into intensity differences that are observed as contrast ***(14)***. Probably the most notable usage of Nomarski microscopy in *C. elegans* was determination of the complete cell lineage ***(2,3,15)***. It has also been used to characterize the morphology of structures that undergo cell fusion ***(2,8,15)*** and to identify mutations in genes that regulate the fusion process ***(16–19)***. An important note when studying cell fusion is the limited contrast of Nomarski imaging, which does not permit accurate resolution of plasma membranes but does allow the observation of nuclear positions that can be indicative of fusion events.

1.2.2. Fluorescence Microscopy

Investigations focused on specialized structures and proteins within a sample must employ an imaging strategy in addition to Nomarski microscopy to label their target of interest. One method involves fluorescently conjugated probes (e.g., chemical dyes linked to antibodies) or proteins fused to fluorescent protein domains, which are visualized by fluorescence microscopy. A fluorescence microscope works via filters and dichroic mirrors, which separate the excitation light (shorter wavelength) from the emission light (longer wavelength). Advances in fluorescent markers, multiple filters, and spectrally resolved detectors allow an experimenter to label multiple structures (typically three but possibly more) in a single sample. This fact is often crucial in proving that cell–cell fusion has occurred by showing the mixing of spectrally distinct dyes from two cells. Fluorescence microscopy combined with immunocytochemical staining of intercellular junctions by the monoclonal antibody MH27 has been a staple technique in recognizing epithelial cell fusions and in characterizing factors that influence the fusion process in *C. elegans* (**Fig. 1A–D; refs. *1,8,9,16–19***). More recently, a transgene marker for the same structure, AJM-1::green fluorescent protein (GFP), has allowed similar observations in live specimens (*see* **Fig. 1A–D; refs. *9,13,20***). Scoring the occurrence of cell fusions using these markers can be undertaken on a standard widefield epifluorescence microscope, including some high-resolution dissecting microscopes. However, disadvantages of widefield fluorescence microscopy include blurring of the image by out-of-focus light and photobleaching or phototoxicity to the specimen.

1.2.3. Confocal and Multiphoton Microscopy

Laser scanning confocal microscopy (LSM) overcomes image blurring due to out-of-focus light by use of an aperture or "pinhole" that prevents light generated outside of the focal plane from reaching the oculars and detector ***(14,21)***. The aperture thus allows optically sectioned images to be acquired in three dimensions (x, y, and z) without image deconvolution. This has proven especially crucial for three-dimensional sample reconstruction to understand the cell structure changes associated with cell fusion in *C. elegans* ***(1,7,9,11,12,19,20,22–24)***.

A disadvantage that LSM shares with widefield fluorescence is that the light path leading from the objective to the focal plane and beyond is excited to fluoresce, thereby generating photobleaching and phototoxicity outside of the field of view. Use of low-intensity laser illumination and/or nonlinear (multiphoton) optics circumvents these disadvantages. Spinning disk confocal microscopes (SDCMs) produce images by scanning the specimen with multiple laser beamlets and collecting light through multiple pinholes (up to 20,000), thereby generating high-frequency, low-intensity illumination, and

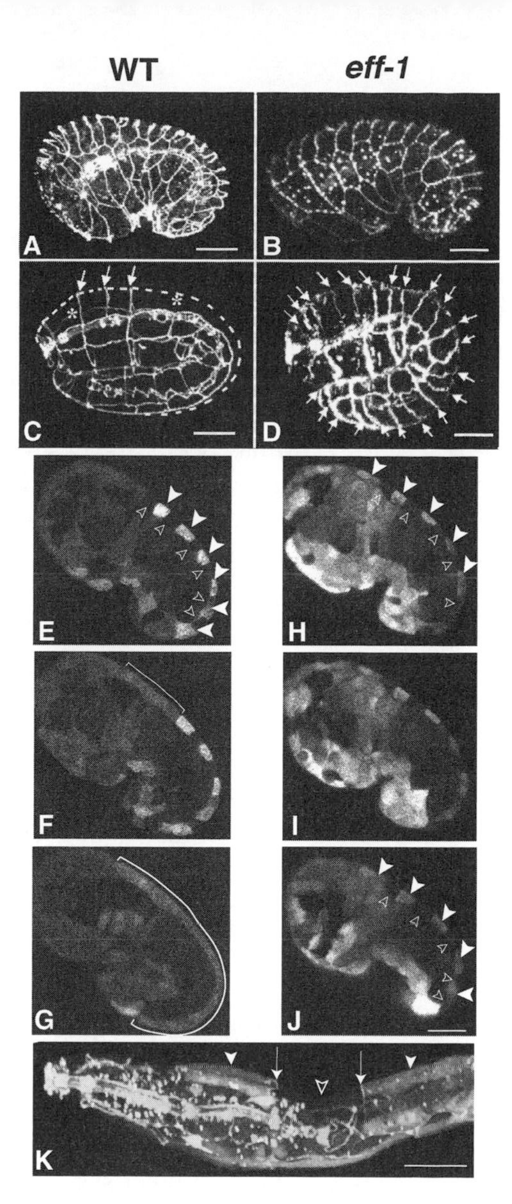

Fig. 1. EFF-1 is necessary for epidermal cell fusion, as assayed by retention of intercellular junctions and lack of membrane permeability. **(A–D)** Adherens junctions in wild-type or *eff-1 (hy21)* mutant embryos labeled with the monoclonal antibody MH27 (**A,C**) or AJM-1::GFP (**B,D**). Embryos in A and B were imaged at the onset of elongation. Embryos in C and D are at the twofold elongation stage. In C, asterisks indicate cells that have fused; arrows indicate unfused cells. In D, cell fusion failure, arrows indicate dorsal junctions. Bar = 10 μm. **(E–K)** Variegated cytoplasmic GFP expression in fusing cells. In E–G, cytoplasmic diffusion between nonfluorescent cells (open arrowheads) and cells labeled with cytoplasmic *lbp-1p::gfp* (white arrowheads) is indicated. Time-lapse optical sections show the development of wild-type (**E–G**) and *eff-1(oj55)* (**H–J**) embryos from comma to 1.5-fold. Cells progressively fuse and mix their cytoplasms during wild-type elongation In F,G, syncytia are indicated by outline. Cytoplasms remain distinct during *eff-1(oj55)* elongation (**I,J**). In K, *eff-1(oj55)* L1 expressing both AJM-1::GFP and *lbp-1p::gfp*. Adherens junctions (arrows) separate three unfused cells with distinct cytoplasms. Bar = 10 μm. (Adapted from **ref. *9*.**)

these instruments capture photons through the use of charge-coupled device cameras with severalfold higher quantum efficiency than the photomultiplier tubes on typical LSMs ***(25)***. The SDCMs have been used to study *C. elegans* embryology, including cell fusion ***(11,26,27)***. Multiphoton microscopy (MPLSM) generates fluorescent signal by the simultaneous absorption of two or three low-energy photons (most probable only at the focal point), thereby limiting photodamage to the plane of the optical section and eliminating the need for a pinhole ***(28)***. The MPLSM has been used in several studies involving live *C. elegans* embryology, including cell fusions ***(9,20,29,30)***. An additional advantage of MPLSM is that thicker specimens can be imaged because of the longer wavelength used for fluorescence excitation (e.g., 900 nm or 1047 lasers ***[30]***).

1.2.4. Multidimensional Imaging

Living specimens change dynamically in both space and time. Specifically, many of the fusing cells of *C. elegans* adopt three-dimensional morphologies that cannot be captured in a single optical section. Multidimensional (multicolor three-dimensional time-lapse) imaging has been essential in delineating the events leading up to and during cell–cell fusion ***(30)***. Thus far, only embryonic *C. elegans* samples have been used for four- and five-dimensional image acquisition ***(11,20,22,30–33)***. Gerlich and Ellenberg have reviewed in detail the specific advantages and disadvantages of using different types of microscopy for multidimensional imaging ***(34)***.

1.3. Conclusions and Insights From Optical Imaging of **C. elegans** *Cell Fusion*

Because of the reproducibility of patterning and the optical compliance of nematode embryos, the dynamic structural intermediates during cell fusions are better understood in *C. elegans* than in most other systems. Syncytium formation in *C. elegans* is marked by several dynamic events that have been observed and measured by light microscopy: cytoplasmic mixing, membrane aperture expansion, and the disappearance of intercellular junctions. Loss of subapical junction markers (e.g., MH27 staining or AJM-1::GFP) has been routinely used to show cell–cell fusion (*see* **Fig. 1A–D; refs.** ***9–11,20,22***). However, disappearance of these junctions occurs at a relatively late stage during a cell fusion event, when two cells are almost completely unified ***(20)***.

Stepping backward chronologically from the final to the initial stages of the fusion process has involved the use of other fluorescent markers. Disappearance of the membranes between cells and widening of a single aperture through the cell–cell interface have been investigated using the membrane-specific dye FM4-64 in laser-permeabilized embryos (**Fig. 2; ref.** ***20***; *see also* **Chapter 14** on ultrastructural imaging in *C. elegans* for details on laser permeabilization

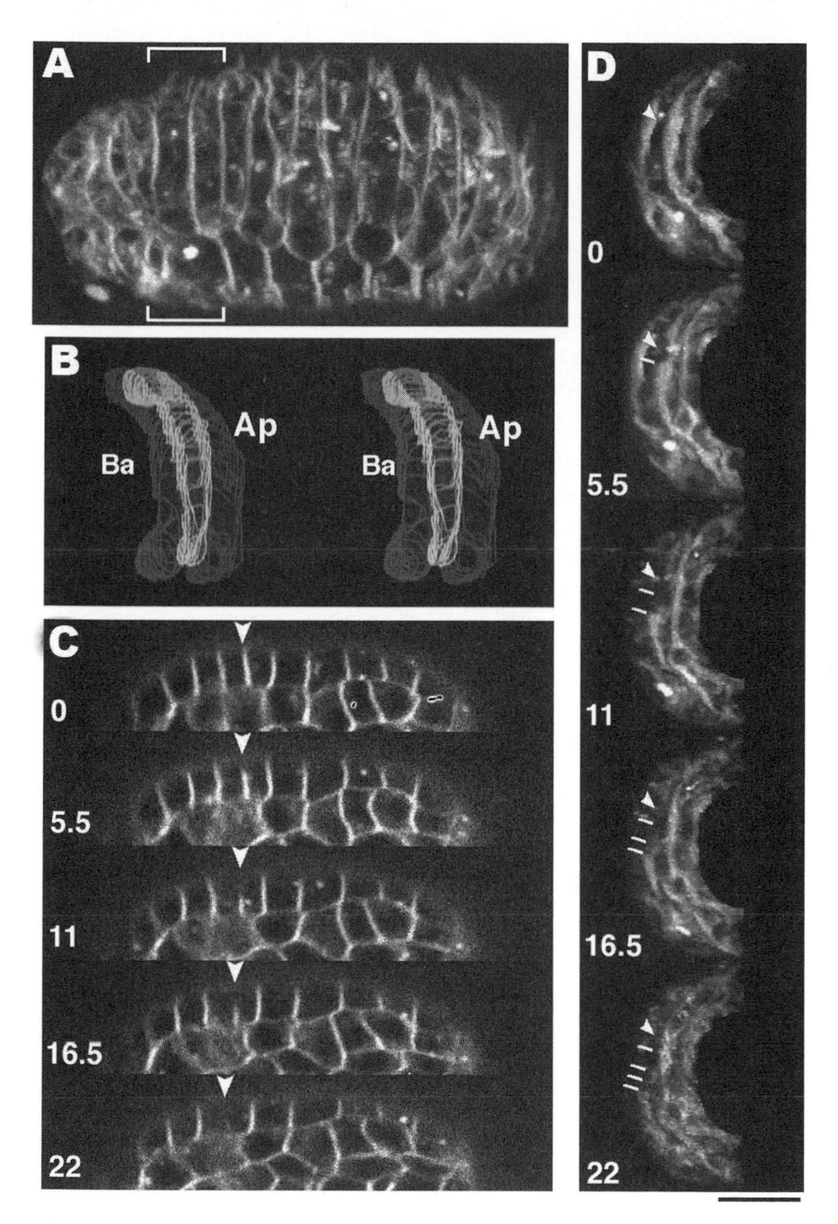

Fig. 2. Three-dimensional reconstruction of fusing cells in the hypodermis of an embryo labeled with FM4-64. **(A)** Dorsolateral three-dimensional reconstruction comprising 30 serial 0.5-μm-spaced multiphoton microscopy sections through half the embryo. Dorsal cells run vertically to the top of the image. Brackets indicate the region isolated in D. **(B)** Computer reconstruction showing the geometry of hyp7 precursors before fusion. Lateral cell borders that fuse appear white in the grayscale stereo image; apical (Ap) and basal (Ba) surfaces are convex and concave, respectively. **(C)** Cell fusion proceeds in an apical to basal direction. Temporal sequence showing the disappearance of lateral membranes (arrowhead) in an optical cross-section through a pair of fusing cells. **(D)** The fusion aperture propagates from a single apical origin. Time sequence of a stereo four-dimensional reconstruction of a fusing cell border from the embryo in A, rotated left 45° about the y-axis. The origin (arrowhead) and progress (tick marks) of the opening in one direction are marked along the length of the cell border. Times shown in C and D are in minutes. Bar = 10μm. (Adapted from **ref.** ***20***.)

of live embryos). However, this method did not resolve the very small pore or pores that are hypothesized to initiate formation of the fusion aperture.

To detect and characterize this very early cytoplasmic connection between fusing cells, soluble GFP has been observed diffusing across the newly permeable membranes at the beginning of cell fusion events (*see* **Fig. 1E–K; refs. *9,11***). Conveniently, several different *C. elegans* promoters (including those of *lbp-1* and *eff-1*) yield variegated expression of cytoplasmic GFP within fields of fusion-fated cells (*see* **Fig. 1E–K**). The rate of equilibration of GFP fluorescence has been used to approximate the initial fusion pore size in yeast mating pairs and has revealed the similarity between naturally occurring and ectopically induced EFF-1–dependent fusion pores in *C. elegans* (**Fig. 3; refs. *11,35***).

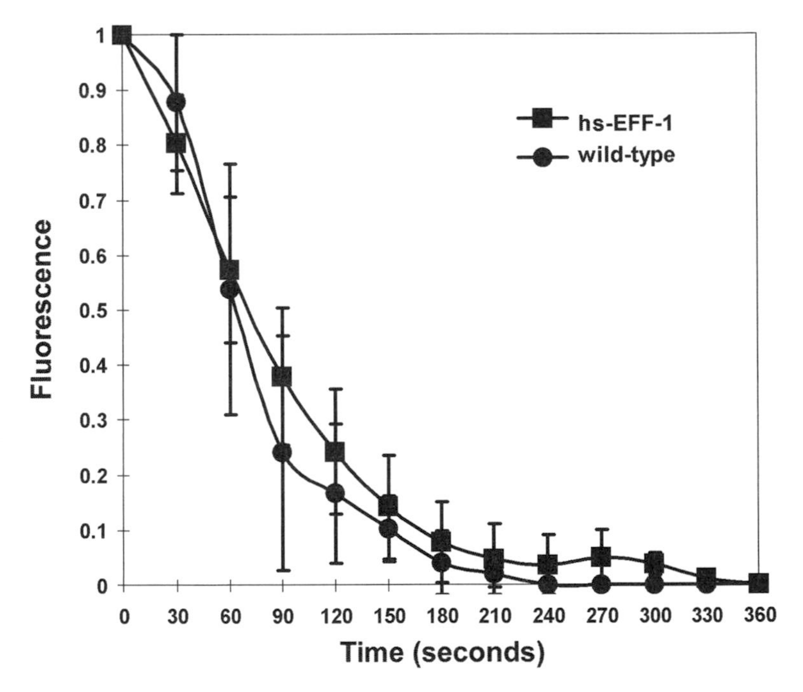

Fig. 3. Permeability of normal and ectopically induced fusion pores. Diffusion of green fluorescent protein across fusing membranes (as shown in **Fig. 1E–G**) was monitored in wild-type hypodermal cell fusions (circles, n = 10) and hs-EFF-1–induced cell fusions at the 100–200 cell stage (squares, n = 9). Mean fluorescence intensities for individual cells were normalized and averaged with initial and final brightness values set to 1 and 0, respectively. Time 0 for each fusion event was the last measurement before the rapid decrease in fluorescence for that bright cell as it fused to a dark neighbor. Plotted data points show the mean value for a group of measurements (normal vs. induced) at each time point after the initiation of permeability. Error bars show standard deviation. (Adapted from **ref. *11***.)

Finally, or actually earliest in the sequence of events, the fusogen molecule that appears to be responsible for forming such pores has been followed during its targeted localization to specific cell interfaces. When a migrating ventral cell first touches its fusion partner in the embryo, fluorescent EFF-1::GFP redistributes from cytoplasmic stores to only the fusion-fated region of the plasma membrane (**Fig. 4; ref.** ***11***). Observation of this localization in several different patterning contexts led to the hypothesis that the homotypic nature of EFF-1–dependent fusion involves stoichiometric homophilic binding between EFF-1 molecules on both cells' membranes ***(11)***.

Headings 2 and **3** of this chapter describe materials and methods that have been used to study cell fusion in embryos and hatched worms via light microscopy. The protocols provided herein will work for either inverted or upright light microscopes, using high-numerical-aperture immersion objectives. The methodology describes the preparation of both fixed specimens and live samples. Protocols for four-dimensional imaging have been published elsewhere ***(34)*** and are therefore not discussed in this chapter.

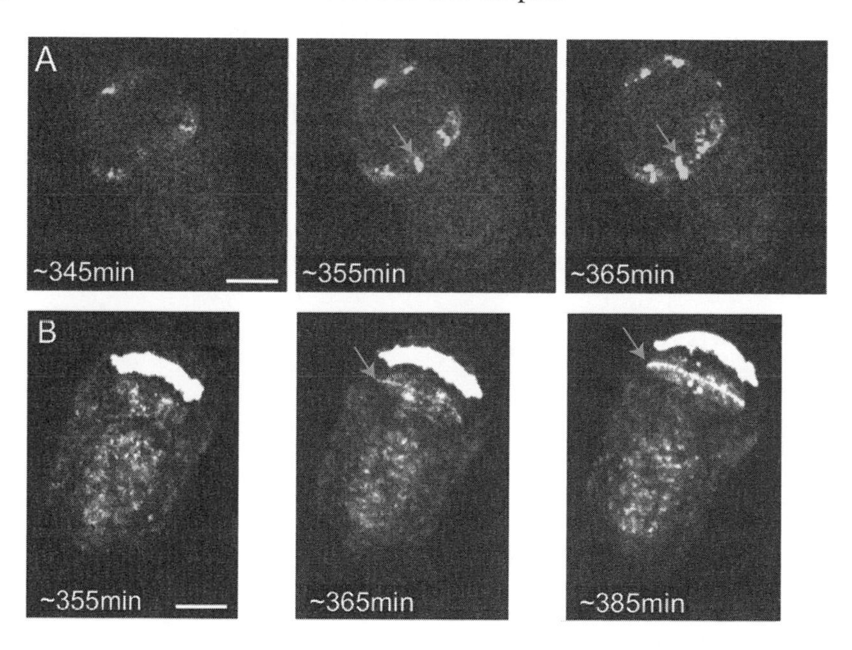

Fig. 4. Dynamics of EFF-1::green fluorescent protein (GFP)–targeted localization. **(A)** Novel contacts between cells already expressing EFF-1::GFP between two ventral cells. EFF-1::GFP junction formed within minutes of initial contact between the cells. **(B)** Dorsal accumulation occurs between two cells with preexisting cell contacts. Arrows indicate EFF-1::GFP accumulation. Anterior is at the top and posterior is at the bottom of each panel. Approximate time of development is shown. Bars = 10 μm. (Adapted from **ref.** ***11***.)

2. Materials

2.1. Collection of Mixed Population of Worms for Fixation (36)

1. Ice.
2. Eppendorf tubes.
3. Pasteur pipette and bulb.
4. Table-top centrifuge.
5. Sterilized M9 buffer: 22 m*M* KH_2PO_4, 42 m*M* Na_2HPO_4, 86 m*M* NaCl, 1 m*M* $MgSO_4$.

2.2. Collection of Large Quantities of Embryos for Fixation (36)

1. Ice.
2. Eppendorf tubes.
3. Pasteur pipette and bulb.
4. Table-top centrifuge.
5. Compound or dissecting microscope.
6. Sterilized M9 buffer (*see* **Subheading 2.1.**).
7. Bleach solution (prepare fresh): 1.0 mL bleach, 0.25 mL 10 N NaOH, 3.75 mL sterile H_2O.
8. Sterilized egg buffer: 118 m*M* NaCl, 48 m*M* KCl, 2 m*M* $CaCl_2$, 2 m*M* $MgCl_2$, 25 m*M* Hepes. Adjust pH to 7.3 with 1 N NaOH.

2.3. Slide Preparation for Fixation Using the Freeze Crack Method (36)

1. Ice.
2. Poly-L-lysine–coated coverslips (*see* **Subheading 2.9.2.**).
3. Poly-L-lysine–coated microscope slides.
4. Dry ice.
5. Razor blade.
6. Pasteur pipette and bulb.
7. Aluminum foil.
8. Table-top centrifuge.
9. Sterilized M9 Buffer (*see* **Subheading 2.1.**).

2.4. Fixation

2.4.1. Finney-Ruvkin Protocol (see ***Note 1; refs.*** **36,37**)

1. Rocker or rotator.
2. Eppendorf tubes.
3. Pasteur pipette and bulb.
4. Fume hood.
5. 37 °C storage.
6. 4 °C storage.
7. Table-top centrifuge.
8. Compound or dissecting microscope.
9. Sterilized M9 buffer (*see* **Subheading 2.1.**).

10. Phosphate buffered saline (PBS) pH 7.4: 1.8 m*M* KH_2PO_4, 4.3 m*M* Na_2HPO_4, 2.7 m*M* KCl, 137 m*M* NaCl.
11. 10% paraformaldehyde, pH 7.2–7.4, in PBS pH 7.4 (*Toxic*: prepare in fume hood by heating PBS to 60 °C while stirring. Adjust pH with NaOH if does not go into solution. Store at 4 °C protected from light.)
12. 2× witches brew: 160 m*M* KCl; 40 m*M* NaCl; 20 m*M* EGTA; 10 m*M* spermidine; 30 m*M* PIPES, pH 7.4; 50% methanol.
13. Tris-Triton buffer: 100 m*M* Tris HCl, pH 7.4; 1% Triton X-100; 1 m*M* EDTA.
14. 40× borate buffer (pH 9.2): 1 *M* H_3BO_3, 0.5 *M* NaOH.

2.4.2. Acetone/Methanol Fixation (see ***Note 1; refs.* 36,38**)

1. 100% acetone.
2. 100% methanol.
3. Ice.

2.5. Blocking Procedure (see ***Note 2; refs.* 36,37**)

1. Fixed specimen.
2. Rocker.
3. Antibody buffer (store at 4 °C): 0.5% TritonX-100, 0.2% mL 0.5 *M* EDTA, 0.1% bovine serum albumin, 0.1% sodium azide, made in PBS.
4. Blocking buffer (store at 4 °C): antibody buffer, 1% bovine serum albumin.

2.6. Antibody Staining of Slide Preparations (36)

1. Fixed specimen that has been blocked.
2. Kimwipes.
3. Flat storage container.
4. Rocker.
5. Primary antibody (*see* **Note 3**).
6. Secondary antibody (if primary not conjugated to a fluorescent molecule).
7. Antibody buffer (*see* **Subheading 2.5.**).
8. Wash buffer: PBS (*see* **Subheading 2.4.1.**), 0.1% Tween-20.
9. Mounting solution.
10. Nail polish.

2.7. Antibody Staining of Tube Preparations (36,37)

1. Fixed specimen that has been blocked.
2. Tabletop centrifuge.
3. Rocker.
4. Primary antibody (*see* **Note 3**).
5. Secondary antibody (if primary not conjugated fluorescent molecule).
6. Antibody buffer (*see* **Subheading 2.5.**).
7. Wash buffer (*see* **Subheading 2.5.**).
8. Mounting solution.
9. Nail polish.

2.8. Mounting Live Worms for Observation (9,20,36,39)

2.8.1. Agar Pad Preparation

1. 3% agarose in distilled water.
2. Pasteur pipette and bulb.
3. Clean microscope slides.
4. Laboratory tape.

2.8.2. Poly-L-Lysine-Coated Coverslips (36)

1. Coverslips.
2. Poly-L-lysine (1 mg/mL in distilled water).

2.8.3. Mounting Live Larva/Adults for Imaging (39)

1. Mouth pipette.
2. Agarose pad (*see* **Subheading 2.8.1.**).
3. Coverslip.
4. Concave slide.
5. Worm pick.
6. Molten Vaseline and fine-tipped paint brush.
7. Levamisole buffer: 0.1 m*M*–0.1 *M* levamisole in M9 (*see* **Note 4**).
8. M9 buffer (*see* **Subheading 2.1.**).

2.8.4. Agar Pad Mount of Live Embryos for Imaging (9,20,39)

1. Mouth pipette.
2. Agarose pad (*see* **Subheading 2.8.1.**).
3. Coverslip.
4. Concave slide.
5. Worm pick.
6. Scalpel.
7. Molten Vaseline and fine-tipped paint brush.
8. 10-mL syringe.
9. Egg Buffer (*see* **Subheading 2.2.**).

2.8.5. Bead Mount of Live Embryos for Imaging

1. Mouth pipette.
2. Concave glass slide for dissection.
3. 22 × 22 mm coverslip.
4. Standard glass slide.
5. Worm pick.
6. Scalpel.
7. Molten Vaseline and fine tipped paint brush.
8. 20-μm-diameter polystyrene beads (PolySciences Inc., Warrington, PA; catalog number 18329).

9. Egg buffer (*see* **Subheading 2.2.**).
10. Bead working solution (store at 4 °C): 33 µL of stock bead slurry mixed into 1 mL of egg buffer containing 1% hydroxymethylcellulose. Resuspend gently before using to disperse settled beads without adding bubbles.

2.8.6. Poly-L-Lysine Mount of Live Embryos (20)

1. Mouth pipette.
2. Poly-L-lysine-coated coverslip.
3. Concave slide.
4. Worm pick.
5. Scalpel.
6. High vacuum grease.
7. Silicone melting point bath oil (Sigma, St. Louis, MO).

3. Methods

3.1. Collection of Mixed Population of Worms for Fixation (36)

1. Prechill M9 on ice.
2. Add 1 mL chilled M9 to a 60-mm NGM plate of worms. (Use a healthy well-fed plate, as worms from a starved plate will autofluoresce.)
3. Allow to sit at room temperatures for 3–5 min.
4. Carefully tilt the NGM plate to one side.
5. Using a Pasteur pipette, remove the liquid from the plate and put into an Eppendorf tube.
6. Place on ice for 5 min.
7. Centrifuge at 1500 rpm (239 *g*) for 2 min.
8. Remove supernatant, being careful not to remove worms.
9. Wash pellet in 1 mL M9 (invert several times).
10. Repeat **steps 7–9** two times.
11. Proceed to fixation procedures (*see* **Subheading 3.4.**).

3.2. Collection of Large Quantities of Embryos for Fixation (36)

1. Prechill M9 buffer and egg buffer on ice.
2. Add 1 mL chilled M9 to a 60-mm NGM plate of worms. (Use a healthy well-fed plate, as worms from a starved plate will autofluoresce.)
3. Allow to sit at room temperatures for 3–5 min.
4. Carefully tilt the NGM plate to one side.
5. Using a Pasteur pipette, remove the liquid from the plate and put into an Eppendorf tube.
6. Place on ice for 5 min.
7. Centrifuge at 1500 rpm (239 *g*) for 2 min.
8. Remove supernatant, being careful not to remove worms.
9. Wash pellet in 1 mL M9 (invert several times).
10. Repeat **steps 7–9** two times.

11. Add 500 µL bleach solution to pellet.
12. Incubate on rotator or rocker for ~5 min. (Remove ~10 µL to look at under microscope to see if worms are lysed and eggs are released. Let the reaction continue until approximately 70% of worms are lysed.)
13. Fill Eppendorf tube with egg buffer.
14. Centrifuge at 1500 rpm (239 *g*) for 2 min.
15. Remove supernatant, being careful not to remove worms.
16. Repeat **steps 13–15** until supernatant is clear.
17. Proceed to fixation protocols (*see* **Subheading 3.4.**).

3.3. Slide Preparation for Fixation Using the Freeze Crack Method (36)

1. Prechill M9 buffer on ice.
2. Add 1 mL chilled M9 to a 60-mm NGM plate of worms. (Use a healthy well-fed plate, as worms from a starved plate will autofluoresce.)
3. Allow to sit at room temperatures for 3–5 min.
4. Carefully tilt the NGM plate to one side.
5. Using a Pasteur pipette, remove the liquid from the plate and put into an Eppendorf tube.
6. Place on ice for 5 min.
7. Centrifuge at 1500 rpm (239 *g*) for 2 min.
8. Remove supernatant, being careful not to remove worms.
9. Wash pellet in 1 mL M9 (invert several times).
10. Repeat **steps 7–9** two times.
11. Resuspend pellet in ~50 µL M9.
12. Using a Pasteur pipette, place 1 drop (~15 µL) of the worms onto a poly-L-lysine–coated microscope slide (*see* **Subheading 3.9.2.**).
13. Gently spread the drops over an area that a coverslip will cover using the side of a Pasteur pipette. (Allow ~70% of liquid to evaporate before applying coverslip.)
14. Place a poly-L-lysine–coated coverslip on top of the worm spread.
15. Gently apply pressure to the four corners and center of the coverslip. (Troubleshoot this step if staining does not work.)
16. Spread dry ice evenly over the bottom of an ice bucket or Styrofoam box. (A rectangular-shaped ice bucket is easiest to get the dry ice to lie flat and uniform.)
17. Place aluminum foil over the even spread of dry ice. (*Optional*: This will prevent the microscope slides from freezing to the dry ice.)
18. Place the microscope slide containing the worms and coverslip on top of the aluminum foil that is on the dry ice.
19. Put a lid over the ice bucket.
20. Incubate for 45 min on dry ice.
21. Remove slide from dry ice and separate the coverslip from the frozen microscope slide using a razor blade. (Do this quickly so as to not allow thawing, or the fixative will not penetrate in later steps.)
22. Proceed to fixation procedures (*see* **Subheading 3.4.**).

3.4 Fixation Procedures

There are numerous fixation protocols that have been used for immunohistochemical studies of *C. elegans* ***(36)***, but two methods have, to our knowledge, been used for most published reports incorporating MH27 staining to monitor intercellular junctions at the site of fusion. The Finney/Ruvkun ***(7–9)*** and methanol/acetone ***(1,7)*** fixation procedures paired with the MH27 antibody have been used to study the disappearance of adherens junctions at fusion sites throughout development. It has been reported that both fixation procedures produce similar results ***(7)***; thus, the choice of protocols will depend on whether other epitopes or molecules will be concurrently stained. Additional protocols ***(36)*** may have to be considered when labeling multiple proteins, because fixatives have differing effects on a protein's conformation and antigen binding (e.g., cold methanol destroys the phalloidin binding site on actin filaments).

3.4.1. Finney/Ruvkun Protocol (see ***Note 1; refs.*** **36,37**)

1. Prechill M9 buffer on ice.
2. Add 1 mL chilled M9 to a 60-mm NGM plate of worms. (Use a healthy well-fed plate, as worms from a starved plate will autofluoresce.)
3. Allow to sit at room temperatures for 3–5 min.
4. Carefully tilt the NGM plate to one side.
5. Using a Pasteur pipette, remove the liquid from the plate and put into an Eppendorf tube.
6. Place on ice for 5 min.
7. Centrifuge at 1500 rpm (239 *g*) for 2 min.
8. Remove supernatant, being careful not to remove worms.
9. Wash pellet in 1 mL M9 (invert several times).
10. Repeat **steps 7–9** two times.
11. Prepare fixative: 1.0 mL cold 2× Witches Brew, 1% formaldehyde, 800 μL distilled water.
12. Add 1.25 mL fixative per tube, and mix well by inverting.
13. Immerse tube into dry ice/ethanol bath to freeze (can be stored at −80 °C).
14. Thaw tube contents in 65 °C water bath until almost all ice has melted.
15. Incubate at 4 °C with occasional agitation for 30 min to overnight.
16. Spin worms at 3000 rpm (956 *g*) for 30 s, and remove supernatant.
17. Wash worms twice with 1× Tris-Triton buffer.
18. Resuspend worms in 1 mL 1× Tris-Triton buffer + 1% β-mercaptoethanol.
19. Incubate at 37 °C for 1.5-2 h with gentle shaking (or rotating).
20. Spin worms at 1500 rpm (239 *g*) for 2 min, and carefully remove supernatant.
21. Resuspend in 1 mL 1× borate buffer.
22. Spin at 1500 rpm (239 *g*), and remove supernatant.
23. Incubate in 1 mL 1× borate buffer plus 10 m*M* DTT for 15 min at room temperature.

24. Spin worms at 1500 rpm (239 *g*), and gently remove supernatant.
25. Resuspend in 1 mL 1× borate.
26. Spin at 1500 rpm (239 *g*), and remove supernatant.
27. Incubate in 1 mL 1× borate buffer plus 0.3% hydrogen peroxide at room temperature for 15 min (wrap tops in parafilm to prevent from opening).
28. Spin at 1500 rpm (239 *g*), and remove supernatant.
29. Resuspend worms in 1 mL 1× borate buffer.
30. Spin at 1500 rpm (239 *g*), and remove supernatant.
31. Proceed to blocking protocol (*see* **Subheading 3.5.**).

3.4.2. Acetone/Methanol Fixation (see ***Note 1; refs.*** **36,38**)

1. Prechill acetone and methanol on ice.
2. Incubate worms in an ice cold methanol for 4 min.
3. Incubate worms in an ice cold acetone for 4 min.
4. Allow to air dry.
5. Proceed to blocking protocol (*see* **Subheading 3.5.**).

3.5. Blocking Procedure (see ***Note 2; refs.*** **36,37**)

1. Incubate fixed specimen in block solution on a rocker for 1 h at room temperature or overnight at 4 °C.
2. Proceed to antibody staining (*see* **Subheadings 3.6.** and **3.7.**)

3.6. Antibody Staining of Slide Preparations **(36)**

1. Dilute antibody in antibody buffer (*see* **Note 3**).
2. Place moistened Kimwipes on the bottom of a flat storage container. If working with fluorescent antibodies, put in a light impenetrable container (easily done with aluminum foil).
3. Remove block solution from fixed sample.
4. Incubate fixed specimen on a rocker with 20–100 μL primary antibody *carefully* applied to the slide surface (directly over the sample) at room temperature for 1.5–2 h at room temperature or overnight at 4 °C.
5. Carefully rinse sample in wash buffer three times over a 30-min period on a rocker. (Do not apply wash buffer directly over worms, or worms will be lost in wash steps.)
6. Dilute secondary antibody in antibody buffer (*see* **Note 3**).
7. Incubate with secondary antibody for 1 h at room temperature on a rocker.
8. Carefully rinse the sample in wash buffer three to five times over a 1-h period.
9. Remove excess wash buffer by tilting microscope slide on its side on top of a Kimwipe. (Do not let slide dry out.)
10. Add mounting media (~10 μL) on top of the worms. (Examples of commercial mounting medium available are Vectasheild, Mowiol, and Aquamount.)
11. Carefully place a coverslip on top of the worms, avoiding bubbles.
12. Seal the edges of the coverslip with nail polish.
13. Store in an air-tight, light-tight container at 4 °C for days or –20 °C for months.

3.7. Antibody Staining of Tube Preparations (36,37)

1. Dilute antibody in antibody buffer (*see* **Note 3**).
2. Centrifuge blocked specimen at 1000 rpm (106 *g*) for 2 min.
3. Remove supernatant, being careful not to remove worms.
4. Incubate fixed specimen on a rocker with 100–200 µL primary antibody at room temperature for 1.5–2 h at room temperature or overnight at 4 °C. (Put in a light impenetrable container if working with fluorescent antibodies.)
5. Incubate fixed specimen in 1-mL wash buffer on a rocker for 10 min.
6. Centrifuge specimen 1000 rpm (106 *g*) for 2 min.
7. Remove supernatant, being careful not to remove worms.
8. Repeat **steps 5–7** two additional times for a total of three washes.
9. Dilute secondary antibody in antibody buffer (*see* **Note 3**).
10. Incubate with 100–200 µL secondary antibody for 1 h at room temperature on a rocker.
11. Incubate fixed specimen in 1 mL wash buffer on a rocker for 15 min.
12. Centrifuge specimen 1000 rpm (106 *g*) for 2 min.
13. Remove supernatant, being careful not to remove worms.
14. Repeat **steps 5–7** three additional times for a total of four washes.
15. Using a Pasteur pipette, transfer ~10–20 µ of the worms onto a clean microscope slide.
16. Add mounting media (~10 µl) on top of the worms (e.g., Mowiol, Vectasheild, or 1. Aquamount).
17. Carefully place a coverslip on top of the worms, avoiding bubbles.
18. Seal the edges of the coverslip with nail polish.
19. Store in an air-tight, light-tight container at 4 °C for days or −20 °C for months.

3.8. Mounting Live Specimens for Observation (9,20,36,39)

Several different means of securing worms or embryos in place for extended high-resolution imaging have been developed. Particularly in the case of embryos, the choice of the mounting method can be critical to optimizing the view of specific cells or the delivery of exogenous fluorescent probes. Both agarose and bead mounts gently squeeze the embryo, and this causes the adoption of either of two predictable orientations (ventral-up or dorsal-up) during the course of embryonic cleavage and early morphogenesis. During the majority of cell fusions, in midmorphogenesis, these "pressure mounts" present a lateral view of the embryo, with the midline conveniently oriented along the plane of sectioning for a confocal microscope. If alternate orientations are desired while observing cell fusions, embryos can be attached to a poly-L-lysine–coated coverslip; such embryos can be observed "unsqueezed" from randomly oriented aspects that may include essentially any angle of view. In addition, poly-L-lysine mounts allow for laser permeabilization of the eggshell without external pressure extruding the embryonic cells. They are therefore the choice for studies involving use of vital dyes or chemical inhibitors that are blocked by the eggshell permeability barrier.

3.8.1. Agar Pad Preparation (39)

1. Securely tape one microscope slide to a flat surface.
2. Apply a second layer of tape on top of the first layer.
3. Melt 3% agarose in the microwave or in a 65 °C water bath.
4. Place a drop of agar onto the microscope slide.
5. Quickly cover the agarose drop with another slide.
6. Press down firmly and evenly.
7. Allow agarose to solidify (do not let agar pad dry out).
8. Gently remove top microscope slide.

3.8.2. Poly-L-Lysine-Coated Coverslips (36)

1. Clean the microscopes and coverslips by wiping with a dry Kimwipe.
2. Place 1 μL poly-L-lysine onto a coverslip.
3. Spread poly-L-lysine over the entire surface using the edge of a different coverslip.
4. Allow to dry for 24 h prior to use (use within 72 h of preparation).

3.8.3. Mounting Live Larvae/Adults for Imaging (39)

1. Place a drop of levamisole solution onto agarose pad (*see* **Note 4**).
2. Place ~100 μL of levamisole buffer into a concave slide.
3. Pick worms into levamisole buffer in concave slide.
4. Using mouth pipette, transfer the worms from the concave slide onto the agarose pad.
5. Cover the worms and agarose pad with ~20 μL of levamisole solution.
6. Place a coverslip over the animals.
7. Backfill with levamisole buffer to ensure that embryos will not dry out.
8. Seal three of four sides with molten Vaseline using a paint brush (leave one side open to replenish the buffer if the mount begins to dry out or if worms start to move).

3.8.4. Agar Pad Mount of Live Embryos for Imaging (9,20,39)

1. Place ~100 μL egg buffer in concave slide.
2. Pick gravid hermaphrodites into egg buffer with worm pick.
3. Cut midsection of mother to release the eggs.
4. Using a mouth pipette, transfer the appropriately staged embryos onto the agar pad with minimal fluid.
5. Cover embryos with a coverslip.
6. Backfill with egg buffer to ensure that embryos will not dry out.
7. Seal edges of coverslip with molten Vaseline using a paint brush.

3.8.5. Bead Mount of Live Embryos for Imaging (J. Murray, Personal Communication)

1. Repeat **steps 1–3** of **Subheading 3.8.4.**
2. Using a mouth pipette, transfer the appropriately staged embryos onto the untreated surface of a standard glass slide, taking care to leave a minimal volume of fluid.
3. Add 3 μL (no more) of bead/egg buffer/hydroxymethylcellulose solution over and around the embryos attached to the coverslip. Avoid forming bubbles.
4. Carefully lower the coverslip onto the drop containing beads and embryos. Allow capillary action to pull the coverslip down onto the embryos and beads, establishing a gentle pressure mount.
5. Seal the edges of the coverslip with molten Vaseline using a paintbrush.

3.8.6. Poly-Lysine Mount of Live Embryos for Imaging (20)

1. Place egg buffer into a concave slide.
2. Transfer worms into egg buffer using a pick.
3. Cut the midsection of a worm with a scalpel or needle to release the embryos.
4. Collect the embryos using a mouth pipette and transfer to poly-L-lysine–coated coverslip (see **Subheading 3.8.2.**) with minimal liquid.
6. Blow air through mouth pipette to attach embryos to poly-L-lysine coverslip.
7. Mount coverslip onto microscope slide using small dots of high-vacuum grease at the four corners of the coverslip.
8. Backfill mount with egg buffer to ensure adequate hydration.
9. Seal edges of coverslip with a thin fillet (~15 μL total) of silicon melting-point bath oil (Sigma) applied using a micropipette.

4. Notes

1. An alternative fixation protocol that also works well with the MH27 antibody (and preserves GFP fluorescence) is 4% paraformaldehyde (PFA). Fixation of embryos, larvae, and adults is performed by incubating the samples in 4% PFA for 10 min, followed by three 5-min washes in PBS ***(36).***
2. If high background signal is observed bound to secondary antibody in the absence of primary antibody, the blocking solution used may have to be varied. There are two alternative blocking solutions: *(a)* antibody buffer + 5%–10% serum or *(b)* antibody buffer + 5% normal goat serum ***(36)***.
3. Titration of each antibody is essential to identify what the optimal working concentration is for that individual antibody. When titrating the antibody concentration, a typical good starting point is 1:20, with successive doubling of the dilution (i.e., 1:40, 1:80, etc.) to optimize the concentration for the experiment in question. Importantly, different fixation procedures can influence the dilution factor of a given antibody's most optimal concentration.
4. Levamisole solutions should be made fresh in order to ensure effective paralysis. Note that there are some mutant *C. elegans* strains that are resistant to levamisole ***(40)***. Alternative immobilization protocols involve 10 m*M* sodium azide treatment ***(41)*** or 1 m*M* aldicarb treatment ***(40)***.

References

1. Podbilewicz, B. and White, J. (1994) Cell fusions in the developing epithelia of *C. elegans*. *Dev. Biol.* **161,** 408–424.
2. Sulston, J., Schierenberg, E., White, J., and Thomson, J. (1983) The embryonic cell lineage of the nematode *Caenorhabditis elegans*. *Dev. Biol.* **100,** 64–119.
3. Sulston, J. and Horvitz, H. (1977) Postembryonic cell lineages of the nematode *Caenorhabditis elegans*. *Dev. Biol.* **56,** 110–156.
4. Shemer, G. and Podbilewicz, B. (2000) Fusomorphogenesis: cell fusion in organ formation. *Dev. Dyn.* **218,** 30–51.
5. Newman, A. P., White, J. G., and Sternberg, P. W. (1996) Morphogenesis of the *C. elegans* hermaphrodite uterus. *Development* **122,** 3617–3623.
6. Kornfeld, K. (1997) Vulval development in *Caenorhabditis elegans*. *Trends Genet.* **13,** 55–61.
7. Sharma-Kishore, R., White, J., Southgate, E., and Podbilewicz, B. (1999) Formation of the vulva in *Caenorhabditis elegans*: a paradigm for organogenesis. *Development* **126,** 691–699.
8. Nguyen, C. Q., Hall, D. H., Yang, Y., and Fitch, D.H. (1999) Morphogenesis of the *Caenorhabditis elegans* male tail tip. *Dev. Biol.* **207,** 86–106.
9. Mohler, W. A., Shemer, G., del Campo, J. J., Valansi, C., Opoku-Serebuoh, E., Scranton, V., Assaf, N., White, J. G., and Podbilewicz, B. (2002) The type I membrane protein EFF-1 is essential for developmental cell fusion. *Dev. Cell* **2,** 355–362.
10. Shemer, G., Suissa, M., Kolotuev, I., Nguyen, K., Hall, D., and Podbilewicz, B. (2004) EFF-1 is sufficient to initiate and execute tissue-specific cell fusion in *C. elegans*. *Curr. Biol.* **14,** 1587–1591.
11. del Campo, J. J., Opoku-Serebuoh, E., Isaacson, A. B., Scranton, V. L., Tucker, M., Han, M., and Mohler, W.A. (2005) Fusogenic activity of EFF-1 is regulated via dynamic localization in fusing somatic cells of *C. elegans*. *Curr. Biol.* **15,** 413–423.
12. Podbilewicz, B., Leikina, E., Sapir, A., Valansi, C., Suissa, M., Shemer, G., and Chernomordik, L. (2006) The *C. elegans* developmental fusogen EFF-1 mediates homotypic fusion in heterologous cells and in vivo. *Dev. Cell* **11,** 471–481.
13. Sapir, A., Choi, J., Leikina, E., Avinoam, O., Valansi, C., Chernomordik, L. V., Newman, A. P., and Podbilewicz, B. (2007) AFF-1, a FOS-1-regulated fusogen, mediates fusion of the anchor cell in *C. elegans*. *Dev. Cell* **12,** 683–698.
14. Inoue, S. and Spring, K. (1997) *Video Microscopy: The Fundamentals*, 2nd ed. Plenum Publishing Corp, New York.
15. Sulston, J. and White, J. (1980) Regulation and cell autonomy during postembryonic development of *Caenorhabditis elegans*. *Dev. Biol.* **78,** 577–597.
16. Alper, S. and Kenyon, C. (2002) The zinc finger protein REF-2 functions with the Hox genes to inhibit cell fusion in the ventral epidermis of *C. elegans*. *Development* **129,** 3335–3348.
17. Alper, S., Kenyon, C. (2001). REF-1, a protein with two bHLH domains, alters the pattern of cell fusion in *C. elegans* by regulating Hox protein activity. *Development* **128,** 1793–1804.

18. Koh, K. and Rothman, J. (2001) ELT-5 and ELT-6 are required continuously to regulate epidermal seam cell differentiation and cell fusion in *C. elegans*. *Development* **128,** 2867–2880.
19. Shemer, G. and Podbilewicz, B. (2002) LIN-39/Hox triggers cell division and represses EFF-1/fusogen–dependent vulval cell fusion. *Genes Dev.* **16,** 3136–3141.
20. Mohler, W. A., Simske, J. S., Williams-Masson, E. M., Hardin, J. D., and White, J. G. (1998) Dynamics and ultrastructure of developmental cell fusions in the *Caenorhabditis elegans* hypodermis. *Curr. Biol.* **8,** 1087–1090.
21. White, J. G., Amos, W. B., and Fordham, M. (1987) An evaluation of confocal versus conventional imaging of biological structures by fluorescence light microscopy. J. *Cell Biol.* **105,** 41–48.
22. Gattegno, T., Mittal, A., Valansi, C., Nguyen, K. C., Hall, D. H., Chernomordik, L. V., and Podbilewicz. B. (2007) Genetic control of fusion pore expansion in the epidermis of Caenorhabditis elegans. *Mol. Biol. Cell* **18,** 1153–1166.
23. Cassata, G., Shemer, G., Morandi, P., Donhauser, R., Podbilewicz, B., and Baumeister, R. (2005) ceh-16/engrailed patterns the embryonic epidermis of *Caenorhabditis elegans. Development* **132,** 739–749.
24. Kontani, K., Moskowitz, I., and Rothman, J. (2005) Repression of cell–cell fusion by components of the *C. elegans* vacuolar ATPase complex. *Dev. Cell* **8,** 787–794.
25. Wang, E., Babbey, C. M., and Dunn, K. W. (2005) Performance comparison between the high-speed Yokogawa spinning disc confocal system and single-point scanning confocal systems. *J. Microsc.* **218,** 148–159.
26. Jansen, L. E., Black, B. E., Foltz, D. R., and Cleveland, D. W. (2007) Propagation of centromeric chromatin requires exit from mitosis. *J. Cell Biol.* **176,** 795–805.
27. Maddox, P. S., Portier, N., Desai, A., and Oegema, K. (2006) Molecular analysis of mitotic chromosome condensation using a quantitative time-resolved fluorescence microscopy assay. *Proc. Natl. Acad. Sci. U.S.A.* **103,** 15097–15102.
28. Denk, W., Strickler, J., and Webb, W. (1990) Two-photon laser scanning fluorescence microscopy. *Science* **248,** 73–76.
29. Mohler, W. and Isaacson, A., (2005) Imaging embryonic development in *Caenorhabditis elegans, in* Imaging in Neuroscience and Development (R. Yuste and A. Konnerth, eds.) Cold Spring Harbor Laboratory Press, Cold Spring Harbor, NY, pp. 119–124.
30. Mohler, W. and White, J. (1998) Multiphoton laser scanning microscopy for four-dimensional analysis of *Caenorhabditis elegans* embryonic development. *Optics Express* **3,** 325–331.
31. Bao, Z., Murray, J., Boyle, T., Ooi., S., Sandel, M., and Waterston, R. (2006) Automated cell lineage tracing in *Caenorhabditis elegans. Proc. Natl. Acad. Sci. U.S.A.* **103,** 2707–2712.
32. Lee, J. Y. and Goldstein, B. (2003). Mechanisms of cell positioning during *C. elegans* gastrulation. *Development* **130,** 307–320.
33. O'Connell, K. F., Caron, C., Kopish, K. R., Hurd, D. H., Kemphues, K.J., Li, Y., and White, J. G. (2001) The *C. elegans* zyg-1 gene encodes a novel regulator of

centrosome duplication with distinct maternal and paternal roles in the embryo. *Cell* **105,** 547–558.

34. Gerlich, D. and Ellenberg, J. (2003) 4D imaging to assay complex dynamics in live specimens. *Nat. Cell Biol.* **5(Suppl),** S14–S19.
35. Nolan, S., Cowan, A.E., Koppel, D., Jin, H., and Grote, E. (2006) Fus1 regulates the opening and expansion of fusion pores between mating yeast. *Mol. Biol. Cell* **17,** 2439–2450.
36. Duerr, J. S. (2006) Immunohistochemistry, in *WormBook (The C. elegans* Research Community, ed.), doi/10.1895/wormbook.1.105.1. http://www.wormbook.org.
37. Finney, M. and Ruvkun, G. (1990) The unc-86 gene product couples cell lineage and cell identity in C. elegans. *Cell* **63,** 895–905.
38. Miller, D. M. and Shakes, D. C. (1995) Immunofluorescence microscopy. *Methods Cell Biol.* **48,** 365–394.
39. Shaham, S. (January 2, 2006) Methods in cell biology, in *WormBook* (The *C. elegans* Research Community, ed.). doi/10.1895/wormbook. http://www.wormbook.org.
40. Lackner, M. R., Nurrish, S. J., and Kaplan, J. M. (1999) Facilitation of synaptic transmission by EGL-30 Gq and EGL-8 PLCß: DAG binding to UNC-13 is required to stimulate acetylcholine release. *Neuron* **24,** 335–346.
41. Chao, M. Y., Komatsu, H., Fukuto, H. S., Dionne, H. M., and Hart, A. C. (2004) Feeding status and serotonin rapidly and reversibly modulate a *Caenorhabditis elegans* chemosensory circuit. *Proc. Natl. Acad. Sci. U.S.A.* **101,** 15512–15517.

14

Ultrastructural Imaging of Cell Fusion in *Caenorhabditis elegans*

Star Ems and William A. Mohler

Summary

Caenorhabditis elegans is a well-established model system particularly suited for studying cell–cell fusion because of its highly predictable and rapid development and its known cell lineage. This chapter focuses on understanding the ultrastructural components of cell fusion through the use of transmission electron microscopy (TEM). Published TEM studies have described the initial demonstration of syncytial cells in the worm, the vesiculation of the bilayers between cells during widening of the normal fusion aperture, and the appearance of microfusion intermediates in the membranes of cells with fusion-defective mutations. Capturing events observed in embryos on the light microscope and preserving the integrity of cellular membranes for examination by TEM require some special considerations that differ from many ultrastructural studies of cells. The principles of different techniques for TEM and details of protocols that have been used to investigate cell fusion in the nematode are discussed in this chapter.

Key Words: Cell fusion; ultrastructure; transmission electron microscopy; *Caenorhabditis elegans*.

1. Introduction

The maximum spatial resolution of conventional or confocal light microscopy (200 nm) *(1)* is insufficient to characterize many subcellular structures. Although fluorescence microscopy is an invaluable tool for live cell imaging, many aspects of the process of cell–cell fusion will be missed without the use of transmission electron microscopy (TEM). The predominant differences between electron and light microscopy lie in the radiation source and lenses. Light microscopy utilizes ultraviolet, visible, or infrared photons as the radiation source paired with optical lenses, whereas electrons and magnetic lenses are used in electron microscopy. Utilizing electrons as the radiation source

From: *Methods in Molecular Biology, vol. 475: Cell Fusion: Overviews and Methods*
Edited by: E. H. Chen © Humana Press, Totowa, NJ

allows the electron microscope, in theory, to produce images with a resolution as low as 3 angstroms (0.3 nm). In reality, the resolution limit is significantly larger (~2 nm) in thin sections of cells or tissues because of damage induced by the electrons interacting with the sample and because contrast is generated by layers of heavy metal stains that bind to the biological material of interest *(2)*. In addition, the axial resolution of standard TEM is often limited by the thickness of sections cut by the diamond knife (typically ~70 nm), although increased resolution in three dimensions can be achieved by electron tomography *(3)*. Even with these limitations, conventional TEM has yielded some key insights into the process of cell fusion in *Caenorhabditis elegans* that could not possibly be made by use of light microscopy alone.

1.1. Ultrastructure of Cell Fusion Events

Thus far, at least eight published papers have presented ultrastructural studies of fused syncytia (embryo *[4]*, pharynx *[5]*, male tail *[6]*, vulva *[7]*, and uterus *[8]*) or actively fusing cells *(9–11)* visualized by TEM in *C. elegans*. Structures that have been visualized by TEM in association with fusion-fated cell–cell interfaces in the worm include intercellular junctions (and associated proteins), membrane pores, membrane vesicles, and conjoined membrane bilayers (**Figs. 1-3**).

Initiation of a cell fusion event is widely thought to involve formation of a pore or pores through the neighboring cell membranes. Ultimately, we hope to visualize fusion pore formation via TEM. However, thus far the existence of such structures has been inferred only by light microscope studies of cytoplasm diffusion across fusing plasma membranes. The initial permeability of the membranes between two fusing hypodermal cells is rather low, typically requiring ~2–4 min for equilibration of green fluorescent protein (GFP) intensity between the cytoplasm of neighboring cells *(12)*. These rates of diffusive equilibration are roughly similar to those measured in fusing yeast cells, where initial fusion pore sizes range from a minimum of 9 nm through a median of 25 nm in estimated diameter *(13)*. No direct TEM evidence for such pores between actively fusing cells has been shown in *C. elegans*. Interestingly, though, TEM images of pharyngeal muscle cells blocked in fusion by a partial loss-of function of the fusogen EFF-1 have revealed "microfusions," cytoplasmic bridges through the double membrane that appear to be less than 50 nm wide (**Fig. 2; ref. *10***). It is possible that these structures represent membrane fusion pores that remain undilated in the mutant cells.

Hypodermal cells undergoing normal cell fusion events have been observed in worm embryos by a combination of time-lapse confocal fluorescence microscopy and correlative TEM on the same specimen *(9)*. Apertures through the fusing cell membranes were visible at optical resolution, progressively

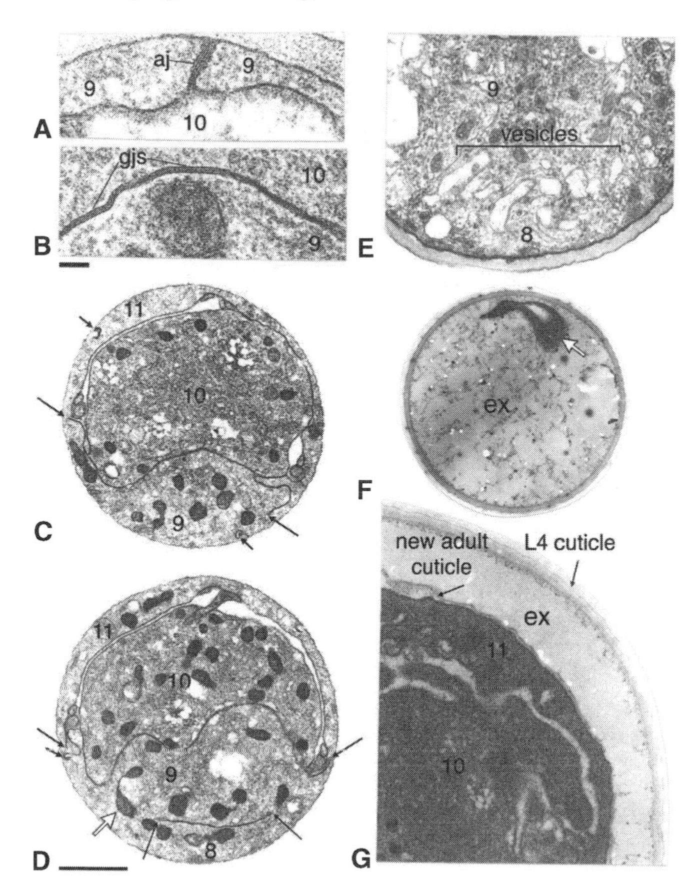

Fig. 1. Ultrastructural features of cell membranes and intercellular junctions in cell fusions during larval male tail tip retraction. Chemical fixation was used to produce serially sectioned electron micrographs. Hypodermal (hyp) cells are labeled by number (8 is hyp8; 9, hyp9; 10, hyp10; 11, hyp11). **(A)** An adherens junction (aj) that seals hyp9 to itself in the very posterior portion of the tail tip. **(B)** Gap junctions (gjs) between hyp9 and hyp10. Bar in B = 10 nm for A and B. **(C,D)** Fusions are noted as breaks in the plasma membranes that tend to be closer to the body wall in posterior C than anterior D sections, suggesting that the fusion began anteriorly and zippered posteriorward. Bar in D = 1 μm for C–E and G but 0.5 μm for F. Short arrows mark curled membrane/adherens remnants. Long arrows indicate remaining broken edges of membrane near presumptive sites of fusion. A mitochondrion spans the newly fused hyp8 and hyp9 cells (open arrow, D). **(E)** In one stage 3 male in which fusions were completed, a collection of membranous vesicles accumulated near the former boundary between hyp8 and hyp9. **(F)** During retraction, the larval cuticle in a stage 3 animal surrounds much extracellular space (ex) at the tail tip with only a remnant of hyp10 tissues (arrow). **(G)** More anteriorly, increased extracellular space primarily lies between old and new cuticle layers, whereas the space between the cells remains small. (Figure reprinted and legend adapted from **ref.** ***6***, with permission of Elsevier.)

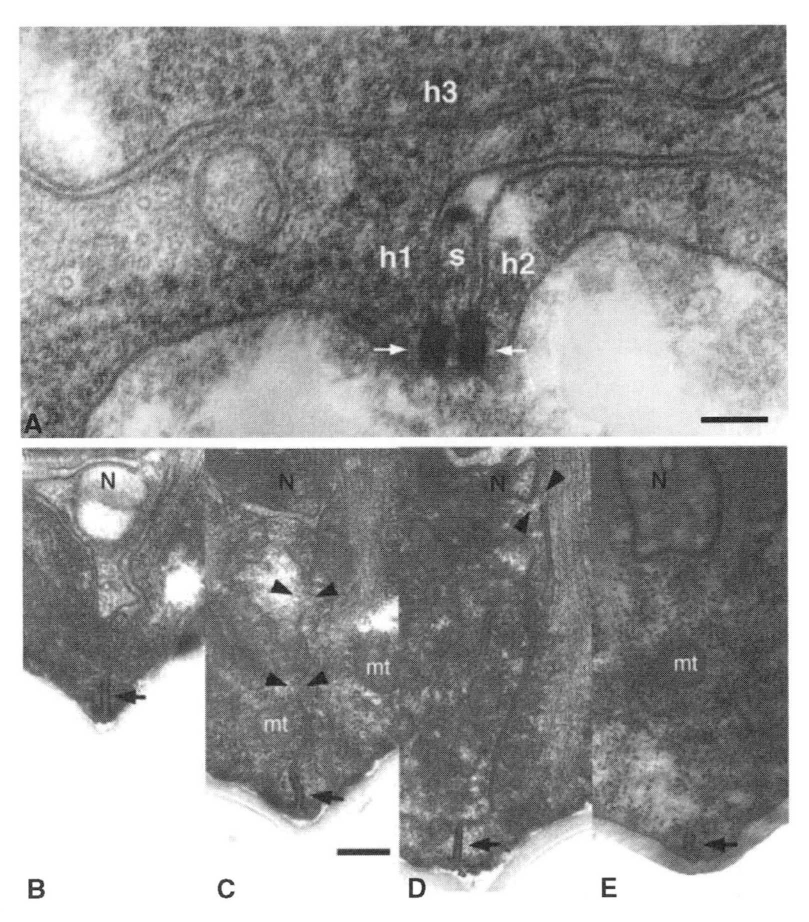

Fig. 2. Transmission electron micrographs of fusion-defective epidermal and pharyngeal muscle cells. **(A)** Transverse thin section of mutant *eff-1 (hy21)* larval stage 4 larva shows normal ultrastructure of the hypodermal membranes, where the separation between apposing plasma membranes that have failed to fuse is generally between 9 and 13 nm, as in wild type. The ultrastructure of the apical junction (arrows) and membranes appears normal. A seam cell process (s) narrows, and three neighboring hypodermal cells (h1, h2, h3) remain unfused along the lateral body wall. **(B–E)** Transmission electron micrograph of pharyngeal muscle pairs showing variable fusion failure in an adult *eff-1 (hy21)* mutant. **(B)** Two cells have failed to fuse, leaving a persistent cell border running from the neuron process (N) to the apical junction (indicated by arrows in B–E). **(C,D)** Above the apical junction, two pairs of *eff-1 (hy21)* mutant cells have formed microfusions (arrowheads), which are so small that no mitochondria (mt) could pass. **(E)** Two wild-type cells fully fused below the neuron process (N) leaving behind a complete apical junction on the plasma membrane of the fused cell pair (Figure reprinted and legend adapted from **ref. *10***, with permission of Elsevier.)

widening from what often appeared to be a single apical site of origin. After chemical fixation of these live-imaged embryos, TEM of the same membranes revealed a single large opening between the two cytoplasms, with a continuous double bilayer of membrane persisting across the remainder of the cell–cell interface (**Fig. 3**). A similar profile, with a large opening breaching an otherwise intact double bilayer, was also seen during hypodermal cell fusions in the developing larval male tail (*see* **Fig. 1; ref. *6***). In the cytoplasm at the outer rim of the widening fusion aperture—the very edge of the septum of conjoined bilayers—membrane vesicles were found clustered near the edge of the retreating membrane. Such vesicular features could be seen in fusing cells of both the embryonic and male tail hypodermis, indicating that removal of the membrane between fusing cells involves active endocytosis of the plasma membranes.

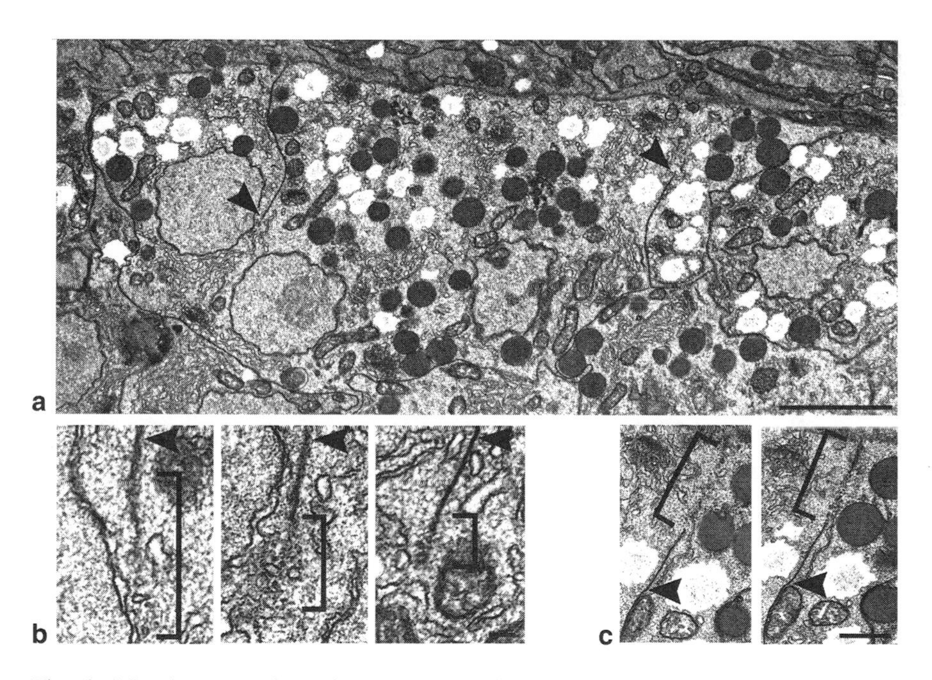

Fig. 3. Membrane vesiculation at the leading edge of the dilating fusion aperture. **(A)** Transmission electron micrograph of an embryo fixed and serially sectioned for correlative electron microscopy after live imaging with multiphoton microscopy. The embryo was fixed while undergoing two visible cell fusion events (arrowheads). Bar = 2 μm. **(B)** Serial cross sections through the leftmost fusing cell border in A are shown at higher magnification. The arrowhead marks the region of intact membrane. The bracket indicates the zone of vesiculation. Spacing between the sections is ~300 nm. Bar = 100 nm. **(C)** Tangential sections through the vesiculating zone of the rightmost fusing border in A. The spacing between sections is ~70 nm. Bar = 500 nm. (Figure reprinted and legend adapted from **ref. *9***, with permission of Elsevier.)

Interestingly, two-channel optical imaging of the process shows that the widening aperture in embryonic hypodermal fusions displaces a marker of subapical intercellular junctions to the basal limit of the cell–cell interface ***(9)***. It should be noted that paired-vesicle structures similar to those seen in TEM of differentiating *Drosophila* myoblasts ***(14)*** were not observed in any of these studies of fusion-fated cells in *C. elegans*. Electron microscopic techniques that have been utilized in *C. elegans* (and that might be of use in the future) to discern these ultrastructural features of cell fusion events are described in this chapter.

1.2. Chemical Fixation

The most common methodology for preparing specimens for TEM involves chemical fixation of the specimen ***(15)***. Typical fixatives include formaldehyde and/or glutaraldehyde as primary fixatives and osmium tetroxide as a secondary fixative and electron-dense stain. After fixation the water in the sample is replaced with an organic solvent, followed by subsequent replacement with resin, which is then polymerized to form a plastic-embedded block that can be thin sectioned. The sections are often poststained with heavy metals (e.g., uranyl acetate) to increase the contrast of the specimen ***(2)***.

Although these methods are widely practiced, there are several general disadvantages or challenges in conventional chemical fixation of worms and embryos and some chemical treatments that can produce artifacts in the plasma membrane structure of animal cells. In *C. elegans*, the action of chemical fixatives is impeded by the impermeability of the cuticle or eggshell that encases the specimen. Transmission electron microscopic studies analyzing the embryo of *C. elegans* have typically used laser ablation ***(16,17)***, enzymatic digestion ***(18)***, or bleaching ***(19)*** to permeabilize the eggshell and vitelline membrane and allow the fixatives to penetrate. Work on hatched animals often involves cutting the larva or adult into pieces ***(20–23)***. Alternatively, microwave treatment during fixation ***(24,25)*** can increase the rate at which fixatives penetrate the eggshell and/or cuticle and thus can provide a faster and more uniform fixation, especially in bulk populations of worms or embryos. However, microwave fixation has been reported to achieve only marginal ultrastructural preservation in most tissues and to be inadequate to resolve very small structures ***(25)***.

Common problems associated with chemical fixation can include leakage of cytoplasmic components, abnormal aggregation of cellular components, and structural rearrangements caused by shrinkage ***(2,26–28)***. Finally, it is important to note that initial fixation with only aldehydes, as occurs in many TEM fixation protocols, has been shown to allow spurious structures to form in the plasma membranes of animal cells ***(29)***. Thus, starting fixation with osmium tetroxide or an osmium/glutaraldehyde mix—or cryopreservation—may be more likely to preserve plasma membranes in a genuine conformation, including any true cell-fusion intermediate structures.

1.3. Cryofixation or Cryo-Electron Microscopy

To minimize artifacts due to handling and chemical fixation, techniques for cryofixation have been developed that involve rapidly freezing the live specimen at temperatures below −140 °C. Ultrarapid cooling promotes the vitrification of water in the specimen, preventing the formation of damaging ice crystals and preserving the ultrastructure of the cells in a near-native state. There are currently two different methods utilized for cryopreservation, high-pressure freezing and plunge freezing. High-pressure freezing involves pressurizing the sample at the same time that it is cooled in liquid nitrogen ***(15)***. Once the specimen is frozen, it is subjected to a freeze substitution procedure wherein the water is substituted with organic solvents and chemical fixatives at temperatures around −80 °C. The temperature is then allowed to slowly increase to permit the chemical fixatives to act while the vitrified specimen is held in its native structure. The sample is warmed to as high as 4 °C for buffer rinses and the beginning of dehydration steps but then is brought back down to −20 °C for final infiltration into resin. Curing is done under ultraviolet light in a "cryobox" over dry ice to keep samples cold until the plastic is hardened. Sample blocks are then stored at room temp before thin sectioning and poststaining ***(2)***.

High-pressure freezing/freeze substitution has been shown to allow fixation without prior permeabilization of worms or embryos, adding a reduction in damaging specimen manipulation to the advantages of cryofixation. In fact, high-pressure freezing/freeze substitution has been used in some of the published images of fusion-fated cell contacts in the worm ***(10)***. However, this method is much more expensive than standard chemical methods and can suffer from some of its own peculiar artifacts ***(15)***. Additional disadvantages of high-pressure freezing/freeze substitution are that preparations can suffer from mass loss, that the precise experimental conditions for high-pressure freezing must be optimized for each sample type in order to prevent ice crystals, and that some material is apparently not suitable for cryofixation ***(15)***. For example, high-pressure freezing/freeze substitution has been noted to work much better with *C. elegans* embryos and larval stage 1 larvae than with older stages ***(19)***.

Even more extreme alternatives to chemical fixation are those combining cryopreservation and cryoimaging. Cells grown directly on electron microscopic grids have been imaged intact, without further fixation or staining, by plunge-freezing the live specimen into liquid ethane or propane and proceeding directly to electron microscopic imaging of the hydrated specimen in an ultracold microscope stage ***(15,26)***. Using unstained specimens requires low doses of electrons in order to prevent sample damage, resulting in images with a lower signal-to-noise level than that observed with staining ***(15)***. However, radiation damage can be significantly reduced at temperatures below −196 °C (liquid nitrogen) or as low as 4 K (liquid helium).

Advantages of cryoimaging methods lie in the potential for increased resolution and the ability to observe both the surface and internal structure of cells. Advances in sample preparation and imaging at ultralow temperatures have produced some extraordinary views of high-resolution native subcellular structure ***(26)***. Because the sample does not have to be embedded or sectioned, high-throughput data can be acquired ***(30)***. Yet, major limitations to the broader use of these techniques include the need for specialized equipment and the difficulty in handling specimens. The predominant disadvantage of plunge freezing is that it is limited to very thin specimens, such as monolayers of cells ***(15,30,31)***. This clearly is a fatal flaw for imaging within worms or embryos.

To our knowledge, cryo-electron microscopic imaging has not been applied to *C. elegans*; all studies of cell fusion described in nematodes to date have used stained, resin-embedded, thin-sectioned specimens, typical of the vast majority of TEM studies. Nevertheless, recent advances in primary nematode cell culture lend optimism to the idea that plunge freezing and unstained cryoimaging may soon become a feasible approach to observing the intact ultrastructure of *C. elegans* cells undergoing fusion in vitro.

1.4. Correlative Use of Light and Electron Microscopy

Added degrees of understanding can be gained by optically imaging a live cell during a dynamic process, followed by fixation and processing for electron microscopy. Swift and careful handling of the sample can ensure that the same structures viewed in the live specimen are immediately fixed and then scrutinized (in the same orientation) at ultrastructural resolution on the electron microscope ***(3)***. In one version of this approach applied to cell fusion, two different contrast agents that highlight plasma membranes were used for the sequential steps on a single specimen: one a vital fluorescent dye (FM4-64) and the other an electron-dense fixative (OsO_4/tannic acid). This method was performed with *C. elegans* embryos by combining automated time-lapse multiphoton confocal imaging with chemical fixation and TEM (*see* **Fig. 3; ref. *9***).

Use of single probes that generate both fluorescence and electron microscopic contrast can allow correlative microscopy to even better target specific molecules of interest. One classic example is the use of antibody-linked fluorescent dyes to induce photoconversion of diaminobenzidine (DAB), generating an electron-dense DAB/osmium precipitate at the sites of localized epitope within a cell ***(32)***. Another correlative optical/electron microscopic method involves cadmium selenide semiconductor quantum dots that are conjugated to reagents (e.g., streptavidin) used in immunocytochemistry ***(33)***. These probes generate bright fluorescent contrast, while their cadmium content can be detected in an energy-filtering electron microscope ***(15,28)***. In a live-imaging correlate of these approaches, genetically encoded tetracysteine tags can be added to protein expression constructs, allowing membrane-permeant biarsenical fluorochromes to bind specifically to the engineered protein of interest. Unlike intrinsically fluorescent proteins (e.g., GFP), these tetracysteine–biarsenical complexes induce very efficient photooxidation of DAB ***(32,34)*** and thereby facilitate the transition from vital fluorescence microscopy to TEM with single-molecule specificity. No published reports have yet described the use of DAB photoconversion, tetracysteine–biarsenical probes, or quantum dot staining in the study of cell fusion in nematodes. However, genetic studies of the worm have begun to reveal molecules critical to the fusion competence of a cell–cell contact, and future studies will undoubtedly involve molecule-specific TEM imaging to improve upon the resolution of their subcellular localization, which until now has been studied only by light microscopy ***(12)***.

1.5. Immunoelectron Microscopy

A much more standard set of methods for immunolabeling of epitopes directly on sections of fixed/embedded specimens has been in use for decades. Typically, a specimen that has been chemically fixed and/or cryopreserved is dehydrated and embedded in a polymerizing resin, often without osmium staining, and then thin sectioned. Antibodies or antigen binding fragments that are covalently linked to gold particles are applied to the surface of sections, incubated, and then washed away. Finally, a poststain such as uranyl acetate may be used to add contrast to cellular features not containing the epitope. Typically, the fixation procedure for immunoelectron microscopy is a compromise between ultrastructural preservation and maintaining the reactivity of the epitope in question (for further details on immunoelectron microscopy methods for worms, *see* **refs. *20,35***).

In only one case, to our knowledge, has immunogold labeling been used to observe fused or fusing cells in the worm. A monoclonal antibody against the subapical junction molecule AJM-1 revealed that intercellular junctions containing this protein remain intact long after the aperture between fused pharyngeal cells has completely dilated (D. Hall, personal communication). This observation agrees with optical observation of the behavior of AJM-1::GFP in fusing embryonic hypodermal cells ***(9)***.

Headings 2 and **3** of this chapter describe materials and methods that have proven useful in the ultrastructural characterization of fusing cells in *C. elegans*. A diverse and well-curated collection of electron microscopic protocols is posted by David Hall (*see* http://www.aecom.yu.edu/wormem/methods.htm). Various protocols that have been used for TEM fixation follow the same general methods, which include sample preparation, fixation, staining, dehydration, embedding, and sectioning. Much of the methodology specific to this species involves bypassing the impermeable eggshell or cuticle to allow efficient chemical fixation and staining as well as strategies for manipulation of specimens to allow the best correlation between light microscope and electron microscopic images of the same sample. In this chapter, we concentrate on the preparative steps leading up to sectioning and imaging. Sample preparation is described separately for embryos and hatched worms (larvae and adult stages).

2. Materials

2.1. Sample Preparation and Aldehyde Fixation for Conventional Electron Microscopy

2.1.1. Embryos

1. Concave slide.
2. Microscope slide.
3. Coverslips.
4. Scalpel.
5. Mouth pipette.
6. Pasteur pipette and bulb.
7. Worm pick.
8. Eyelash brush.
9. Coumarin-440 dye ablation laser (Photonic Instruments, Inc., St. Charles, IL).
10. Sealed plastic dish.
11. Kimwipes.
12. Razor blades.
13. Sterilized egg buffer: 118 m*M* NaCl, 48 m*M* KCl, 2 m*M* $CaCl_2$, 2 m*M* $MgCl_2$, 25 m*M* Hepes. Adjust pH to 7.3 with 1 N NaOH.
14. Agarose fixative: 3 ml 2% agarose (1.5% low gel temperature agarose + .05% high gel temperature agarose), 5 μL 1 *M* $MgCl_2$, 400 μL 0.2 *M* Na cacodylate (pH 7.4), 400 μL sucrose (68.46 g/100 mL), 500 μL 25% glutaraldehyde (*see* **Note 1**).
15. Dissecting microscope.
16. Nine-well glass plate.
17. 0.2 *M* sodium cacodylate.

2.1.2. Larvae and Adults (10,20)

1. Concave slide.
2. Scalpel.
3. Mouth pipette.
4. Pasteur pipette.
5. Worm pick.
6. Eyelash brush.
7. Sealed plastic dish.

8. Sterilized M9 buffer: 22 m*M* KH_2PO_4, 42 m*M* Na_2HPO_4, 86 m*M* NaCl, 1 m*M* $MgSO_4$.
9. 0.2 *M* Hepes.
10. 8% ethanol in M9 buffer.
11. Aldehyde fixative solution: 2.5% glutaraldehyde, 2% paraformaldehyde, in 0.2 *M* Hepes.
12. Razor blades.
13. Dissecting microscope.

2.2. Secondary Osmium Fixation and Staining

1. 0.2 *M* sodium cacodylate (0.1 *M* Hepes, pH 7.4 for worms).
2. 0.1 *M* sodium cacodylate.
3. 1% OsO_4 + 1% $K_3Fe(CN)_6$ in 0.2 *M* sodium cacodylate (for worms: 1% OsO_4, 0.5% $K_3Fe[CN]_6$ in 0.1 *M* Hepes).
4. 0.2% tannic acid in 0.1 *M* sodium cacodylate (*embryos only*).
5. 0.1 *M* sodium acetate.
6. 1% uranyl acetate in 0.1 *M* sodium acetate.
7. 2% agarose (3% SeaPlafue agarose for worms).
8. Distilled water.

2.3. Dehydration

1. 30% ethanol.
2. 50% ethanol.
3. 70% ethanol.
4. 90% ethanol.
5. 95% ethanol.
6. 100% ethanol.

2.4. Embedding

1. 50% mixture of propylene oxide diluted in dehydration agent.
2. 100% propylene oxide.
3. 2:1 mixture of propylene oxide:resin.
4. 1:2 mixture of propylene oxide:resin.
5. Resin.
6. Embedding mold.
7. 60 °C oven.

2.5. Alternative Protocol: Correlative Confocal Imaging and Aldehyde-Free Fixation for Transmission Electron Microscopy (9)

1. Concave slide.
2. Microscope slide.
3. Coverslips.
4. Scalpel.
5. Mouth pipette.

6. Pasteur pipette.
7. Worm pick.
8. Eyelash brush.
9. Poly-L-lysine (1 mg/mL of high molecular weight).
10. Embryo culture medium: 65% Schneider's insect cell medium, 35% ES-cell certified fetal bovine serum, 1× lipid concentrate, 1× basal medium Eagle (BME) vitamin concentrate (Invitrogen, Carlsbad, CA; **ref. *36***).
11. Coumarin-440 dye ablation laser (Photonic Instruments).
12. Confocal or fluorescence microscope.
13. Second fixation materials (*see* **Subheading 2.2.**).
14. Dehydration materials (*see* **Subheading 2.3.**), substituting acetone for ethanol.
15. Embedding preparation materials (*see* **Subheading 2.4.**).
16. Embedding materials (*see* **Subheading 2.5.**).
17. Hydrofluoric acid.

3. Methods

3.1. Sample Preparation

3.1.1. Embryo Preparation and First Fixation

1. Prepare agarose fixative pads on microscope slide (*see* **Note 2**).
2. Place egg buffer into a concave slide.
3. Transfer worms into egg buffer using a pick.
4. Cut the midsection of a worm with a scalpel or needle to release the embryos.
5. Collect the embryos using a mouth pipette and transfer to the agar pad.
6. Arrange the embryos on the agar pad according to the developmental stage of the embryo, and cover the pad with a glass coverslip.
7. Observe embryos using differential interference contrast optics on a compound microscope fitted with a coumarin-dye ablation laser. When an embryo reaches the developmental stage of interest, permeabilize the egg shell with ablation laser (*see* **Note 3**).
8. Place the slide in a humidified chamber at room temperature for 2 h for fixation to complete (*see* **Note 4**).
9. Remove coverslip by pipetting egg buffer around the base of the pad, and gently slide the coverslip off.
10. Cut up the agarose pad to place individual eggs in separate wells of a nine-well glass plate (*see* **Note 5**).
11. Rinse two times in egg buffer for 30 min each, followed by an additional 2-min wash.
12. Wash in 0.2 *M* sodium cacodylate pH 7.4.

    a. Two times for 30 min each.
    b. Once for 5 min.
    c. Once for 10 min.
    d. Two times for 60 min each.

13. Proceed to second fixation section (*see* **Subheading 3.2.**)

3.1.2. Worm Preparation and First Fixation

1. Collect worms in M9 buffer.
2. Wash several times (about three) in chilled M9 buffer (*see* **Note 6**).
3. Anesthetize worms in 8% ethanol in M9 for 5 min in a glass well slide.
4. Cut open animals with clean razor blade.
5. Incubate with aldehyde fixative for 2 h on ice.
6. Rinse three times in 0.2 *M* Hepes for 10 min each (*see* **Note 6**).
7. Proceed to second fixation section (*see* **Subheading 3.2.**).

3.2. Second Fixation

1. Stain in 1% OsO_4 + 1% $K_3Fe(CN)_6$ in 0.2 *M* sodium cacodylate, pH 7.4, for 90 min (45 min for correlative electron microscopy).
2. Wash in 0.1 *M* sodium cacodylate, pH 7.4 (for worms wash in 0.1 *M* Hepes).

   a. Two times for 30 min each.
   b. Once for 5 min.
   c. Two times 10 min each.

3. Stain in 0.2% tannic acid in 0.1 *M* sodium cacodylate for 15 min (described for embryos only).
4. Wash in 0.1 *M* sodium acetate.

   a. Two times for 30 min each.
   b. Once for 5 min.
   c. Once for 10 min.
   d. Once for 30 min.

5. Stain in 1% uranyl acetate in 0.1 *M* sodium acetate for 3 h (1 h for worms; *see* **Note 7**).
6. Wash in 0.1 *M* sodium acetate.

   a. Two times for 30 min each (for worms, three washes for 5 min each).
   b. Two times for 10 min each.

7. Wash in 0.1 *M* Hepes (for worms, use distilled water for three washes for 1 min each).

   a. Once for 30 min.
   b. Two times for 5 min each.

8. Overlay sample with agarose (embryo in 2% agarose and worm pieces in 3% SeaPlafue agarose).
9. Cut individual samples embedded in agar into blocks (*see* **Note 8**).
10. Rinse in distilled water.
11. Proceed to dehydration section (*see* **Subheading 3.3.**).

3.3. Dehydration

1. Ethanol washes.

   a. 30% ethanol for 6 min.

   b. 50% ethanol for 10 min.
   c. 70% ethanol for 10 min (*see* **Note 9**).
   d. 90% ethanol for 6 min.
   e. 95% ethanol for 6 min.
   f. Three washes with 100% ethanol for 20 min each.

2. Proceed to preparation for embedding (*see* **Subheading 3.4.**).

3.4 Embedding

1. Wash in 50% mixture of propylene oxide diluted in dehydration agent for 20 min.
2. Wash two times in 100% propylene oxide for 8 min each (three times for worms).
3. Wash in 2:1 mixture of propylene oxide:resin for 30 min (3 h for worms).
4. Wash in 1:2 mixture of propylene oxide:resin for 60 min (5 h for worms).
5. Wash two times in resin (room temperature) for 30 min each (3 h for worms; for embryos, *see* **Note 10**).
6. Cover with fresh resin and store for 24 h.
7. Transfer samples to embedding mold, and orient to desired aspect for trimming and cutting.
8. Cover with fresh resin and polymerize in 60 °C oven for 3 days.
9. Proceed to sectioning sample.

3.5. Alternative Protocol: Embryo Preparation for Correlative Confocal Imaging and Aldehyde-Free Fixation for Transmission Electron Microscopy (9)

1. Place egg buffer into a concave slide.
2. Transfer worms into egg buffer using a pick.
3. Cut the midsection of a worm with a scalpel or needle to release the embryos.
4. Collect the embryos using a mouth pipette and transfer to poly-L-lysine–coated coverslip (*see* **Note 11**) with minimal liquid.
5. Blow air through mouth pipette to firmly attach embryos to poly-L-lysine coverslip.
6. Mount coverslip onto microscope slide using small dots of vacuum grease at the four corners of the coverslip.
7. Backfill mount with egg buffer or embryo culture medium containing vital dye (e.g., FM4-64), and seal edges with ~15 μL of silicone oil (Sigma, St. Louis, MO; catalog number M-6884).
8. Permeabilize embryos using ablation laser to admit vital dye through eggshell.
9. Acquire three-dimensional stacks of confocal optical sections over time. These multidimensional images can be referred to in selecting TEM sections for high-resolution analysis (*see* **Note 12**).
10. When an event of interest (e.g., fusion aperture) is observed, quickly remove coverslip and attached embryos from slide and immerse the coverslip (embryo-side up) in fixative. All subsequent steps (starting in **Subheading 3.2.**) leading up to sectioning are performed with the embryos attached to coverslip glass by moving the coverslip between solutions.

11. At the embedding step (*see* **Subheading 3.4.**), invert the embedding capsule containing resin to make a tight seal over the region of the coverslip containing embryos. Polymerize resin with the coverslip attached to the block face.
12. Dissolve coverslip glass from block surface in hydrofluoric acid. (Alternatively, pop coverslip glass off of block face using a drop of liquid nitrogen.)
13. Proceed to sectioning sample. The flat face of the resin block presents the embryo in precisely the same orientation as it was viewed during confocal imaging, allowing for direct correlation between optical sections and thin TEM sections.

4. Notes

1. This solution is toxic; use in fume hood. Make agarose fixative mixture immediately before mounting embryos, and store in 47 °C waterbath until use. The pad should be thick, so use triple thickness tape when preparing slides.
2. Melt the agarose by placing it in a 47 °C waterbath. Place a drop of warmed agarose on top of a glass slide that has been taped to the bench. Avoid the introduction of air bubbles. Place another slide on top of the agarose drop and press down firmly. Allow the agarose to solidify. Carefully remove the upper slide.
3. Aim for an area of the eggshell near the polar body. Prestaining the eggshells of embryos with Trypan blue can improve ease of permeabilization. It is important to note the properties of the embryo after permeabilization. You may need to experiment with small variations in sucrose concentration in the agarose fixative to determine what concentration to use that will not cause the embryo to collapse or swell after being ablated.
4. A humidified chamber can be a flat, air-tight container with damp paper towels placed on the bottom.
5. It is important to trim away excess agar with a clean razor blade.
6. Centrifuge between washes at 1500 rpm for 2 min using a table-top centrifuge. Carefully remove 90% of the liquid in between washes with a Pasteur pipette.
7. Skip this step for correlation electron microscopy.
8. Blocks should be larger than the pipette tip but small enough to fit into embedding mold.
9. This can be stored in 70% ethanol overnight at 4 °C.
10. Trim agarose from around the egg and transfer to separate wells of a nine-well plate.
11. Poly-L-lysine coverslip coating: Place 1 μL poly-L-lysine onto the coverslip and spread across the surface using a different coverslip. Store in a closed container for 24 h prior to use.
12. When all embryos are permeabilized *before* light microscopy, typically only a single embryo will be selected for TEM observation after fixation (for detailed protocol for four-dimensional imaging, *see* **ref. *37***).

References

1. Andreasen, A. and Ren, H. (2003) Extending the resolution of light microscopy and electron microscopy digitized images with reference to cellular changes after in vivo low oxygen exposure. *J. Neurosci. Methods* **122,** 157–170.

2. Frey, T. G., Perkins, G. A., and Ellisman, M. H. (2006) Electron tomography of membrane-bound cellular organelles. *Annu. Rev. Biophys. Biomol. Struct.* **35,** 199–224.
3. Koster, A. J. and Klumperman, J. (2003) Electron microscopy in cell biology: integrating structure and function. *Nat. Rev. Mol. Cell Biol.* **4,** SS6–SS10.
4. Sulston, J. E., Schierenberg, E., White, J. G., and Thomson, J. N. (1983) The embryonic cell lineage of the nematode *Caenorhabditis elegans. Dev. Biol.* **100,** 64–119.
5. Albertson, D. G. and Thomson, J. N. (1976) The pharynx of *C. elegans. Philos. Trans. R. Soc. Lond. Biol.* **275,** 299–325.
6. Nguyen, C. Q., Hall, D. H., Yang, Y., and Fitch, D. H. (1999) Morphogenesis of the *Caenorhabditis elegans* male tail tip. *Dev. Biol.* **207,** 86–106.
7. Sharma-Kishore, R., White, J. G., Southgate, E., and Podbilewicz, B. (1999) Formation of the vulva in *Caenorhabditis elegans*: a paradigm for organogenesis. *Development* **126,** 691–699.
8. Newman, A. P., White, J. G., and Sternberg, P. W. (1996) Morphogenesis of the *C. elegans* hermaphrodite uterus. *Development* **122,** 3617–3623.
9. Mohler, W. A., Simske, J. S., Williams-Masson, E. M., Hardin, J. D., and White, J. G. (1998) Dynamics and ultrastructure of developmental cell fusions in the *Caenorhabditis elegans* hypodermis. *Curr. Biol.* **8,** 1087–1090.
10. Shemer, G., Suissa, M., Kolotuev, I., Nguyen, K., Hall, D., and Podbilewicz, B. (2004) EFF-1 Is sufficient to initiate and execute tissue-specific cell fusion in *C. elegans. Curr. Biol.* **14,** 1587–1591.
11. Gattegno, T., Mittal, A., Valansi, C., Nguyen, K. C., Hall, D. H., Chernomordik, L. V., and Podbilewicz. B. (2007) Genetic control of fusion pore expansion in the epidermis of *Caenorhabditis elegans. Mol. Biol. Cell* **18,** 1153–1166.
12. del Campo, J. J., Opoku-Serebuoh, E., Isaacson, A. B., Scranton, V. L., Tucker, M., Han, M., and Mohler, W. A. (2005) Fusogenic activity of EFF-1 is regulated via dynamic localization in fusing somatic cells of *C. elegans. Curr. Biol.* **15,** 413–423.
13. Nolan, S., Cowan, A. E., Koppel, D., Jin, H., and Grote, E. (2006) Fus1 regulates the opening and expansion of fusion pores between mating yeast. *Mol. Biol. Cell* **17,** 2439–2450.
14. Doberstein, S. K., Fetter, R. D., Mehta, A. Y., and Goodman, C. S. (1997) Genetic analysis of myoblast fusion: blown fuse is required for progression beyond the prefusion complex. *J. Cell Biol.* **136,** 1249–1261.
15. Lucic, V., Forster, F., and Baumeister, W. (2005) Structural studies by electron tomography: from cells to molecules. *Annu. Rev. Biochem.* **74,** 833–865.
16. Priess, J. R. and Hirsh, D. I. (1986) *C. elegans* morphogenesis: the role of the cytoskeleton in elongation of the embryo. *Dev. Biol.* **117,** 156–173.
17. Kirkham, M., Müller-Reichert, T., Oegema, K., Grill, S., and Hyman, A. (2003) SAS-4 is a *C. elegans* centriolar protein that controls centrosome size. *Cell* **112,** 575–587.
18. Rappleye, C. A., Paredez, A. R., Smith, C. W., McDonald, K. L., and Aroian, R. V. (1999) The coronin-like protein POD-1 is required for anterior–posterior axis formation and cellular architecture in the nematode *Caenorhabditis elegans. Genes Dev.* **13,** 2838–2851.

19. Rostaing, P., Weimer, R. M., Jorgensen, E. M., Triller, A., and Bessereau J. L. (2004) Preservation of immunoreactivity and fine structure of adult *C. elegans* tissues using high-pressure freezing. *J. Histochem. Cytochem.* **52,** 1–12.
20. Hall, D. H. (1995) Electron microscopy and 3D image reconstruction, in *Caenorhabditis elegans: Modern Biological Analysis of an Organism* (H. F. Epstein and D. C. Shakes, eds.), vol. 48. Academic Press, New York, pp. 395–436.
21. Ward, S., Thomson, N., White, J. G., and Brenner, S. (1975) Electron microscopical reconstruction of the anterior sensory anatomy of the nematode *Caenorhabditis elegans. J. Comp. Neurol.* **160,** 313–338.
22. White, J. G., Southgate, E., Thomson, J. N., and Brenner, S. (1976) The structure of the ventral nerve cord of *Caenorhabditis elegans. Philos. Trans. R. Soc. Lond. Biol. Sci.* **275,** 327–348.
23. White, J. G., Southgate, E., Thomson, J. N., and Brenner, S. (1986) The structure of the nervous system of the nematode *Caenorhabditis elegans. Philos. Trans. R. Soc. Lond. Biol.* **314,** 1–340.
24. Jones, J. T. and Gwynn, I. (1991) Method for rapid fixation and dehydration of nematode tissue for transmission electron microscopy. *J. Microsc. 164*, 43–51.
25. Paupard, M. C., Miller, A., Grant, B., Hirsh, D., and Hall, D. H. (2001) Immuno-EM localization of GFP-tagged yolk proteins in *C. elegans* using microwave fixation. *J. Histochem. Cytochem.* **49,** 949–956.
26. Subramaniam, S. and Milne, J. L. (2004) Three-dimensional electron microscopy at molecular resolution. *Annu. Rev. Biophys. Biomol. Struct.* **33,** 141–155.
27. Sartori Blanc, N., Studer, D., Ruhl, K., and Dubochet, J. (1998) Electron beam-induced changes in vitreous sections of biological samples. *J. Microsc.* **192,** 194–201.
28. Leapman, R. D. (2004) Novel techniques in electron microscopy. *Curr. Opin. Neurobiol.* **14,** 591–598.
29. Hasty, D. L. and Hay, E. D. (1978) Freeze-fracture studies of the developing cell surface. *J. Cell Biol.* **78,** 756–768.
30. Al-Amoudi, A., Norlen, L. P. O., and Dubochet, J. (2004) Cryo-electron microscopy of vitreous sections of native biological cells and tissues. *J. Struct. Biol.* **148,** 131–135.
31. Richter, T., Biel, S., Sattler, M., Wenck, H., Wittern, K., Wiesendanger, R., and Wepf, R. (2007) Pros and cons: cryo-electron microscopic evaluation of block faces versus cryo-sections from frozen-hydrated skin specimens prepared by different techniques. *J. Microsc.* **225,** 109–207.
32. Deerinck, T. J., Martone, M. E., Lev-Ram, V., Green, D. P., Tsien, R. Y., Spector, D. L., Huang, S., and Ellisman, M. H. (1994) Fluorescence photooxidation with eosin: a method for high resolution immunolocalization and in situ hybridization detection for light and electron microscopy. *J. Cell Biol.* **126,** 901–910.
33. Nisman, R., Dellaire, G., Ren, Y., Li, R., and Bazett-Jones, D. P. (2004) Application of quantum dots as probes for correlative fluorescence, conventional, and energy-filtered transmission electron microscopy. *J. Histochem. Cytochem.* **52,** 13–18.

34. Gaietta, G., Deerinck, T., Adams, S., Bouwer, J., Tour, O., Laird, D., Sosinsky, G., Tsien, R., and Ellisman, M. (2002) Multicolor and electron microscopic imaging of connexin trafficking. *Science* **296,** 503–507.
35. Selkirk, M. E., Yazdanbakhsh, M., Freedman, D., Blaxter, M. L., Cookson, E., Jenkins, R. E., and Williams, S. A. (1991) A proline-rich structural protein of the surface sheath of larval *Brugia* filarial nematode parasites. *J. Biol. Chem.* **266,** 11002–11008.
36. Shelton, C. A. and Bowerman, B. (1996) Time-dependent responses to glp-1–mediated inductions in early *C. elegans* embryos. *Development* **122,** 2043–2050.
37. Mohler, W. A. and Squirrell, J. M. (2000) Multiphoton imaging of embryonic development, in *Imaging Neurons: A Laboratory Manual* (A. Konnerth, ed.). Cold Spring Harbor Laboratory Press, Cold Spring Harbor, New York, pp. 21.1–21.11.

15

Live Imaging of *Drosophila* Myoblast Fusion

Brian E. Richardson, Karen Beckett, and Mary K. Baylies

Summary

Myoblast fusion requires a number of cellular behaviors, including cell migration, recognition, and adhesion, as well as a series of subcellular behaviors, such as cytoskeletal rearrangements, vesicle trafficking, and membrane dynamics, leading to two cells becoming one. With the discovery of fluorescent proteins that can be introduced and studied within living cells, the possibility of monitoring these complex processes within the living embryo is now a reality. Live imaging, unlike imaging techniques for fixed embryos, allows the opportunity to visualize and measure the dynamics of these processes in vivo. This chapter describes the development and use of live imaging techniques to study myoblast fusion in *Drosophila*.

Key Words: *Drosophila*; myoblast fusion; muscle development; fluorescent proteins; live imaging.

1. Introduction

There are 30 individual muscles per hemisegment of the *Drosophila* embryo. Formation of these individual body wall muscles depends on the specification and fusion of two myoblast cell types, founder cells and fusion-competent myoblasts (discussed in greater detail in **Chapter 5**). Each founder cell contains the necessary information to direct the formation of a specific muscle. Founder cells can be identified by expression of identity genes, such as the transcriptional regulators *even-skipped, apterous, slouch*, and *Krüppel*. The combination of identity genes expressed by a particular founder cell is thought to regulate the final morphology of the specific muscle. Fusion-competent myoblasts, in contrast, are thought to be naive cells. Upon fusion to a founder cell, fusion-competent myoblasts become reprogrammed to the founder cell's specific developmental program, as witnessed by each newly incorporated fusion-competent myoblast nucleus expressing the founder cell's particular combination of identity genes ***(1–5)***. Myoblast fusion is

From: *Methods in Molecular Biology, vol. 475: Cell Fusion: Overviews and Methods*
Edited by: E. H. Chen © Humana Press, Totowa, NJ

a reiterative process; depending on the particular muscle, body wall muscles in *Drosophila* embryos arise from between 2 and 25 fusion events ***(6)***.

Fusion occurs during stages 12 to 15 (7.5–13h after egg laying [AEL]). Founder cells/myotubes and fusion-competent myoblasts are arranged in multiple cell layers prior to and during the fusion process (**Fig. 1; ref.** ***7***). As fusion commences, the mesoderm is arranged with the founder cells occupying both the most external and most internal positions, with multiple layers of fusion-competent myoblasts found in between (*see* **Fig. 1A**). As germband retraction and dorsal closure proceed during stages 13 and 14, the ventral founder cells/myotubes and fusion-competent myoblasts move externally to lie underneath the epidermis and central nervous system (*see* **Fig. 1B,C**). While some fusion-competent myoblasts contact the founder cells/myotubes and are responsible for the initial fusion events, the remaining fusion-competent myoblasts are located more internally and must migrate to find their fusion partners. An appreciation of these cell arrangements and movements is essential for the analysis of myoblast fusion.

Fusion does not start concurrently in all muscles (*see* **Fig. 1D; ref.** ***7***). For example, fusion begins during stage 12 (7.5–9.5h AEL) in the dorsal DA1 muscle but does not begin until stage 13 (9.5–10.5h AEL) in the ventral VA2 muscle. However, for all muscles examined to date, the majority of fusion events occur during stage 14 (10.5–11.5h AEL; *see* **Fig. 1D**), making this a particularly useful stage for the analysis of the cellular biology underlying myoblast fusion.

Although genetic analysis has revealed a number of genes required for the fusion process (reviewed in **Chapter 5**), their precise function during fusion has been hampered by the lack of direct, cellular assays. For example, while many of the known genes encode regulators of the actin cytoskeleton ***(8)***, the impact of cytoskeletal rearrangements on fusion were unclear. Furthermore, the arrangements of founder cells and fusion-competent myoblasts ***(7)*** have implicated a critical role of migration in the fusion process that remains to be analyzed. Finally, the site of fusion has only been implied by localization of proteins implicated in fusion and not directly located. To address these issues, we developed live imaging techniques for *Drosophila* to further our understanding of the myoblast fusion process. This chapter deals with the methodology and considerations underlying this approach, including collection and mounting of the embryos for live imaging, as well as the imaging itself and techniques for processing and presenting the data.

2. Materials

2.1. Embryo Collection

1. Embryo laying chamber: 100-mL plastic beaker, punched with holes to allow air exchange and to prevent condensation.
2. Embryo collection plates, attached to the laying chamber with a rubber band: Microwave 1500mL of ddH_2O, 50g of granulated sugar and 45g of agar until an

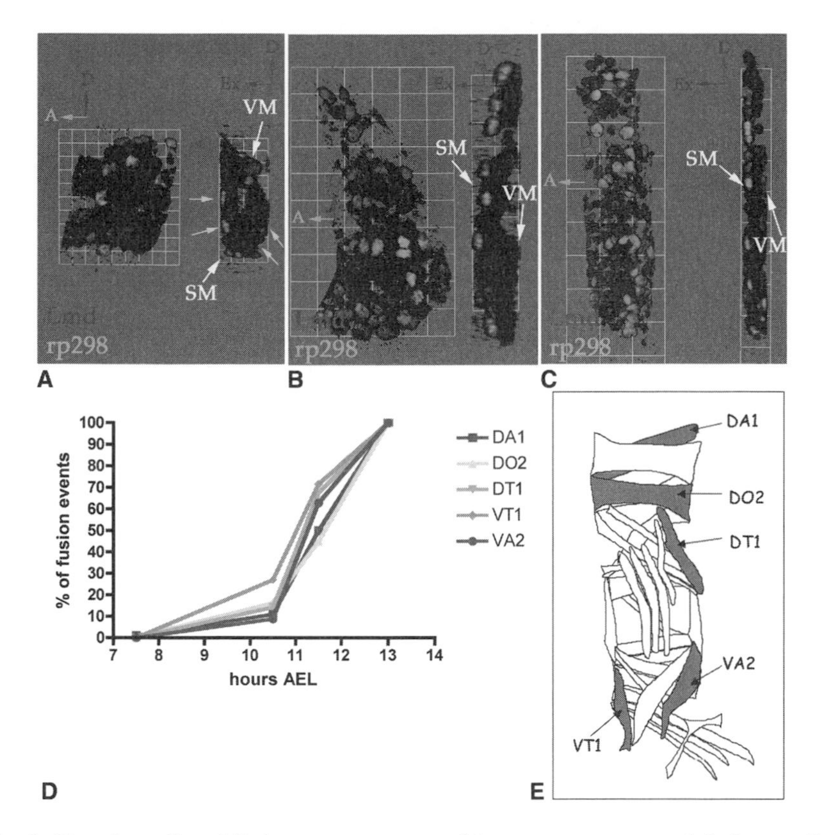

Fig. 1. Founder cell and fusion-competent myoblast arrangements and fusion profiles of individual muscles. A–C, stages 12 (A), 13 (B), and 14 (C) *rp298-lacZ* embryos were stained with antibodies against ß-gal to label founder cell/myotube nuclei (green) and Lameduck (Lmd) to label fusion-competent myoblasts (blue). Three-dimensional renderings of single mesodermal hemisegments at stages 12 (A, 1 grid unit = 5.7 μm), 13 (B, 1 grid unit = 10.9 μm), and 14 (C, 1 grid unit = 14.1 μm) are shown. Each panel shows an external view (left) and a side view rotated 90° clockwise (right). Red arrows point to dorsal, green arrows point to anterior, and blue arrows point to external. SM, somatic mesoderm; VM, visceral mesoderm. **(A)** At stage 12, the somatic mesoderm folds into two layers so that the founder cells (green) are concurrently the most external and internal cells (yellow arrows, A), with the fusion-competent myoblasts (blue) in between. **(B)** At stage 13, the internal founder cells and fusion-competent myoblasts have moved externally to underlay the overlaying epidermis (not labeled). The founder cells (green) appear to rest on top of the fusion-competent myoblasts (blue) at this stage, and the cells are tightly packed together. **(C)** By stage 14, the number of *rp298-lacZ*–expressing nuclei (green) have increased because of fusion. The fusion-competent myoblasts (blue) have separated from one another. **(D)** Wild-type stages 12–15 embryos were stained with antibodies against Eve (DA1), Runt (DO2) or Slouch (DT1, VT1, and VA2) in combination with phalloidin to assist accurate staging. The number of nuclei for each muscle and stage were counted in 50 hemisegments (A2–4). Graph shows the percentage of fusion events that occur during each stage for each muscle during the course of fusion (7.5–13 h after egg laying [AEL]). The mean number of nuclei observed for each muscle at stage 15 is 100% and a single nucleus is 0%. **(E)** Schematic showing which muscles were analyzed for wild-type fusion profiles (blue). (Adapted from **ref. 7**.) (*See Color Plates*)

even solution is formed. Add 500 mL of cold apple juice. Cool to 65 °C, and add 40 mL of 10% Tegosept in 100% ethanol. Pour into 35-mm Petri dishes; makes approximately 200.
3. Yeast paste: Water plus dry baker's yeast, stirred to make a paste. Store at 4 °C.
4. Paint brush.
5. Dechorionating baskets: Cut the top off of a 15-mL Falcon tube at approximately the 12-mL line, and cut the center out of the tube cap.
6. Nitex membrane, screwed onto the dechorionating basket with the cap.
7. 50% Clorox bleach.
8. Small Petri dish.

2.2. Embryo Mounting

1. For inverted microscopy: Uncoated 35-mm glass bottom microwell dish (MatTek Cultureware, Ashland, MA).
2. For upright microscopy: air-permeable Teflon membrane mounted on Perspex frame (designed by E. Wieschaus; **refs.** ***9,10***).
3. Microscope slides.
4. Embryo hook.
5. 18 × 18 mm coverslips.
6. 22 × 40 mm coverslips.
7. Technau glue: Double-sided Scotch tape dissolved in heptane.
8. Halocarbon oil 700 (Halocarbon Products Corp., River Edge, NJ).

2.3. Fluorescent Proteins

When selecting a fluorescent protein for live imaging studies, important factors include brightness of signal, spectral properties, folding time, and cytotoxicity (for review, *see* **ref.** ***11***). Brightness in particular is of the utmost importance in the developing musculature, which extends several cell layers and up to 25 μm below the surface of the embryo. Fluorescent proteins that we have used successfully to label the forming musculature of the embryo are listed in **Table 1**.

The protein of choice for most purposes continues to be green fluorescent protein (GFP). Green fluorescent protein is small, nontoxic, and easily visualized with most fluorescent microscopes. Mutation of the original sequence has provided derivatives with enhanced properties, such as increased thermostability, increased expression, and optimized excitation peaks ***(12)***. For colabeling studies requiring the use of multiple fluorescent proteins, yellow fluorescent proteins (YFPs) and red fluorescent proteins (RFPs) have both proven useful. We have not encountered a cyan fluorescent protein (CFP) that is bright enough to visualize myogenesis. Yellow fluorescent proteins require newer filter sets to distinguish their emission signal from GFP. The original RFP, DsRed, had problems with slow maturation times and contamination of the GFP signal ***(13)***.

Table 1
Fluorescent Proteins Useful for Live Imaging of Myoblast Fusion

Protein	Localization	Excitation/emission	Source/reference
EGFP	Cytoplasm	488 nm/509 nm	Clontech ***(24,25)***
EYFP	Cytoplasm	514 nm/527 nm	Invitrogen ***(24)***
GFP::actin	Actin cytoskeleton	488 nm/509 nm	***(16)***
GFP::moesin	Actin cytoskeleton	488 nm/509 nm	***(26)***
GFP::α-tubulin	Microtubules	488 nm/509 nm	***(18)***
NLS::EGFP	Nucleus	488 nm/509 nm	***(25)***
NLS::DsRed.T4	Nucleus	556 nm/586 nm	***(15)***
H2B:YFP	Nucleus	514 nm/527 nm	***(27)***
Src::GFP	Cell membrane	488 nm/509 nm	***(9)***
mCD8::GFP	Cell membrane	488 nm/509 nm	***(17)***

EGFP, enhanced green fluorescent protein; EYFP, enhanced yellow fluorescent protein; GFP, green fluorescent protein; NLS, nuclear localization signal; DsRed, *Discosoma* Red; H2B, human histone 2B; mCD8, mouse CD8. Excitation and emission represent approximate maximum values.

However, we have had success with DsRed.T4, which does not produce green fluorescence and has substantially improved maturation time ***(14,15)***.

Although the proteins listed in **Table 1** have been quite useful in our studies, fluorescent proteins with increased brightness, distinct spectral properties, and other advantages are being developed constantly and should be considered in any new studies. Thanks to focused engineering, a wide range of fluorescent proteins across all spectra of light are now available. These include mPlum (far-red), mCherry and mStrawberry (red), Venus (yellow), Emerald (green) and Cerulean (cyan), which represent a diverse reagent set for multicolor live imaging ***(11)***.

The fusion of fluorescent proteins to other proteins or sequences is useful for marking specific cell structures during fusion. For example, we have used a GFP::actin fusion protein in our studies of the role of the actin cytoskeleton in fusion (**Figs. 2–4; ref.** ***16***), as well as a nuclear localization signal fused to DsRed.T4 to track subsets of myoblast nuclei through the fusion process (*see* **Fig. 4; ref.** ***15***). Src::GFP and mCD8::GFP are both useful as cell membrane markers ***(9,17)***. A GFP::α-tubulin fusion protein can be used to examine microtubule dynamics ***(18)***.

2.4. Expression of Fluorescent Proteins

The galactose (GAL4)/upstream activating sequence (UAS) system is a widely used system for expressing a variety of proteins, including fluorescent protein, in a tissue-specific manner ***(19)***. As mentioned, brightness of signal is critical, necessitating the use of strongly expressing GAL4 lines. Combining

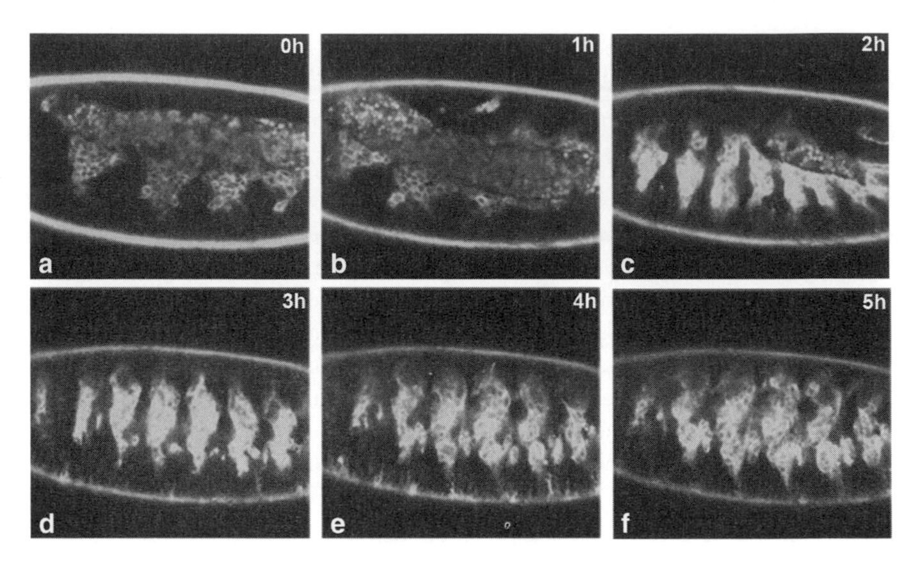

Fig. 2. All of *Drosophila* myogenesis can be visualized using live imaging. Lateral view of live *twi-GAL4; Dmef2-GAL4* × *UAS-GFP::actin* embryo. These GAL4 lines strongly express throughout myogenesis. Each image **(A–F)** represents a single frame from a time-lapse sequence over the course of 5 h, covering stages 12–15. During this time, most of myoblast fusion is completed. Images are single optical slices.

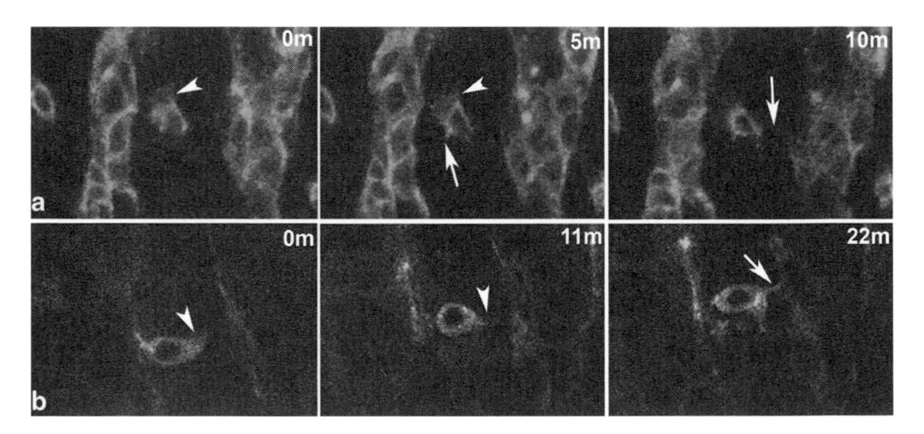

Fig. 3. Dynamic actin bases behaviors of myoblasts. Lateral views of live *twist* promoter *–GFP::actin* embryos. Actin labeling is concentrated at the cell cortices and in cellular extensions such as lamellipodia and filopodia. Such behaviors appear critical for myoblast fusion. The nucleus is evident as a nonlabeled structure in the middle of the cells. Each panel represents a single time point from a time-lapse sequence over 10 min. (A) or 22 min. (B) Images are single optical slices. **(A)** Founder cells for a segment border muscle prior to fusion extends lamellipodia (arrowheads) and filopodia (arrows). **(B)** Fusion-competent myoblast extends lamellipodia (arrowheads) and a filopodium (arrow) prior to fusion.

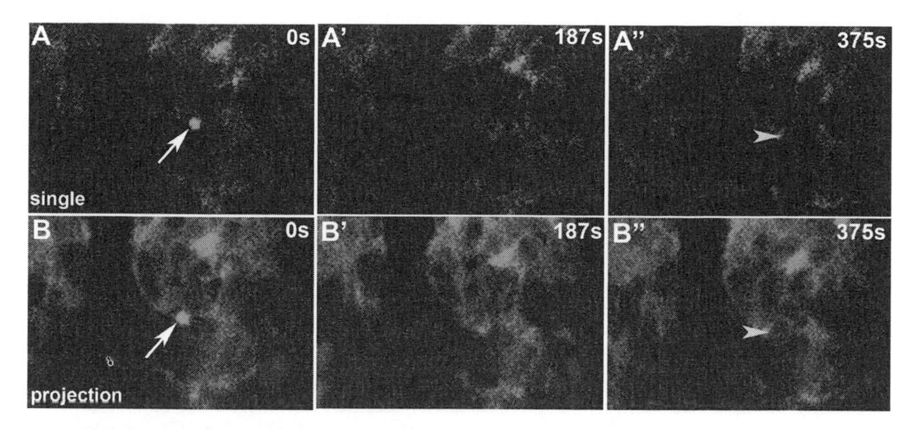

Fig. 4. Single myoblast fusion event. Lateral view of live *twist* promoter *–GFP::actin, apME580-NLS::dsRed.T4* embryo. Each row of panels represents a time point from a time-lapse sequence over 375 s. **(A)** Single optical slice. In this sequence, an actin focus (arrow) forms at the site of adhesion between a fusion-competent myoblast and an *apterous*-labeled myotube. This focus resolves, followed by fusion and addition of an additional labeled nucleus (arrowhead) to the myotube. **(B)** Same sequence as in A but as an optical projection displaying 9 μm of the Z-axis. (*See Color Plates*)

different UAS-fluorescent protein lines under control of a single GAL4 allows labeling of different cell structures, such as actin cytoskeleton and nucleus, with different fluorescent colors. However, the GAL4/UAS system does not allow multicolor labeling of different cell types, such as founder cells versus fusion-competent myoblasts. An alternative approach is to create direct enhancer and/or promoter fusions to the coding regions of fluorescent proteins. This allows simultaneous transgenic expression of different fluorescent proteins in two or more cell types, allowing multicolor live imaging. We also observe that fluorescence signal is often visible sooner using enhancer/promoter fusions than a comparable GAL4/UAS combination, presumably because of lag time for GAL4 to bind and activate UAS.

Reagents that we have found useful for driving expression of fluorescent proteins during myoblast fusion are listed in **Table 2**. Specifically, we have found the *twist* promoter *(20)* to be a useful way to drive expression throughout the mesoderm during all stages of myogenesis, both as a GAL4 construct and fused to GFP::actin (*see* **Figs. 2–4**). Specific enhancers of founder cell identity genes, such as *apterous even-skipped* and *slouch* ***(2,21–23)***, are useful for visualizing founder cells/myotubes as fusion proceeds (*see* **Fig. 4**). We are currently lacking promoters or enhancers for fusion-competent myoblast-specific expression, which would facilitate analysis of the contribution of this cell type to the fusion process.

Table 2
Promoters and Enhancers Useful for Driving Fluorescent Proteins During Myoblast Fusion

Promoter/enhancer	Tissue expression	Reference
twi promoter	Somatic mesoderm	***(20)***
Dmef2 promoter	Somatic mesoderm	***(28)***
apME580	Founder cell subset	***(2)***
eveMHE	Founder cell subset	***(23)***
slouch/S59	Founder cell subset	***(21,22)***
5053	Founder cell subset	***(29)***
Mhc promoter	Muscle	***(30)***

twi, twist; apME580, apterous mesodermal enhancer 580; Dmef2, drosophila myocyte enhancing factor 2; eveMHE, even-skipped mesoderm and heart enhancer; Mhc, myosin heavy chain.

2.5. Fluorescent Balancer Chromosomes

Analysis of mutants using live imaging presents the additional challenge of correctly separating homozygous and heterozygous embryos. This is especially important because the process being studied often takes place before the final mutant phenotype is apparent. Correct identification can be achieved by negative selection against a fluorescently labeled balancer chromosome. In our hands, the CTG (Cyo, *twi-GAL4, UAS-2XEGFP*) and TTG (TM3, *twi-GAL4, UAS-2XEGFP*) balancers created by A. Michelson's group can be easily identified under a standard dissecting scope fitted for fluorescence microscopy ***(24)***.

3. Methods

3.1. Embryo Collection

1. Collect embryos of the appropriate age on an apple juice/agar plate.
2. Wash embryos into an embryo basket using water and a paintbrush.
3. Dechorionate embryos for 3 min in 50% Clorox bleach in a Petri dish, and then rinse well with water.

3.2. Embryo Mounting

1. Transfer embryos to a microscope slide with a paintbrush, and cover them with Halocarbon 700 oil.
2. Select embryos of the appropriate age and genotype (using negative selection against fluorescent balancers, if necessary).
3. For an upright microscope, apply a thin layer of Technau glue onto an air-permeable Teflon membrane stretched over a Perspex frame. Place embryos onto the glue, which will help prevent rolling and drifting (*see* **Note 1**). Cover embryos with a drop of Halocarbon 700 oil. Place an 18 × 18 mm coverslip on either side of the embryos, and "bridge" them with a 22 × 40 mm coverslip ***(9,10)***.

4. For an inverted microscope, apply a thin layer of Technau glue onto a glass bottom dish. Place embryos onto the glue, which will help prevent rolling and drifting (*see* **Note 1**). Cover embryos with a drop of Halocarbon 700 oil.

3.3. Confocal Microscopy

Although it is possible to visualize fluorescent proteins using standard epifluorescence microscopy, *Drosophila* embryos display a high level of autofluorescence that can obscure the signal of interest. Furthermore, fluorescent signal is significantly dampened in internal cell layers such as the mesoderm. Fortunately, the thin optical sections and relative strength of confocal microscopy can be used to minimize the effects of these problems. Our standard setup for live imaging is a Zeiss LSM510 scanning confocal system mounted on an Axiovert 100 M microscope with a 63× 1.2 NA C-Apochromat water objective. Green fluorescent protein and YFP can be detected using standard fluorescein isothiocyanate settings, and RFP can be detected using standard rhodamine settings. Newer systems feature more advanced filters and completely filterless setups, allowing further spectral separation required for the use of newer fluorescent proteins.

A sufficient signal-to-noise ratio is critical for proper analysis of confocal data. To reduce noise, it is possible to perform frame averaging and decrease the laser scan speed. However, both of these approaches increase photobleaching and phototoxicity (*see* **Note 2**) and decrease temporal resolution because of longer scan times. We find a frame averaging of four frames and a scan speed of 7 (pixel time of 2.56 μs) are a good balance with our system. Laser strength can be raised to increase signal strength, although this also increases photobleaching (*see* **Note 2**). We generally set our 488-nm laser (which excites GFP) to 25%. Finally, the confocal pinhole can be opened to allow more light to be imaged. However, this increases the thickness of the optical slice and decreases axial resolution. We set the pinhole to capture 1–1.5 μm per optical slice for live imaging. When performing multicolor imaging, it is essential to set pinholes so that the same size of optical slice is captured for each emission spectrum.

One must consider the type of movie desired before beginning an imaging sequence. For example, imaging the entirety of myoblast fusion requires approximately 6 h of scanning, making photobleaching a significant hurdle. Therefore, sampling every 5–10 min is advisable, as it is sufficient to capture overall tissue dynamics while avoiding drastic photobleaching. For more in-depth analysis of fusion dynamics, more frequent sampling becomes necessary. Myoblast migration and fusion events occur quite rapidly; therefore, sampling every 1–2 min is necessary to gain an appreciation of these processes. Finally, the analysis of subcellular dynamics during an individual migration or fusion event requires even more rapid imaging; we have found relevant cytoskeletal changes to occur on the order of every 5–10 s. Such temporal resolution stretches the limits of our current setup, but newer confocal systems such as the Zeiss LSM 5 Live have been specifically designed for high-speed laser scanning imaging and will increase the temporal resolution of future live imaging studies. Although spinning disc confocal microscopy already exists as a tool for high-speed imaging, we have found its optical resolution insufficient for visualizing myoblast migration and fusion, presumably because the cells are too internal in the embryo.

A final but critical consideration when imaging myoblast migration and fusion is the analysis of multiple optical slices along the Z-axis of the embryo. Tissue and cell movement can bring cells of interest in and out of a single optical slice, confusing interpretation of true migration and fusion events. Confocal software allows one to scan several optical slices sequentially over time to create a three-dimensional time sequence, also known as four-dimensional imaging. However, scanning multiple optical slices does have disadvantages, such as decreased temporal resolution and increased photobleaching. Therefore, it is desirable to visualize the fewest number of optical slices while still being able to follow all relevant cell movements.

3.4. Processing Software

The Zeiss LSM software allows confocal images to be exported in .avi or .tif formats. The .avi files can be played on PC computers, while Apple Quicktime can be used to compile .tif files into an image sequence. The LSM software can also be used to create optical projections of three-dimensional data. Other confocal software has similar functions. Alternatively, confocal data files can be directly imported into third-party software programs such as Volocity (Improvision, Waltham, MA) for additional three-dimensional reconstruction analysis and presentation options.

4. Notes

1. Even with specific cell markers and proper technique, several experimental artifacts are possible when imaging myoblast fusion. The yolk cells of the embryo are autofluorescent and often present near the mesoderm, especially during the earliest stages of fusion (*see* **Fig. 2**). Conversely, near the end of myogenesis, autofluorescent macrophages are present near the musculature and should not be confused with myoblasts. Unintended movement can cause multiple problems during live imaging. Insufficient gluing of the embryo during mounting often results in embryo rolling or lifting, which can be mistaken for tissue or cell movement. Cell and tissue movement between optical slices can also be confused with true cell migration or fusion events. However, proper use of four-dimensional imaging can circumvent this problem, as discussed.

2. Three papers defining new GAL4 lines expressed in PCMS were published: one focused on an FCM enhancer derived from the Dmef2 enhancer ***(31)*** and two focused on enhancer regions isolated from the sticks-and-stones gene ***(32,33)***.
3. All fluorophores, including those present on fluorescent proteins, are eventually destroyed by exposure to light. This problem can be minimized in several ways, such as reducing laser intensity, reducing frame averaging, increasing the scan speed, or increasing the frame delay. Also, selecting the correct promoter or enhancer to drive expression will allow continual synthesis of new fluorescent protein in the cell, replenishing the supply as old protein is extinguished.

Acknowledgments

We thank members of Baylies' laboratory, Owen Richardson, and especially Kat Hadjantonakis for stimulating discussions and advice. We also recognize the valuable input from Julia Kaltschmidt during our early days of filming. This work is supported by NIH grants GM56989 and GM78318 to M.B.

References

1. Baylies, M. K., Bate, M., and Ruiz Gomez, M. (1998) *Cell* **93,** 921–927.
2. Capovilla, M., Kambris, Z., and Botas, J. (2001) *Development* **128,** 1221–1230.
3. Frasch, M. (1999) *Curr. Opin. Genet. Dev.* **9,** 522–529.
4. Beckett, K. and Baylies, M. K. (2006) *Dev. Biol.* **299,** 176–192.
5. Carmena, A. and Baylies, M. (2006) *Muscle Development in Drosophila* (Sink, H., ed.). Landes Bioscience, New York, pp. 79–89.
6. Bate, M. (1990) *Development* **110,** 791-804.
7. Beckett, K. and Baylies, M. K. (2007) 3D analysis of founder cell and fusion competent myoblast arrangements outlines a new model of myoblast fusion. *Dev. Biol.* **309,** 113–125.
8. Chen, E. H. and Olson, E. N. (2004) *Trends Cell Biol.* **14,** 452–460.
9. Kaltschmidt, J. A., Davidson, C. M., Brown, N. H., and Brand, A. H. (2000) *Nat. Cell Biol.* **2,** 7–12.
10. Haseloff, J., Dormand, E., and Brand, A. H. (1998) *Methods Mol. Biol.* **122,** 241–259.
11. Shaner, N. C., Steinbach, P. A., and Tsien, R. Y. (2005) *Nat. Methods* **2,** 905–909.
12. Zacharias, D. A. and Tsien, R. Y. (2006) *Methods Biochem. Anal.* **47**, 83–120.
13. Baird, G. S., Zacharias, D. A., and Tsien, R. Y. (2000) *Proc. Natl. Acad. Sci. U.S.A.* **97,** 11984–11989.
14. Bevis, B. J. and Glick, B. S. (2002) *Nat. Biotechnol.* **20,** 83–87.
15. Barolo, S., Castro, B., and Posakony, J. W. (2004) *Biotechniques* **36,** 436–440, 442.
16. Verkhusha, V. V., Tsukita, S., and Oda, H. (1999) *FEBS Lett.* **445,** 395–401.
17. Lee, T. and Luo, L. (1999) *Neuron* **22,** 451–461.
18. Grieder, N. C., de Cuevas, M., and Spradling, A. C. (2000) *Development* **127,** 4253–4264.

19. Brand, A. H. and Perrimon, N. (1993) *Development* **118,** 401–415.
20. Baylies, M. K. and Bate, M. (1996) *Science* **272,** 1481–1484.
21. Schnorrer, F. and Dickson, B. J. (2004) *Dev Cell* **7,** 9–20.
22. Knirr, S., Azpiazu, N., and Frasch, M. (1999) *Development* **126,** 4525–4535.
23. Halfon, M. S., Carmena, A., Gisselbrecht, S., Sackerson, C. M., Jimenez, F., Baylies, M. K., and Michelson, A. M. (2000) *Cell* **103,** 63–74.
24. Halfon, M. S., Gisselbrecht, S., Lu, J., Estrada, B., Keshishian, H., and Michelson, A. M. (2002) *Genesis* **34,** 135–138.
25. Barolo, S., Carver, L. A., and Posakony, J. W. (2000) *Biotechniques* **29,** 726, 728, 730, 732.
26. Dutta, D., Bloor, J. W., Ruiz-Gomez, M., VijayRaghavan, K., and Kiehart, D. P. (2002) *Genesis* **34,** 146–151.
27. Bellaiche, Y., Gho, M., Kaltschmidt, J. A., Brand, A. H., and Schweisguth, F. (2001) *Nat. Cell Biol.* **3,** 50–57.
28. Ranganayakulu, G., Schulz, R. A., and Olson, E. N. (1996) *Dev. Biol.* **176,** 143–148.
29. Schnorrer, F., Kalchhauser, I., and Dickson, B. J. (2007) *Dev. Cell* **12,** 751–766.
30. Chen, E. H. and Olson, E. N. (2001) *Dev. Cell* **1,** 705–715.
31. Beckett, K., Rochlin, K. M., Duan, H., Nguyen, H. T., and Baylies, M. K. (2008) Expression and functional analysis of a novel Fusion Competent Myoblast specific GAL4 driver. Gene Expr. Patterns **8,** 87–91.
32. Stute, C., Kesper, D., Holz, A., Buttgere, D., and Renkerwitz-Ashl, R. (2006) Establishment of cell type specific Galy-driver lines for the mesoderm of drosophila. *Dros. Inf. Serv.* **89,** 111–115.
33. Kocherlakota, K. S., Wu, J. M., McDermott, J., and Abmayr, S. M. (2008) Analysis of the Cell Adhesion Molecule Sticks-and-Stones Reveals Multiple Redundant Functional Domains, Protein-Interaction Motifs and Phosphorylated Tyrosines That Direct Myoblast Fusion in Drosophila melanogaster. *Genetics* **178,** 1371–1383.

16

Ultrastructural Analysis of Myoblast Fusion in *Drosophila*

Shiliang Zhang and Elizabeth H. Chen

Summary

Myoblast fusion in *Drosophila* has become a powerful genetic system with which to unravel the mechanisms underlying cell fusion. The identification of important components of myoblast fusion by genetic analysis has led to a molecular pathway toward our understanding of this cellular process. In addition to the application of immunohistochemistry and live imaging techniques to visualize myoblast fusion at the light microscopic level, ultrastructural analysis using electron microscopy remains an indispensable tool to reveal fusion intermediates and specific membrane events at sites of fusion. In this chapter, we describe conventional chemical fixation and high-pressure freezing/freeze substitution methods for visualizing fusion intermediates during *Drosophila* myoblast fusion. Furthermore, we describe an immunoelectron microscopic method for localizing specific proteins relative to the fusion apparatus.

Key Words: Cell fusion; myoblast fusion; *Drosophila*; electron microscopy; chemical fixation; high-pressure freezing/freeze substitution; immunoelectron microscopy.

1. Introduction

The development of multinucleated skeletal muscle is a fascinating process that requires fusion of mononucleated myoblasts. Myoblast fusion is a conserved cellular process that occurs in multicellular organisms ranging from insects to humans. The somatic muscle in the fruit fly *Drosophila* is functionally equivalent to vertebrate skeletal muscle, yet the fly musculature is much simpler and takes only a short time to develop *(1)*. These features, together with the great genetic tools available for *Drosophila*, make it an ideal system in which to study myoblast fusion in vivo.

Myoblast fusion in *Drosophila* embryo occurs between two types of muscle cells: muscle founder cells and fusion-competent myoblasts *(2,3)*. Muscle

From: *Methods in Molecular Biology, vol. 475: Cell Fusion: Overviews and Methods*
Edited by: E. H. Chen © Humana Press, Totowa, NJ

founder cells determine the position, orientation, and size of the future muscle fibers, whereas fusion-competent myoblasts migrate toward, adhere to, and fuse with founder cells to generate multinucleated muscle fibers. One commonly used technique to monitor myoblast fusion is imaging fixed or live embryos with light microscopy. While antibodies against structural proteins including muscle myosin heavy chain and β3-tubulin are often used to label either mature muscle fibers in wild-type embryos or fusion-defective myoblasts in mutants *(4,5)*, the sites of fusion in founder cells or fusion-competent myoblasts are marked by antibodies against proteins required for fusion that are localized or recruited to the sites of cell attachment (*see* **Chapter 5; refs.** *2,3*). These immunohistochemical studies have provided a wealth of information about the function of genes required for myoblast fusion. Besides examining protein expression and localization in fixed embryos, the cellular dynamics of the somatic musculature can be monitored live with fluorescent proteins driven by appropriate mesodermal promoters (*see* **Chapter 5; ref.** *6*).

The advantages of light microcopy are the simplicity of the technique and the ability to monitor the fusion process live. However, the maximum resolution of light microscopy (200 nm), limited by the wavelength of the light, far exceeds the thickness of cell membrane (3–10 nm). Therefore, another type of microscopy with much higher resolution is required to visualize fusion-related intracellular organelles or ultrastructural events occurring at the plasma membrane. Electron microscopy utilizes an electron beam with a far smaller wavelength than light. As a result, the resolution of a standard transmission electron microscope is about 0.2 nm. To date, electron microscopic analyses have provided significant information on the ultrastructural fusion intermediates leading to myoblast fusion.

1.1. Ultrastructural Analysis of Myoblast Fusion in Drosophila

The first electron microscopic study of *Drosophila* myoblast fusion was published in 1997 *(7)*. In this landmark paper, Doberstein et al. revealed several fusion intermediates at the ultrastructural level, including paired vesicles with electron-dense margins, rare electron-dense plaques, and multiple membrane discontinuities (fusion pores) along the apposing myoblast membranes. Although subsequent electron microscopic work from several groups verified the presence of these fusion intermediates *(8–11)*, two issues are worth noting. First, the presence of clusters of prefusion vesicles in wild-type embryos (**Fig. 1**) is far less frequent than those shown in **Figures 2A** and **3** of Doberstein et al. *(7)* and Zhang and Chen, unpublished observation. Second, the number and morphology of fusion pores reported in wild-type embryos prepared by the conventional chemical fixation method require a reevaluation by an independent sample preparation method (*see* **Subheading 1.3.**) that better preserves the lipid bilayers of the plasma membrane.

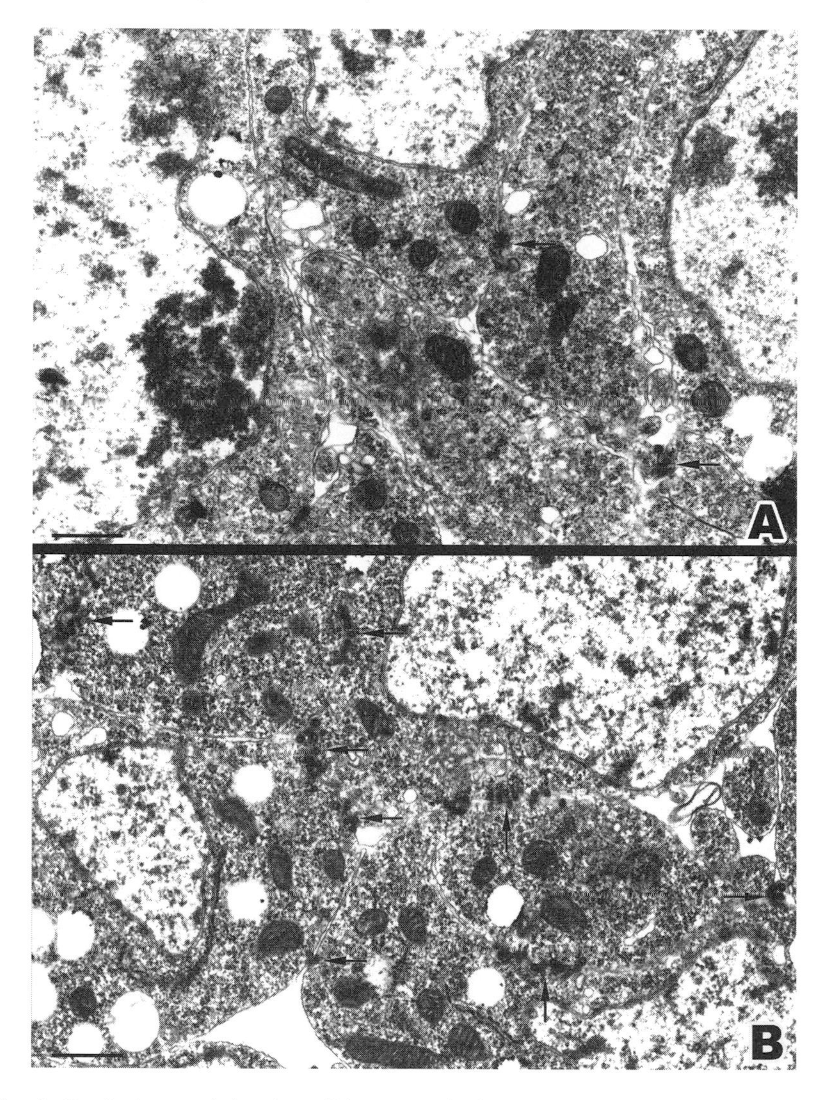

Fig. 1. Prefusion vesicles in wild-type and *sltr* mutant embryos revealed by conventional electron microscopy. A typical cluster of myoblasts in the ventral muscle group VL1–4 is shown in both panels. **(A)** Prefusion vesicles (arrows) in an early stage 13 wild-type embryo. Note the scarcity and low number of prefusion vesicles at cell contact sites. By the end of stage 13, myoblast fusion in this group of muscles is complete in wild-type embryos, and prefusion vesicles are no longer observed (not shown). **(B)** Prefusion vesicles (arrows) are accumulated in an early stage 14 *sltr* mutant embryo. Note the increased frequency and number of prefusion vesicles compared with the wild type. Scale bars = 500 nm.

Besides revealing the fusion intermediates localized at the plasma membrane, electron microscopic analyses have also provided information on the origin and trafficking of fusion-related intracellular organelles. For example, prefusion vesicles have been observed budding off from the Golgi apparatus or associating with the microtubules, suggesting that these vesicles are of exocytic origin and are perhaps transported by the microtubule cytoskeleton to the plasma membrane *(9)*.

At the molecular level, fusion-related proteins can be localized relative to the ultrastructural fusion intermediates by immunoelectron microscopy at a resolution that cannot be achieved by light microscopy. To date, there has been only one published immunoelectron microscopic study of *Drosophila* myoblast fusion *(9)*. This study revealed a correlation between actin-enriched foci at cell contact sites and the directional targeting of the Golgi-derived prefusion vesicles.

The significance of these electron microscopic and immunoelectron microscopic studies is underscored by the ultrastructural phenotypes of different fusion mutants that block the fusion process at various stages. For example, some mutants have been found with no prefusion vesicles (*mbc, rols7/ants*; **refs.** *7,12*) or accumulated vesicles (*blow, sltr*; **refs.** *7,9*). Others form few or no pores (*Drac1*^{G12V}; **ref.** *7*) or multiple pores without completing the fusion process (*Dwip*; **ref.** *10*). Interestingly, while membrane discontinuities along adhering myoblasts have been observed in embryos of both mutant alleles of *sltr/Dwip* prepared by the conventional chemical fixation method *(9,10)*, fusion pores are not seen in *sltr* mutant embryos prepared by the high-pressure freezing/freeze substitution method *(9)*. This apparent discrepancy likely reflects the intrinsic differences between the two sample preparation methods and calls for caution when using the conventional chemical fixation method to examine lipid bilayers during myoblast fusion.

1.2. Conventional Chemical Fixation of Drosophila Embryos

Most of the electron microscopic studies of *Drosophila* myoblast fusion to date have relied on the conventional chemical fixation method. Briefly, embryos are fixed at room temperature with a series of chemicals: first with glutaraldehyde to cross-link protein molecules and later with osmium tetroxide to preserve lipids *(13)*. The fixed embryos are then dehydrated with organic solvent such as ethanol before being embedded in a resin, which can be polymerized into a hardened block for subsequent sectioning. Thin sections are cut on an ultramicrotome and later stained with heavy metals such as lead and uranium to give contrast between different cellular structures *(13)*.

Despite the relative ease and cost efficiency of this method, the major limitation is the slow diffusion (from seconds to minutes) of chemical fixatives into cells, especially cells within thick tissues that have diffusion barriers. The cylindrical *Drosophila* embryos are approximately 200 μm in diameter and 500 μm in length surrounded by the vitelline membrane, which is a natural diffusion barrier. In addition, the embryonic mesoderm is "protected" by the overlaying epidermal cells, making it more difficult for the chemical fixatives to reach. Thus, diffusion of chemicals into the fusing myoblasts can take a considerable amount of time, thus affecting the preservation of cellular structures ***(14,15)***.

Another limitation for this method is the selectivity of the cross-linking reactions of chemical fixatives. For example, the primary fixative glutaraldehyde cross-links only proteins, not lipids. As a result, the lipid bilayers of the plasma membrane are less well preserved before the later application of osmium tetroxide ***(13)***. Insufficient preservation can cause ultrastructural artifacts, including loss of continuity in the membranes, leakage of some cytoplasmic components, and distortion or disorganization of cytoskeletal organelles (**Fig. 2; refs.** ***14,16,17***). These artifacts may confound our interpretation of gene functions during myoblast fusion.

1.3. High-Pressure Freezing/Freeze Substitution of Drosophila *Embryos*

An alternative method to conventional chemical fixation is the high-pressure freezing and freeze substitution method ***(14–16)***. Samples are frozen at below −140 °C under high pressure to immobilize cellular structures. Subsequently, the frozen water (amorphous at this low temperature) in the sample is substituted with organic solvent containing chemical fixatives at approximately −80 °C. The major advantage of the high-pressure freezing/freeze substitution method is the ultrarapid speed (between 20 and 50 ms) of immobilization of all molecules, thus allowing near-native preservation of cellular ultrastructure ***(14–17)***. In addition, the high pressure applied during the freezing process makes it possible to preserve thick tissues of up to a few hundred micrometers, circumventing the problem of slow/inadequate fixation of internal tissues by the conventional chemical fixation method ***(17)***. Moreover, proteins and lipid molecules are chemically fixed at low temperatures (approximately −50 °C by glutaraldehyde and −30 °C by osmium) when their thermal energy is low, thus avoiding structural distortions caused by room temperature fixation ***(13)***. Taken together, high-pressure freezing/freeze substitution ensures instant and adequate fixation of the entire cellular architecture, with especially marked improvement on the morphology of lipid bilayers of the plasma membrane over chemically fixed samples (*see* **Fig. 2**). Thus, it is an optimal method for studying membrane fusion events in whole embryos.

Even though high-pressure freezing/freeze substitution is the preferred method for ultrastructural analysis of myoblast fusion, the cost of a high-pressure freezing

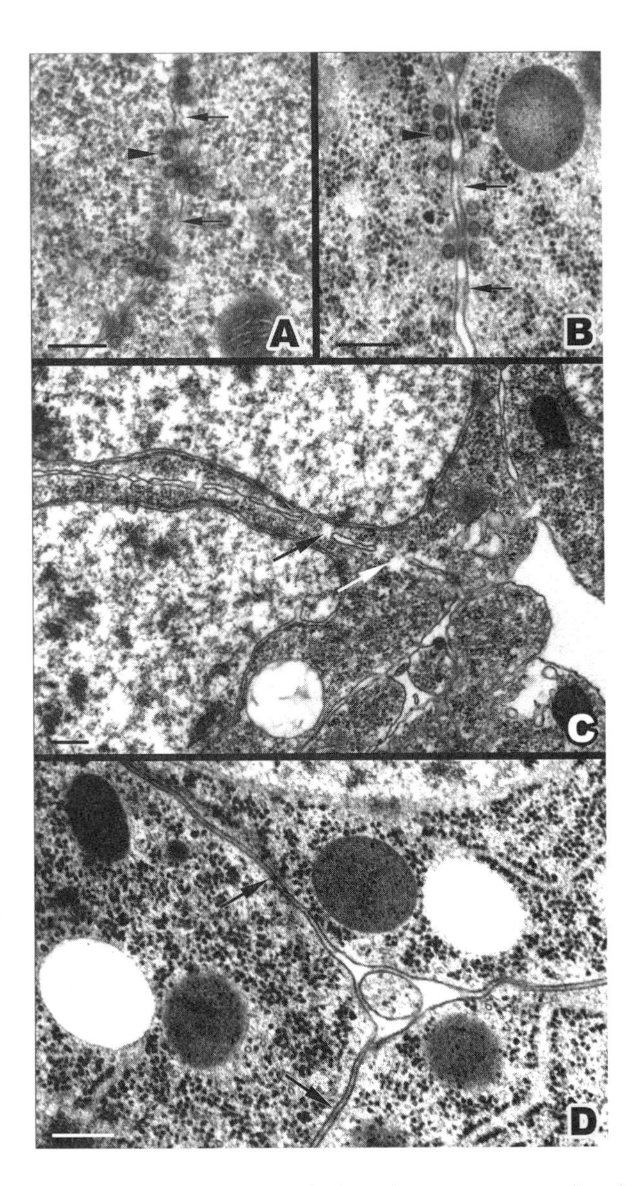

Fig. 2. Comparison of membrane morphology between conventional chemical fixation and high-pressure freezing/freeze substitution electron microscopy. All electron micrographs are taken from *sltr* mutant embryos at early stage 14. Samples in A and C are prepared by chemical fixation, and those in B and D are prepared by high-pressure freezing/freeze substitution. **(A,B)** Prefusion vesicles at myoblast membrane contact sites. Note that prefusion vesicles with electron-dense margins (arrowhead) are mostly paired (a few paired vesicles in B are out of focus). While the plasma membrane (arrows) is not well preserved by chemical fixation **(A)**, it appears smooth and intact in embryos prepared by high-pressure freezing/freeze substitution **(B)**. **(C)** Membrane discontinuities/ruptures (arrow) in a stage 14 embryo prepared by chemical fixation. **(D)** The smooth and intact plasma membrane (arrow) in a stage 14 embryo prepared by high-pressure freezing/freeze substitution. Scale bars = 200 nm.

unit is high, and only a limited number of such units are currently available. For ultrastructural studies on structures other than the plasma membrane, the conventional chemical fixation method remains a valid approach. Therefore, we describe methods for both conventional chemical fixation and high-pressure freezing/freeze substitution of *Drosophila* embryos in this chapter.

1.4. Immunoelectron Microscopy of Drosophila *Embryos*

While light microscopy is routinely used in visualizing fusion-related proteins in the myoblasts, the resolution is insufficient to pinpoint these proteins relative to the fusion intermediates. In addition, it is difficult to unambiguously localize cell type–specific proteins at sites of fusion to either side of the closely apposed plasma membranes with light microscopy. The use of immunoelectron microscopy, in combination with immunohistochemistry at the light microscopic level, will help to circumvent these problems.

Two different approaches are utilized for immunoelectron microscopy, preembedding and postembedding labeling ***(15)***. The former approach involves staining whole embryos with an antibody prior to processing for electron microscopy, whereas the latter method involves embedding the embryos in a resin, followed by sectioning and antibody staining of the thin sections. The advantage of the preembedding labeling is that, once labeled, the embryos can be processed following standard electron microscopic preparation procedures. This method is most suitable for labeling antigens present on the surface of the embryo, where antibodies can readily detect the antigens without penetrating into deep layers of the embryo ***(18)***. Although treatment of embryos with detergent or organic solvent opens up spaces for antibody penetration and is commonly used in immunohistochemistry for light microscopy, such treatment interferes with the preservation of ultracellular structure and thus is not recommended for immunoelectron microscopy ***(15)***. Comparatively, the advantage of postembedding is the improved preservation of the ultrastructure of embryos and the possibility of staining internal tissues. However, many thin sections may need to be stained and screened if the antigen is not widely expressed.

In principle, postembedding labeling is more advantageous for studies of myoblast fusion, since myoblasts residing in deeper layers of the embryo are difficult for antibodies to approach. However, because of the limited number of fusion events at a given time point in wild-type embryos ***(9,19)*** and the specific and transient expression of many fusion-related proteins at sites of fusion ***(2,3)***, hundreds of thin sections may need to be stained to ensure the presence of antigen on at least one or a few sections. An alternative and perhaps more practical approach is to modify the preembedding protocol (e.g., cutting the embryo open to allow antibody penetration to the mesoderm) and subsequently screen for positive signals on thin sections, as demonstrated by Kim et al. ***(9)*** in their localization of actin relative to the prefusion vesicles.

No matter which method is used, one should keep in mind that the quality of the antibody is the most critical factor for the success of an immunoelectron microscopic experiment. In general, a higher concentration of antibody should be applied to electron microscopic samples than embryos for immunohistochemistry. However, some antibodies that work well for light microscopy may not do so for immunoelectron microscopy, mainly because of the loss of antigenicity of proteins during the harsh treatment of samples prepared for electron microscopy.

2. Materials

2.1. Conventional Chemical Fixation in Drosophila *Embryos*

2.1.1. Embryo Collection

1. Empty Tri-Stir plastic beaker (Ted Pella, Inc., Redding, CA, catalog number 12904). Punch a few holes in the bottom and along the side with a 20-gauge needle to allow air flow.
2. Egg collection plates: mix 300 mL of H_2O, 100 mL of 50% grape juice, 17 g of agar, and 12 g of sucrose. Microwave the mixture in a flask at 100% power until boiling (~2.5 min). Heat at 10% power and swirl intermittently until all crystals disappear. Cool down to 50°–60 °C and pour into 60-mm Petri dish (BD Falcon, Franklin Lakes, NJ, catalog number 35-3002). Makes approximately 140 plates. Store the plates at 4 °C and warm them up to room temperature before use.
3. Yeast paste: Add water to dry baker's yeast in a beaker and mix with a spatula to make a wet paste. Store at 4 °C.

2.1.2. Embryo Fixation

1. Scintillation vial.
2. Primary fixative:

   Solution A (total volume is 10 mL; *see* **Note 1**):

   a. 5.0 mL 50% glutaraldehyde in water (Sigma, St. Louis, MO, catalog number G7651).
   b. 2.5 mL 0.4 *M* sodium cacodylate, pH 7.4 (Sigma, catalog number C4945).
   c. 1.0 mL 100% acrolein (PolySciences, Warrington, PA, catalog number 00016).
   d. 1.5 mL distilled H_2O.

Add 10 mL heptane to 10 mL of Solution A and shake vigorously. Let the phases separate and shake again. Repeat intermittently for about 10 min. Withdraw the top heptane phase for embryo fixation. The actual amount of glutaraldehyde and acrolein that goes into the heptane phase is fairly low.

Solution B (total volume is 10 mL; *see* **Note 2**):

a. 1.6 mL 50% glutaraldehyde in water (Sigma, catalog number G7651).
b. 5.0 mL 16% paraformaldehyde (Electron Microscopy Science, Hatfield, PA, catalog number 15710).

c. 2.5 mL 0.4 M sodium cacodylate, pH 7.4 (Sigma, catalog number C4945).
d. 0.9 mL distilled H_2O.

Add 10 mL heptane to Solution B and follow the procedure described above to make equilibrated heptane with fixatives.

3. Small paintbrush.
4. Dechorionating basket: use a 50-mL centrifuge tube (BD Falcon, catalog number 35-2070) from which the conical end has been cut off and a large hole cut in the cap. Place a square of Nitex mesh between the threads of the tube and screw on the cap.
5. 50% bleach diluted in water.
6. Fluorescent dissecting microscope (for picking green fluorescent protein [GFP]–negative homozygous embryos if the balancer has a GFP marker).
7. Nutator or shaker.
8. 0.1 *M* sodium cacodylate (pH 7.4).
9. Double sticky tape.
10. Microscope slide (Fisher Scientific, Pittsburg, PA, catalog number 12-550-14).
11. Microscope slide with a clear silicone rubber well (*see* **Note 3**).
12. Pasteur pipettes (Fisher Scientific, catalog number 13-678-6A).
13. Dissecting microscope with light source from the bottom (this type of microscope makes the amnioserosa clearly visible).
14. Sharpened tungsten needle (Electron Microscopy Sciences, catalog number 62091-01).
15. Eppendorf tubes.
16. Secondary fixative: 2% glutaraldehyde in 0.1 *M* sodium cacodylate buffer (pH 7.4).

2.1.3. Embryo Postfixation

1. 1% reduced osmium tetroxide (*see* **Note 4**):

Tube A:

a. 0.5 mL 0.4 M sodium cacodylate buffer (pH 7.4).
b. 1.5 mL distilled H_2O.
c. 40 mg potassium ferrocyanide (Sigma, catalog number 3289).
d. Vortex to mix.

Tube B:

a. 0.5 mL 0.4 M sodium cacodylate buffer (pH 7.4).
b. 0.5 mL distilled H_2O.
c. 1.0 mL 4% osmium tetroxide (Ted Pella, catalog number 18459).

Right before use, mix the contents of tube A and tube B, add 80 µL 0.3 *M* $CaCl_2$, and rotate the mixture slowly on a Nutator. The solution will turn dark brown.

2. 1% osmium tetroxide in 0.1 *M* sodium cacodylate (pH 7.4).
3. Ice.
4. Foil.

5. Nutator or shaker.
6. 0.25% tannic acid diluted in distilled water.
7. 1% uranyl acetate diluted in distilled water (*see* **Note 5**).

2.1.4. Embryo Dehydration

1. Ethanol series: 10%, 30%, 50%, 70%, 95%, and 100%.
2. Nutator or shaker.

2.1.5. Embryo Embedment

1. Glass scintillation vials.
2. Propylene oxide (Ted Pella, Cat. # 18601).
3. Disposable plastic pipettes: 5 mL and 25 mL.
4. Two EPON stock solutions (*see* **Note 6**):

EPON A:

a. 71.92 g Eponate 12 resin.
b. 100 g DDSA.

EPON B:

a. 116 g Eponate 12 resin.
b. 121 g NMA.

The reagents are available in the Eponate 12 kit (Ted Pella, catalog number 18010).

5. 1:1 EPON:propylene oxide (3 mL EPON A, 3 mL EPON B, 6 mL propylene oxide).
6. EPON: Mix the needed volume of EPON A and EPON B at a ratio of 1:1 and add 1.5% DMP-30 as a catalyst. DMP-30 is also available in the Eponate 12 kit.
7. Nutator or shaker.
8. Waste container.
9. Fine forceps (Electron Microscopy Sciences, catalog number 72700-D).
10. Embedding mold (Ted Pella, catalog number 105).
11. Oven.

2.1.6. Sectioning

1. Ultramicrotome.
2. Knife maker.
3. Ultra glass knife strips (Electron Microscopy Sciences, catalog number 71012).
4. GKB plastic troughs (Ted Pella, catalog number 123-3).
5. Nail polish (Electron Microscopy Sciences, catalog number 72180).
6. Transmission electron microscopy razor blades (Tousimis, Rockville, MD, catalog number 7250).
7. Microscope slides (Fisher Scientific, catalog number 12-550-14).
8. Syringe.
9. Hot plate.

10. Staining solution: 1 g of toluidine blue O, 1 g of sodium tetraborate decahydrate, 30 mL of 100% ethanol, 70 mL of distilled H_2O.
11. Inverted microscope.
12. Diamond knife (Diatome, Biel, Switzerland, ultra 45°).
13. Fine forceps.
14. Dust-Off FGSA kit (Electron Microscopy Sciences, catalog number 70705).
15. Filter paper (Whatman, catalog number 1001 090).
16. Slot grids (Ted Pella, catalog number 1GC12H) or slot with formvar/carbon (Electron Microscopy Sciences, catalog number FCF2010-Cu).
17. Grid storage box.

2.1.7. Staining

1. 5% uranyl acetate diluted in distilled water (*see* **Note 5**).
2. Syringe (Becton Dickinson, Franklin Lakes, NJ, catalog number 309644).
3. Syringe driven filter unit (Millipore, catalog number SLGV033RS).
4. Glass Petri dishes.
5. Parafilm.
6. Fine forceps.
7. Small weighing boat.
8. Sato's lead (Sato, 1968): Weigh out 0.1 g of lead nitrate, 0.1 g of lead citrate, 0.1 g of lead acetate, and 0.2 g of sodium citrate. Add 8.2 mL of degassed distilled water to the above mixture of chemicals in a 15-mL Falcon tube and shake vigorously for 1 min. The solution looks very milky. Add 1.8 mL of freshly made 4% NaOH. The solution becomes clear except for some large white grains at the bottom of the tube. Filter the solution with a 0.22-µm syringe driven filter unit. It is ready for use (*see* **Note 7**).

2.1.8. Microscopy

1. Transmission electron microscope.
2. Liquid nitrogen.

2.2. High-Pressure Freezing and Freeze Substitution in Drosophila *Embryos*

2.2.1. Embryo Dechorionation and Staging

1. Small paintbrush.
2. Dechorionating basket (*see* **Subheading 2.1.2.**).
3. 50% bleach diluted in water.
4. Embryo collection plates (*see* **Subheading 2.1.2.**).
5. Fluorescent dissecting microscope for staging embryos.

2.2.2. High-Pressure Freezing

1. High-pressure freezing device and accessories.

2. Yeast paste: add 10% methanol to dry baker's yeast in the beaker and mix well with a spatula (*see* **Note 8**).
3. Liquid nitrogen.
4. Filter paper.
5. Specimen holders: the top holder and the bottom holder.
6. Forceps.

2.2.3. Freeze Substitution

1. Freeze substitution device.
2. Aluminum foil.
3. Cryovials.
4. Pencil for labeling cryovials.
5. Osmium fixatives in acetone: measure 25 mL of dry acetone in 50-mL disposable centrifuge tube and chill on dry ice. Quickly dissolve 0.25 g of crystalline OsO_4 in the chilled acetone using glassware with a screw cap. Uranyl acetate crystal (25 μg) is dissolved in 0.25 mL dry methanol to make a 10% uranyl acetate solution. Add this solution to the OsO_4/acetone container and mix. As a result, this fixative contains about 98% acetone, 1% methanol, 1% OsO_4, and 0.1% uranyl acetate. Aliquot this fixative into cryovials. Cap the cryovials and immediately immerse them in liquid nitrogen to freeze the solution. Keep the cryovials upright during freezing so that most of the fixative stays at the bottom of the tubes. Store until ready to use (*see* **Note 9**).

2.3. Immunoelectron Microscopy in **Drosophila** *Embryos*

2.3.1. Embryo Prefixation

1. 50% bleach diluted in water.
2. Small paintbrush.
3. Dechorionating basket (*see* **Subheading 2.1.2.**).
4. Embryo collection plates.
5. Fluorescent dissecting microscope.
6. 0.1 *M* sodium phosphate buffer (pH 7.2): mix 68.4 mL of 1 *M* Na_2HPO_4 (141.96 g Na_2HPO_4 per liter) and 31.6 mL of 1 *M* NaH_2PO_4 (119.98 g NaH_2PO_4 per liter). Dilute the mixture with distilled H_2O up to 1 L. This solution can be stored at room temperature for months.
7. Scintillation vials.
8. Fixation solution:

    a. 5 mL 16% paraformaldehyde.
    b. 25 μL 8% glutaraldehyde.
    c. 5 mL phosphate buffer (pH 7.2).
    d. 10 mL heptane.
    e. Mix right before use.

9. Nutator or shaker.
10. PBS+: PBS contains 1% normal goat serum, 50 mM glycine, 1 mg/mL BSA, 0.02% NaN_3, 0.1% gelatin.

11. Double sticky tape.
12. Microscope slide.
13. Microscope slide with a clear silicone rubber well (*see* **Note 3**).
14. Pasteur pipettes.
15. Dissecting microscope with light source from the bottom.
16. Sharpened tungsten needle.
17. Eppendorf tubes.

2.3.2. Antibody Staining

1. PBS++: PBS contains 1% normal goat serum, 50 m*M* glycine, 1 mg/mL BSA, 0.02% NaN_3, 0.1% gelatin, 0.01%–0.02% saponin.
2. Nutator.
3. Primary antibody.
4. Immunoelectron microscopy grade secondary antibody (Nanogold).

2.3.3. Embryo Postfixation

1. 0.1 *M* sodium phosphate buffer (pH 7.2).
2. Fixation solution: 2% formaldehyde, 1% glutaraldehyde in phosphate buffer.
3. Nutator.

2.3.4. N-Propyl-Gallate (NPG) Silver Enhancement

1. High-speed centrifuge.
2. 15 mL Falcon centrifuge tubes.
3. Gum Arabic stock: Dissolve 50 g of Gum Arabic (Sigma, catalog number 51200-250G) in 100 mL of distilled water under constant agitation over several days. Centrifuge at 12,000 *g* for 4–5 h. Aliquot the solution into Falcon centrifuge tubes (5 mL each tube) and store at −20 °C (*see* **Note 10**).
4. Medium-sized weighing boat.
5. Nutator.
6. PH meter.
7. Stirring plate.
8. Freshly prepare the following solutions:

    a. 1 *M* Hepes (Sigma, catalog number 54457-50G-F), pH 6.8. (Adjust pH with NaOH. If too much NaOH has been added, remake the solution instead of readjusting with HCl.)
    b. 50 m*M* Hepes, pH 5.8, containing 200 m*M* sorbitol (Sigma, catalog number 85529).
    c. 20 m*M* Hepes, pH 7.4 containing 250 m*M* sodium thiosulfate (Sigma, catalog number S7026).
    d. 10 mg of NPG (Sigma, catalog number 02370) dissolved in 250 µL of 100% ethanol first. Bring volume up to 5 mL with distilled water.
    e. 36 mg of silver lactate (Sigma, catalog number 85210) in 5 mL of distilled water. Make this solution right before use and store in the dark.

9. 24-well plate.

2.3.5. Embryo Postfixation

1. 0.1 *M* sodium phosphate buffer (pH 7.2).
2. 0.1% osmium tetroxide in phosphate buffer.

2.3.6. Embryo Dehydration

1. Ethanol series: 50%, 70%, 95%, and 100%.
2. Nutator.

2.3.7. Staining

1. 5% uranyl acetate diluted in distilled water (*see* **Note 5**).
2. Syringe (Becton Dickinson, catalog number 309644).
3. Syringe driven filter unit (Millipore, catalog number SLGV033RS).
4. Glass Petri dishes.
5. Parafilm.
6. Fine forceps.

3. Methods

3.1. Conventional Chemical Fixation in Drosophila *Embryos*

The protocol for conventional chemical fixation in *Drosophila* embryos is modified from Lin et al. ***(20)*** and McDonald et al. ***(15)***.

3.1.1. Embryo Collection

1. Set up the collection beakers 2–3 days ahead of use.
2. Start a fresh collection plate in the late afternoon of the day before harvest. On the day of harvest, collect all embryos (0–16 h) laid overnight with a small paintbrush to the dechorionating basket.

3.1.2. Embryo Fixation

1. Rinse the embryos well with distilled water and dechorionate the embryos for 2 min with 50% of fresh bleach.
2. Rinse the embryos with PBS or distilled water to remove the residual bleach. Try to remove as much liquid as possible by blotting the bottom of the Nitex mesh with Kimwipes, but do not let the embryos completely dry out. If you need to pick up homozygous mutant embryos, follow **step 3**. Otherwise, move to **step 4**.
3. Transfer the bleached embryos from the mesh onto a fresh egg collection plate. Try to distribute the embryos evenly on the plate. If the balancer chromosome is marked by GFP, homozygous mutant embryos can be identified with a fluorescent dissecting microscope by their absence of GFP expression.
4. Move the embryos with a small paintbrush into a scintillation vial containing heptane previously equilibrated with fixatives. Fix for 20–30 min with rotation. Depending on which cellular ultrastructure you are interested in visualizing, different fixatives (Solution A or Solution B) can be used for optimal preservation (*see* **Notes 1** and **2**).

5. Rinse the embryos three times with fresh heptane to remove residual glutaraldehyde, paraformaldehyde, and/or acrolein.
6. Transfer the embryos with a glass pipette and slowly place them on a glass slide so that a monolayer of embryos forms. The embryos should eventually clump together into a single tight monolayered group. Wait until almost all the heptane has evaporated before starting the next step.
7. Gently lay a piece of double sticky tape down on the embryos (*see* **Note 11**).
8. Carefully peel the tape off the glass slide. The embryos are now attached to one side of the tape. Invert the tape with embryos on top, and place the tape in the previously made silicone rubber well on a glass slide.
9. Immediately cover the embryos with 0.1 *M* sodium cacodylate buffer (pH 7.4).
10. Carefully poke at one end of the fixed embryos with a sharpened tungsten needle. Once the embryos are popped out of the vitelline membrane, they will float on the aqueous solution. During this step, you can pick embryos at the desired stage based on the shape of the amnioserosa (**Fig. 3;** *see* **Note 12**).
11. Transfer the embryos at desired stages with a glass pipette to an Eppendorf tube containing fresh 2% glutaraldehyde in 0.1 *M* sodium cacodylate buffer (pH 7.4). Fix the embryos for 1 h with rotation at room temperature.

3.1.3. Embryo Postfixation

1. Rinse the embryos two to three times for 5 min each with 0.1 *M* sodium cacodylate buffer (pH 7.4).
2. Postfix the embryos with 1% osmium tetroxide in 0.1 *M* sodium cacodylate buffer (pH 7.4) or 1% reduced osmium tetroxide for 1 h at 4 °C with rotation in the fume hood (*see* **Note 4**). Keep the Eppendorf tubes covered with foil.
3. Rinse the embryos two to three times for 5 min each in 0.1 *M* sodium cacodylate buffer (pH 7.4).
4. Rinse the embryos two to three times for 5 min each in distilled H_2O.
5. *Optional*: Fix the embryos for 5 min in 0.25% tannic acid (*see* **Note 13**). Rinse two to three times for 5 min each with distilled water.
6. At room temperature stain the embryos with 1% uranyl acetate in distilled water for 1 h in the dark (*see* **Note 5**).

3.1.4. Embryo Dehydration

1. Rinse the embryos two times for 5 min each with distilled H_2O.
2. Dehydrate the embryos in an ethanol series for 5 min each: 10%, 30%, 50%, 70%, 95%, and 100% (three times). Use a *new* unopened bottle of 100% ethanol for each experiment (*see* **Note 14**).

3.1.5. Embryo Embedment

1. Transfer the embryos to a glass scintillation vial. Transit through propylene oxide from 100% ethanol by three changes of 100% of propylene oxide, 5 min each.
2. Add 1:1 EPON:propylene oxide, and rotate 30 min at room temperature.

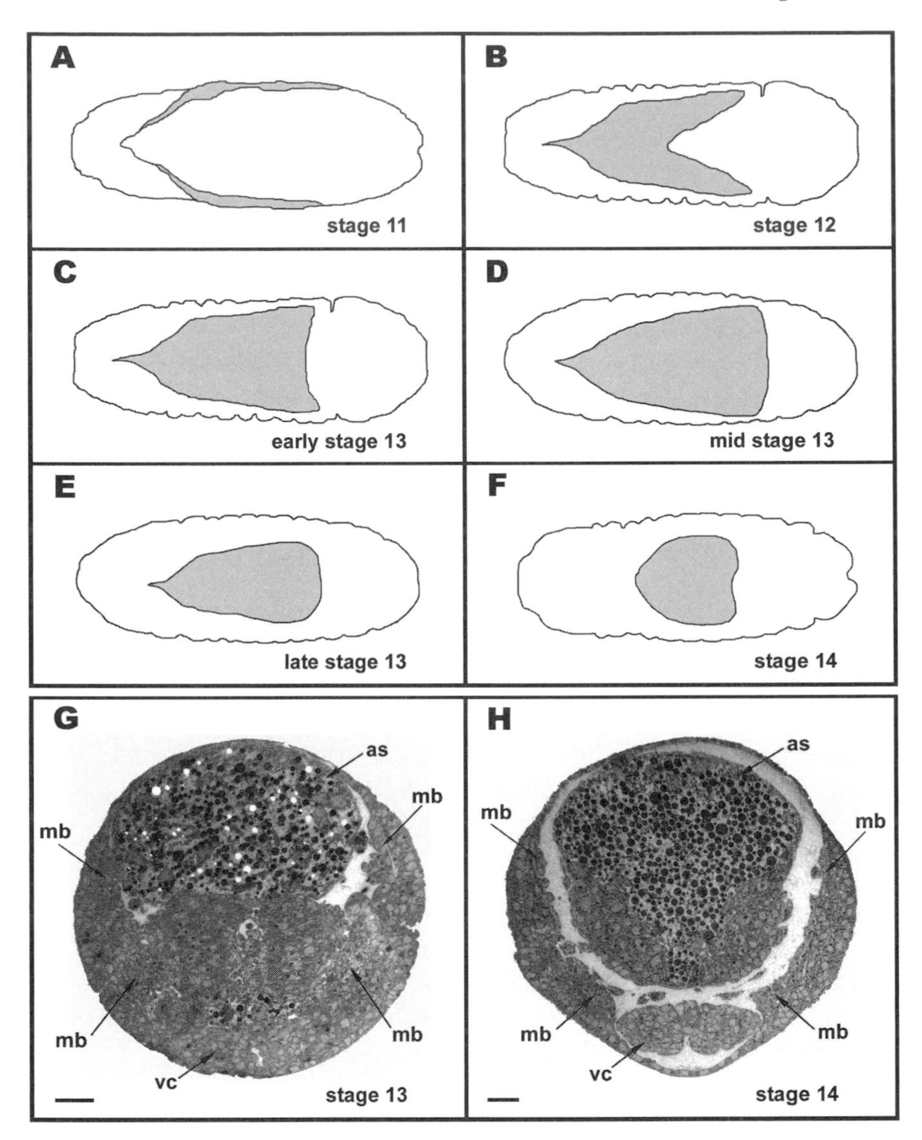

Fig. 3. Embryo staging during myoblast fusion. **(A–F)** Schematic drawings of stages 11–14 wild-type embryos. In all panels, a dorsal view of the embryo is shown, and anterior is to the left. The amnioserosa is marked in grey. Note that the specific shapes of the amnioserosa can be used to further stage embryos during stage 13. **(G,H)** Transverse sections of the abdominal segments of stages 13 and 14 wild-type embryos. as, amnioserosa; vc, ventral cord; mb, myoblasts. Scale bars = 20 μm.

3. Add EPON (6 mL EPON A, 6 mL EPON B, 180 μL DMP-30) and rotate overnight at room temperature.
4. Change EPON twice on the next day before embedding. Allow at least 2 h between each change.
5. Embed the embryos with fresh EPON in embedding mold. Orient and line up the embryos at one end of the block. Bake in a 70°–80 °C oven for 24–48 h.

3.1.6. Sectioning

1. Use a glass knife maker to make glass knives following the manual. Make a "boat" for each glass knife with a GKB plastic trough and nail polish.
2. Perform coarse sectioning with the glass knife. Sections will float on the water within the boat. Put a drop of water on a glass slide. Transfer the sections from the "boat" to the water drop on the slide. Place the slide on a hotplate. Stain the sections with 1% toluidine blue O and 1% sodium tetraborate decahydrate.
3. Examine the sections under an inverted microscope to gauge the quality of fixation and infiltration of the resin. Verify the stage of the embryo and determine which segments have been sectioned.
4. Continue with serial thin sectioning with a diamond knife if the sample is from the desired embryo segment and stage. The ideal thickness for the section is about 50–70 nm. Carefully transfer the sections onto a slot with supporting film. Store the slot in the grid storage box.

3.1.7. Staining

1. Line the bottom of glass Petri dish with parafilm. Add drops of 5% filtered uranyl acetate onto the parafilm. Stain the grid for 10–15 min with 5% uranyl acetate.
2. Gently rinse the grid with degassed distilled H_2O and place it back to the grid storage box until dry.
3. Stain with Sato's lead as described by Sato ***(21)*** for 3 min. This is done in a covered glass Petri dish in the presence of NaOH pellets. Do not stain more than five grids at one time.
4. Carefully rinse the grid with degassed distilled H_2O and place it back into the grid storage box.

3.1.8. Microscopy

1. Screen the sections under a transmission electron microscope (*see* **Note 15**). Find the myoblasts or myotubes at low magnification (*see* **Fig. 3**).
2. Change to higher magnification to observe the fusion intermediates during myoblast fusion.

3.2. High-Pressure Freezing and Freeze Substitution in Drosophila *Embryos*

The protocol for high-pressure freezing and freeze substitution in *Drosophila* embryos is modified from McDonald ***(14)*** and McDonald et al. ***(15)***.

3.2.1. Embryo Dechorionation and Staging

1. Follow **Steps 1–3** in **Subheading 3.1.2.**
2. Stage both wild-type and mutant embryos by the shape of their amnioserosa under the fluorescent dissecting microscope (*see* **Fig. 3**).

3.2.2. High-Pressure Freezing

1. Place about five embryos at the desired stage on a slot containing a small amount of yeast paste in 10% methanol. Remove the excess moisture with filter paper. Mix the embryos with just enough yeast paste, and place embryos with yeast into the bottom of a specimen holder containing a small amount of yeast paste. Place the top holder on the bottom, and squeeze gently together with forceps (*see* **Note 16**).
2. Carefully place this sandwich into the specimen carrier to be inserted into the high-pressure freezing device. Make sure the specimen carrier and specimen holders match very well.
3. Freeze the samples in the high-pressure freezer and maintain the frozen sandwich in liquid nitrogen until processing with freeze substitution. Separate top and bottom specimen holders if you are ready for freeze substitution.

3.2.3. Freeze Substitution

1. Remove the frozen samples from the storage container in liquid nitrogen and transfer them to a cryovial containing frozen fixatives. Leave the cap of the cryovial slightly loose to release the liquid nitrogen that got in the tube during transfer. Avoid warming up the cryovials during the entire procedure.
2. Move all individual cryovials to the freeze substitution device and run the appropriate program.

3.2.4. Embryo Embedment

1. Remove the cryovials from the device when the freeze substitution program is complete. Rinse each sample two to three times with fresh 100% acetone at room temperature and embed in the EPON resin. *See* **Subheading 3.1.5.** for the subsequent steps of embryo embedment.

3.3. Immunoelectron Microscopy in Drosophila Embryos

The protocol of immunoelectron microscopy in *Drosophila* embryos is modified from Burry et al. ***(22)***, Jongens et al. ***(23)***, and Tepass ***(18)***.

3.3.1. Embryo Prefixation

1. Follow **Steps 1–3** in **Subheading 3.1.2.**
2. Fix the embryos in a scintillation vial containing 10 mL of heptane and 10 mL of fixatives in phosphate buffer (pH 7.2) at room temperature with vigorous shaking for 30 min (*see* **Note 17**).
3. Briefly wash the embryos twice with phosphate buffer (pH 7.2). Transfer the embryos with a plastic pipette to a glass slide. Use a piece of double sticky tape

to transfer to another slide with a silicone rubber well. Immediately cover the embryos with PBS+.

4. Devitellinize the embryos with a sharpened tungsten needle under a dissecting microscope with light source from the bottom. Stage the embryos (*see* **Fig. 3**), and cut off the anterior 20% of the embryos to help the penetration of antibodies to the myoblasts.

3.3.2. Antibody Staining

1. Transfer the devitellinized embryos with a glass pipette to an Eppendorf tube. Wash the embryos three times for 5 min each with PBS++.
2. Dilute the primary antibody in PBS++ at a desired concentration, and leave the embryos in the antibody solution at room temperature for 2 h. Shake lightly on a Nutator or shaker.
3. Wash the embryos five times for 20 min each in PBS++.
4. Incubate the embryos in PBS++ containing the secondary Nanogold antibody at 1:200 dilution at room temperature for 2 h. Shake lightly on a Nutator or shaker.

3.3.3. Embryo Postfixation

1. Wash the embryos three times for 5 min each with phosphate buffer (pH 7.2).
2. Postfix the embryos in a fixative containing 2% formaldehyde and 1% glutaraldehyde in phosphate buffer overnight at 4 °C with constant shaking.

3.3.4. N-Propyl-Gallate Silver Enhancement

1. On the second day, perform NPG silver enhancement before osmium postfixation.
2. Pour 5 mL of Gum Arabic stock and 2 mL of 1 *M* Hepes (pH 6.8) into a medium-sized plastic weighing boat. Slowly mix them with a stir bar on a stirring plate for at least 15 min and make sure no bubbles appear.
3. Wash the embryos three times for 5 min each with 50 mM Hepes (pH 5.8) containing 200 m*M* sorbitol. During the washes, prepare the enhancement solution as described in **steps 4** and **5**.
4. Add 1.5 mL of freshly made NPG solution to the weighing boat under constant agitation.
5. Add 1.5 mL of freshly made silver lactate solution to the above weighing boat under constant agitation. After adding silver lactate, solution in the weighing boat should be kept in the dark and used as soon as possible.
6. Transfer the washed embryos from **step 3** to a new 24-well plate. Remove the Hepes buffer, and add 1 mL of NPG silver solution from **step 5**. Cover the plate and keep it in the dark for 10 min.
7. Stop the reaction by washing three times for 5 min each with 20 m*M* Hepes (pH 7.4) containing 250 m*M* sodium thiosulfate.

3.3.5. Embryo Postfixation

1. Rinse the embryos in phosphate buffer three times for 5 min each. Transfer the embryos to an Eppendorf tube.

2. Postfix the embryos in 0.1% OsO_4 in phosphate buffer for 30 min at 4 °C with constant shaking.
3. Rinse the embryos three times for 5 min each with phosphate buffer.

3.3.6. Embryo Dehydration

1. Dehydrate the embryos in an ethanol series for 5 min each: 50%, 70%, 95%, and 100% (three times). Use a *new* unopened bottle of 100% ethanol for each experiment (*see* **Note 13**).
2. Transfer the embryos to a glass scintillation vial. Transit through propylene oxide from the 100% ethanol by two changes of 100% of propylene oxide 5 min each.

3.3.7. Staining

1. Line the bottom of a glass Petri dish with parafilm. Add drops of 5% filtered uranyl acetate onto the parafilm. Stain the grid for 10–15 min with 5% uranyl acetate.
2. Gently rinse the grid with degassed distilled H_2O, and place it back to the grid storage box until dry.
3. No lead staining is applied for immunoelectron microscopy (*see* **Note 18**).

3.3.8. Microscopy

Screen the sections under a transmission electron microscope (*see* **Note 19**).

4. Notes

1. Solution A is sufficient for visualizing prefusion vesicles. Acrolein is used as an intermediate in the manufacture of acrylic acid. It is extremely toxic to humans from inhalation and dermal exposure. It is supplied in vials capped by a rubber stopper and must be taken out from the vial with a syringe and needle. All these fixatives must be used in a fume hood.
2. Follow Solution B if more detailed intracellular structures, such as the Golgi, microtubules, or actin filaments, need to be visualized.
3. The silicone rubber well should be prepared at least 1 day ahead, because the acetic acid present in the commercially available liquid silicone rubber prevents its polymerization. Once the liquid is out of the container, it takes about 1 day for acetic acid to evaporate from the hardening silicone rubber. Acetic acid can degrade the ultrastructure if it gets into the aqueous solution used in the next step to cover embryos.
4. Reduced osmium tetroxide can result in better preservation of the ultrastructure. Osmium tetroxide is highly toxic and is a rapid oxidizer. Exposure to the vapor can cause severe chemical burns to the eyes, skin, and respiratory tract. Wear nitrile gloves (osmium can penetrate latex gloves) and eye protection. Do not open any vials of osmium tetroxide outside of the fume hood. Place unused 4% stock in a dark glass container with a screw cap. Make sure the container is closed tightly and sealed with parafilm to prevent vapor leakage. Keep the stock at 4 °C.

5. Uranyl acetate is a heavy metal poison. For better staining, filter the uranyl acetate with a 0.22-μm filter before use. Uranyl acetate helps to increase membrane contrast.
6. Store EPON A and EPON B in separate bottles. Keep them dry and store at 4 °C until the day of use. Warm up to room temperature before opening the bottles.
7. Usually, the staining solution is ineffective after 3 days. For good staining, always use freshly made lead staining solution in each experiment.
8. Yeast paste made up with H_2O is also usable. However, 10% methanol gives embryos some additional cryoprotection without negative side effects.
9. Acetone and methanol are the most popular fixatives for high-pressure freezing/freeze substitution. Methanol replaces water much faster than acetone. However, McDonald ***(14)*** reported that acetone dehydration results in better preservation of the ultrastructure for *Drosophila* embryos.
10. Prepare Gum Arabic stock several days before use. The aliquots can be stored in −20 °C for months.
11. Do not exert excessive pressure the embryos so that they are not squished. If too much heptane is left on the slide, some embryos may not stick to the tape well, and it would be difficult to devitellinize them. However, if the embryos are left too dry, the quality of fixation will decrease.
12. To visualize fusion intermediates, embryos at stages 12, 13, and/or 14 should be used. The easiest way to distinguish these stages is to observe the shape of the amnioserosa through a dissecting microscope with a light source from the bottom (*see* **Fig. 3**). It is also possible to distinguish the stages under a fluorescent dissecting microscopy (*see* **Subheading 3.2.1.**).
13. Tannic acid fixation can improve the resolution of the ultrastructure, especially for microtubule and actin filaments.
14. If water is present after dehydration, the resin will not polymerize properly and the embedded samples will be impossible to section.
15. Get properly trained before using the transmission electron microscope. Take care not to burn the supporting film by abruptly increasing the voltage of the filament or switching from high to low magnifications without decondensing the electron beam.
16. Any excess yeast squeezed out needs to be carefully but quickly removed. Dry yeast paste around the embryos will affect the freezing process.
17. These fixatives contain a low concentration of glutaraldehyde. Equilibrated heptane with these fixatives does not fix the embryos well. Thus, we choose to fix the embryos with a mixture of heptane and fixatives with high-speed shaking.
18. Lead staining causes increased contrast in the sample. Thus, it may be difficult to distinguish between lead staining versus gold staining if the former were applied.
19. Because low osmium tetroxide and no lead staining are applied to the samples, it would be a bit harder to identify the myoblasts under the transmission electron microscope. In general, the resolution of the ultrastructure is not as high for immunoelectron microscopy as it is for regular electron microscopy.

Acknowledgments

We thank Michael Delannoy, Richard Fetter, Kent McDonald, and Guofeng Zhang for valuable advice on conventional chemical fixation and high-pressure freezing/freeze substitution; Ulrich Tepass for immunoelectron microscopy; and members of Chen's laboratory for comments. Work in Chen's laboratory is supported by the NIH, the Packard Foundation, and the Searle Scholars Program.

References

1. Baylies, M. K., Bate, M., and Ruiz Gomez, M. (1998) Myogenesis: a view from *Drosophila. Cell* **93**, 921–927.
2. Abmayr, S. M., Balagopalan, L., Galletta, B. J., and Hong, S. J. (2003) Cell and molecular biology of myoblast fusion. *Int. Rev. Cytol.* **225**, 33–89.
3. Chen, E. H. and Olson, E. N. (2004) Towards a molecular pathway for myoblast fusion in *Drosophila. Trends Cell Biol.* **14**, 452–460.
4. Erickson, M. R., Galletta, B. J., and Abmayr, S. M. (1997) *Drosophila* myoblast city encodes a conserved protein that is essential for myoblast fusion, dorsal closure, and cytoskeletal organization. *J. Cell Biol.* **138**, 589–603.
5. Paululat, A., Holz, A., and Renkawitz-Pohl, R. (1999) Essential genes for myoblast fusion in *Drosophila* embryogenesis. *Mech. Dev.* **83**, 17–26.
6. Chen, E. H. and Olson, E. N. (2001) Antisocial, an intracellular adaptor protein, is required for myoblast fusion in *Drosophila. Dev. Cell* **1**, 705–715.
7. Doberstein, S. K., Fetter, R. D., Mehta, A. Y., and Goodman, C. S. (1997) Genetic analysis of myoblast fusion: blown fuse is required for progression beyond the prefusion complex. *J. Cell Biol.* **136**, 1249–1261.
8. Schroter, R. H., Lier, S., Holz, A., Bogdan, S., Klambt, C., Beck, L., and Renkawitz-Pohl, R. (2004) kette and blown fuse interact genetically during the second fusion step of myogenesis in *Drosophila. Development* **131**, 4501–4509.
9. Kim, S., Shilagardi, K., Zhang, S., Hong, S. N., Sens, K. L., Bo, J., Gonzalez, G. A., and Chen, E. H. (2007) A critical function for the actin cytoskeleton in targeted exocytosis of prefusion vesicles during myoblast fusion. *Dev. Cell* **12**, 571–586.
10. Massarwa, R., Carmon, S., Shilo, B. Z., and Schejter, E. D. (2007) WIP/WASp-based actin-polymerization machinery is essential for myoblast fusion in *Drosophila. Dev. Cell* **12**, 557–69.
11. Estrada, B., Maeland, A. D., Gisselbrecht, S. S., Bloor, J. W., Brown, N. H., and Michelson, A. M. (2007) The MARVEL domain protein, Singles Bar, is required for progression past the pre-fusion complex stage of myoblast fusion. *Dev. Biol.* **307**, 328–339.
12. Rau, A., Buttgereit, D., Holz, A., Fetter, R., Doberstein, S. K., Paululat, A., Staudt, N., Skeath, J., Michelson, A. M., and Renkawitz-Pohl, R. (2001) rolling pebbles (rols) is required in *Drosophila* muscle precursors for recruitment of myoblasts for fusion. *Development* **128**, 5061–5073.
13. Hayat, M. A. (2000) *Principles and Techniques of Electron Microscopy: Biological Applications*. Cambridge University Press, Cambridge, England.

14. McDonald, K. L. (1994) Electron microscopy and EM immunocytochemistry. *Methods Cell Biol.* **44**, 411–444.
15. McDonald, K. L., Sharp, D. J., and Rickoll, W. (2000) Preparation of this sections of Drosophia for examination by transmision electron microscopy. In *Drosophila Protocols* (W. Sullivan, M. Ashburner, and R.S. Hawley, eds.). Cold Spring Harbor Laboratory Press, Cold Spring Harbor, NY, pp. 245–272.
16. McDonald, K. L. and Auer, M. (2006) High-pressure freezing, cellular tomography, and structural cell biology. *Biotechniques* **41**, 137–143.
17. Lucic, V., Forster, F., and Baumeister, W. (2005) Structural studies by electron tomography: from cells to molecules. *Annu. Rev. Biochem.* **74**, 833–865.
18. Tepass, U. (1996) Crumbs, a component of the apical membrane, is required for zonula adherens formation in primary epithelia of *Drosophila. Dev. Biol.* **177**, 217–225.
19. Beckett, K. and Baylies, M. K. (2007) 3D analysis of founder cell and fusion competent myoblast arrangements outlines a new model of myoblast fusion. *Dev. Biol.* **309**, 113–125.
20. Lin, D. M., Fetter, R. D., Kopczynski, C., Grenningloh, G., and Goodman, C. S. (1994) Genetic analysis of Fasciclin II in *Drosophila*: defasciculation, refasciculation, and altered fasciculation. *Neuron* **13**, 1055–1069.
21. Sato, T. (1968) A modified method for lead staining of thin sections. *J Electron Microsc. (Tokyo)* **17**, 158–159.
22. Burry, R. W., Vandre, D. D., and Hayes, D. M. (1992) Silver enhancement of gold antibody probes in pre-embedding electron microscopic immunocytochemistry. *J. Histochem. Cytochem.* **40**, 1849–1856.
23. Jongens, T. A., Ackerman, L. D., Swedlow, J. R., Jan, L. Y., and Jan, Y. N. (1994) Germ cell–less encodes a cell type–specific nuclear pore–associated protein and functions early in the germ-cell specification pathway of *Drosophila. Genes Dev.* **8**, 2123–2136.

17

A Genomic Approach to Myoblast Fusion in *Drosophila*

Beatriz Estrada and Alan M. Michelson

Summary

We have developed an integrated genetic, genomic, and computational approach to identify and characterize genes involved in myoblast fusion in *Drosophila*. We first used fluorescence-activated cell sorting to purify mesodermal cells both from wild-type embryos and from 12 variant genotypes in which muscle development is perturbed in known ways. Then, we obtained gene expression profiles for the purified cells by hybridizing isolated mesodermal RNA to Affymetrix GeneChip arrays. These data were subsequently compounded into a statistical metaanalysis that predicts myoblast subtype-specific gene expression signatures that were later validated by in situ hybridization experiments. Finally, we analyzed the myogenic functions of a subset of these myoblast genes using a double-stranded RNA interference assay in living embryos expressing green fluorescent protein under control of a muscle-specific promoter. This experimental strategy led to the identification of several previously uncharacterized genes required for myoblast fusion in *Drosophila*.

Key Words: Cell–cell fusion; myoblast; mesoderm; myogenesis; muscle development; *Drosophila*; genomics; gene expression profiling.

1. Introduction

Cell fusion is necessary for the development of different human tissues, including bone, placenta, and muscle. During myogenesis, mononucleated myoblasts fuse with each other to form nascent, multinucleated functional myofibers. Thus, normal muscle growth and regeneration of injured tissue require the fusion of new myoblasts. Also, there are several myopathies related to defects in myoblast fusion ***(1–3)***. However, the molecular mechanisms underlying these pathologies are largely unknown.

Studies performed with mammalian cells in vitro and in *Drosophila* embryos have demonstrated that myoblast fusion involves an ordered set of specific cellular events. First, myoblasts recognize and adhere. Then alignment occurs

From: *Methods in Molecular Biology, vol. 475: Cell Fusion: Overviews and Methods*
Edited by: E. H. Chen © Humana Press, Totowa, NJ

through the parallel apposition of the membranes of elongated myoblasts with myotubes or other myoblasts. Finally, membrane union takes place between the aligned plasma membranes in small areas of cytoplasmic continuity, with vesiculation of the excess plasma membrane in the fusion area *(2)*. Genetic analysis combined with light and electron microscopy of *Drosophila* fusing myoblasts have assigned the function of specific proteins to these particular cellular processes, providing an entry point for studying the molecular mechanisms underlying myoblast fusion *(4)*. Despite recent progress in this area, there are still many unanswered questions related to the molecular basis of muscle fusion, including the role of the cytoskeleton in cell shape changes that occur in fusing myoblasts, the identification and functional analysis of molecules responsible for the actual fusion of muscle cell membranes, and the mechanisms that govern the invariant size of each muscle, as determined by the number of fusion events that occur during formation of a particular myofiber.

Compared with vertebrate systems, *Drosophila* offers several advantages as an experimental organism for studying muscle development. First, the relatively short generation time of *Drosophila* allows myogenesis to be analyzed in vivo in a more rapid manner. Second, *Drosophila* offers outstanding genetic resources for studying muscle development. Third, this model organism has a smaller number of genes, which gives the advantage of circumventing potential functional redundancies inherent to mammalian genomes. Fourth, many of the key components responsible for myogenesis at the molecular level are well conserved between flies and humans so that knowledge about fly muscle development is highly relevant to human biology and disease.

In an attempt to better understand the molecular mechanisms underlying myoblast fusion, we have undertaken an integrated genetic, genomic, and computational strategy to determine which genes are expressed during and are essential for *Drosophila* myoblast fusion. We hypothesized that a functional genomic approach would reveal fusion genes that were not identified in previous forward genetic screens *(1,5,6)*. Specifically, we used fluorescence-activated cell sorting (FACS) to purify embryonic myoblasts marked with green fluorescent protein (GFP). Included in the analysis were embryos derived from wild-type and from genetically modified strains in which mesoderm development was perturbed in predictable and informative ways (**Figs. 1** and **2**). The RNA from these cell populations was hybridized to Affymetrix GeneChip arrays, and pairwise comparisons were made to detect genes that are differentially expressed between the mesoderm and the rest of the embryo, as well as between the mutant mesodermal cells and their wild-type counterparts. Then, we combined the results from all the microarray data and subjected the resulting compendium of expression profiles to a statistical metaanalysis that was designed to reveal candidate genes expressed in each of the two fusing myoblast populations, founder cells and fusion-competent myoblasts *(7)*. Finally, we validated hundreds of these

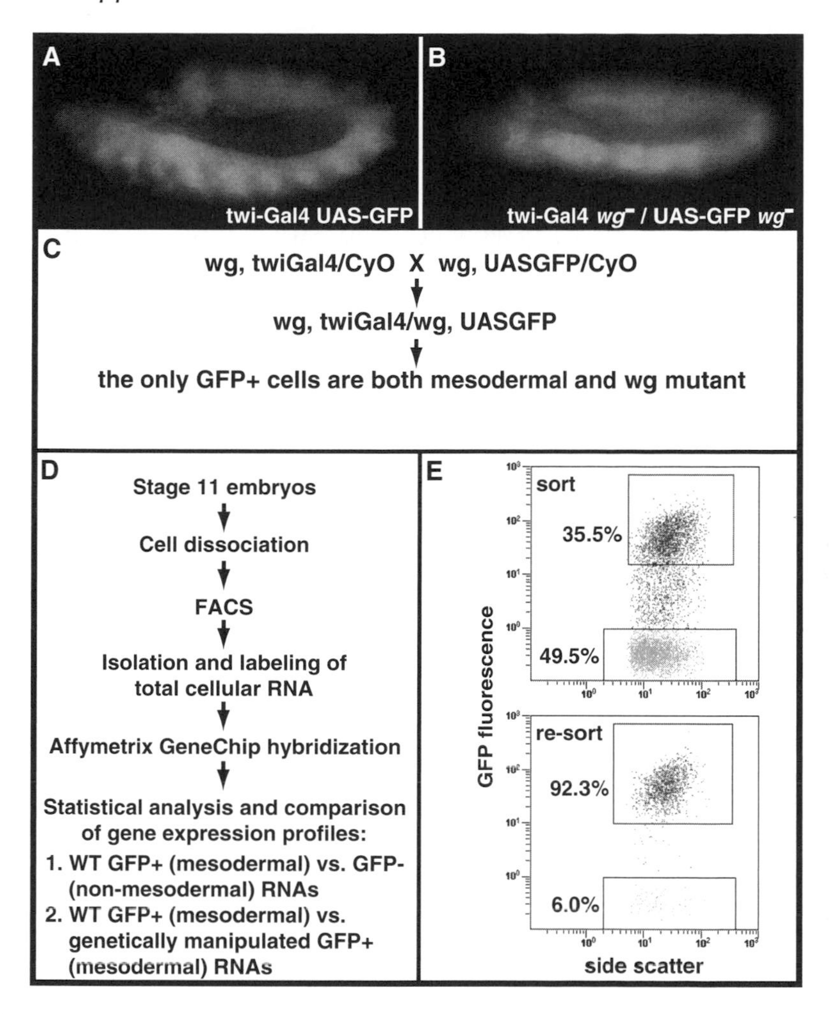

Fig. 1. Experimental strategy to obtain gene expression signatures of purified *Drosophila* embryonic mesodermal cells. **(A)** Transgenic, stage 11, *Drosophila* embryo expressing Gal4 under the control of the *twi* promoter and green fluorescent protein (GFP) under the upstream activation sequence (UAS) regulatory sequence ***(21)***, resulting in GFP-positive mesodermal cells. **(B)** Transgenic, stage 11, *wingless (wg)* mutant embryo with GFP-positive mesodermal cells. **(C)** Genetic crossing scheme to obtain homozygous mutant GFP-positive mesodermal cells. Strains bearing independently generated recombinant chromosomes having the mutant gene of interest (e.g., *wg*) and either the *twiGal4* or *UASGFP* transgenes are crossed. **(D)** A representative fluorescence-activated cell sorting (FACS) experiment to obtain total RNA from mesodermally purified cells. **(E)** Representative FACS scatter plots before **(top panel)** and after **(bottom panel)** the separation of GFP-positive and -negative cell populations. Top, upper box: GFP-positive sort in blue; top, lower box: GFP-negative sort in green. Bottom, upper box: in blue are shown the re-sorted GFP-positive cells to verify that purity obtained from the primary sort was greater than 90%. (*See Color Plates*)

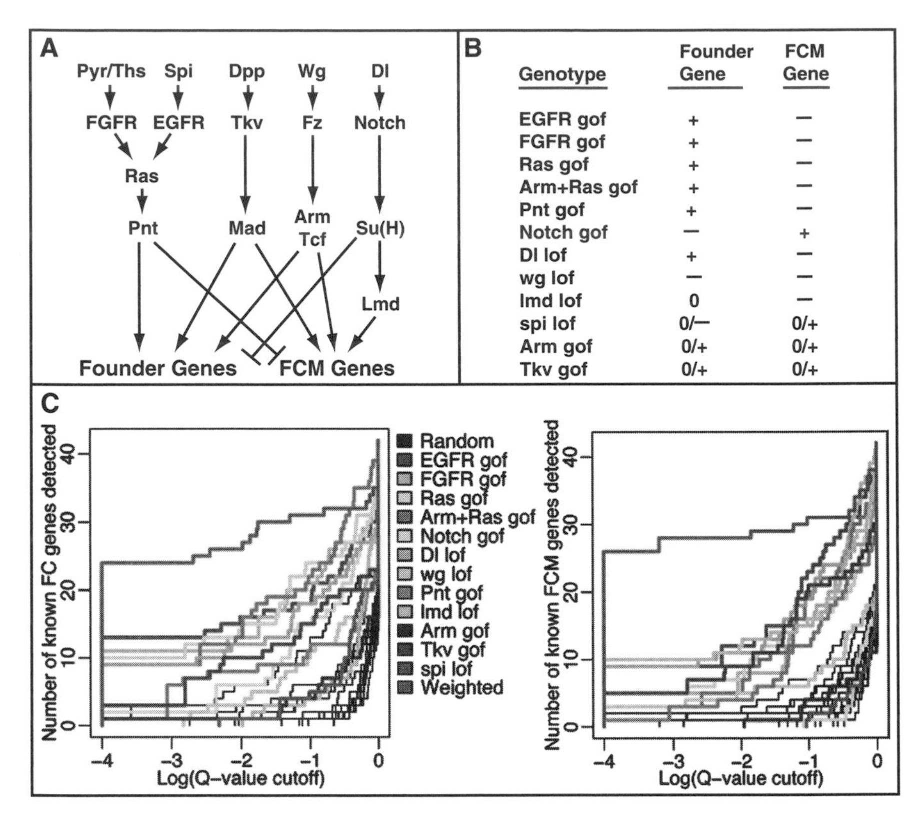

Fig. 2. Statistical metaanalysis of an expression profiling compendium to predict myoblast-specific gene signatures. **(A)** Summary of the signaling pathways and transcription factors that positively and negatively regulate the expression of myoblast subpopulation-specific genes (founder genes and fusion-competent myoblast [FCM] genes). **(B)** Expected gene expression changes for founder or FCM genes in each of the 12 genetic perturbations used in our study *(6)*. **(C)** Detection curves showing the number of genes from the training set (see below) detected, as a function of q-value (predicted measurement of false-positive rate in Log scale), for FC genes (left) and FCM genes (right). In both panels, the predictive value of individual genotype/wild-type comparisons (various colors as indicated on the figure) are compared to randomly generated rankings (thin black lines) and to composite rankings derived from a combination of all datasets (gray). To avoid introducing biases for or against any genotype, the training sets were composed of known genes from the literature as well as the mesodermally enriched genes that had been verified by in situ hybridization in this study to be founder cell or FCM genes. (*See Color Plates*)

predictions by gene-specific *in situ* hybridization and assessed the functions of a selected group of myoblast genes using double-stranded (ds) RNA interference (RNAi) analysis of muscle development in whole living embryos *(6)*. This integrated approach uncovered several previously uncharacterized genes that are involved in myoblast fusion (**Fig. 3E,F** and data not shown), with many more candidates remaining to be analyzed by RNAi.

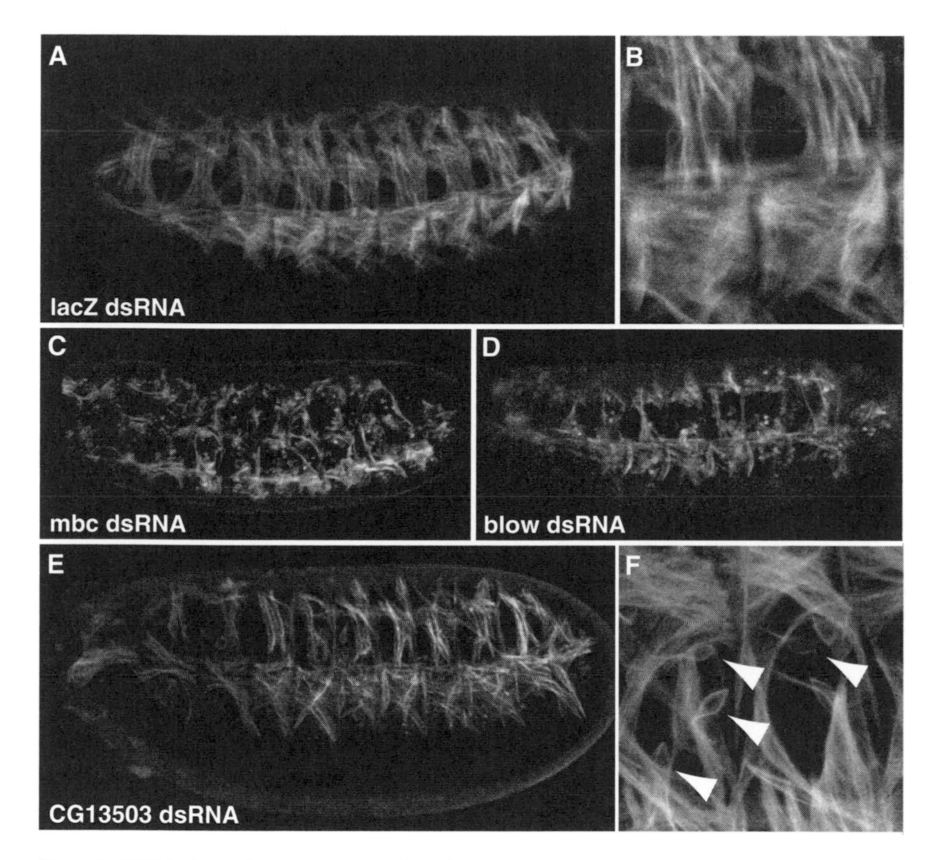

Fig. 3. RNA interference analysis of genes necessary for myoblast fusion. Live transgenic embryos expressing tau–green fluorescent protein under the control of the myosin heavy chain promoter (20) were injected with negative control *lacZ* double stranded (ds) RNA **(A,B)**, positive control *myoblast city (mbc,* **C**) or *blown fuse (blow,* **D**) dsRNA and gene-specific test dsRNA **(E,F)**. In E and F, RNA interference for CG13503, *solas*, shows abundant unfused myoblasts (arrowheads) indicating that this gene is required for myoblast fusion. Because this gene is the homolog of the vertebrate WH2 actin-binding protein, and its expression is restricted to fusion-competent myoblasts, we hypothesized that it is required for the unique cytoskeletal rearrangements occurring in fusion-competent myoblasts during myoblast fusion *(6)*.

2. Materials

2.1. Generating Appropriate Fly Strains

The following *Drosophila* stocks were used to obtain both wild-type and genetically modified mesodermal cells expressing GFP:

1. Twi-Gal4, UAS-2EGFP *(8)*.
2. Twi-Gal4 alone (*see* **Note 5; ref.** ***9***).
3. UAS-λTop (constitutively activated EGFR) ***(10)***.
4. UAS-Dof, UAS-λ-Htl (constitutively activated Heartless FGFR together with Downstream of FGFR/Heartbroken/Stumps) (***11,12***).
5. UAS-Ras1^{Act} (activated Ras) ***(13)***.
6. UAS-Pnt^{P2VP16} (activated Pointed) ***(14)***.
7. UAS-Tkv^{QD} (activated Thick veins Decapentaplegic receptor) ***(15)***.
8. UAS-Arm^{S10} (activated Armadillo) ***(16)***.
9. UAS-Arm^{S10}; UAS-Ras1^{Act}.
10. Twi-Gal4, *wg*^{CX4}/CyO.
11. *wg*^{IG22}, UAS-2EGFP.
12. UAS-N^{intra} ***(17)***.
13. Twi-Gal4, *lmd*^{1}/TM3, *ftz-lacZ*.
14. UAS-2EGFP, *lmd*^{2}/TM3, *ftz-lacZ*.
15. Twi-Gal4, *Dl*^{X}/TM3, *ftz-lacZ*.
16. UAS-2EGFP, *Dl*^{X}/TM3, *ftz-lacZ*.

2.2. Collecting the Embryos

1. Fly incubator.
2. Yeast paste.
3. Molasses plates (cook 378 g of molasses with 65 g of bacteriological agar in 2 L of water for 45 min; when it cools down to 55°–60 °C, add 30 mL of Tegosept: 80 g of methyl 4-hydroxybenzoate and 750 mL of reagent alcohol).
4. Fly cages (400 mL tripour plastic beakers perforated with a hot 25- gauge needle).

2.3. Preparation of Embryonic Single Cell Suspensions for Flow Cytometry

1. Plastic squeeze bottles.
2. Camel hair brush.
3. 70-μm strainer (Falcon).
4. Weighing balance.
5. Glass Dounce homogenizers (VWR 62400-620; Wheaton Scientific, Millville, NJ) with loose-fitting pestle (clearance 0.0035–0.005 inch)
6. Desktop clinical centrifuge.
7. Hemocytometer (optional).
8. 5-mL round bottom plastic tubes.

9. 40-μm nylon mesh (www.smallparts.com).
10. Sterile plastic transfer pipettes.
11. Solutions: 50% bleach, 0.01% Triton X-100, 8% fetal bovine serum (FBS; Sigma, St. Louis, MO) in Schneider's medium (Gibco).

2.4. Sorting Green Fluorescent Protein-Positive and -Negative Cells by Flow Cytometry

1. Fluorescent-activated cell sorter (we used EPICS Altra with Hyper Sort Option from Beckman Coulter, Fullerton, CA).
2. 5-mL round bottom plastic tubes.
3. Carry-on cooler to transport cells on ice.
4. Solutions: Seecof saline (6 m*M* Na_2HPO_4, 3.67 m*M* KH_2PO_4, 106 m*M* NaCl, 26.8 m*M* KCl, 6.4 m*M* $MgCl_2$, 2.25 m*M* $CaCl_2$, pH 6.8). Sterilize by filtering; do not autoclave. ("RNAlater" from Ambion, Austin, TX.)

2.5. Obtaining Total RNA From Sorted Cells

1. Beckman J6-M centrifuge with JS 5.2 rotor (or equivalent centrifuge).
2. 2-mL Eppendorf tubes.
3. Bioanalyzer microfluidics platform (Agilent, Santa Clara, CA); alternatively, use a spectrophotometer and RNA electrophoresis gel to check the quantity and quality of the total RNA.
4. Solutions: Schneider's *Drosophila* Medium (Gibco), Trizol Reagent (Life Technologies, Inc., Gaithersburg, MD), isopropanol, RNAse free water (e.g., from Ambion).

2.6. Hybridization on Affymetrix Chips

For GeneChip *Drosophila* genome arrays, the hybridizations are done at a microarray core facility that will provide all the reagents, apparatus, and experimental expertise required for the procedure.

2.7. Data Analysis

Data analyses involved the use of the statistical programming language R (http://www.r-project.org), the Bioconductor suite of bioinformatics R packages (http://www.bioconductor.org), and the goldenspike R package (http://www.elwood9.net/spike).

2.8. Validation of the Predicted Gene Expressions

2.8.1. Reagents Used to Generate Gene Specific Digoxigenin-Labeled Antisense RNA Probes for Embryonic In Situ Hybridizations

1. T7 and SP6 RNA polymerase (Roche, Indianapolis, IN).
2. Digoxigenin-labeled nucleotide mix (Roche).
3. RNAse inhibitor (Roche).

2.8.2. Template DNA Used to Generate Gene Specific Digoxigenin-Labeled Antisense RNA Probes for Embryonic In Situ Hybridizations

1. *Drosophila Gene Collection* (DGC1 and 2; http://www.fruitfly.org/DGC/index.html).
2. *Drosophila* embryonic primary cDNA.

2.8.3. Antibodies

1. Anti-digoxigenin-AP (Roche).
2. Rabbit anti-Lmd (from Duan et al. ***[18]***) was used at 1:1000. Rabbit anti-β-galactosidase (Promega, Madison, WI) was used at 1:500.

2.8.4. Fly Stocks Used in Embryonic In Situ Hybridizations

The following stocks were used to determine gene expression patterns in mutant backgrounds:

1. Twi-Gal4, UAS-Ras1^{Act}.
2. *Dl*^{X}/TM3, *ftz-lacZ.*
3. *lmd*^{1}/TM3, *ftz-lacZ.*
4. The enhancer trap line *rP298-lacZ* was used to test for localization of gene expression to founder cells ***(19)***.

2.9. Functional Analysis of Newly Identified Myoblast Genes

RNA interference assay:

1. MEGAscript RNAi kit (Ambion).
2. *Drosophila* embryonic primary cDNA.
3. DEPC-treated 1× injection buffer (5 mM KCl, 0.1 m*M* PO_4, pH 7.8).
4. MHC–tau-GFP embryos ***(20)***.
5. Microinjector, needle puller, or commercially available microinjection needles.
6. Molasses plates.
7. 50% bleach.
8. Halocarbon oil (series 700, from VWR, part number 700-1).

3. Methods

3.1. Generating Appropriate Fly Strains

Specific gene promoter-driven GFP or Gal4-UAS ***(21)*** system-based strategies are some of the possibilities to express GFP (or other fluorescent protein) in the desired population of cells. Choose a strain of flies that expresses the highest levels of GFP (or other fluorescent protein). Engineering several copies of the reporter gene or using di-cistronic versions of GFP will increase the intensity ***(8)***. This strategy maximizes the fluorescence intensity and provides a higher recovery of cells with the greatest purity (*see* **Note 2**).

3.2. Collecting the Embryos

1. Maintain flies in large plastic cages (400-mL tripour beakers) at 25 °C, feeding them using molasses plates streaked with yeast paste. Put 700–1000 flies per cage. More flies prevent sufficient ventilation and results in the flies sticking to the sides of the cage.
2. Change molasses plates (with yeast) one time a day for 2–3 days.
3. Change the plate, and allow flies to prelay for one period of 2h on fresh food (prelaying for two periods of 2h improves the yield of the first collection if it is not initially adequate; *see* **Note 3**).
4. Change the plate every 2h and let the embryos age to the appropriate stage for cell dissociation. To obtain myoblasts that have already been specified but have not yet undergone fusion, use stage 11 embryos (approximately 5.5 7.5h after egg laying at 25 °C; *see* **Note 4**).

3.3. Preparation of Embryonic Single Cell Suspensions for Flow Cytometry

Single cell suspensions are obtained by homogenizing embryos with a Dounce tissue grinder (colloquially, "douncing") as follows (protocol modified from **ref.** ***22***):

1. To prepare the eggs for douncing, start approximately 20–30min before the time assigned for the actual dissociation of the eggs.
2. Wash eggs and yeast from plates using dII_2O from a squeeze bottle, loosen eggs with a camel hair brush, and pour the eggs through a 70-μm strainer (the weight of the strainer should be measured beforehand). Wash the materials in the strainer until the yeast is removed.
3. Blot the strainer with filter paper to remove excess liquid, and weigh the strainer to obtain the wet-weight of the collected eggs.
4. Immerse the strainer in 50% (v/v) bleach to cover the eggs and dechorionate them for 5min.
5. During dechorionation, fill the douncer with 7-mL Schneider's medium and keep on ice.
6. After dechorionation, wash the bleach completely from the eggs in the strainer with 0.01% (v/v) Triton X-100 from a squeeze bottle, rinse with dH_2O and with a final rinse in Schneider medium, and then brush the eggs into the douncer. *Important*: Try not to put more than 0.03g of embryos per douncer, as it reduces the yield of single cell suspension. Divide the total weight of embryos by 0.03 to estimate the number of douncers needed.
7. Use a loose pestle to gently but firmly dounce to the bottom. Give seven strokes. Push up and down only, without rotating the pestle. From now on, keep the tubes on ice.
8. Transfer dounced materials from two douncers (7 + 7mL) into one conical centrifuge tube (15-mL Falcon tube). Spin at 40*g* (418rpm in a desktop clinical

centrifuge) for 5 min to pellet the tissue and cell debris, clumps, and vitelline membranes. Single cells and yolk are in the supernatant.

9. Transfer the supernatant to a clean tube and spin at 380 *g* for 10 min (1255 rpm in a desktop clinical centrifuge) to pellet single cells, and then discard the supernatant. Resuspend the cells with 8% FBS in Schneider's medium (1 mL per tube).
10. *Optional*: Draw 40 µL of the cell suspension, and count the cells using a hemocytometer.
11. Add another 1–2 mL of 8% FBS in Schneider's medium once the cells are resuspended.
12. Sieve the cells through a 40-µm nylon mesh into 5-mL round bottom Falcon tube as follows. Cut the mesh in approximately 7 × 7 cm pieces. Hold the tube and the mesh with one hand, and hold the mesh with your thumb to create a funnel to which you will apply the cell suspension with a sterile plastic transfer pipette. Pipette 1 mL of cells into the funnel-shaped mesh, and then sieve the cells through the mesh. Repeat the pipetting and sieving of the cells to sieve the rest of the volume of cells from that tube.
13. *Optional*: Rinse the original tube with another 1 mL of 8% FBS in Schneider's medium, and rinse the mesh with the same solution. Discard the mesh.
14. Repeat the entire process for each tube of cells.
15. Fill the tubes to 4–4.5 mL with 8% FBS in Schneider's medium. Maintain the tubes on ice, and take to the sorting facility.

3.4. Sorting Green Fluorescent Protein-Positive and -Negative Cells by Flow Cytometry

At the flow cytometry facility, specify the following (*see* **Notes 1** and **4**):

1. Use Seecof saline as the FACS running buffer (*Drosophila* cells have a higher osmolarity than mammalian cells ***(23)***.
2. Turn on the cooler so that the cells being sorted are kept at 4 °C.
3. Run the cell sorter between 5000 and 10,000 cells per second to maximize cell viability.
4. Calibrate the FACS for GFP-positive and GFP-negative windows using cells derived from a negative control strain (*see* **Note 5**).
5. Check that the purity of the GFP-positive cells is at least 90%. This can be achieved by re-sorting a small aliquot of both populations of sorted cells before continuing with the bulk of the cells *see* **Fig. 1E**).
6. Prepare labeled 5-mL round bottom tubes with 1.5 mL of RNAlater solution to collect the cells while they are being sorted.
7. After one collecting tube is full, ask the FACS operator to invert that tube to mix the cells with the RNAlater solution. After this step, all tubes must be maintained on ice.

3.5. Isolating Total RNA From Sorted Cells

1. When the FACS runs from the two consecutive 2-h egg collections are finished, dilute the cells in RNAlater with Schneider's medium (dilution is necessary

because RNAlater is very viscous and would prevent later pelleting of the cells by centrifugation). Place ~30 mL of sorted cells (up to six tubes) in a 50 mL Falcon tube and fill it with Schneider's medium. Sorted cell suspensions are then diluted with Schneider's medium so that RNAlater is no more than 20% of the total volume.

2. Centrifuge at 4400 rpm (4900 *g*) in a Beckman J6-M centrifuge with the JS 5.2 rotor (or equivalent centrifuge) for 15 min; decant the supernatant.
3. Resuspend the cells in 200–300 μL of Trizol and transfer to a 2 mL Eppendorf tube.
4. Precipitate the RNA in isopropanol (follow the Trizol manufacturer's instructions for total RNA isolation). Keep RNA at −80 °C until it is analyzed on a Bioanalyzer.
5. Resuspend the RNA in 20–50 μL of RNAse-free water from Ambion. The RNA needs to be at least at 430 ng/μL to start the first strand synthesis of cDNA at the Affymetrix facility (assuming that 3 μg of labeled RNA will be hybridized on each chip; *see* **Note 4**).

3.6 Hybridization of Affymetrix GeneChips

Total cellular RNA (2.5–3 μg) from each genotype is labeled in one round of linear amplification and used for hybridization to a single Affymetrix GeneChip using standard methods recommended by the manufacturer (http://www.affymetrix.com/support/technical/manual/expression_manual.affx). Each RNA sample is independently labeled and hybridized in triplicate (*see* **Note 6**).

3.7. Data Analysis

3.7.1. Method

To find genes with differential expression between two conditions or samples, we chose an analysis method that showed optimized detection of spiked-in control RNAs *(24)*. This method involved the calculation of multiple expression summaries per probe set, using the Tukey-Biweight and median polish summary methods and testing for differential expression using a regularized t-statistic metric as described elsewhere *(25)*.

3.7.2. Meta-Analysis

To identify genes that were specifically expressed in the two different populations of myoblasts, founder cells and fusion-competent myoblasts, we pooled all the data from the expression profiling experiments for all 12 different genotypes and performed a statistical meta-analysis. This strategy allowed us to rank all the genes based on the similarity of their expression patterns with the canonical founder cell (or fusion-competent myoblast) patterns of expression based on our prior observations *(6)*. To perform this meta-analysis, we devised a metric that is basically a sum of the t-statistics for each genotype-to-wild-type comparison, with each term multiplied by 1 if the expected response is increased expression in the genotype or by –1 if the expected response is downregulation (*see* **Fig. 2B**). Also, each term is multiplied by a weighting factor that was chosen to optimize the detection of a training set of known founder cells (or fusion-competent myoblast) genes. The resulting summation gives us an overall metric for how well each gene follows the canonical founder cells (or fusion-competent myoblast) expression pattern (*see* **Fig. 2C**).

3.7.3. Computational Analysis

A detailed description of the computational analysis is not provided here because of space limitations. However, it can be found in the main text and Supplementary Experimental Procedures, Method E, of Estrada et al. ***(6)***.

3.8. Validation of the Predicted Gene Expression Patterns

3.8.1. Validation

Validations were done by embryonic in situ hybridizations both in wild type and the following mutant backgrounds: Constitutively activated Ras (Ras1Act) and Delta (*D1X*) mutant embryos produce an expansion of FCs at the expense of FCMs ***(13,26–28)***. Thus, the expression of any gene which is expanded in these mutants compared to the wild-type expression is considered to be expressed in FCs. lameduck mutant embryos (lmd1) fail to develop FCMs; thus the expression of any gene which is reduced in these mutants compared to the wild type expression is considered to be expressed in FCMs.

Gene-specific digoxigenin-labeled antisense RNA probes were synthesized using cDNA clones obtained from the *Drosophila* Gene Collection (DGC1 and 2, http://www.fruitfly.org/DGC/index.html). For genes without an available cDNA, gene-specific polymerase chain reaction (PCR) primers were designed and 0–18h embryonic RNA was used for standard first-strand cDNA synthesis followed by PCR amplification. A microtiter plate method was used for parallel synthesis of multiple probes (http://www.fruitfly.org/about/methods/RNAinsitu.html).

3.8.2. Antibody Staining

Antibody staining was carried out as described elsewhere ***(13)*** on *Dl* or *lmd* mutant embryos where we used the β-galactosidase staining from a *lacZ*-marked TM3 balancer chromosome to identify the homozygotes. β-Galactosidase staining of the enhancer trap line *rP298-lacZ* was used to test for localization of gene expression to founder cells ***(19)***.

3.9. Functional Analysis of Newly Identified Genes

Gene segments for dsRNA synthesis were selected to be 300–700 bp in length and common to all predicted splice variants of the targeted gene. To avoid off-target RNAi effects ***(29,30)***, regions were chosen that lack any consecutive 18 bp of identity to another predicted gene in the *Drosophila melanogaster* genome. These sequences were PCR amplified from primary embryonic cDNA using primers that incorporated T7 promoters on both ends. Purified PCR product was transcribed in vitro and purified using the MEGAscript RNAi kit, precipitated, resuspended, and diluted to 2 mg/mL in DEPC-treated 1× injection buffer ***(31)***. Dechorionated MHC–tau-GFP embryos ***(20)*** were injected midventrally during the syncytial blastoderm stage, and then allowed to develop to stages 16 to 17 before assessment. Each gene was initially injected and scored blindly, with negative control (*lacZ* dsRNA) and positive control (*mbc* or *blow* dsRNA) injections performed in parallel (*see* **Fig. 3**). Only embryos that developed robust GFP expression and that lacked obvious major morphological defects (typically 60%–80% of those injected) were included in the analysis.

4. Notes

1. Talk to the people who run the FACS facility about sorting *Drosophila* cells. In many cases, this will be the first time that they will be dealing with *Drosophila*, so they need to follow your instructions about the running buffer and the sorting speed (*see* **Subheading 3.4.**).
2. One of the limiting steps in this genomic approach is gathering sufficient RNA to hybridize to the Affymetrix GeneChips, especially for mutants where only one fourth of the collected embryos has the desired genotype and expresses the fluorescent protein marker for cell sorting (*see* **Fig. 1B,C**). It is important to do a pilot FACS run at the sorting facility (*see* **Fig. 1E**) in order to establish optimum conditions for achieving adequate separation of the GFP-positive and -negative cell populations such that sufficient quantities of high-quality total RNA will be obtained (*see* **Subheading 3.1**).
3. To maximize the number of embryos collected, aim to collect at dawn or dusk when female flies are most active at laying eggs. You can also change their circadian clock by appropriately controlling the experimental light cycle to suit your own convenience (*see* **Subheading 3.2.3.**).
4. A typical experiment involves the collection of 350–1500 mg (wet weight) of embryos from 7000–12,000 flies (*see* **Subheading 3.2.**), the sorting of $1–3 \times 10^7$ total cells from the initial suspension, and the isolation of $1.6–8 \times 10^6$ GFP-positive cells (*see* **Subheading 3.4.**). At least six independent cell collections were pooled for gene expression profiling of a given genotype (*see* **Subheading 3.2.**). Approximately 1 μg of total RNA was obtained from 1×10^6 cells (*see* **Subheading 3.5.**).
5. To calibrate the FACS and optimize the separation of GFP-positive and GFP-negative cells, use a strain of flies that is related to the study strain except that

it should lack GFP expression. We used Twi-Gal4 alone as our negative control. Yolk autofluorescence of *Drosophila* embryos needs to be differentiated from GFP expression (*see* **Subheadings 2.1** and **3.4**).

6. We used three technical replicates of pooled embryo collections to optimize statistical significance. Although biological replicates—that is, independent embryo collections for RNA isolation and for each labeling reaction and hybridization—are ideal for statistical power, this approach was neither practical nor cost-effective (in terms of required FACS time) for obtaining sufficient precisely staged RNA for each chip hybridization. Thus, we pooled multiple RNA collections from different cell sorting runs to minimize the variation inherent among collections and to obtain adequate RNA from each genotype (*see* **Subheading 3.6.**).

Acknowledgments

We thank Stephen S. Gisselbrecht and Sung E. Choe for comments on the manuscript. This work was funded by the Howard Hughes Medical Institute and the National Institutes of Health.

References

1. Chen, E. H. and Olson, E. N. (2005) Unveiling the mechanisms of cell–cell fusion. *Science* **308(5720),** 369–373.
2. Horsley, V. and Pavlath, G. K. (2004) Forming a multinucleated cell: molecules that regulate myoblast fusion. *Cells Tissues Organs* **176(1–3),** 67–78.
3. Pomerantz, J. and Blau, H. M. (2004) Nuclear reprogramming: a key to stem cell function in regenerative medicine. *Nat. Cell. Biol.* **6(9),** 810–816.
4. Chen, E.H. and Olson, E. N. (2004) Towards a molecular pathway for myoblast fusion in Drosophila. *Trends Cell Biol.* **14(8),** 452–460.
5. Dworak, H. A. and Sink, H. (2002) Myoblast fusion in *Drosophila. BioEssays* **24,** 591–601.
6. Estrada, B., Choe, S. E., Gisselbrecht, S. S., et al. (2006) An integrated strategy for analyzing the unique developmental programs of different myoblast subtypes. *PLoS Genet.* **2(2),** e16.
7. Baylies, M. K. and Michelson, A. M. (2001) Invertebrate myogenesis: looking back to the future of muscle development. *Curr. Opin. Genet. Dev.* **11,** 431–439.
8. Halfon, M. S., Gisselbrecht, S., Lu, J., Estrada, B., Keshishian, H., and Michelson, A. M. (2002) New fluorescent protein reporters for use with the *Drosophila* Gal4 expression system and for vital detection of balancer chromosomes. *Genesis* **34,** 135–138.
9. Greig, S. and Akam, M. (1993) Homeotic genes autonomously specify one aspect of pattern in the *Drosophila* mesoderm. *Nature* **362,** 630–632.
10. Queenan, A. M., Ghabrial, A., and Schüpbach, T. (1997) Ectopic activation of *torpedo/Egfr*, a *Drosophila* receptor tyrosine kinase, dorsalizes both the eggshell and the embryo. *Development* **124,** 3871–3880.
11. Michelson, A. M., Gisselbrecht, S., Buff, E., and Skeath, J. B. (1998) Heartbroken is a specific downstream mediator of FGF receptor signalling in *Drosophila. Development* **125,** 4379–4389.

12. Vincent, S., Wilson, R., Coelho, C., Affolter, M., and Leptin, M. (1998) The *Drosophila* protein Dof is specifically required for FGF signaling. *Mol. Cell* **2,** 515–525.
13. Carmena, A., Gisselbrecht, S., Harrison, J., Jiménez, F., and Michelson, A. M. (1998) Combinatorial signaling codes for the progressive determination of cell fates in the *Drosophila* embryonic mesoderm. *Genes Dev.* **12,** 3910–3922.
14. Halfon, M. S., Carmena, A., Gisselbrecht, S., et al. (2000) Ras pathway specificity is determined by the integration of multiple signal-activated and tissue-restricted transcription factors. *Cell* **103(1),** 63–74.
15. Nellen, D., Burke, R., Struhl, G., and Basler, K. (1996) Direct and long-range action of a DPP morphogen gradient. *Cell* **85(3),** 357–368.
16. Pai, L.-M., Orsulic, S., Bejsovec, A., and Peifer, M. (1997) Negative regulation of Armadillo, a Wingless effector in *Drosophila. Development* **124,** 2255–2266.
17. Lieber, T., Kidd, S., Alcamo, E., Corbin, V., and Young, M. W. (1993) Antineurogenic phenotypes induced by truncated Notch proteins indicate a role in signal transduction and may point to a novel function for Notch in nuclei. *Genes Dev.* **7,** 1949–1965.
18. Duan, H., Skeath, J. B., and Nguyen, H. T. (2001) *Drosophila* Lame duck, a novel member of the Gli superfamily, acts as a key regulator of myogenesis by controlling fusion-competent myoblast development. *Development* **128,** 4489–4500.
19. Nose, A., Isshiki, T., and Takeichi, M. (1998) Regional specification of muscle progenitors in *Drosophila*: the role of thc *msh* homcobox gene. *Development* **125,** 215–223.
20. Chen, E. H. and Olson, E. N. (2001) Antisocial, an intracellular adaptor protein, is required for myoblast fusion in *Drosophila. Dev. Cell* **1,** 705–715.
21. Brand, A. H. and Perrimon, N. (1993) Targeted gene expression as a means of altering cell fates and generating dominant phenotypes. *Development* **118,** 401–415.
22. Donady, J. J. and Fyrberg, E. A. (1977) Mass culturing of *Drosophila* embryonic cells in vitro. *Tissue Culture Assoc. Manual* **3,** 685–687.
23. Singleton, K. and Woodruff, R. I. (1994) The osmolarity of adult *Drosophila* hemolymph and its effect on oocyte-nurse cell electrical polarity. *Dev. Biol.* **161(1),** 154–167.
24. Choe, S. E., Boutros, M., Michelson, A. M., Church, G. M., and Halfon, M. S. (2005) Preferred analysis methods for Affymetric GeneChips revealed by a wholly-defined control dataset. *Genome Biol.* **6,** R16.
25. Baldi, P. and Long, A. D. (2001) A Bayesian framework for the analysis of microarray expression data: regularized t -test and statistical inferences of gene changes. *Bioinformatics* **17(6),** 509–519.
26. Ruiz-Gomez, M., Coutts, N., Suster, M. L., Landgraf, M., and Bate M. (2002) Myoblasts incompetent encodes a zinc finger transcription factor required to specify fusion-competent myoblasts in *Drosophila. Development* **129(1),** 133–141.
27. Carmena, A., Buff, E., Halfon, M. S., et al. (2002) Reciprocal regulatory interactions between the Notch and Ras signaling pathways in the *Drosophila* embryonic mesoderm. *Dev. Biol.* **244,** 226–242.

28. Bour, B. A., Chakravarti, M., West, J. M., and Abmayr, S. M. (2000) *Drosophila* SNS, a member of the immunoglobulin superfamily that is essential for myoblast fusion. *Genes Dev.* **14,** 1498–1511.
29. Ma, Y., Creanga, A., Lum, L., and Beachy, P. A. (2006) Prevalence of off-target effects in *Drosophila* RNA interference screens. *Nature* **443(7109),** 359–363.
30. Kulkarni, M. M., Booker, M., Silver, S. J., et al. (2006) Evidence of off-target effects associated with long dsRNAs in *Drosophila melanogaster* cell-based assays. *Nat. Methods* **3(10),** 833–838.
31. Kennerdell, J. R. and Carthew, R. W. (1998) Use of dsRNA-mediated genetic interference to demonstrate that *frizzled* and *frizzled 2* act in the wingless pathway. *Cell 95*, 1017–1026.

18

Mesenchymal Cell Fusion in the Sea Urchin Embryo

Paul G. Hodor and Charles A. Ettensohn

Summary

Mesenchymal cells of the sea urchin embryo provide a valuable experimental model for the analysis of cell–cell fusion in vivo. The unsurpassed optical transparency of the sea urchin embryo facilitates analysis of cell fusion in vivo using fluorescent markers and time-lapse three-dimensional imaging. Two populations of mesodermal cells engage in homotypic cell–cell fusion during gastrulation: primary mesenchyme cells and blastocoelar cells. In this chapter, we describe methods for studying the dynamics of cell fusion in living embryos. These methods have been used to analyze the fusion of primary mesenchyme cells and are also applicable to blastocoelar cell fusion. Although the molecular basis of cell fusion in the sea urchin has not been investigated, tools have recently become available that highlight the potential of this experimental model for integrating dynamic morphogenetic behaviors with underlying molecular mechanisms.

Key Words: Sea urchin; embryo; gastrulation; primary mesenchyme cells; secondary mesenchyme cells; blastocoelar cells; cell fusion; time-lapse microscopy; fluorescent dyes.

1. Introduction

During sea urchin gastrulation, two populations of mesoderm cells (primary [PMC] and secondary [SMC] mesenchyme cells) undergo homotypic cell–cell fusion. One important advantage of studying cell–cell fusion in the transparent sea urchin embryo is that powerful optical imaging approaches can be employed to analyze this process in vivo. In addition, a growing understanding of signaling pathways and gene regulatory networks that underlie cell specification and morphogenesis during early sea urchin development *(1,2)* provides a rich context for linking cell behaviors (including cell fusion) to genetic mechanisms. The genome of a representative sea urchin, *Strongylocentrotus purpuratus*, has recently been sequenced and annotated, and tools for gene discovery and perturbation of gene function are available *(3,4)*.

From: *Methods in Molecular Biology, vol. 475: Cell Fusion: Overviews and Methods*
Edited by: E. H. Chen © Humana Press, Totowa, NJ

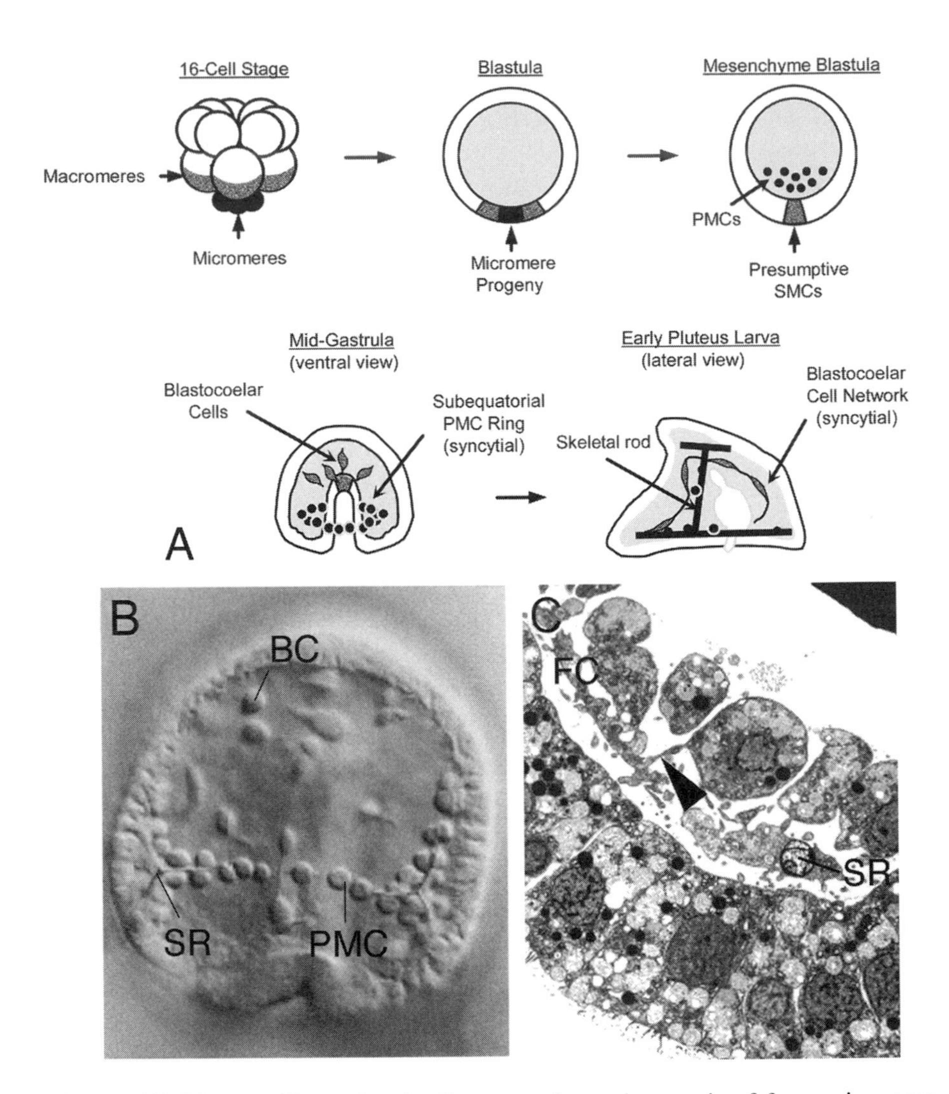

Fig. 1. **(A)** Diagram illustrating the lineage and morphogenesis of fusogenic mesenchymal cell populations in the sea urchin embryo. Primary mesenchyme cells (PMCs) are derived from the micromeres of the 16-cell stage embryos (black cells). Blastocoelar cells and other secondary mesenchyme cells (SMCs) are derived from the vegetal part of the macromeres (stippled region). The PMCs ingress at the mesenchyme blastula stage, and presumptive SMCs fill the vacated region of the epithelium. By the midgastrula stage, the PMCs have formed a subequatorial ring pattern and have fused to form a single syncytial network. Blastocoelar cells begin to ingress from the tip of the archenteron at this stage. For simplicity, other SMC subtypes (pigment cells and circumesophageal muscle cells) are not shown. **(B)** Living embryo at the late gastrula stage, viewed with differential interference contrast optics. The PMCs are arranged in a subequatorial ring pattern, and fusion is complete. BC, blastocoelar cell; PMC, primary mesenchyme cell; SR, skeletal rudiment. **(C)** Transmission electron micrograph of a late embryo showing the PMC filopodial cable (FC) and the thin stalks (arrow) that join PMC cell bodies to the filopodial cable. SR, skeletal rod. (Figure reprinted and legend adapted from **ref.** ***12*** with permission from Elsevier.)

1.1. Primary Mesenchyme Cell Fusion

The mesoderm of the sea urchin embryo consists of several distinct populations of cells. The best studied of these is the primary mesenchyme. Primary mesenchyme cells are descendants of the micromeres, four small cells that form at the vegetal pole of the embryo at the fourth cleavage division (**Fig. 1A**). The progeny of the micromeres become incorporated into the epithelial wall of the embryo during cleavage and blastula stages. At the onset of gastrulation, most of the micromere descendants undergo epithelial–mesenchymal transition (EMT) and ingress into the blastocoel cavity; after ingression these cells are referred to as *primary mesenchyme cells*. After a brief period of quiescence, PMCs extend filopodia and migrate along the inner wall of the blastocoel. By the midgastrula stage, the cells accumulate in a characteristic ringlike pattern near the equator of the embryo (the subequatorial PMC ring) and begin to secrete the biomineralized endoskeleton of the late embryo and larva (*see* **Fig. 1A,B**).

The specification, differentiation, and morphogenesis of PMCs have been investigated intensively ***(5)***. A large collection of expressed sequence tags (ESTs) derived from gastrula-stage PMCs has been used to identify genes that control the specification and morphogenesis of these cells ***(6–10)***. These and many other recent studies have led to the characterization of a complex gene regulatory network that underlies the specification and differentiation of the large micromere–PMC lineage ***(5,11)***.

Microinjection of fluorescent tracers and time-lapse three-dimensional confocal imaging have been used to analyze PMC fusion in vivo (**Fig. 2; ref. *12***). These studies have shown that fusion begins about 2h after ingression, during the PMC migratory phase, and that all PMCs fuse with at least one partner by the midgastrula stage, when the subequatorial ring has formed (4–5h post-ingression). Heterochronic cell transplantation experiments (**Figs. 3** and **4**) and cell culture studies demonstrate that the capacity to fuse is autonomously programmed in the large micromere–PMC lineage and independent of extrinsic signals. Surprisingly, PMCs remain fusion competent long after fusion is normally complete ***(12)***.

1.2. Blastocoelar Cell Fusion

Secondary mesenchyme cells constitute a second population of mesodermal cells (*see* **Fig. 1A**). Like PMCs, SMCs undergo EMT, but they ingress later in gastrulation in several asynchronous waves. Most SMCs give rise to pigment cells and blastocoelar cells, two highly migratory cell types, while other SMCs undergo much more limited movements after ingression and form the circumesophageal musculature of the larva ***(13)***. Of the three classes of SMCs, cell–cell fusion has been documented only in the case of blastocoelar cells. It seems likely that esophageal myoblasts also undergo fusion, based on what is known of myogenesis in many other organisms, but this has not yet been clearly demonstrated in the sea urchin ***(13)***.

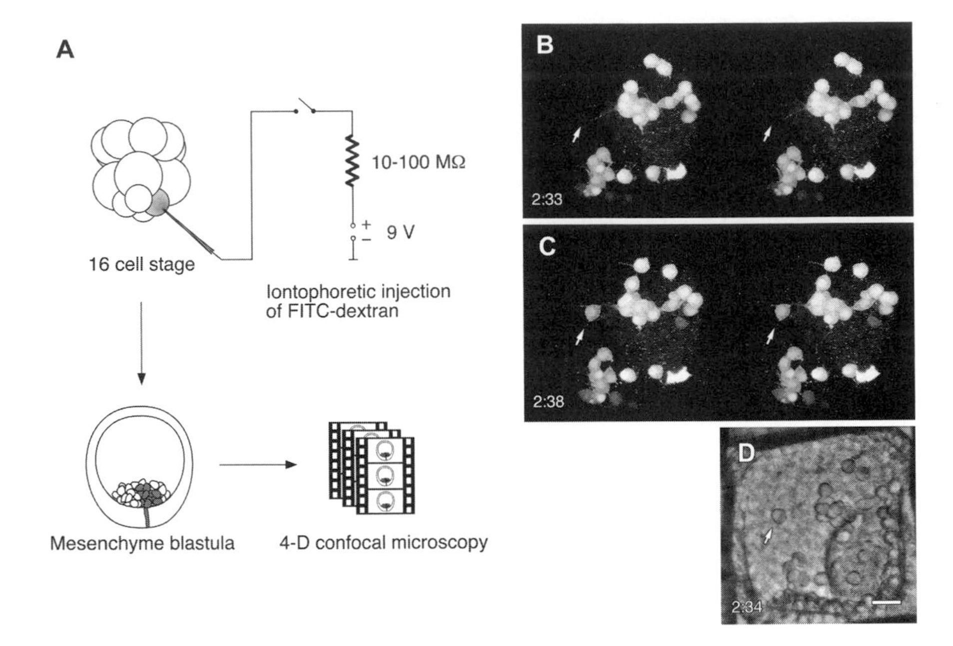

Fig. 2. Analysis of primary mesenchyme cell (PMC) fusion by microinjection of fluorescent dextran and time-lapse microscopy. **(A)** Experimental method. Single micromeres were iontophoretically injected with fluorescein dextran at the 16-cell stage. Development to the mesenchyme blastula stage produced an embryo with one quarter of the PMCs labeled. Fusion was monitored by time-lapse, three-dimensional laser scanning confocal microscopy. **(B–D)** In vivo fusion dynamics within the PMC population. In B and C, confocal stereo pairs show migrating PMCs labeled with FD10, which are undergoing fusion. In D, a bright-field image of the same embryo is shown. The arrows indicate a PMC that is unlabeled in B but acquires FD10 label in C by fusion with a PMC cluster to its right. Note that fusion occurs through filopodia. Times from PMC ingression are indicated in hours: minutes. Scale bar = 20 μm. (Figure reprinted and legend adapted from **ref. *12*** with permission from Elsevier.)

Blastocoelar cells have been described as fibroblastlike because of their mesodermal origin, mesenchymal morphology, and location within the extracellular matrix (ECM) of the blastocoel cavity ***(14)***. They migrate within the blastocoelar ECM during gastrulation and become arranged in a branched, reticular network concentrated around the gut (especially in the vicinity of the hindgut), along the skeletal rods, and within the larval arms. Only a few studies have addressed the early specification of blastocoelar cells ***(15–18)***, although several molecular markers have been identified that are expressed selectively by these cells ***(19–23)***.

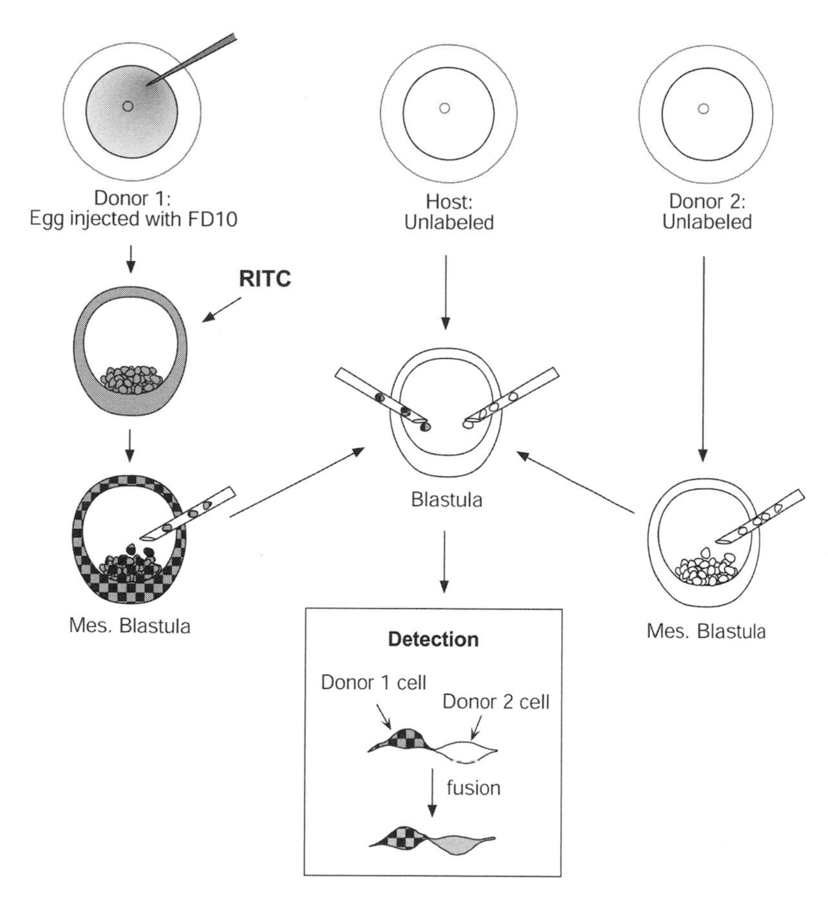

Fig. 3. Analysis of primary mesenchyme cell (PMC) fusion by transplantation of fluorescently labeled cells. This diagram shows a method used to monitor PMC fusion after heterochronic transplantation into young host embryos. Donor 1 embryos were first labeled with FD10 by injection into the egg and then double-labeled with rhodamine B isothiocyanate (RITC) at the mesenchyme (Mes.) blastula stage. The PMCs from these embryos were transplanted together with unlabeled PMCs from Donor 2 embryos into an early blastula host. Fusion was monitored by the transfer of FD10 to unlabeled PMCs. The RITC label was not transferred upon fusion and was used to identify the original, double-labeled donor cells. (Figure reprinted and legend adapted from **ref.** ***12*** with permission from Elsevier).

1.3. Contrasts in Mesenchymal Cell Fusion

There are intriguing differences in the morphology of cell–cell fusion as it occurs in PMCs and blastocoelar cells. The PMCs fuse primarily via their filopodia, which join together to form a cablelike filopodial chain ***(12,24–26)***. The bodies of the PMCs, which contain the cell nuclei, remain spherical in

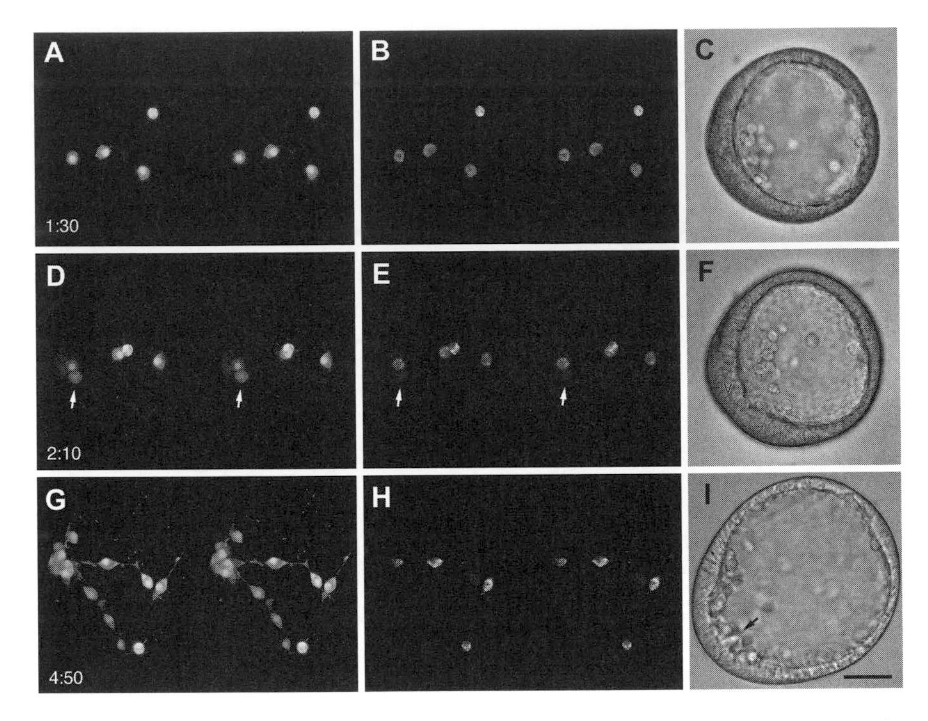

Fig. 4. Primary mesenchyme cell (PMC) fusion in an early embryonic environment. Approximate times after PMC ingression in sibling donor embryos are shown in hours:minutes. **(A,D,G)** Confocal stereo pairs showing FD10 labeling. **(B,E,H)** Confocal stereo pairs showing RITC labeling. **(C,F,I)** Transmitted light images of the same embryo recorded at the same times. In A–C, 4 double-labeled PMCs and 10–15 unlabeled PMCs were transplanted into an early blastula embryo several hours before the ingression of host PMCs. In D–F, after about 2 h, fusion is revealed by the transfer of FD10 to unlabeled donor cells (arrows). In G–I, later, about 15 donor PMCs can be seen joined in a syncytium while host PMCs are just beginning to ingress (I, arrow). Scale bar = 40 μm. (Figure reprinted and legend adapted from **ref.** ***12*** with permission from Elsevier.)

shape and are joined to the filopodial cable by narrow stalks (*see* **Fig. 1C; ref.** ***27***). The PMCs that undergo fusion in vitro also retain distinct, spherical cell bodies, although under these conditions the morphology of the filopodial cable is altered ***(12,28)***. In contrast, when blastocoelar cells fuse, the bodies of the partners often merge completely into a large, fibroblastic, multinucleated cell.

Another fascinating feature of mesenchymal fusion is that although PMCs and blastocoelar cells sometimes make direct filopodial contacts with one another at developmental stages when both cell types are fusogenic, these contacts never result in heterotypic cell fusion (**Fig. 5; ref.** ***12***).

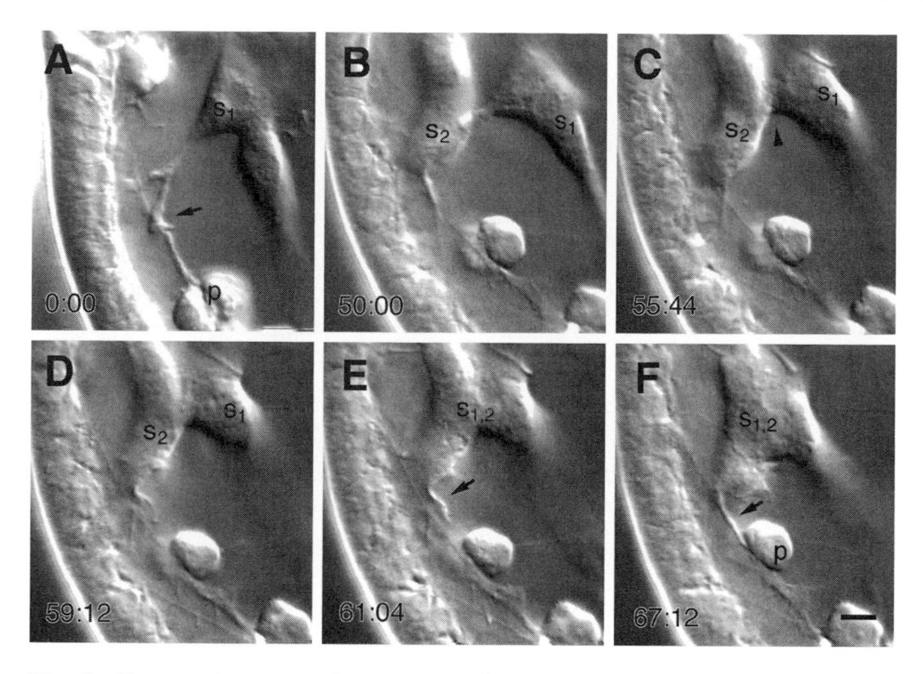

Fig. 5. Contacts between primary mesenchyme cells (PMCs) and blastocoelar cells during gastrulation, concurrent with blastocoelar cell fusion. Six images of a time-lapse recording of a late gastrula viewed with differential interference contrast optics. Times indicated are in minutes:seconds. **(A)** Two blastocoelar cells fuse (s_1), while one cell engages in continuous filopodial contact (arrow) with the PMC filopodial cable (p). Another blastocoelar cell moves into the focal plane (s_2 in **B**). Fusion between s_1 and s_2 results in coalescence of cell bodies (**C**, arrowhead). A filopodium extended by s_2 makes direct contact with PMCs during this time (**E,F**, arrow). The PMCs are capable of homotypic fusion throughout this period of development but do not fuse with blastocoelar cells despite these cell–cell contacts ***(12)***. Scale bar = 5 µm. (Figure reprinted and legend adapted from **ref. *12*** with permission from Elsevier).

Presumably, there are differences in the underlying mechanisms of fusion in PMCs and blastocoelar cells that maintain this strict cell type specificity. Mesenchymal cell fusion during sea urchin gastrulation therefore provides a unique example of two intermingled populations of cells, both highly fusogenic, that remain distinct by employing different mechanisms of fusion.

In this chapter, we describe methods for studying mesenchymal cell fusion in vivo. Our emphasis is on PMC fusion, but we also outline ways in which the same methods can be used to analyze the fusion of blastocoelar cells. Related protocols, including detailed methods for gamete collection, fertilization, micromanipulation, and microinjection, can be found elsewhere

(29–32). Space limitations prevent us from providing a detailed description of a relatively new method of tagging cells in vivo that may prove useful in future studies of cell fusion, namely, expression of fluorescent reporters in transgenic embryos ***(33)***. Several DNA regulatory elements have been identified that drive the expression of reporter genes such as *GFP* selectively in PMCs or blastocoelar cells ***(34–40)***. Because expression of transgenes in sea urchin embryos is mosaic, the reporter gene is expressed in only a subset of the targeted cells, and GFP protein is subsequently transferred from cell to cell via fusion ***(34,35)***.

2. Materials

2.1. Analysis of Primary Mesenchyme Cell Fusion by Microinjection of Fluorescent Dyes Into Progenitor Blastomeres

2.1.1. Embryo Culture

1. Gravid adult sea urchins (*Lytechinus variegatus*) (*see* **Note 1**).
2. 0.5 *M* KCl.
3. 10-cc syringe with needle.
4. Several 100-mL plastic beakers (*see* **Note 2**).
5. Glass culture bowls, 2–8 inches in diameter (*see* **Note 2**).
6. Seawater.
7. Pasteur pipettes and rubber bulbs.
8. Light microscope (stereomicroscope).
9. Temperature-controlled waterbath or incubator (18°–25 °C; *see* **Note 3**).

2.1.2. Immobilization of Embryos in Kiehart Chambers

1. Poly-L-lysine hydrobromide, MW = 150–300 × 10^3 (Sigma, St. Louis, MO, catalog number P1399). Prepare 50-mL of a 1 mg/mL solution of poly-L-lysine in deionized water, and store 1.5-mL aliquots at −80 °C.
2. Warming oven (60 °C).
3. Double-sided Scotch tape, No. 665 (3M Company) (*see* **Note 4**).
4. Glass coverslips, 22 × 22 mm.
5. Diamond scribe.
6. Silicone high-vacuum grease (Dow Corning Corp., Midland, MI; *see* **Note 5**).
7. Kiehart microinjection chambers (*see* **Note 6**).
8. Pasteur pipettes and rubber bulbs.
9. Seawater.

2.1.3. Iontophoretic Injection of Micromeres

1. Iontophoresis apparatus.
2. Filamented glass capillaries (Item #1B100F-6, 1-mm diameter, 6-inch length, World Precision Instruments, Sarasota, FL).

3. Pipette puller (e.g., Sutter P-97 Horizontal Pipette Puller, Sutter Instruments, Novato, CA).
4. Fluorescein isothiocyanate dextran, MW = 10,000 (FD10). FD10 should be prepared as a 10% (wt/vol) solution in sterile-filtered (0.2-μm) deionized water and stored at 4 °C.
5. Epifluorescence microscope with appropriate filter sets.
6. Model MP-2 micromanipulator with micropipette holder (Narishige Scientific Instrument Laboratory, Tokyo, Japan).
7. Mouth pipette.
8. 35-mm plastic dishes.
9. Light-safe, humid box.
10. Temperature-controlled waterbath or incubator (18°–25 °C).

2.1.4. Four-Dimensional Confocal Microscopy

1. 22 × 50 mm glass coverslips.
2. Poly-L-lysine hydrobromide, MW = 150–300 × 103 (Sigma #P1399).
3. Warming oven (60 °C).
4. Nylon cloth, 47-μm mesh (Small Parts, Inc., Miami Lakes, FL).
5. Watchmakers' forceps.
6. Confocal microscope with software for time-lapse recording and three-dimensional image reconstruction.

2.2. Investigating Fusion by Transplantation of Primary Mesenchyme Cells

2.2.1. Injecting Eggs With Fluorescein Dextran

1. Protamine sulfate (Sigma #P-4020). A 1% solution of protamine sulfate should be prepared by adding 0.4 g protamine sulfate to 40 mL deionized water in a 50-mL plastic screw-top tube. Incubate the solution on a tilter until completely dissolved (~1 h). The solution can be used for several months if stored at 4 °C. Before each use, warm the solution to room temperature and ensure that the protamine sulfate is dissolved completely
2. 60-mm plastic tissue culture dishes (lids only).
3. Diamond scribe.
4. Glass coverslips.
5. Silicone high-vacuum grease (Dow Corning; *see* **Note 5**).
6. Pasteur pipettes and rubber bulbs.
7. 15-mL conical, glass centrifuge tubes.
8. Clinical tabletop centrifuge.
9. Mouth pipette.
10. Filamented glass capillaries (World Precision Instruments, Sarasota, FL).
11. Epifluorescence microscope with appropriate filter sets.
12. Micromanipulator (e.g., Leitz joystick micromanipulator)
13. Microinjector (e.g., Picospritzer II, Parker Instrumentation, General Valve Division, Fairfield, NJ).

14. 35-mm plastic dishes.
15. Light-safe, humid box.
16. Temperature-controlled waterbath or incubator (18°–25 °C).
17. Seawater.

2.2.2. Labeling Embryos With Rhodamine B Isothiocyanate

1. Rhodamine B isothiocyanate (RITC).
2. Dimethylsulfoxide (DMSO).
3. Vortexer.

2.2.3. Primary Mesenchyme Cell Transplantation

1. Nonfilamented glass capillary tubes, OD = 0.86 mm, ID = 0.51 mm (Drummond Scientific Co. Broomall, PA).
2. Sigmacote silicone solution (Sigma, St. Louis, MO).
3. Pipette puller.
4. Micropipette beveler (K. T. Brown-Type Micropipette Beveler, Model BP-10, Sutter Instruments, Novato, CA).
5. Kiehart chambers and accessories (*see* **Subheading 2.1.2.**).
6. Model MP-2 micromanipulator with micropipette holder (Narishige Scientific).
7. Micrometer syringe.
8. Plastic tubing (Clay Adams Intramedic polyethylene tubing, PE-60).
9. Dow Corning 200 fluid (polydimethylsiloxane).
10. Microscope.
11. Vibration isolation table.

2.2.4. Microscopic Examination

1. Poly-L-lysine hydrobromide, MW = 150–300 × 10^3 (Sigma #P1399).
2. Glass microscope slides.
3. Glass coverslips.
4. Double-sided Scotch tape, No. 665 (3M Company; *see* **Note 4**).
5. Confocal microscope.

3. Methods

3.1. Analysis of Primary Mesenchyme Cell Fusion by Microinjection of Fluorescent Dyes Into Progenitor Blastomeres

3.1.1. Embryo Culture

1. Obtain or prepare seawater. Several formulas for artificial seawater have been described. We routinely use Instant Ocean (Aquarium Systems, Inc., Mentor, OH) dissolved in deionized water following the manufacturer's instructions or natural seawater collected at the Marine Biological Laboratory (MBL; Woods Hole, MA). The MBL also has a formula for artificial seawater:

NaCl	24.72 g
KCl	0.67 g
$CaCl_2.2H_2O$	1.36 g
$MgCl_2.6H_2O$	4.66 g
$MgSO_4.7H_2O$	6.29 g
$NaHCO_3$	0.18 g

Dissolve the salts in ~800 mL deionized or distilled water and then bring to a final volume of 1 L. Adjust the pH to 8.3.

2. Induce spawning of adult sea urchins (L. variegatus) by intracoelomic injection of 0.5 M KCl.
3. Collect eggs in a beaker filled with seawater at room temperature and use within a few hours. Collect sperm dry and store at 4 °C for up to 1 wk.
4. Pipette an aliquot of eggs into a glass finger bowl filled with seawater to a depth of 0.5–1.0 cm. Ensure that no more than 25%–50% of the surface area of the dish is covered when the eggs settle, as embryos will not develop normally if overcrowded.
5. Fertilize the eggs by pipetting a few drops of a suspension of sperm (diluted in seawater) into the bowl and mixing gently. Confirm that fertilization has occurred by monitoring elevation of the fertilization envelope using a light microscope.
6. Incubate the developing embryos at 18°–25 °C.

3.1.2. Immobilization of Embryos in Kiehart Chambers

At the 8-cell stage, transfer embryos into a Kiehart chamber ***(41)***.

1. Coat a glass coverslip, covering the surface with 1 mg/mL poly-L-lysine solution, removing the excess liquid after 5–10 sec, and drying the coverslip in a warm oven. Rinse the coverslip in deionized water for several seconds and dry it again. Attach the coverslip to the top of the Kiehart chamber with silicone grease.
2. Attach a spacer consisting of double-sided Scotch tape (No. 665) with the side parallel to the edge of the coverslip.
3. When the embryos have reached the 16-cell stage, use a mouth pipette to place a few of them into the space between the tape and the edge of the coverslip.
4. Press down a coverslip fragment across the double-sided tape such that the embryos are slightly compressed between the fragment and the top coverslip.
5. Immediately close the chamber with a second coverslip attached to the bottom, and fill it with seawater (*see* **Note 7**).

3.1.3. Iontophoretic Injection of Micromeres

A simple iontophoresis apparatus can be custom built and consists of a 9-V battery connected with the positive pole to a resistor and through a push-button switch to a platinum wire inserted into the injection needle. A selector switch allows the value of the resistor to be chosen between 10 and 100 MW. The role of the resistor is to set an upper limit of the electric current in the range of 1–0.1 nA.

1. Prepare injection needles by pulling filamented glass capillaries with a micropipette puller. Backfill the needles with 10% fluorescein isothiocyanate dextran MW = 10,000 (FD10). The liquid will be drawn into the needle along the internal filament by capillary forces. Keep the needles affixed to a strip of modeling clay in a humid box (*see* **Note 8**).
2. Place the Kiehart chamber containing 16-cell-stage embryos onto the stage of an epifluorescence microscope. Mount an injection needle onto the micromanipulator in a horizontal position, to fit into the space between the top coverslip of the chamber and the coverslip fragment, in the region where the embryos are immobilized. Insert the platinum electrode of the iontophoresis apparatus into the back of the needle. Move the tip of the needle into the vicinity of the embryos. Connect the negative pole of the battery to the microscope stage with an alligator clamp, thus closing the circuit through the metal frame of the Kiehart chamber and the seawater.
3. Perform the dye injection by viewing the embryos under dim light, using a combination of a fluorescein filter set and bright-field illumination. First test the needle opening by viewing the needle tip in seawater and pressing the push-button switch for a few seconds. A slow stream of fluorescent dye should appear immediately at the tip of the needle (*see* **Note 9**).
4. Insert the tip of the needle into one of the micromeres of a 16- or 28-cell-stage embryo. Press the push-button switch for a few seconds, watching the transfer of dye into the cell. The dye will rapidly diffuse throughout the cell. Stop when the fluorescence is clearly visible (*see* **Note 10**).
5. After the labeling is completed, immerse the Kiehart chamber into a dish containing seawater and gently disassemble it, keeping track of the labeled embryos. Using a mouth pipette, transfer labeled embryos individually to 35-mm tissue culture dishes containing seawater. Culture the embryos in the dark until they reach the mesenchyme blastula stage.

3.1.4. Four-Dimensional Confocal Microscopy

The principal difficulty in imaging sea urchin embryos over time is their mobility. A special microchamber needs to be built, in which embryos have to be strictly immobilized, while at the same time keeping them healthy, such that they can further develop to the pluteus stage.

1. Start with a 22 × 50 mm #1 glass coverslip, cleaned by briefly flaming over a gas burner.
2. Apply high vacuum grease with a syringe in a thick square frame 20-mm wide. The frame defines a large culture chamber for the embryo.
3. Apply two thin parallel strips of vacuum grease in the center of the chamber, 1–2 mm apart. These will hold the microchamber in which the embryo will be immobilized.
4. Add several drops of a 1 mg/mL poly-L-lysine solution over the space between the grease strips. Allow to dry at 60 °C.
5. Rinse the chamber several times with deionized water, and then fill it with seawater.

6. Cut out a 2 × 4 mm nylon mesh piece with 47 μm opening size. With a pair of watchmaker's forceps, pull out a single thread in each direction, making a square chamber in the center of the mesh, with an opening of 130 μm (*see* **Note 11**).
7. Mouth pipette a single labeled embryo onto the glass surface of the chamber, between the two grease strips. The embryo will attach to the coated glass surface. Orient it with the vegetal pole down toward the glass.
8. Lay the mesh piece over the embryo, fitting the embryo into the microchamber and attaching the sides of the mesh to the grease strips.
9. Close the microchamber with a coverslip fragment similar in size to the nylon mesh, pressing it down to attach it to the vacuum grease. At this step it is possible to make fine adjustments to the orientation of the embryo. When the microchamber is closed, the embryo will be pressed against the two glass surfaces on its vegetal and animal poles and against the nylon square on its sides. The polylysine on the large coverslip will hold it immobilized. Because of compression, the shape of the embryo will become somewhat cuboidal (*see* **Note 12**).
10. Close the large chamber with a 22 × 22 mm #1.5 coverslip by attaching it to the outer grease frame. There should be no air bubbles left in the chamber, and the grease seal should be checked for any leaks.
11. View mounted embryos by laser-scanning confocal microscopy. Record alternative bright-field and fluorescence images every 5–10 min over several hours (*see* **Note 13**).
12. Examine the four-dimensional movie of the fluorescence stacks. Use the bright-field images as a spatial reference for labeled cells relative to the embryonic morphology (*see* **Note 14**).

3.1.5. Analysis of Secondary Mesenchyme Cell Fusion

The methods described above (*see* **Subheadings 3.1.1.–3.1.4.**) can also be used to analyze blastocoelar cell fusion. The only change in the protocol is that dye is microinjected into macromeres (the progenitors of the blastocoelar cells) rather than micromeres. As their name implies, macromeres are larger cells than micromeres, and fluorescent dextran can be delivered using either iontophoresis (described in **Subheading 3.1.3.**) or pressure injection (below in **Subheading 3.2.1**). Subsequent time-lapse microscopy is carried out as described above.

3.1.6. Use of Photoactivatable Cell Markers

As an alternative to microinjection into individual progenitor cells, caged cell markers can be microinjected into the fertilized egg and locally photoactivated in specific cells. The advantage of this procedure is that it is much easier to inject fertilized eggs, which are very large cells, than the smaller blastomeres of cleavage stage embryos. This approach, however, requires a means of focal irradiation with light of the wavelength appropriate for photoactivation (usually ultraviolet or blue light). The two most common approaches are to employ (1) a conventional epifluorescence microscope with a pinhole aperture placed in the conjugate focal plane between the mercury lamp and the specimen or (2) a scanning confocal microscope equipped with an ultraviolet laser. A variety of photoactivatable cell markers are available commercially, although the widely used caged, fluorescent dextrans formerly produced by Molecular Probes, Inc. have recently been discontinued.

3.2. Investigating Fusion by Transplantation of Primary Mesenchyme Cells

This approach allows the experimenter to observe fusion between PMCs double labeled with FD10 and RITC and unlabeled PMCs. Upon fusion, FD10 transfers to unlabeled fusion partners, while RITC is used to track the cells that were labeled initially.

3.2.1. Injecting Eggs With Fluorescein Dextran

1. Coat the lids (*see* **Note 15**) of 60-mm plastic dishes by pouring 1% protamine sulfate (PS) solution into the lid (enough to cover the entire surface) and allowing the solution to stand for 2 min. Pour the PS solution back into its original tube, as it can be reused many times. Rinse the PS-coated lids for 5–10 min in several changes of deionized water and air dry. Using a diamond scribe, cut small fragments of coverslips (~ 1.0 × 0.3 cm) and mount one on each dish near the center, attaching it with a very small dab of silicone stopcock grease. This coverslip fragment will be used to break open the tip of the microinjection needle (*see* number 6 below). Dried, coated dishes should be stored in a dust-free container at room temperature and can be used for several weeks.
2. Using a Pasteur pipette, transfer unfertilized *L. variegatus* eggs into a 15-mL conical, glass centrifuge tube. Dejelly the eggs by washing them six to eight times (brief washes) with seawater. After each wash, collect the eggs by gentle centrifugation using a clinical centrifuge (30 s, half speed).
3. Fill a PS-coated dish with seawater. Using a mouth pipette and a stereomicroscope, carefully place a single file of dejellied eggs near the center of the dish. Eggs should adhere firmly within a few seconds and should not move when the dish is moved gently back and forth.
4. Fertilize the row of immobilized eggs by adding a few drops of diluted sperm to the seawater directly above the eggs. Allow 2–5 min for fertilization envelopes to elevate and harden.
5. Prepare the injection apparatus. Pull filamented glass needles and fill them with FD10 as described in **Subheading 3.1**. Instead of using a platinum electrode, connect the needle to a micromanipulator and a microinjection apparatus. Several different configurations of microscopes, micromanipulators, and pressure injectors can be used **(29)**. We use a fixed-stage, inverted, epifluorescence microscope, a free-standing micromanipulator with a joystick and course and fine movements in all three axes, and a picospritzer.

6. Place the dish containing the fertilized eggs on the stage of the epifluorescence microscope, and bring the injection needle into view. Use a combination of bright-field and fluorescence (fluorescein filter set) illumination to monitor injection. Break the tip of the needle by touching it gently to the edge of the coverslip fragment. Apply positive pressure, and confirm that a stream of fluorescent dye is released from the tip of the needle. Then move the needle near the row of eggs and inject them in sequence by inserting the needle through the plasma membrane and allowing the dye to enter for 1–2 seconds (*see* **Note 16**).
7. After injection, place the injection dish in a light-safe, humid container and allow the embryos to develop in the dark at 18°–25 °C until they reach the mesenchyme blastula stage.

3.2.2. Labeling Embryos With Rhodamine B Isothiocyanate

1. Prepare RITC solution immediately before use. Dissolve 1 mg of RITC in 20 mL of dimethyl sulfoxide, add 5 mL of seawater, and vortex. Make a 1:100 dilution in seawater of this concentrated solution. Place the diluted solution into a 35-mm tissue culture dish.
2. Collect FD10-labeled, mesenchyme blastula stage embryos by gentle centrifugation, and transfer them into the staining solution. Label them with RITC for 10 min (*see* **Note 17**).
3. Transfer the double-labeled embryos with a mouth pipette into a fresh dish of seawater.

3.2.3. Primary Mesenchyme Cell Transplantation

1. Coat glass capillary tubes with Sigmacote silicone solution to prevent adherence of cells to the glass. Aspirate the silicone solution into each capillary tube, and then remove the excess by touching one tip of the tube to absorbant paper. Allow the capillaries to air dry (a warm oven speeds this process). Rinse each coated capillary by drawing deionized water back and forth through it several times. Allow the washed capillary tubes to air dry, and store them in a dust-free container.
2. Pull siliconized capillaries using the pipette puller.
3. Bevel the tip of each needle using the needle beveler. The tip diameter of the beveled needles should be 10–15 μm (about 2× the diameter of a PMC).
4. Mount embryos into a Kiehart chamber, as described in **Subheading 3.1.2**. When multiple groups of embryos are used, each group can be mounted into a separate Kiehart chamber. Alternatively, double chambers can be used, in which two separate coverslip fragments each immobilize separate groups of embryos. The PMC donor embryos should be at the mesenchyme blastula stage, while recipients can be at any developmental stage.
5. Mount a beveled needle onto the Narishige micromanipulator (*see* **Note 18**).
6. Insert the needle into the blastocoel of an unlabeled donor embryo, and bring its tip into the vicinity of the PMCs. Slightly tap the needle holder to dislodge the PMCs from the wall of the blastocoel. Draw the PMCs into the needle by applying negative pressure. The needle can be loaded in this way with PMCs from multiple embryos.

7. Move the needle containing labeled PMCs to the group of recipient embryos, and insert it into the blastocoel of an embryo. Inject PMCs into the blastocoel, and remove the needle. The position of injected PMCs within the blastocoel is not critical. Multiple embryos can be injected in sequence.
8. Repeat transferring of PMCs from double-labeled embryos to the same recipients that received unlabeled PMCs.
9. Disassemble the Kiehart chamber containing recipient embryos in a tissue culture dish containing seawater.

3.2.4. Microscopic Examination

Fusion between double-labeled and unlabeled PMCs can be studied by four-dimensional microscopy, as described in **Subheading 3.1.4.** Alternatively, embryos can be mounted on slides for individual pictures at several time points.

1. Immobilize one or more embryos between a poly-L-lysine–coated coverslip and a glass slide, using double-sided Scotch tape spacers (*see* **Note 4**). Although this method allows for some movement of the embryos over time compared with immobilization in a nylon mesh, it is suitable for taking pictures and is much easier to perform.
2. View the embryo on an epifluorescence microscope, collecting bright-field and fluorescence images using fluorescein and rhodamine filter sets. Alternatively, image the embryos with a confocal microscope.

4. Notes

1. Several species of sea urchins are commonly used for developmental studies. *Lytechinus variegatus* embryos are ideal for imaging cell fusion and other morphogenetic processes because they are highly transparent and develop at room temperature.
2. It is critical that all glass and plastic items used for collecting gametes and culturing embryos be clean and completely detergent free. It is advisable to keep "embryoware" separate from other laboratory supplies and to wash these items with water only.
3. The rate of embryonic development is slower at cooler temperatures. For example, the time required for *L. variegatus* embryos to develop to the mesenchyme blastula stage is 19–20 h at 18 °C and 9–10 h at 25 °C.
4. Scotch #665 double-sided tape is only slightly thinner than an *L. variegatus* embryo. It is therefore very convenient to use as a spacer. For other species of sea urchins with larger or smaller embryos, it may be necessary to use more than one layer of tape or small pieces of plastic films as spacers.
5. A convenient means of delivering vacuum grease is to load it into a 10-cc syringe from the back. The syringe can be used without a needle to dispense small dabs of grease.
6. These chambers are not available commercially and must be manufactured locally by a machine shop. Aluminum or stainless steel can be used. Diagrams and specifications can be found elsewhere ***(41)***.

7. Once the embryos are compressed between the two glass surfaces, it is important to complete the assembly of the chamber quickly to prevent the embryos from desiccating. For the same reason, after the chamber is filled, one should periodically add a drop of seawater with a Pasteur pipette to the open edge of the chamber.
8. It is important to use a precision instrument for pulling the capillaries. This ensures that the injection needles have a reproducible tip shape across many experiments. Needles should be prepared on the day they will be used.
9. The appropriate size of the needle opening must be determined by trial and error. If the opening is too large, seawater can be drawn into the needle between pulses of the picospritzer, causing a delay in the appearance of the dye when positive pressure is applied, or the dye can flow outward without the application of an electric current. In such cases the needle must be discarded. If the opening is too small, or the tip is completely closed, the needle tip should be gently touched to the edge of the Scotch tape in the chamber and then re-tested by applying the electric current.
10. There is a tradeoff between having a strong enough fluorescent signal in the micromeres and maintaining a normal development of injected cells. Overlabeled cells usually lyse within a few minutes. Otherwise the fluorescent dextran appears to be quite inert and even brightly labeled cells develop normally, as long as they are not exposed to light.
11. The exact type of nylon mesh for building the microchamber will depend on the manufacturer and also on the sea urchin species used in the experiment. The dimensions of the chamber must be such that an embryo fits in snugly.
12. Compression of the embryo is critical for keeping it immobilized for several hours. The embryo is still able to gastrulate and form skeletal rods, although occasionally the orientation of the rods is abnormal.
13. Because of exposure to light during image capture, the fluorescence signal bleaches over time. To reduce the impact of bleaching on the recording, adjust the level of illumination as low as practically possible and decrease image sizes and scan times. Also use nonlinear amplification to boost low-level confocal signals, which will compensate for decreasing signal intensities. In our particular setting, we used a Bio-Rad MRC600 microscope equipped with a krypton/argon laser and 20× (NA = 0.80) and 40× (NA = 1.00) plan apochromatic oil immersion objectives. The laser light intensity was set to 3%. Bright-field stacks consisted of 5 image planes 15 μm apart, and fluorescence stacks consisted of 20 planes 4 μm apart. Nonlinear amplification was set to +4. Individual image sizes were 256 × 256 pixels.
14. A variety of software applications can be used to generate the movie. On the MRC600, pseudo stereo image pairs can be generated from each fluorescence image stack by projecting them into a single image twice, with a pixel shift between consecutive planes of +0.5 and –0.5, respectively. The stereo pairs can be viewed side by side in monochrome mode or can be pseudo-colored in red and green, overlaid, and viewed with stereo glasses.
15. Lids are used because they are shallower than the bottom part of the dish and therefore do not obstruct the injection needle, which is brought down at a descending angle from the side.

16. The egg membrane is flexible and can easily be deformed without the needle entering the cytoplasm. It may be necessary to move the needle sharply into the eggs to successfully puncture the membrane. Once the needle is inside the egg cytoplasm, one can visually observe rapid diffusion of the fluorescent dye throughout the cytoplasm.
17. It is very easy to overlabel embryos with RITC such that their development will be impaired. Practice the RITC labeling in advance with a few batches of uninjected embryos, and examine their development compared with unlabeled controls. Adjust the dye concentration and labeling time as needed.
18. In contrast to the injection setup, avoid any air bubbles in the tubing and syringe. In the case of pressure injection, the presence of air bubbles is beneficial because it dampens sharp changes in dye flow. For cell transplantations, however, it is important that the flow of water and cells immediately responds to changes in pressure applied to the syringe by the experimenter.

References

1. Ben-Tabou de-Leon, S. and Davidson, E. H. (2007) Gene regulation: gene control network in development. *Ann. Rev. Biophys. Biomol. Struct.* **36,** 191–212.
2. Levine, M. and Davidson, E. H. (2005) Gene regulatory networks for development. *Proc. Natl. Acad. Sci. U.S.A.* **102,** 4936–4942.
3. Cameron, R. A., Rast, J. P., and Brown, C. T. (2004) Genomic resources for the study of sea urchin development. *Methods Cell Biol.* **74,** 733–757.
4. Sea Urchin Genome Sequencing Consortium (2006) The genome of the sea urchin *Strongylocentrotus purpuratus. Science* **314,** 941–952.
5. Wilt, F. H. and Ettensohn, C. A. (2007) The morphogenesis and biomineralization of the sea urchin larval skeleton, in Handbook of Biomineralization (E. Bauerlein, ed.). Wiley-VCH Press, New York, pp. 183–210.
6. Cheers, M. S. and Ettensohn, C. A. (2005) P16 is an essential regulator of skeletogenesis in the sea urchin embryo. *Dev. Biol.* **283,** 384–396.
7. Ettensohn, C. A., Illies, M. R., Oliveri, P., and De Jong, D. L. (2003) Alx1, a member of the Cart1/Alx3/Alx4 subfamily of Paired-class homeodomain proteins, is an essential component of the gene network controlling skeletogenic fate specification in the sea urchin embryo. *Development* **130,** 2917–2928.
8. Illies. M. R., Peeler, M. T., Dechtiaruk, A. M., and Ettensohn, C. A. (2002) Identification and developmental expression of new biomineralization proteins in the sea urchin, *Strongylocentrotus purpuratus. Dev. Genes Evol.* **212,** 419–431.
9. Livingston, B. T., Killian, C. E., Wilt, F., Cameron, A., Landrum, M. J., Ermolaeva, O., Sapojnikov, V., Maglott, D. R., Buchanan, A. M., and Ettensohn, C. A. (2006) A genome-wide analysis of biomineralization-related proteins in the sea urchin, *Strongylocentrotus purpuratus. Dev. Biol.* **300,** 335–348.
10. Zhu, X., Mahairas, G., Illies, M., Cameron, R. A., Davidson, E. H., and Ettensohn, C. A. (2001) A large-scale analysis of mRNAs expressed by primary mesenchyme cells of the sea urchin embryo. *Development* **128,** 2615–2627.

11. Oliveri, P. and Davidson, E. H. (2004) Gene regulatory network controlling embryonic specification in the sea urchin. *Curr. Opin. Genet. Dev.* **14,** 351–360.
12. Hodor, P. G. and Ettensohn, C. A. (1998) The dynamics and regulation of mesenchymal cell fusion in the sea urchin embryo. *Dev. Biol.* **199,** 111–124.
13. Burke, R. D. and Alvarez, C. M. (1988) Development of the esophageal muscles in embryos of the sea urchin *Strongylocentrotus purpuratus. Cell Tissue Res.* **252,** 411–417.
14. Tamboline, C. R. and Burke, R. D. (1992) Secondary mesenchyme of the sea urchin embryo: ontogeny of blastocoelar cells. *J. Exp. Zool.* **262,** 51–60.
15. Kominami, T. and Takata, H. (2003) Specification of secondary mesenchyme–derived cells in relation to the dorso-ventral axis in sea urchin blastulae. *Dev. Growth Differ.* **45,** 129–142.
16. Ruffins, S. W. and Ettensohn, C. A. (1993) A clonal analysis of secondary mesenchyme cell fates in the sea urchin embryo. *Dev. Biol.* **160,** 285–288.
17. Ruffins, S. W. and Ettensohn, C. A. (1996) A fate map of the vegetal plate of the sea urchin *(Lytechinus variegatus)* mesenchyme blastula. *Development* **122,** 253–263.
18. Sweet, H. C., Gehring, M., and Ettensohn, C. A. (2002) LvDelta is a mesoderm-inducing signal in the sea urchin embryo and can endow blastomeres with organizer-like properties. *Development* **129,** 1945–1955.
19. Howard-Ashby, M., Materna, S. C., Brown, C. T., Chen, L., Cameron, R. A., and Davidson, E. H. (2006) Gene families encoding transcription factors expressed in early development of *Strongylocentrotus purpuratus. Dev. Biol.* **300,** 90–107.
20. Miller, R. N., Dalamagas, D. G., Kingsley, P. D., and Ettensohn, C. A. (1996) Expression of *S9* and actin *CyIIa* mRNAs reveals dorso-ventral polarity and mesodermal sublineages in the vegetal plate of the sea urchin embryo. *Mech. Dev.* **60,** 3–12.
21. Rottinger, E., Besnardeau, L., and Lepage, T. (2004) A Raf/MEK/ERK signaling pathway is required for development of the sea urchin embryo micromere lineage through phosphorylation of the transcription factor Ets. *Development* **131,** 1075–1087.
22. Shoguchi, E., Tokuoka, M., and Kominami, T. (2002) In situ screening for genes expressed preferentially in secondary mesenchyme cells of sea urchin embryos. *Dev. Genes Evol.* **212,** 407–418.
23. Sweet, H. C., Hodor, P. G., and Ettensohn, C. A. (1999) The role of micromere signaling in Notch activation and mesoderm specification during sea urchin embryogenesis. *Development* **126,** 5255–5265.
24. Okazaki, K. (1960) Skeleton formation of sea urchin larvae. II. Organic matrix of the spicule. *Embryologia* **5,** 283–320.
25. Okazaki, K. (1965) Skeleton formation of sea urchin larvae. V. Continuous observation of the process of matrix formation. *Exp. Cell Res.* **40,** 585–596.
26. Gustafson, T. and Wolpert, L. (1967) Cellular movement and contact in sea urchin morphogenesis. *Biol. Rev.* **42,** 441–498.
27. Gibbins, J. R., Tilney, L. G., and Porter, K. R. (1969) Microtubules in the formation and development of the primary mesenchyme in *Arbacia punctulata.* I. The distribution of microtubules. *J. Cell Biol.* **41,** 201–226.

28. Okazaki, K. (1975) Spicule formation by isolated micromeres of the sea urchin embryo. *Am. Zool.* **15,** 567–581.
29. Cheers, M. S. and Ettensohn, C. A. (2004) Rapid microinjection of fertilized eggs. *Methods Cell Biol.* **74,** 287–310.
30. Foltz, K. R., Adams, N. L., and Runft, L. L. (2004) Echinoderm eggs and embryos: procurement and culture. *Methods Cell Biol.* **74,** 39–74.
31. Jaffe, L. A. and Terasaki, M. (2004) Quantitative microinjection of oocytes, eggs, and embryos. *Methods Cell Biol.* **74,** 219–242.
32. Sweet, H., Amemiya, S., Ransick, A., Minokawa, T., McClay, D. R., Wikramanayake, A., Kuraishi, R., Kiyomoto, M., Nishida, H., and Henry, J. (2004) Blastomere isolation and transplantation. *Methods Cell Biol.* **74,** 243–271.
33. Arnone, M. I., Dmochowski, I. J., and Gache, C. (2004) Using reporter genes to study cis-regulatory elements. *Methods Cell Biol.* **74,** 621–652.
34. Arnone, M. I., Bogarad, L. D., Collazo, A., Kirchhamer, C. V., Cameron, R. A., Rast, J. P., Gregorians, A., and Davidson, E. H. (1997) Green fluorescent protein in the sea urchin: new experimental approaches to transcriptional regulatory analysis in embryos and larvae. *Development* **124,** 4649–4659.
35. Arnone, M. I., Martin, E. L., and Davidson, E. H. (1998) Cis-regulation downstream of cell type specification: a single compact element controls the complex expression of the *CyIIa* gene in sea urchin embryos. *Development* **125,** 1381–1395.
36. Harkey, M. A., Klueg, K., Sheppard, P., and Raff, R. A. (1995) Structure, expression, and extracellular targeting of PM27, a skeletal protein associated specifically with growth of the sea urchin larval spicule. *Dev. Biol.* **168,** 549–566.
37. Makabe, K. W., Kirchhamer, C. V., Britten, R. J., and Davidson, E. H. (1995) Cis-regulatory control of the SM50 gene, an early marker of skeletogenic lineage specification in the sea urchin embryo. *Development* **121,** 1957–1970.
38. Klueg, K. M., Harkey, M. A., and Raff, R. A. (1997) Mechanisms of evolutionary changes in timing, spatial expression, and mRNA processing in the msp130 gene in a direct-developing sea urchin, *Heliocidaris erythrogramma. Dev. Biol.* **182,** 121–133.
39. Martin, E. L., Consales, C., Davidson, E. H., and Arnone, M. I. (2001) Evidence for a mesodermal embryonic regulator of the sea urchin *CyIIa* gene. *Dev. Biol.* **236,** 46–63.
40. Yamasu, K. and Wilt, F. H. (1999) Functional organization of DNA elements regulating SM30alpha, a spicule matrix gene of sea urchin embryos. *Dev. Growth Differ.* **41,** 81–91.
41. Kiehart, D. (1982) Microinjection of echinoderm eggs: apparatus and procedures. *Methods Cell Biol.* **25,** 13–31.

19

Sperm–Egg Fusion Assay in Mammals

Naokazu Inoue and Masaru Okabe

Summary

As representatives of the 60 trillion cells that make a human body, a sperm and an egg meet, recognize each other, and fuse to create a new generation. Thus, gamete fusion is an extremely important process that must transpire without error to launch life activity. This may drive the fusion mechanism to evolve into an unfailing and steady process. At the same time, fusion must be restricted to occur only between two gametes of the same species. However, the molecular bases of the fusion event in fertilization have not yet been clarified. In this chapter, we describe the methods to evaluate fusion by staining the swollen sperm nuclei after fertilization.

Key Words: Sperm; egg; fusion; fertilization; Izumo.

1. Introduction

Fertilization is well known as one of the most typical cell–cell fusion processes in vivo. Among the many ejaculated sperm journeying into the uterus, and subsequently to the ampulla of the oviduct, only a few bump into eggs. At the final stage of the long journey, a selected sperm penetrates into the zona pellucida (ZP) that surrounds the eggs and fuses with the egg plasma membrane. Until recently, the molecular basis for fertilization was poorly understood, especially the final sperm–egg fusion process.

To identify factors involved in sperm–egg fusion, we utilized an antimouse sperm monoclonal antibody OBF13 that specifically inhibits the fusion process, and we succeeded in finding a novel fusion factor that we named *Izumo* after the Japanese shrine dedicated to marriage *(1)*. On the egg plasma membrane, the fusion-related protein CD9 was discovered rather serendipitously by three groups *(2–4)*. In those studies, the assessment of fusion was performed by either the Hoechst prestain or poststain method described below. At present, Izumo and CD9 are the only factors that have been proved to be indispensable

From: *Methods in Molecular Biology, vol. 475: Cell Fusion: Overviews and Methods*
Edited by: E. H. Chen © Humana Press, Totowa, NJ

through gene-manipulated animals. Elucidation of the real central factor of the fusion machinery is yet to come.

2. Materials

2.1. Preparation of Oocytes

1. Female mouse 8 weeks old (or older) of an appropriate strain (10 weeks old or older female if hamster).
2. Hypodermic needle, 30 gauge, 1/2 inch syringe, 1 mL sterile disposable.
3. Pregnant mare's serum gonadotropin (PMSG; Sigma, St. Louis, MO, catalog number G-4877), human chorionic gonadotropin (hCG; Sigma, catalog number C-1063) for superovulation.
4. Hyaluronidase type IV-S (Sigma, catalog number H-4272).
5. 30- or 60-mm no surface coated plastic dish for bacteria (Iwaki, Holliston, MA, numbers 1000-035, 1010-060).
6. Watchmaker's forceps (Fontax 5C).
7. Egg-handling pipette (finely drawn capillary tube (Funakoshi 1-40-7500) with mouth pipette (Sigma, catalog number A5177).
8. Stereoscopic microscope (Olympus SZX12).
9. CO_2 incubator (37 °C, 5%CO_2, Asshi 4020).
10. Modified kSOM medium (for culture of mouse eggs; **Table 1**).
11. FHM medium (for collection of mouse eggs; *see* **Table 1**).
12. BWW medium (for culture of hamster eggs; *see* **Table 1**), modified BWW medium (containing 3 mg/mL of human serum albumin [HAS; Sigma catalog number A-1653]).
13. Mineral oil (Sigma, catalog number M-8410).

2.2. Removal of the Zona Pellucida

1. Piezo-manipulator PMAS-CT150 (Prime Tech LTD, Japan).
2. Stereoscopic microscope.
3. Egg-handling pipette.
4. Acidic Tyrode's solution (Sigma, catalog number T-1788).
5. Borosilicate glass tube (Sutter Instrument Co., Novato, CA, B100-75-10).
6. Sutter P97 puller to make a 10-μm diameter capillary.
7. 12.5% PVP in Hepes-CZB (*see* **Table 1**).
8. Mineral oil.

2.3. Collection of Sperm

1. 10-week-old (or older) appropriate strain male mouse.
2. Human sperm from healthy male donors.
3. TYH medium (for capacitation of mouse sperm; *see* **Table 1**).
4. Modified BWW medium (for capacitation of human sperm; *see* **Table 1**).
5. Watchmaker's forceps (Fontax 5C).

Table 1
Composition of Media

Reagent	Source	Modified kSOM	FHM	TYH	BWW	Hepes-CZB
NaCl	Sigma S-5886	2.775 g	2.775 g	3.488 g	2.455 g	2.38 g
KCl	Sigma P-5405	0.093 g	0.093 g	0.178 g	0.178 g	0.18 g
KH_2PO_4	Sigma P-5655	0.0238 g	0.0238 g	0.081 g	0.081 g	0.08 g
$MgSO_4$	Sigma M-2643	0.012 g	0.012 g	0.072 g	0.072 g	0.071 g
Sodium lactate (60% [w/v])	Sigma L-7900	1.18 mL	1.18 mL	—	1.54 mL	1.85 mL
Sodium pyruvate	Sigma P-4562	0.011 g	0.011 g	0.028 g	0.014 g	0.0176 g
Glucose	Sigma G-6152	0.018 g	0.018 g	0.5 g	0.5 g	0.5 g
Glutamine (200 m*M* stock)	Sigma G-7513	2.5 mL	2.5 mL	—	—	2.5 mL
Bovine serum albumin	Sigma A-3311	0.5 g	0.5 g	2.0 g	—	1.5 g
EDTA 0.2 Na	Sigma E-6635	—	—	—	—	0.0205 g
EDTA 0.4 Na (10 m*M*	Dai-ichi 1.3 mL × 10	0.5 mL	0.5 mL	—	—	—
$NaHCO_3$	Sigma S-5761	1.05 g	0.168 g	1.053 g	1.5 g	0.21 g
Hepes	Sigma H-6147	—	2.38 g	—	—	2.6 g
$CaCl_2/2H_2O$	Sigma C-7902	0.126 g	0.126 g	0.126 g	0.126 g	0.125 g
Essential amino acid	GibcoBRL 11130-051	10 mL	10 mL	—	—	—
Nonessential amino acid	GibcoBRL 11140-050	5 mL	5 mL	—	—	—
Penicillin-streptomycin	GibcoBRL 15140-148	2.5 mL	2.5 mL	2.5 mL	2.5 mL	2.5 mL
Phenol Red 0.5%	Dai-ichi 1.3 mL × 10	1.0 mL	1.0 mL	0.5 mL	0.5 mL	1.0 mL
1.0 N NaOH	Nacalai tesque 315-11	—	Adjust to pH 7.4	—	—	Adjust to pH 7.4
Water	GibcoBRL 15230-162	Adjust to 500 mL	Adjust to 500 mL	Adjust to 500 mL	Adjust to 500 mL	Adjust to 500 mL

These medium were filtrated with 0.2 μm filter, divided into 10 mL, and stored at –20°C until used.

6. Straight-blade Vannas scissors (Natume MB54-1).
7. Stereoscopic microscope.
8. CO_2 incubator.
9. 30- or 60-mm nontreated plastic dish.
10. Mineral oil.

2.4. Staining of Fused Sperm

1. Hoechst 33342 (Invitrogen H-3570).
2. CO_2 incubator.
3. 0.25% glutaraldehyde for fixation.
4. Stereoscopic microscope.
5. Ix-70 fluorescent microscope (Olympus).
6. Egg-handling pipette.
7. Mineral oil.
8. Mixture of solid paraffin and Vaseline (mixing ratio is 1:9).

3. Methods

3.1. Preparation of Oocytes

This procedure is described in more detail by Nagy et al. *(5)*.

1. Intraperitoneal injections of 5 IU PMSG and 5 IU hCG at a 48-h interval to 8-week-old (or older) female mice (injection 30 IU PMSG and 30 IU hCG at 72-h interval to 10-week-old or older female hamster).
2. Sacrifice the mice at 13–15h after hCG injection (sacrifice the hamster at 17h after hCG).
3. Dissect the oviducts.
4. Transfer the oviduct to a mineral oil–covered 30- or 60- mm plastic dish.
5. Newly ovulated oocytes, surrounded by cumulus cells, are found in the ampulla of oviduct.
6. Place one oviduct beside a mineral oil–covered 100-μL drop of FHM medium prepared on 60-mm plastic dish.
7. Use watchmaker's forceps to grasp the oviduct and supplementary forceps to tear the oviduct close to where the oocytes are located.
8. Release the clutch of cumulus cells into a 100μL drop of FHM medium.
9. Allow eggs to incubate in 35 IU/mL hyaluronidase solution until the cumulus cells are completely removed (it may take approximately 5min).
10. Wash eggs by pipetting in and out and subsequently transferring eggs into fresh drops of FHM medium (repeat at least three times).
11. Place up to 50 eggs in a 50-μL drop of modified kSOM at 37°C under 5% CO_2 in air until removal of the zona pellucida is performed.

3.2. Removal of the Zona Pellucida

Two different methods for the preparation of zona pellucida–free eggs are described. Acidic Tyrode's method is a quick and easy method to remove zona pellucida and is often used by researchers. However, a drawback of this method is that a part of the dissolved zona pellucida was found to be readsorbed on the egg plasma membrane *(6)*. In natural conditions, the acrosome intact sperm do not bind to egg plasma membrane, but with acidic Tyrode's method we cannot eliminate a massive adhesion between acrosome intact sperm and eggs with readsorbed zona pellucida on their plasma membrane. Because this "false binding" can be eliminated by mechanical zona pellucida removal methods as shown in **Figures 1** and **2** *(6)*, we always use this method to perform sperm–egg fusion assay. The other available mechanical removal method is reported by another group *(7)*.

3.2.1. Acidic Tyrode's Method

1. Prepare eggs and remove cumulus cells as described in **Subheading 3.1.**
2. Transfer cumulus-free eggs into a 50-μL drop of mineral oil–covered acidic Tyrode's solution prepared on a plastic dish.
3. To remove remnant medium, transfer eggs into a second 50 μL of acidic Tyrode's solution.
4. Repeatedly pipette the eggs in and out until the zona pellucida are dissolved under the stereoscopic microscope (it may be finished within 30 s).
5. Wash zona pellucida–free eggs at least three times by transferring eggs into fresh new drops of TYH medium to remove remnant acidic Tyrode's solution. (For human sperm, modified BWW medium is required instead of TYH medium.)

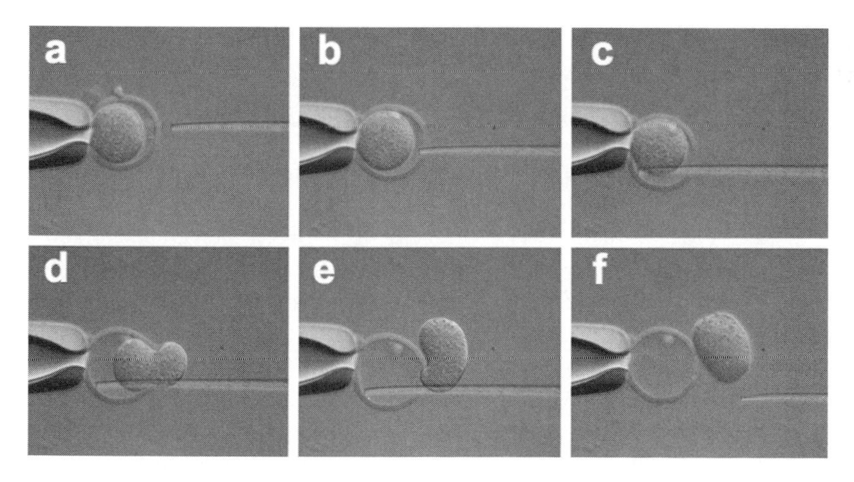

Fig. 1. Preparation of zona pellucida–free eggs using a piezo-micromanipulator. Eggs were freed from the cumulus cells and placed in a drop of FHM. A pipette, attached to a piezo-driven micromanipulator, was used to form a slit in the zona pellucida **(A,B)**. The egg then was flushed out through the slit by rapidly introducing medium into the perivitelline space from behind the eggs **(C–F)**.

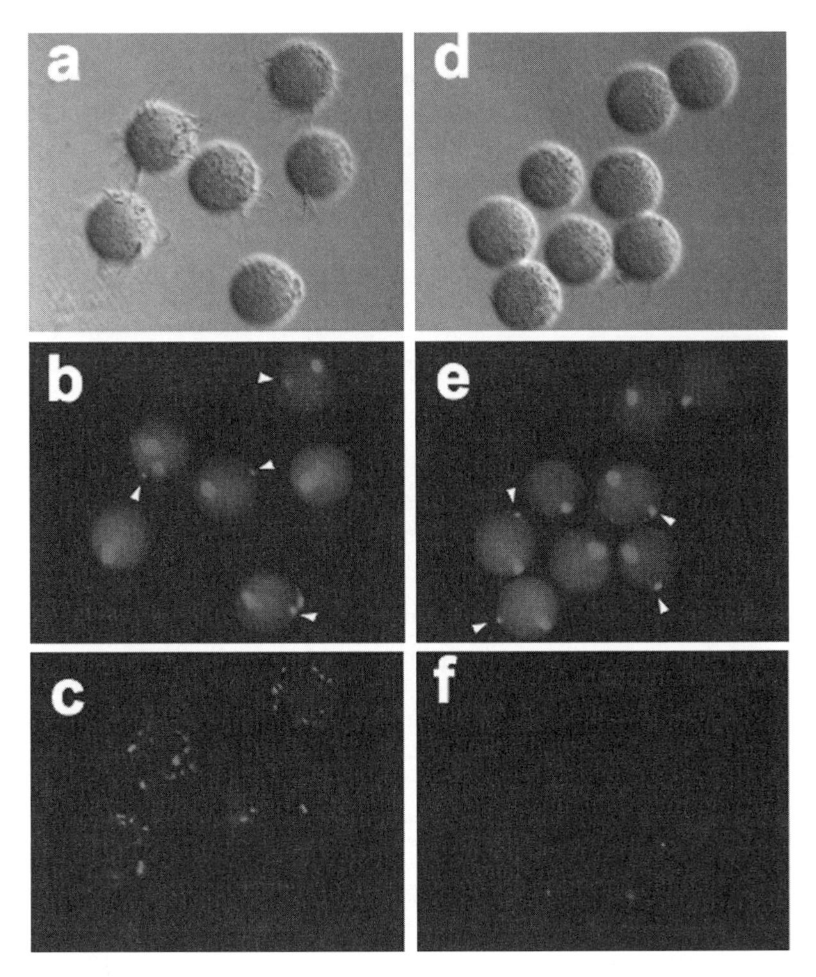

Fig. 2. An example of the Hoechst "preloading" method: comparison of sperm binding and fusing abilities to denuded egg plasma membrane (PM) by acidic Tyrode's (acid PM) or manipulator (piezo PM). "Green sperm" from *Acr-EGFP* transgenic mice *(8)* were capacitated for 2 h and then mixed with eggs preloaded with Hoechst 33342 that were prepared with acidic Tyrode's solution (in A–C) and a piezo-manipulator (in D–F). After 30 min of incubation, eggs were fixed and visualized by fluorescence microscopy to assay sperm PM binding and fusion. **(A,D)** Bound sperm (Hoffman modulation contrast optics). Considerably more sperm were present on the acid PM than on the piezo PM. **(B,E)** Fused sperm are stained with Hoechst (arrowheads) because of the dye transfer from the egg. Similar numbers of fused sperm were seen with both types of egg preparation. The larger egg nucleus was also stained. **(C,F)** Acrosome intact sperm had "green fluorescent" acrosomes. The majority of sperm bound to eggs prepared with the acidic Tyrode's solution were acrosome intact, while few sperm prepared by the piezo-micromanipulator that bound to eggs were acrosome intact.

6. Incubate in TYH medium for more than 1 h at 37 °C under 5% CO_2 in air to allow surface proteins to recover.

3.2.2. Mechanical Method Using a Piezo-Manipulator

1. Prepare eggs and remove cumulus cells as described in **Subheading 3.1** (**see Fig. 1; ref.** ***6***).
2. Prepare the drilling pipette from a borosilicate glass tube by pulling using a Sutter P97 puller to a diameter of 10 µm according to an appropriate textbook ***(5)***.
3. Add a few microliters of mercury to the tip of the pipette to enhance the drilling efficiency.
4. Prepare several 6-µL drops of mineral oil–covered 12.5% PVP in Hepes-CZB and FHM media prepared on the top of a 60-mm plastic dish.
5. Equip the pipette to the piezo-driven micromanipulator and first soak the pipette wall with 12.5% PVP in Hepes-CZB.
6. Make an approximately 30-µm slit in the zona pellucida by applying a piezo-pulse (*see* **Fig. 1**).
7. Flush out the oocyte from the slit by rapidly introducing medium into the perivitelline space from behind the eggs (*see* **Fig. 1**).
8. Place up to 50 zona pellucida–free eggs into a 50 µL drop of TYH medium and incubate at 37 °C under 5% CO_2 in air until use.

3.3. Collection of Sperm

3.3.1. Human Sperm

1. Collect sperm from healthy male donors by masturbation.
2. Liquefy for 30–60 min at 37 °C under 5% CO_2 in air and divide into 0.5-mL aliquots.
3. Place 0.5-mL aliquots to the bottom of 2 mL of modified BWW medium.
4. Incline the tubes to an angle of 30 ° and incubate at 37 °C under 5% CO_2 in air for 1 hr.
5. Take out approximately 1.0 mL of the upper part of the medium containing motile sperm into another 1.5-mL tube.
6. Centrifuge for 5 min at 500 *g* at room temperature.
7. Discard supernatant and resuspend the sperm in 1 mL of modified BWW medium.
8. Repeat **steps 6** and **7** twice more. Eventually resuspend in a mineral oil–covered 400-µL drop of modified BWW medium prepared on 60-mm plastic dish at 37 °C under 5% CO_2 in air until use.

3.3.2. Mouse Sperm

1. Dissect the cauda epididymis from 12-week-old (or older) mice according to textbook ***(5)***.
2. Place the epididymis beside a 100-µL drop of TYH medium covered by mineral oil in a 30- or 60-mm plastic dish.

3. Use watchmaker's forceps to grasp the cauda epididymis and make a cut at the proximal cauda epididymis (mature sperm are stored) with straight-blade Vannas scissors.
4. After squeezing the sperm out from the cut area, hold the swarm of sperm by sticking them to the tip of supplementary forceps.
5. Introduce the swarm of sperm into a 200-μL drop of TYH medium.
6. At 1 h after incubation, check the sperm motility by observing well-dispersed sperm.
7. Cultivate sperm for an additional 1 h in TYH medium at 37 °C under 5% CO_2 in air to induce capacitation and spontaneous acrosome reaction before insemination.

3.4. Staining of Fused Sperm

Fusion assessment can be performed in two different ways.

3.4.1. Hoechst "Preloading" Method

1. Prepare the zona pellucida–free eggs as described in **Subheading 3.2**.
2. Introduce eggs into a 50-μL drop of Hoechst 33342 (1 μg/mL) in TYH medium (up to 50 eggs per spot), and allow to stand for 10 min at 37 °C under 5% CO_2 in air.
3. Transfer the eggs into another fresh 50-μL drop of TYH medium covered with mineral oil.
4. Incubate dye-loaded eggs for 15 min at 37 °C under 5% CO_2 in air to discharge excess dye.
5. Repeat **steps 3** and **4** three more times, and subject to fusion assay.
6. Cultivate sperm as described in **Subheading 3.3**. to induce capacitation.
7. Transfer the eggs into a 50-μL drop of FHM medium containing 0.25% glutaraldehyde for fixation after 30 min of incubation with 2×10^5 mouse sperm.
8. Allow to stand for 5 min at room temperature.
9. Wash the sperm-bound eggs by transferring eggs into fresh drops of FHM medium several times
10. Observe under a fluorescence microscope (ultraviolet excitation light). With this method, only nuclei of fused sperm are stained by the dye transferred into sperm after membrane fusion (*see* **Fig. 2** and **3**; **Note 1**).

3.4.2. Hoechst "Poststaining" Method: Observation of Swollen Sperm

1. Prepare the zona pellucida–free eggs as described in **Subheading 3.2.**
2. Cultivate sperm as described in **Subheading 3.3.** to induce capacitation.
3. Prepare a 100-μL drop of TYH medium covered by mineral oil prepared on 60-mm plastic dish.
4. Introduce zona pellucida–free eggs into a drop of TYH medium.
5. Incubate zona pellucida–free eggs with 2×10^5 (for mouse) or 1×10^6 (for human) sperm for 6 h in TYH or modified BWW medium at 37 °C under 5% CO_2 in air,

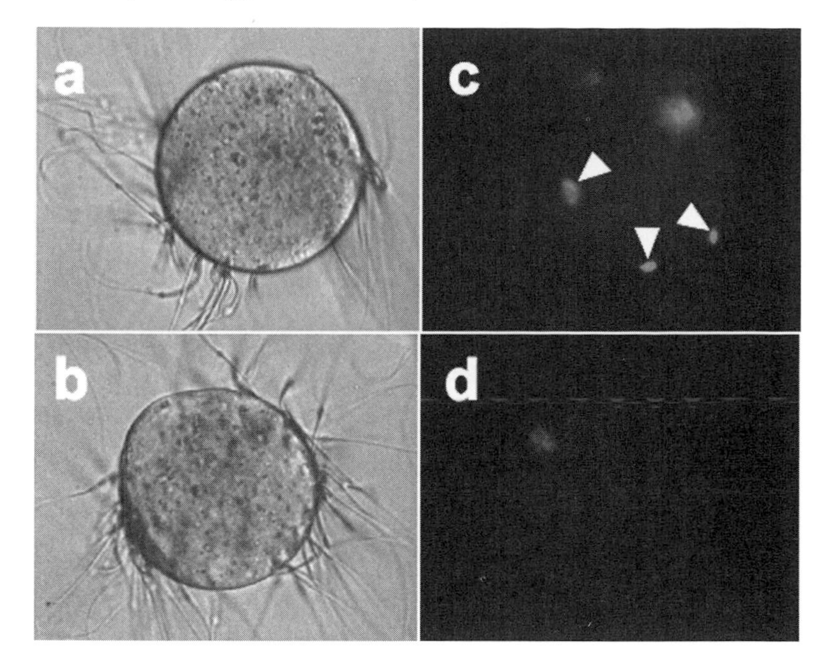

Fig. 3. An example of the Hoechst "preload" method: fusion assay of Izumo knockout sperm. Fused sperm were stained by the egg that was preloaded with Hoechst 33342. The arrowheads show the fused sperm. In comparison with a few *Izumo* +/– sperm, which successfully fused with eggs **(A,C)**, *Izumo* –/– sperm never fused with eggs **(B,D).** This defect was limited to the fusion process, because pups from *Izumo* –/– sperm could be obtained by intracytoplasmic sperm injection. (Reprinted from **ref.** ***1*** with permission of Nature Publishing Group.)

respectively. Fused sperm heads launch to swell during this incubation period; some enlarged sperm heads can be seen under a phase contrast microscope.

6. Wash the sperm bound eggs by pipetting and transferring into fresh new drops to remove weakly bound sperm.
7. Incubate the eggs with 1 µg/mL of Hoechst 33342 for 10 min at 37 °C under 5% CO_2 to stain swollen sperm nucleus.
8. Wash the eggs several times by transferring them into fresh TYH medium.
9. To observe sperm-fused eggs, apply four small dabs of Vaseline mix (vaseline: solid paraffin = 9:1) on a slide glass by injecting out from a syringe without needle.
10. Place a few eggs into a 1-µL drop of FHM medium, and cover the eggs with a cover glass.
11. Gently press the eggs with the cover glass to flatten the eggs under the stereoscopic microscope to make the observation easier (*see* **Note 2**).
12. Observe under a fluorescence microscope (ultraviolet excitation light).

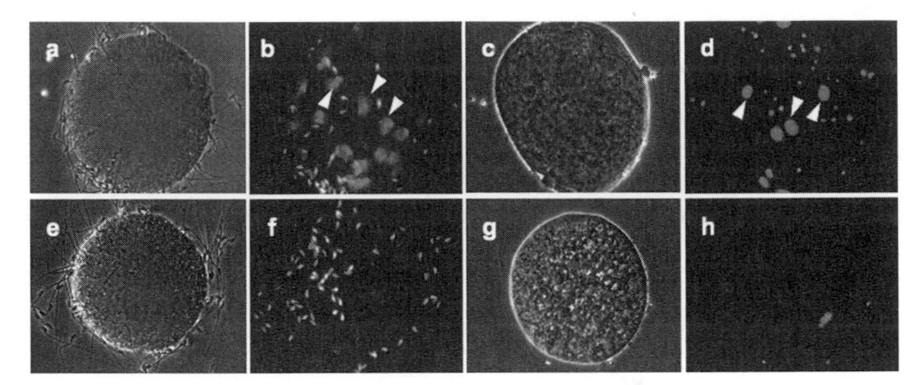

Fig. 4. Exemplification of Hoechst "poststain" method: involvement of Izumo in the xeno-species fusion system. Six hours after the insemination of zona pellucida–free hamster eggs with *Izumo* +/– **(A,B)** and –/– **(E,F)** mouse sperm, sperm heads were stained by adding Hoechst 33342 to the medium **(B,F)**. The sperm–egg binding was strong enough to resist the repeated pipetting. Human sperm were also added with 25 μg/mL of antihuman Izumo **(G,H)** or control IgG **(C,D)** to zona pellucida–free hamster eggs. No fusion was observed in the presence of anti-Izumo antibody. The arrowheads indicate the swelling sperm head after staining with Hoechst 33342 **(B,D)**. (Reprinted from **ref. *1*** with permission of Nature Publishing Group.)

Fused sperm (enlarged heads) will be stained with the dye as in **Figure 4**. (For human sperm, the use of modified BWW medium is required instead of TYH medium; *see* **Note 3.**)

4. Notes

1. The Hoechst "preloading" method is not applicable when hamster eggs are used. They seem to pump out the Hoechst dye from the cytosol. As a result, hamster eggs are not able to accumulate enough dye for fusion assay (*see* **Subheading 3.4.1.**).
2. For the observation of swollen sperm, the cover glass is pressed down to squeeze the eggs such that the sperm become more visible, but this has to be done carefully. The eggs burst easily with excess amount of pressurization (*see* **Subheading 3.4.2.**).
3. The mixing of gametes of xeno species is ethically restricted in many ways. Please follow the ethical laws in the countries where the experiments are pursued (*see* **Subheading 3.4.2.**).

References

1. Inoue, N., Ikawa, M., Isotani, A., and Okabe M. (2005) The immunoglobulin superfamily protein Izumo is required for sperm to fuse with eggs. *Nature* **434,** 234–238.
2. Le Naour, F., Rubinstein, E., Jasmin, C., Prenant, M., and Boucheix, C. (2000) Severely reduced female fertility in CD9-deficient mice. *Science* **287,** 329–321.

3. Miyado, K., Yamada, G., Yamada, S., Hasuwa, H., Nakamura, Y., Ryu, F., Suzuki, K., Kosai, K., Inoue, K., Ogura, A., Okabe, M., and Mekada, E. (2000) Requirement of CD9 on the egg plasma membrane for fertilization. *Science* **287,** 321–324.
4. Kaji, K., Oda, S., Shikano, T., Ohnuki, T., Uematsu, T., Sakagami, J., Tada, N., Miyazaki, S., and Kudo, A. (2000) The gamete fusion process is defective in eggs of Cd9-deficient mice. *Nat. Genet.* **24,** 279–282.
5. Nagy, A., Gertsenstein, M., Vintersten, K., and Behringer, R. (2002) *Manipulating the Mouse Embryo*, 3rd ed. Cold Spring Harbor Laboratory Press, Cold Spring Harbor, NY.
6. Yamagata, K., Nakanishi, T., Ikawa, M., Yamaguchi, R., Moss, S, B., and Okabe, M. (2002) Sperm from the calmegin-deficient mouse have normal abilities for binding and fusion to the egg plasma membrane. *Dev. Biol.* **250,** 348–357.
7. Ziyyat, A., Rubinstein, E., Monier-Gavelle, F., Barraud, V., Kulski, O., Prenant, M., Boucheix, C., Bomsel, M., and Wolf, J. P. (2006) CD9 controls the formation of clusters that contain tetraspanins and the integrin alpha 6 beta 1, which are involved in human and mouse gamete fusion. *J. Cell Sci.* **119,** 416–424.
8. Nakanishi, T., Ikawa, M., Yamada, S., Parvinen, M., Baba, T., Nishimune, Y., and Okabe, M. (1999) Real-time observation of acrosomal dispersal from mouse sperm using GFP as a marker protein. *FEBS Lett.* **449,** 277–283.

20

Quantitative Assays for Cell Fusion

Jessica H. Shinn-Thomas, Victoria L. Scranton, and William A. Mohler

Summary

Cell fusion would seem to be obviously recognizable upon visual inspection, and many studies employ a simple microscopic fusion index to quantify the rate and extent of fusion in cell culture. However, when cells are not in monolayers or when there is a large background of multinucleation through failed cytokinesis, cell–cell fusion can only be proven by mixing of cell contents. Furthermore, determination of the microscopic fusion index must generally be carried out manually, creating opportunities for unintended observer bias and limiting the numbers of cells assayed and therefore the statistical power of the assay. Strategies for making assays dependent on fusion and independent of visual observation are critical to increasing the accuracy and throughput of screens for molecules that control cell fusion. A variety of in vitro biochemical and nonbiochemical techniques have been developed to assay and monitor fusion events in cultured cells. In this chapter, we briefly discuss several in vitro fusion assays, nearly all based on systems of two components that interact to create a novel assayable signal only after cells fuse. We provide details for the use of one example of such a system, intracistronic complementation of β-galactosidase activity by mutants of *Escherichia coli lacZ*, which allows for either cell-by-cell microscopic assay of cell fusion or quantitative and kinetic detection of cell fusions in whole populations *(1)*. In addition, we describe a combination of gene knock-down protocols with this assay to study factors required for myoblast fusion.

Key Words: Cell fusion; *lacZ*; β-galactosidase; complementation; fluorescence histochemistry; protein interactions; small interfering RNA.

1. Introduction

Cell fusion occurs in a diverse range of cell types and organisms; however, the basic process is similar regardless of cell type. Plasma membranes merge by the joining of lipid bilayers, and soluble contents mix between the two cytoplasmic compartments *(2–4)*. Quantitative fusion assays are based on signals resulting from either lipid mixing or cytoplasm mixing, often involving detection of a signal that depends on direct interaction between two essential components. Many studies have used a system involving two-color mixing of spectrally distinct

From: *Methods in Molecular Biology, vol. 475: Cell Fusion: Overviews and Methods*
Edited by: E. H. Chen © Humana Press, Totowa, NJ

fluorescent proteins, dyes, or both to convincingly demonstrate fusion events in culture *(5–7)*. Such two-color cells can only arise from fusion events, and they exclude the background of multinucleated cells from failed cytokinesis. However, like the fusion index based on counting of nuclei, such dye-mixing assays still require a concerted effort by the microscopist to resolve and count fused cells, and the mixed-color signal of a fused cell cannot be easily distinguished from aggregates of cells by flow cytometry or fluorimetry. In this chapter, we focus on assays in which cell–cell fusion generates a novel signal that is fusion dependent and that can be quantified in mass cultures without a requirement for cell counting. Examples include photochemical, biochemical, and genetic interactions.

1.1. Assays Using Membrane and Cytoplasmic Dyes

1.1.1. Fluorescence Dequenching Assays

Several lipid and aqueous dyes allow monitoring of either of the two major events during fusion, membrane lipid mixing and cytoplasm mixing. Fluorescent lipid probes such as octadecyl rhodamine B (R18; Invitrogen, Carlsbad, CA) and PKH26 (Sigma-Aldrich, St. Louis, MO) have been used to detect membrane mixing by observing the transfer of these dyes, often from loaded red blood cells to the lipid membranes of experimental cell lines. More important for the question of population-based assays, these reagents have also been used to determine the rapid kinetics of fusion by spectrofluorometric monitoring of fluorescence dequenching, as both membrane-soluble and aqueous-soluble dyes are diluted by fusion of heavily dye-loaded cells with unlabeled partners (**Fig. 1; refs.** ***8–13***).

Note that this system is based on only a single component, the dye, and that it is not entirely specific to cell fusion events, as dequenching can also occur during lysis of labeled cells. Moreover, it is typically employed only when one cell type (e.g., the heavily dye-loaded red blood cell) is an “inert” target for active fusion by the unlabeled cells, as in assays of virus-induced cell fusion. In such cases, the physiological effects of the heavy labeling of the target cell appear not to greatly perturb the cell fusion reaction.

1.1.2. Förster Resonance Energy Transfer-Based Mixing Assays

Förster resonance energy transfer (FRET) arises upon close interaction between two different fluorescent probes with overlapping excitation and emission spectra. FRET has been exploited to detect cell–cell fusions by incorporating a fluorescent donor probe in the plasma membrane of one population of cells and a fluorescent acceptor probe in the membrane of a separate population. The distance between even closely apposed cell surfaces is too great to permit efficient energy transfer. Therefore, FRET signal does not appear because of simple aggregation of cells. After fusion between cells of the separately labeled populations, however, the two fluorochromes commingle within the same membrane, allowing detectable FRET

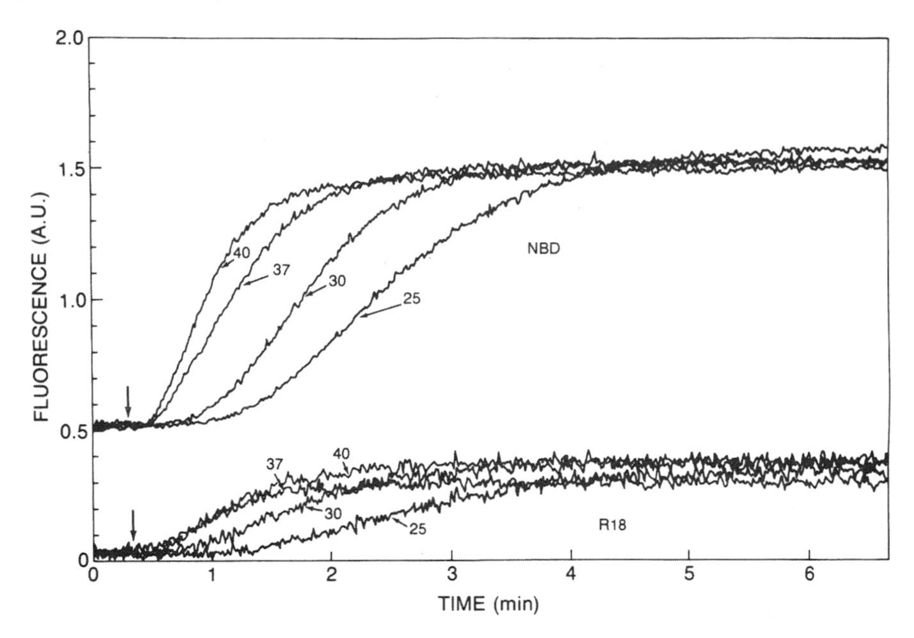

Fig. 1. Temperature dependence of hemagglutinin (HA)–induced cell fusion measured by dequenching of two different probes. Red blood cells (RBCs) were double labeled with R18 (lipid probe) and NBD-taurine (cytoplasmic probe). Complexes were formed with loaded RBCs and GP4F cells (NIH3T3 cells expressing high levels of HA) and injected into a cuvette containing PBS, pH 7.4, prewarmed to different temperatures (arrows) followed by lowering media pH to 5.0 to induce fusion (bold arrows). Spectrofluorometric measurements were taken for NBD (upper curves) and R18 (lower curves). (Reproduced from **ref. *11*,** with permission of The Rockefeller University Press.)

to occur (**Fig. 2**). The increase in acceptor probe emission upon excitation of the donor probe is proportional to the number of cell fusion events ***(14)***. Lipophilic dyes (DiI and DiO) have recently been used to yield fusion-dependent FRET quantified by microscopy or fluorescence-activated cell sorting (FACS) analysis ***(15)***. It is conceivable that pairs of tightly binding cytoplasmic dyes or fluorescent proteins could also be used for FRET-based fusion assays, but we are not aware of any examples where this has been put into practice.

1.2. Assays Via Fusion-Induced Gene Activation

Reporter genes such as *lacZ*, luciferase, chloramphenicol acetyltransferase (CAT), or green fluorescent protein (GFP) are employed in a variety of molecular and biochemical assays. Widely used cell fusion assays make use of a two-part system in which such a reporter gene is driven by a prokaryotic promoter. In these assays, one population of cells expresses T7 RNA polymerase and a separate population of cells harbors a reporter gene under the control of the T7

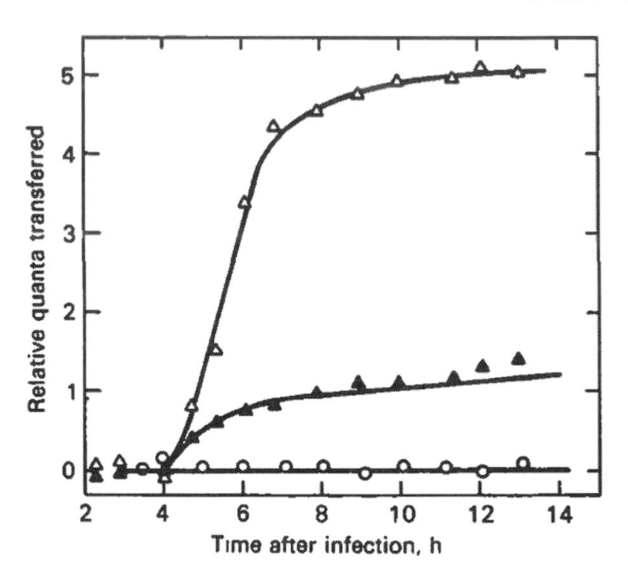

Fig. 2. Förster resonance energy transfer (FRET) analysis of cell fusion of virally infected cells using fluorescent probes. Cell populations were labeled with either F18 or R18, infected with HSV-I wild-type virus or an HSV-I mutant (*syn20*) that promotes extensive fusion, harvested, mixed in equal proportions, and seeded in culture dishes. R18 fluorescence was measured in mixed cells for mock infected (○),wild-type infected (▲), and *syn20* infected (Δ) cells. The emission spectrum of F18 overlaps the absorption spectrum of R18, similar to the overlap of fluorescein and rhodamine, respectively. The increase in fluorescence for *syn20* infected cells began at a time when fusion was observed to occur in parallel experiments using a Coulter counter assay to monitor fusion. (Reproduced from **ref. *14*,** with permission of the Company of Biologists.)

promoter, which is silent in the absence of T7 polymerase. Cell fusion between cells of these two distinct populations results in polymerase-dependent activation of reporter gene expression. These assays can be manipulated to detect various molecular and cellular fusion requirements with a quantitative readout of fusion by reporter activity measurements ***(16–25)***. More recently a high-throughput cell fusion assay has been developed using a viral transcriptional transactivator (tat) and a long terminal repeat (LTR) promoter. The adaptability of this system could be harnessed to study a variety of cell fusion processes ***(26)***.

Reporter gene activation via recombinase-induced gene rearrangement has also been shown to yield a robust fusion-dependent signal ***(27–29)***. A reporter construct held silent by termination sequences or a shift of reading frame in one cell population is activated by a recombinase enzyme (e.g., Cre or FLP) expressed in a distinct population. Upon fusion, site-specific recombination to produce a constitutively active reporter gene should occur only in syncytia formed by fusion of the two cell types (**Fig. 3**). Because recombination is a single, binary event

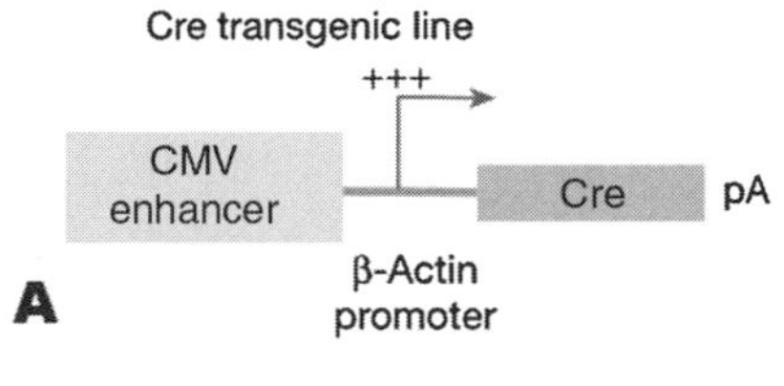

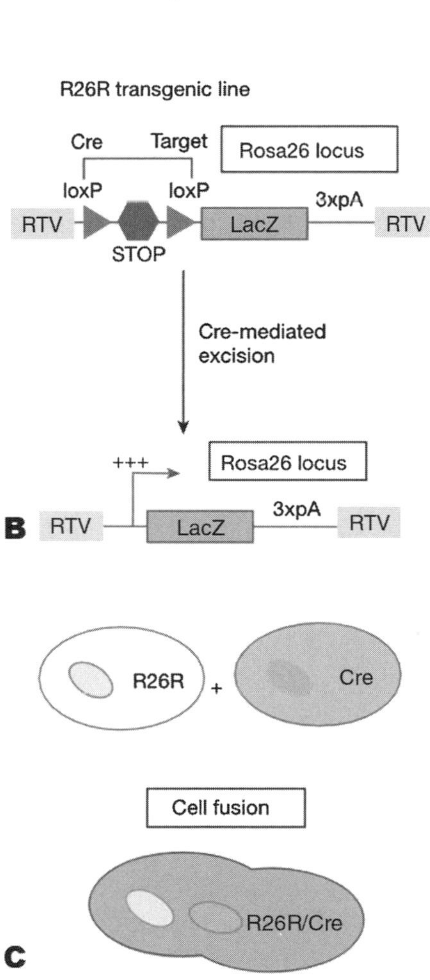

Fig. 3. Schematic of the Cre/lox method used to detect cell–cell fusion. **(A,B)** Representation of transgenes expressed in cell lines. **(C)** When a cell expressing Cre recombinase (A) fuses with a cell bearing the *lacZ* reporter (or other reporter gene; B) the floxed stop cassette is excised and the reporter is expressed in the fused cell. Expression of the reporter is analyzed to quantify the extent of cell fusion in a culture. (Adapted from **ref.** ***27***, with permission of Macmillan Publishers Ltd.)

with perduring and irreversible effects on reporter activity, it is possible (but never reported) that such assays could betray a background signal due to shedding or transcytosis of recombinase from cell to cell rather than true cell fusion.

1.3. Assays Via Fusion-Induced Biochemical Complementation

Another class of assays employs pairs of reporter genes that encode complementary parts of a signal-generating complex, with each component constitutively expressed in separate cell populations. Assayable activity, either enzymatic or autofluorescent, is generated only upon mixing of the two components by fusion within a coculture of complementary cells. Because the formation of active complexes is concentration dependent, significant signal is created within fused cells but not within lysates of cocultures that have not undergone fusion *(1)*. The potential for very rapid association of complementary components suggests that these assays may produce measurable signal more rapidly after cell fusion than systems that involve activating new transcription of reporter genes. Thus, biochemical complementation schemes probably lie intermediate in response time between strictly photonic (dequenching or FRET) and expression-activating (polymerase or recombinase) assays of cell fusion.

1.3.1. lacZ *Complementation Fusion Assays*

Escherichia coli β-galactosidase is normally a homotetrameric complex of monomers encoded by the *lacZ* gene. Classic genetic studies showed that complementary pairs of fragments of the monomer (each individually inactive) can produce active enzyme when coexpressed as long as each fragment contains domains lacking from the other (**Fig. 4**). This phenomenon, originally described as intracistronic complementation, involves formation of an active hetero-octomeric complex *(30)*. A pair of deletion mutants of *lacZ* (Δμ and Δω) were found to produce particularly strong fusion-dependent enzyme activity in mouse myoblasts *(1)*. Complementation by *lacZ* fragments has also been adapted to quantitative studies of cell fusion during yeast mating (E. Grote, personal communication). β-galactosidase activity can be detected using a variety of substrates that yield chromogenic, fluorescent, or chemiluminescent cleavage products; these products are assayable by histochemistry or flow cytometry or in whole-culture lysates (discussed in **Heading 3; refs. *1,31–34***). Applications of such assays to the question of myoblast fusion have involved time-course and dose–response studies of physicochemical fusion inhibitors, temporal correlation of adhesion molecule expression with fusion competence, and quantification of fusion rates in normal and knockout mutant myoblasts ***(35–37)***. **Headings 2** and **3** of this chapter include detailed descriptions of β-galactosidase complementation protocols in a mouse myoblast cell line, including chromogenic and fluorescence histochemistry, as well as chemiluminescence assays in combination with small interfering (si)RNA knockdown of genes.

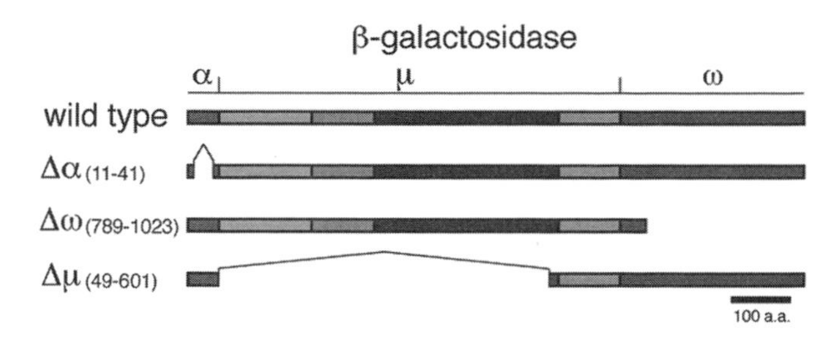

Fig. 4. Schematic representation of wild-type lacZ protein and deleted polypeptides Δα, Δω, and Δμ. Shaded regions represent distinct domains of the β-galactosidase monomer as portrayed by Jacobson et al. *(42)*. Subscripts indicate the ranges of amino acids removed by the deletions. (Adapted from **ref. *1*,** with permission of the National Academy of Sciences, U.S.A.)

1.3.2. Other Possible Fusion Assays Based on Biochemical Complementation

Recently *lacZ* complementation and a number of other two-component protein fragment complementation schemes (e.g., bipartite versions of DHFR, β-lactamase, luciferase, and derivatives of GFP) have been used to assay for protein–protein interactions in live cells ***(32,38,39)***. In principle, any of these reporter pairs should be, like *lacZ*, a viable candidate for creation of a cell-fusion assay system. A major concern in designing such a system must be to optimize the affinity of the interacting fragments to enhance the rate of association and generation of fusion-induced signal. As many of these pairs have been inversely optimized for minimal intrinsic affinity, to reduce the background in protein–protein interaction assays, this may require redesigning existing complementation pairs, possibly though the addition of "third-party" high-affinity interacting domains. One such assay has recently been created through adaptation of bimolecular fluorescence complementation (BiFC) using affinity-enhanced fragments of a fluorescent protein (H. Lin and T. Kerppola, personal communication). The availability of multiple spectrally distinct BiFC pairs could allow for design of complex multiple-readout fusion assays, including measurement of competition for fusion partners between cells of distinct genotypes or states of differentiation (**Fig. 5 refs. *39–41***).

2. Materials

2.1. Cell Culture of C2F3 Myoblasts

1. Growth medium (GM): Dulbecco's Modified Eagle's Medium, 5% fetal bovine serum, 15% calf serum, 1% penicillin/streptomycin.

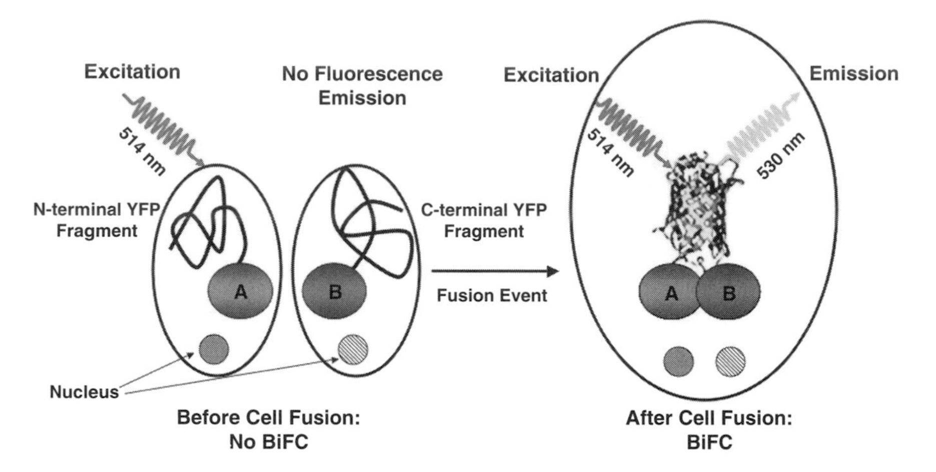

Fig. 5. Schematic representation of cell–cell fusion assay using bimolecular fluorescence complementation (BiFC) by halves of a split YFP fluorophore. N- and C- terminal fragments of YFP are linked to high-affinity interacting domains A and B, respectively, and expressed in separate cell populations. In the absence of cell fusion, the fluorophore halves remain nonfunctional. Following fusion of complementary cells, a functional fluorophore is reconstituted that emits fluorescence upon excitation with an appropriate wavelength. (Adapted from **ref. *43***; http://creativecommons.org/licenses/by/3.0/.)

2. Differentiation medium (DM): Dulbecco's Modified Eagle's Medium, 5% horse serum, 1% penicillin/streptomycin.

2.2. Indigogenic Histochemistry

1. Wash buffer: PBS, pH 7.4 (1.06 M KH_2PO_4, 155.17 *M* NaCl, 2.97 *M* Na_2HPO_4).
2. Fixative: 4% paraformaldehyde in PBS, pH 7.4.
3. Stain: stock solution of 40 mg/mL conventional X-gal (5-bromo-4-chloro-3-indolyl β-D-galactopyranoside; Sigma, St. Louis, MO in dimethylformamide (stored at −20 °C and protected from light). Stock solution is diluted to 1 mg/mL in staining buffer.
4. Staining buffer: 1 m*M* $MgCl_2$, 5 m*M* $K_3Fe(CN)_6$, 5 m*M* $K_4Fe(CN)_6$, 2 m*M* $MgCl_2$ in PBS, pH7.4 (protect from light).

2.3. Fluorescence Histochemistry

1. Type I collagen (Sigma): 0.1% in 0.1 *M* acetic acid.
2. Wash buffer: PBS, pH 7.4 (1.06 *M* KH_2PO_4, 155.17 *M* NaCl, 2.97 *M* Na_2HPO_4).
3. Fixative: 4% paraformaldehyde in PBS, pH 7.4.
4. Detection: stock solution of 50 mg/mL Fast Red Violet-LB Salt (Sigma); stock solution of 50 mg/mL "5-6-X-gal" (5-bromo-6-chloro-3-indolyl *β*-D-galactopyranoside in dimethylformamide; Fluka, St. Louis, MO).

2.4. Chemiluminescence Assays for β-Galactosidase

1. Wash buffer: PBS, pH 7.4 (1.06 *M* KH_2PO_4, 155.17 *M* NaCl, 2.97 *M* Na_2HPO_4).
2. Galacto-Light Plus System (Applied Biosystems, Foster City, CA), which contains lysis solution: 100 m*M* KH_2PO_4 (pH7.8), 0.2% Triton X-100; chemiluminescent substrate: Galacton-Plus 100× concentration; reaction buffer diluent: 100 m*M* sodium phosphate (pH 8.0), 1 m*M* $MgCl_2$; light emission accelerator containing Sapphire-II enhancer.
3. Luminometer (MGM Instruments, Hamden, CT; Optocomp II used for data shown here).

2.5. β-Galactosidase Assay in Conjunction With Small Interfering RNA of Target Genes

1. Growth medium without antibiotics: Dulbecco's Modified Eagle's Medium, 5% fetal bovine serum, 15% calf serum.
2. Dulbecco's Modified Eagle's Medium.
3. Differentiation medium (DM): Dulbecco's Modified Eagle's Medium, 5% horse serum, 1% penicillin-streptomycin.
4. Opti-MEM (Invitrogen), a Modified Eagle's Minimum Essential Medium, buffered with Hepes and sodium bicarbonate (2.4 g/L) and supplemented with hypoxanthine, thymidine, sodium pyruvate, L-glutamine, trace elements, and growth factors.
5. siRNA duplexes (Dharmacon, Chicago, IL) for genes of interest.
6. Lipofectamine (Invitrogen).

3. Methods

3.1. Cell Culture

1. For cell line maintenance and passage, individual cell lines C2F3-462 (Δμ) and C2F3-787 (Δω) (*see* **Fig. 4**) are seeded into GM as appropriate to keep confluency below 50% to prevent differentiation induced by contact (*see* **Note 1**). Cells are typically split 1:10 every other day. Cells can be split as low as 1:40 if necessary.
2. Differentiation is induced by switching to DM when cells have reached 80%–100% confluency. Differentiation Medium is changed daily.
3. For lacZ complementation assays, a total of 400,000 cells/well are co-plated in a 24-well dish and grown for 24 h in 500 μL of GM per well. A 3:1 ratio of cell number (C2F3-462:C3F3-787) is typically used to compensate for a slower cell-doubling rate in C2F3-462. At 100% confluency, cells are switched to DM, and fresh medium is added every day for the next 96 h. Control plates containing an equal number of cells from the individual cell lines are recommended, as there is some variability in the rate of myotube formation between cell lines. One should make sure that both individual strains are forming myotubes at the time of lysate collection (*see* **Note 2**).

3.2. Indigogenic Histochemistry for β-Galactosidase

1. Cells are washed twice with PBS, pH 7.4.
2. Fix cells in 4% paraformaldehyde for 5 min at room temperature.

3. Wash cells with PBS, pH 7.4, twice (5 min per wash).
4. Stain cells with 1 mg/mL X-Gal in staining solution overnight at 37 °C.
5. Examine by transmitted light microscopy for blue cells. X-Gal staining can be most easily observed with phase-contrast rings removed from the microscope condenser.

3.3. Fluorescence Histochemistry for β-Galactosidase

1. Cells are cultured on sterilized collagen-coated glass coverslips. Number 1.5 thickness coverslips are baked at ~200 °C to sterilize them and then allowed to cool. Coverslips are soaked in 0.1% Type I collagen in 0.1 *M* acetic acid overnight at 4 °C. Coverslips are then rinsed with sterile water and placed in 35-mm dishes, and cells are plated in coculture and induced to fuse with DM.
2. Cells are washed twice briefly in PBS, pH 7.4, and fixed in ice cold 4% paraformaldehyde for 5 min.
3. Cells are washed with PBS, pH 7.4, twice (5 min per wash). Cells may be stored at 4 °C before staining. Additionally, immunofluorescence labeling for molecules of interest may be done before Fluor-X-gal staining (*see* **Note 3**). Antibodies must be compatible with paraformaldehyde fixation, as ethanol or methanol fixation can destroy complemented β-galactosidase. Keep dishes at 4 °C during immunostaining.
4. Dilute the stock solutions of Fast Red Violet LB and 5-6-X-gal together in PBS, pH 7.4, to a final concentration of 100 μg/mL and 25 μg/mL, respectively. Filter final solution through a 0.45-μm syringe filter to remove precipitating particles. Cells are incubated in this Fluor-X-gal staining solution for 30 min at 37 °C.
5. Wash cells with PBS, pH 7.4, for 30 min at room temperature.
6. Mount coverslips on slides in PBS, pH 7.4, and seal with nail polish. Glycerol mounting media should be avoided, as they may lead to bleeding of the localized fluorescent reaction product.
7. Fluor-X-gal product emits with a peak at ~560 nm and is best viewed with a tetramethyl Rhodamine isothiocyanate filter set, although it can also be imaged through fluorescein isothiocyanate or Texas Red filter sets. Because of its broad emission spectrum, Fluor-X-gal stain is best combined with Cy5-labeled antibodies for dual-channel imaging (**1**).

3.4. Chemiluminescence Detection of β-Galactosidase

1. Co-plated cells are lysed within culture wells with 50 μL lysis solution by incubation at 37 °C for 10 min. Cells are then scraped off the plate with a pipette and triturated several times in the well before a 20-μL sample is removed and placed in a luminometer tube or 96-well luminometer assay plate.
2. 70 μL of Reaction Buffer (Galacton-Plus diluted 100× in Reaction Buffer Diluent) is added to each sample and incubated at room temperature for 1 h.
3. 100 μL of Accelerator is added to each tube. Individual samples are typically read on a luminometer within 1–2 s of Accelerator addition. Luminescence continues for up to 1 h, and staging of reaction start times and Accelerator addition times can therefore be used to process large numbers of samples on an automated multiwell luminometer.

3.5. β-Galactosidase Assay in Conjunction With Small Interfering RNA of Target Genes

1. For lacZ complementation assays, a total of 400,000 cells/well are co-plated in a 24-well dish and grown for 24 h in 500 μL of GM per well.
2. Before transfection, briefly wash cells twice with PBS, pH 7.4, and replenish wells with 400 μL Dulbecco's Modified Eagle's Medium.
3. Dilute siRNA to approximately 200 ng/well (*see* **Note 4**) in 50 μL of Opti-MEM. Dilute Lipofectamine 4 μL per well (*see* **Note 4**) to a total volume of 55 μL in Opti-MEM. Combine siRNA solution and Lipofectamine solution and mix gently. Incubate at room temperature for 20 min (*see* **Note 4**).
4. Add 105 μL of siRNA–Lipofectamine mixture to each well.
5. Change cells to DM after 24 h and daily thereafter.
6. Proceed with chemiluminescence detection protocol (*see* **Section 3.4.**) 96 h post-transfection. Results from a series of control and siRNA knock-down cultures are shown in **Figure 6**.

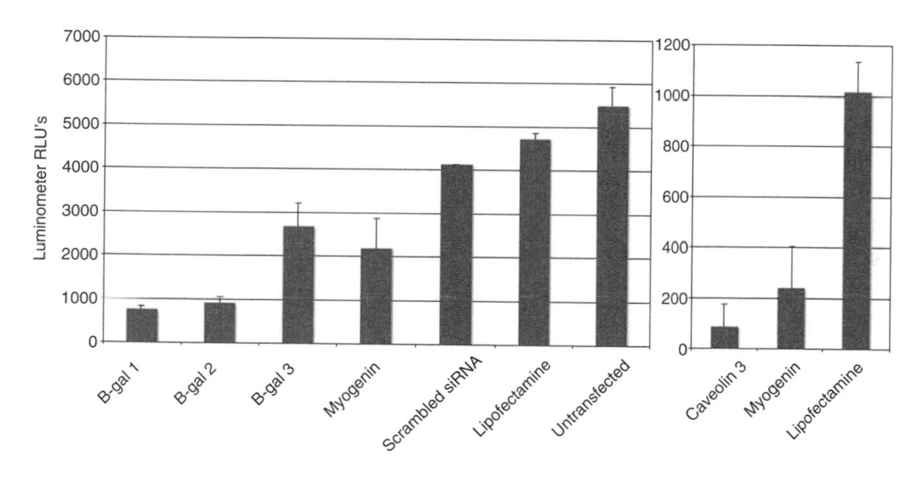

Fig. 6. Luminescence readings for small interfering (si)RNA knock-downs in C2F3 *lacZ* complementation cells. Luminescence recordings were taken 96 h after transfection and transfer to DM media. β-Galactosidases 1, 2, and 3 are control siRNA duplexes against *lacZ* that show high, medium, and low silencing capabilities, respectively. Scrambled siRNA is a nonspecific negative control duplex. Myogenin and caveolin-3 SMART pools each comprise a combination of three predicted high-efficiency siRNA duplexes for the target gene. Right and left panels show separate experiments, each with a matched negative control culture to correct for variation in the reporter response from one experiment to the next. Knock-down of myogenin and caveolin-3 has been previously reported to decrease myoblast fusion ***(44,45)***.

4. Notes

1. Myoblast fusion is greatly affected by cell contact and density. Therefore, it is extremely important to avoid cell–cell contact until you are prepared to proceed with the experiment. Individual cell lines should be passaged to maintain less than 50% confluency. Likewise, it is important to plate equal numbers of cells in all wells of an assay group to prevent any variation in rates of cell fusion.
2. The C2F3-462 cell line appears to have a slight delay in its cell cycle and in the formation of myotubes compared with C2F3-787. The ratio of cocultured plating can be adjusted to increase the number of myotubes formed in a 96-h period and to optimize the level of β-galactosidase expression in control cultures.
3. If planning to proceed with immunofluorescence in combination with Fluor-X-Gal, block fixed cells with PBS, pH 7.4, and 10% horse serum prior to adding first-layer antibodies.
4. As with any transfection, the amounts of siRNA or Lipofectamine and the time of transfection incubation can be adjusted empirically or per manufacturer's instructions.

Acknowledgment

This work was supported by grants from the Patterson Trust and the National Institutes of Health (HD43156) to W.A.M.

References

1. Mohler, W. A. and Blau, H. M. (1996) Gene expression and cell fusion analyzed by lacZ complementation in mammalian cells. *Proc. Natl. Acad. Sci. U.S.A.* **93,** 12423–12427.
2. Blumenthal, R., Clague, M. J., Durell, S. R., and Epand, R. M. (2003) Membrane fusion. *Chem. Rev.* **103,** 53–69.
3. Chen, E. H. and Olson, E. N. (2005) Unveiling the mechanisms of cell–cell fusion. *Science* **308,** 369–373.
4. Jahn, R., Lang, T., and Sudhof, T. C. (2003) Membrane fusion. *Cell* **112,** 519–533.
5. Hu, C., Ahmed, M., Melia, T. J., Sollner, T. H., Mayer, T., and Rothman, J. E. (2003) Fusion of cells by flipped SNAREs. *Science* **300,** 1745–1749.
6. Giraudo, C. G., Hu, C., You, D., Slovic, A. M., Mosharov, E. V., Sulzer, D., Melia, T. J., and Rothman, J. E. (2005) SNAREs can promote complete fusion and hemifusion as alternative outcomes. *J. Cell Biol.* **170,** 249–260.
7. Podbilewicz, B., Leikina, E., Sapir, A., Valansi, C., Suissa, M., Shemer, G., and Chernomordik, L. V. (2006) The *C. elegans* developmental fusogen EFF-1 mediates homotypic fusion in heterologous cells and in vivo. *Dev. Cell* **11,** 471–481.
8. Morris, S. J., Sarkar, D. P., White, J. M., and Blumenthal, R. (1989) Kinetics of pH-dependent fusion between 3T3 fibroblasts expressing influenza hemagglutinin and red blood cells. Measurement by dequenching of fluorescence. *J. Biol. Chem.* **264,** 3972–3978.
9. Kemble, G. W., Bodian, D. L., Rose, J., Wilson, I. A., and White, J. M. (1992) Intermonomer disulfide bonds impair the fusion activity of influenza virus hemagglutinin. *J. Virol.* **66,** 4940–4950.

10. Bagai, S. and Lamb, R. A. (1996) Truncation of the COOH-terminal region of the paramyxovirus SV5 fusion protein leads to hemifusion but not complete fusion. *J. Cell Biol.* **135,** 73–84.
11. Sarkar, D. P., Morris, S. J., Eidelman, O., Zimmerberg, J., and Blumenthal, R. (1989) Initial stages of influenza hemagglutinin-induced cell fusion monitored simultaneously by two fluorescent events: cytoplasmic continuity and lipid mixing. *J. Cell Biol.* **109,** 113–122.
12. Fischer, C., Schroth-Diez, B., Herrmann, A., Garten, W., and Klenk, H. D. (1998) Acylation of the influenza hemagglutinin modulates fusion activity. *Virology* **248,** 284–294.
13. Chernomordik, L. V., Frolov, V. A., Leikina, E., Bronk, P., and Zimmerberg, J. (1998) The pathway of membrane fusion catalyzed by influenza hemagglutinin: restriction of lipids, hemifusion, and lipidic fusion pore formation. *J. Cell Biol.* **140,** 1369–1382.
14. Keller, P. M., Person, S., and Snipes, W. (1977) A fluorescence enhancement assay of cell fusion. *J. Cell Sci.* **28,** 167–177.
15. Huerta, L., Lopez-Balderas, N., Larralde, C., and Lamoyi, E. (2006) Discriminating in vitro cell fusion from cell aggregation by flow cytometry combined with fluorescence resonance energy transfer. *J. Virol. Methods* **138,** 17–23.
16. Nussbaum, O., Broder, C. C., and Berger, E. A. (1994) Fusogenic mechanisms of enveloped-virus glycoproteins analyzed by a novel recombinant vaccinia virus–based assay quantitating cell fusion–dependent reporter gene activation. *J. Virol.* **68,** 5411 2542.
17. Feng, Y., Broder, C. C., Kennedy, P. E., and Berger, E. A. (1996) HIV-1 entry cofactor: functional cDNA cloning of a seven-transmembrane, G protein–coupled receptor. *Science* **272,** 872–877.
18. Okuma, K., Nakamura, M., Nakano, S., Niho, Y., and Matsuura, Y. (1999) Host range of human T-cell leukemia virus type I analyzed by a cell fusion–dependent reporter gene activation assay. *Virology* **254,** 235–244.
19. Russell, C. J., Kantor, K. L., Jardetzky, T. S., and Lamb, R. A. (2003) A dual-functional paramyxovirus F protein regulatory switch segment: activation and membrane fusion. *J. Cell Biol.* **163,** 363–374.
20. Yi, Y., Isaacs, S. N., Williams, D. A., Frank, I., Schols, D., De Clercq, E., Kolson, D. L., and Collman, R. G. (1999) Role of CXCR4 in cell–cell fusion and infection of monocyte-derived macrophages by primary human immunodeficiency virus type 1 (HIV-1) strains: two distinct mechanisms of HIV-1 dual tropism. *J. Virol.* **73,** 7117–7125.
21. Broer, R., Boson, B., Spaan, W., Cosset, F. L., and Corver, J. (2006) Important role for the transmembrane domain of severe acute respiratory syndrome coronavirus spike protein during entry. *J. Virol.* **80,** 1302–1310.
22. He, B., McAllister, W. T., and Durbin, R. K. (1995) Phage RNA polymerase vectors that allow efficient gene expression in both prokaryotic and eukaryotic cells. *Gene* **164,** 75–79.
23. Dutch, R. E. and Lamb, R. A. (2001) Deletion of the cytoplasmic tail of the fusion protein of the paramyxovirus simian virus 5 affects fusion pore enlargement. *J. Virol.* **75,** 5363–5369.

24. Paterson, R. G., Russell, C. J., and Lamb, R. A. (2000) Fusion protein of the paramyxovirus SV5: destabilizing and stabilizing mutants of fusion activation. *Virology* **270,** 17–30.
25. Waning, D. L., Schmitt, A. P., Leser, G. P., and Lamb, R. A. (2002) Roles for the cytoplasmic tails of the fusion and hemagglutinin-neuraminidase proteins in budding of the paramyxovirus simian virus 5. *J. Virol.* **76,** 9284–9297.
26. Bradley, J., Gill, J., Bertelli, F., Letafat, S., Corbau, R., Hayter, P., Harrison, P., Tee, A., Keighley, W., Perros, M., Ciaramella, G., Sewing, A., and Williams, C. (2004) Development and automation of a 384-well cell fusion assay to identify inhibitors of CCR5/CD4-mediated HIV virus entry. *J. Biomol. Screen.* **9,** 516–524.
27. Alvarez-Dolado, M., Pardal, R., Garcia-Verdugo, J. M., Fike, J. R., Lee, H. O., Pfeffer, K., Lois, C., Morrison, S. J., and Alvarez-Buylla, A. (2003) Fusion of bone-marrow–derived cells with Purkinje neurons, cardiomyocytes and hepatocytes. *Nature* **425,** 968–973.
28. Harris, R. G., Herzog, E. L., Bruscia, E. M., Grove, J. E., Van Arnam, J. S., and Krause, D. S. (2004) Lack of a fusion requirement for development of bone marrow–derived epithelia. *Science* **305,** 90–93.
29. Reinecke, H., Minami, E., Poppa, V., and Murry, C. E. (2004) Evidence for fusion between cardiac and skeletal muscle cells. *Circ. Res.* **94,** e56–e60.
30. Ullmann, A., Jacob, F., and Monod, J. (1967) Characterization by in vitro complementation of a peptide corresponding to an operator-proximal segment of the beta-galactosidase structural gene of *Escherichia coli. J. Mol. Biol.* **24,** 339–343.
31. Lojda, Z. (1979) *Enzyme Histochemistry: A Laboratory Manual.* Springer-Verlag, Berlin.
32. Rossi, F. M., Blakely, B. T., Charlton, C. A., and Blau, H. M. (2000) Monitoring protein–protein interactions in live mammalian cells by beta-galactosidase complementation. *Methods Enzymol.* **328,** 231–251.
33. Fiering, S. N., Roederer, M., Nolan, G. P., Micklem, D. R., Parks, D. R., and Herzenberg, L. A. (1991) Improved FACS-Gal: flow cytometric analysis and sorting of viable eukaryotic cells expressing reporter gene constructs. *Cytometry* **12,** 291–301.
34. Blakely, B. T., Rossi, F. M., Tillotson, B., Palmer, M., Estelles, A., and Blau, H. M. (2000) Epidermal growth factor receptor dimerization monitored in live cells. *Nat. Biotechnol.* **18,** 218–222.
35. Mohler, W. A. (1996) *Analysis of Mouse Myoblast Fusion.* Stanford University, Stanford, CA.
36. Charlton, C. A., Mohler, W. A., Radice, G. L., Hynes, R. O., and Blau, H. M. (1997) Fusion competence of myoblasts rendered genetically null for N-cadherin in culture. *J. Cell Biol.* **138,** 331–336.
37. Charlton, C. A., Mohler, W. A., and Blau, H. M. (2000) Neural cell adhesion molecule (NCAM) and myoblast fusion. *Dev. Biol.* **221,** 112–119.
38. Remy, I. and Michnick, S. W. (2007) Application of protein-fragment complementation assays in cell biology. *Biotechniques* **42,** 137, 139, 141 passim.
39. Hu, C. D., Chinenov, Y., and Kerppola, T. K. (2002) Visualization of interactions among bZIP and Rel family proteins in living cells using bimolecular fluorescence complementation. *Mol. Cell* **9,** 789–798.

40. Hu, C. D. and Kerppola, T. K. (2003) Simultaneous visualization of multiple protein interactions in living cells using multicolor fluorescence complementation analysis. *Nat. Biotechnol.* **21,** 539–545.
41. Jach, G., Pesch, M., Richter, K., Frings, S., and Uhrig, J. F. (2006) An improved mRFP1 adds red to bimolecular fluorescence complementation. *Nat. Methods* **3,** 597–600.
42. Jacobson, R. H., Zhang, X. J., DuBose, R. F., and Matthews, B. W. (1994) Three-dimensional structure of beta-galactosidase from E. coli. *Nature* **369,** 761–766.
43. Bhat, R. A., Lahaye, T., and Panstruga, R. (2006) The visible touch: in planta visualization of protein–protein interactions by fluorophore-based methods. *Plant Methods* **2,** 12.
44. Gunther, S., Mielcarek, M., Kruger, M., and Braun, T. (2004) VITO-1 is an essential cofactor of TEF1-dependent muscle-specific gene regulation. *Nucleic Acids Res.* **32,** 791–802.
45. Galbiati, F., Volonte, D., Engelman, J. A., Scherer, P. E., and Lisanti, M. P. (1999) Targeted down-regulation of caveolin-3 is sufficient to inhibit myotube formation in differentiating C2C12 myoblasts. Transient activation of p38 mitogen-activated protein kinase is required for induction of caveolin-3 expression and subsequent myotube formation. *J. Biol. Chem.* **274,** 30315–30321.

21

Fusion Assays and Models for the Trophoblast

Sascha Drewlo, Dora Baczyk, Caroline Dunk, and John Kingdom

Summary

A healthy syncytium in the placenta is vital to a successful pregnancy. The trophoblast builds up the natural barrier between the mother and the developing fetus and is the site of gas, nutrition, and waste exchange. An inadequate formation of this tissue leads to several pathologies of pregnancy, which may result in fetal death during the second trimester or iatrogenic preterm delivery due to intrauterine growth restriction, preeclampsia, or abruption.

Cytotrophoblastic cells fuse constantly with the overlying syncytiotrophoblast/syncytium to maintain the function of the trophoblast. Syncytin-1 is the only molecule known to directly induce fusion in the placental trophoblast. Many other proteins, such as gap junctions (e.g., connexin 40) and transcription factors, play a role in the molecular pathways directing the trophoblast turn over. Despite the significance of this process for successful placentation, the mechanisms regulating its activity remain poorly understood.

In this chapter we present several different model systems that can be utilized to investigate the regulation of the cell fusion process in the trophoblast. We describe cell-based assays as well as tissue-related protocols. We show how fusion can be monitored in (1) BeWo cells as a trophoblast cell line model, (2) HEK239 using syncytin-1 as a fusion molecule, and (3) a floating villi explant model. Furthermore, we will present strategies to inhibit fusion in the different models. These techniques represent powerful tools to study the molecular mediators of cell fusion in the trophoblast.

Key Words: Cell fusion; trophoblast; syncytiotrophoblast; fusion assay; syncytin-1.

1. Introduction

Cell fusion is an important mechanism in human syncytiotrophoblast development and maintenance. Investigations of the fusion process in placental tissue are a challenge because of its complexity. Here we describe cell- and tissue-based models on this topic. These models are useful tools

From: *Methods in Molecular Biology, vol. 475: Cell Fusion: Overviews and Methods*
Edited by: E. H. Chen © Humana Press, Totowa, NJ

to study trophoblast fusion because of the controlled conditions of the cell culture environment and the flexibility of their application in different approaches. In the second part of this chapter we describe a floating placental villi model, which is useful to investigate the fusion process in the placental tissue in vitro.

1.1. Cell Lines and Model Systems

Cell fusion takes place only under particular circumstances. These conditions vary across experimental models. The trophoblastic cell line BeWo (choriocarcinoma) fuses spontaneously and can be enhanced after a treatment with forskolin or cyclic adenosine monophosphate (cAMP; **refs. *1,2***). In contrast, HEK293 cells do not fuse spontaneously. However, their fusion can be induced under certain conditions. Syncytin-1, a placental fusion molecule, can be transfected into cell lines, and fusion will take place as long as the protein is expressed and the partner cell lines provide the complementary receptor (RDR, ATB0; **ref. *3***). Certain prerequisites are necessary, but the initial fusion process is initiated by the interaction of syncytin-1 and its receptor.

Various external factors, for example, GCM1, have been shown to mediate syncytin-1 expression and cell-cell fusion in trophoblast model systems ***(4,5)***. The most common cell-based model used to study trophoblast fusion is a panel of choriocarcinoma cell lines: BeWo, JAR, and JEG3. BeWo and JAR cells were shown to fuse upon stimulation with forskolin ***(1,2)***, an agent that increases the levels of intracellular cAMP. Forskolin was later shown to increase syncytin-1 mRNA levels in JEG3, JAR, and BeWo cells ***(6–9)***. However, levels of syncytin-1 mRNA in JEG3 cells (whether stimulated with forskolin or not) are low compared with those in JAR and BeWo cells ***(6–9)***.

Fusion is generally monitored by measuring the disappearance of desmoplakin immunostaining and/or by measuring the release of a product of the syncytiotrophoblast: human chorionic gonadotropin (hCG; **ref. *5***). An alternative method described here is the two-color–based assay, which is fast, reliable, and easy to use and shows only real fusion events in double fluorescent syncytia.

We also show how to generate truncated and wild-type protein expressing mutants of syncytin-1. Human syncytin-1 can be subcloned in a mammalian expression system and syncytin-1–driven fusion monitored. The assays can be effectively used to assess interaction pathways (fusion regulation) and other molecular mediators of cell fusion (**Fig. 1**).

1.2. Primary Trophoblast Cells

Another cellular model to study trophoblast fusion is the cultivation of primary isolates of trophoblast cells from early or late trimester placental

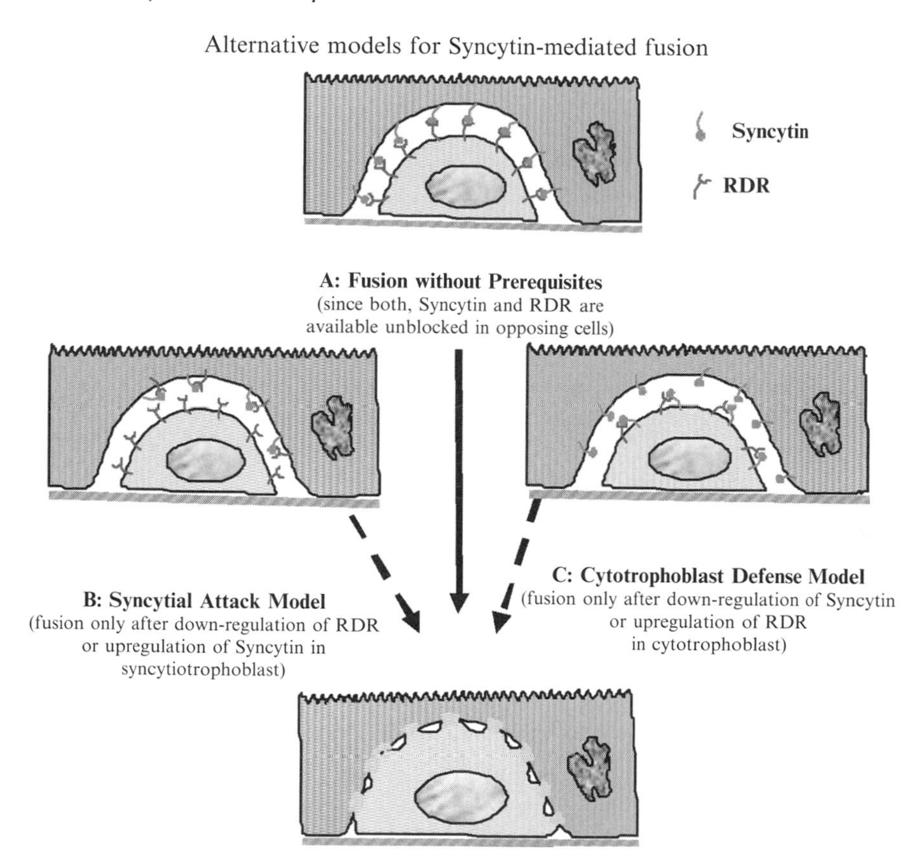

Fig. 1. Examples of how trophoblast fusion may be regulated through local availability of syncytin-1 and its receptor (RDR). The arrows indicate the transition of a nonfusogenic state to initiation of fusion. Intercellular contacts between syncytin-1 and receptor leading to fusion are depicted symbols. Expression of syncytin-1 or receptor on the apical syncytiotrophoblast membrane is considered irrelevant for fusion and therefore is not depicted in this.

material. Primary trophoblast cells are the ultimate, but not easy to handle, fusion model. The heterogeneous cell population, the limited access to tissue, the small amount of cells obtainable during purification, and their limited life span in culture make them difficult to work with. It seems that these primary cells undergo terminal differentiation processes, which have not been fully understood yet. Nevertheless these cells serve as a useful tool to study the fusion process, because they are more relevant to the tissue than immortalized cells. However, because of inconsistency of the procedure, we do not present protocols for this material here.

1.3. Villous Tissue Explants

Villous explants of placental material can be kept in culture for several days and are an excellent in vitro model ***(10)***. Denudation of the villi with trypsin leads to a reformation of the syncytium in culture over a 72-h period. Cytotrophoblastic cell fusion can be monitored, stimulated, or blocked (**Fig. 2; ref *10***).

2. Materials

2.1. Fusion Assay Using BeWo Cells and CTgreen and CTorange

1. F12-K media + 2 m*M* L-glutamine (ATCC, Manassas, VA).
2. Fetal bovine serum, nondeactivated (Gibco, Grand Island, NY).
3. CytoTracker Orange CMTMDR (Molecular Probes, Eugene, OR, number C2927) and CellTracker Green (Molecular Probes, number C2925): aliquot after resolving in dimethyl sulfoxide (DMSO; 5 m*M*) and store at −20 °C. Keep solutions in the dark to maintain stability of the dye (handle with gloves only).
4. Use forskolin in the concentration range of 25 to 100 μ*M* (prepare in DMSO, a 50-m*M* stock solution), and store at −20 °C (handle with gloves only).
5. cAMP at 20 μ*M* (100 m*M* stock in water); store at −20 °C.
6. Fluorescent microscope.

2.2. HEK239 Fusion Assay and Construct Design

1. Plasmid DNA: plasmid cytomegalovirus (pCMV)–syncytin; plEGFP-N1 (Clontech, Mountain View, CA).
2. Lipofectamine 2000 (Invitrogen, Carlsbad, CA) or calcium transfection kit (Promega, Mannheim, Germany).

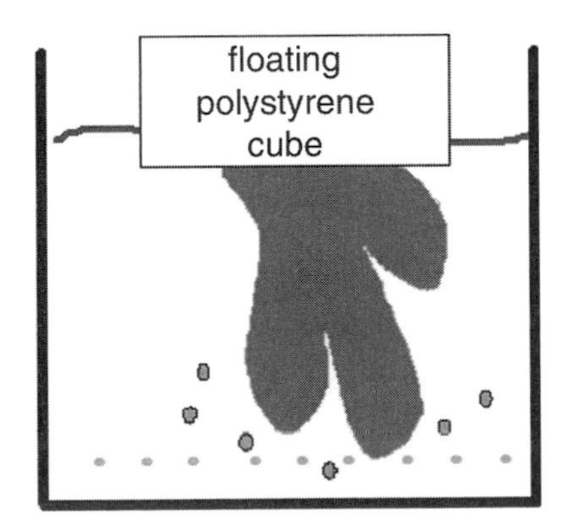

Fig. 2. Scheme of the floating villi model.

3. Hoechst 3342 (Sigma-Aldrich, St. Louis, MO).
4. Fluorescence microscope.

2.3. Construction of Syncytin-1 Expression Constructs and Mutants

1. All plasmids need to be tested by restriction analysis and sequencing.
2. RNA isolated from various human choriocarcinoma cell lines.
3. cDNA synthesis kit, Pwo polymerase.
4. Taq polymerase.
5. Syncytin-1 (GenBank, accession number AF206181).
6. Vector pGEM-T (Promega), pVSV-G (pantropic retroviral expression system; BD Clontech, Heidelberg, Germany; **Fig. 3**).

2.4. Inhibition of Induced Fusion in Cell-Based Models

2.4.1. Inducing Fusion in Cell-Based Models

1. Small interfering (si)RNA against GCM1, 50 n*M* per well (Qiagen, Valencia, CA; sequence in **Table 1**).
2. Fluorescent siRNA; for example, Cy3-labeled siRNA controls.
3. CytoTracker dyes (*see* **Subheading 2.1.**).
4. Lipofectamine 2000 and OptiMem (Invitrogen).
5. Forskolin in EtOH (10 m*M* stock) and cyclic AMP in ddH_2O (100 m*M* stock); store at −20 °C (Sigma).
6. Toxicity of siRNA and antisense oligonucleotides is monitored with human interferon-α ELISA kit (PBL Biomedical Laboratories, Piscataway, NJ).

Table 1
Immunohistochemical Reagents

Antigen	Clone	Titer (μg/mL)	Source
Human cytokeratin-7	OV-TL 12/30	0.5	DAKO (Carpinteria, CA)
Ki-67	Ki-S5	0.2	DAKO (Carpinteria, CA)
Human chorionic gonadotropin-.4	INN-hFSH-132	1.5	Accurate Chemical & Scientific Corp. (Westbury, CT)
Human leukocyte antigen-G (HLA-G)	MEM-G/1	0.1	Cederlane (Hornby, Canada)
Syncytin-1	RP69	1:100–1:1000	***(3,14)***
Syncytin-1	6A2B2	1:100–1:500	Blond et al. ***(11)*** (2000), Biomerieux, France

2.4.2. Inhibition of Fusion Using Syncytin-1 Soluble Mutants

1. pCMV–syncytin-1 plasmid (*see* **Subheading 2.2.**).
2. HEK239 cell line.

2.4.3. Production and Concentration of Soluble Syncytin-1 Variant 515delTM

1. Amicon 30kDa spin off column (Millipore).
2. OptiMem (Invitrogen).
3. Fetal calf serum (FCS).
4. Calcium transfection kit (Promega).

2.4.4. Peptide Inhibition of Fusion

Peptide HRB-1 can be synthesized by a specialized company like Aplagen GmbH (Baesweiler, Germany; **ref. *12***). The peptide should be C-terminally amidated and N-terminally acetylated to enhance stability in the media and purified by the supplier.

1. Peptide sequence of HRB-1: SGIVTEKVKEIRDRIQRRAEELRNTGPWGL (411–440).
2. Dissolve peptide in phosphate buffered saline (PBS): use at 3μg/mL, and store at −80°C.
3. Phosphate buffered saline.

2.5. Cell Surface Biotinylation of Syncytin-1 Variant: Investigation of Cell Surface Proteins

1. Biotinylation kit: BK101 Immunoprobe Biotinylation Kit (Roche, Basel, Switzerland).
2. Lysis buffer: 20m*M* Tris-HCl, pH 7.5; 150m*M* NaCl; 1% Triton X-100; 0.5% deoxycholate; 0.05% SDS; and protein inhibitor cocktail (Complete Mini, Roche; add fresh). Lysis buffer can be stored in the refrigerator until use.
3. MACS streptavidin microbeads, MACS columns (Miltenyi Biotech, Bergisch Gladbach, Germany).
4. Sample loading buffer (Laemmli loading dye) 3× stock: 2.4mL 1*M* Tris-Cl, pH 6.8; 3mL 20% SDS; 3mL 100% glycerol; 1.6mL β-mercaptoethanol; 10mL bromophenol blue 0.006g (store 4°C)

2.6. Fusion in a Placental Tissue-Based System: Explant Villi Model

2.6.1. Floating Villous Explant Culture

1. Fresh placental tissue in PBS.
2. Styrofoam pieces, 1 cm^3, sterile.
3. Dulbecco's Modified Eagle's Medium (DMEM)/F12 with 1% liquid media supplement, 100 units/mL penicillin, 100 units/mL streptomycin, 2m*M* l-glutamine, 100μg/mL gentamicin, and 2.5μg/mL Fungizone (Gibco) and no FCS; ITS

(1.0 mg/ml insulin from bovine pancreas, 0.55 mg/mL human transferrin [substantially iron free], 0.5 μg/mL sodium selenite, 50 mg/mL bovine serum albumin and 470 μg/mL linoleic acid [Sigma, St Louis, MO]).

2.6.2. Syncytial Denudation and Syncytia Reformation

1. Trypsin.
2. DMEM/F12 + 2 m*M* glutamine.
3. Gentamycin/Fungizone 100×.
4. Penicillin/streptomycin 100×.
5. ITS solution (all solutions from Gibco).

2.6.3. GCM1 Antisense siRNA Strategy to Inhibit Fusion in Explants for Tissue Use

1. Media as in **Subheading 2.3.** and siRNA (*see* **Table 1**).
2. Interferon alpha ELISA (PBL Biomedical Laboratories).

2.6.4. Histology and Immunohistochemistry

1. Ethyl alcohol, methanol, PBS.
2. Antibodies (*see* **Table 1**).
3. 10 m*M* sodium citrate, pH 6.0.

3. Methods

3.1. Inducing Fusion in Cell-Based Models

The standard models to investigate trophoblast fusion in cell-based assays are the desmoplakin and the two-color fluorescence fusion assays. The desmoplakin assay has often been used to show syncytium formation and is dependent on the disappearance of desmoplakin, when the membranes are lost between fusing cells ***(12)***.

An easier and faster way to demonstrate cell fusion is the two-color fluorescence assay ***(6)***. Two different fusion-competent cell populations are labeled, each with different color dyes, and the fluorescence of the cell population is evaluated. Double fluorescent syncytia reflecting "real" cell fusion events are scored. The following protocol is for fusion assay using BeWo cells and CytoTracker Green and Orange dyes.

1. BeWo cells (passage numbers 1–20) are grown to 70% confluency in F12-K medium containing 10% FCS, 1% penicillin/streptomycin, and L-glutamine (5% CO_2).
2. Trypsinize 15 cm^2 BeWo cells, remove medium, wash 2 × 4 mL Hanks buffered salt solution (HBSS), incubate in 2 mL trypsin/EDTA 5–15 min at 37 °C, resuspend cells in 10 mL F12-K + FCS 10%, and put into plastic tube. Centrifuge and resuspend cells in 12 mL F12-K + FCS 10%.
3. Thaw CytoTracker Green dye (CTgr) and CytoTracker Orange dye (CTor) in a dark, dry place according to the manufacturer's recommendations.

4. Count sample of cells in Neubauer chamber or cell counter, and calculate total number of cells.
5. Divide cells equally into two tubes.
6. Centrifuge 5 min at 800 *g*, and remove supernatant. Resuspend cells in serum-free F12 medium (washing solution): dilute to 1×10^6 cells/mL.
7. Dim the light on a clean bench. Wear gloves when handling dyes in next step!
8. Add CTgr (5 m*M* in DMSO) to tube labeled "one" (2 μL/mL ≥ 10 μ*M* final concentration) and CTor (5 m*M* in DMSO) to tube labeled "two" (2 μL/mL ≥ 10 μ*M* final concentration). Incubate cells for 20 min at 37 °C in the dark. Put CTgr and CTor dyes back into the freezer.
9. Wash stained cells: centrifuge 5 min at 800 g, resuspend cells in 2 × 5 mL F12 + FCS 10%
10. Repeat washing twice (in total, wash three times). Count cells again during the last wash and use Trypan blue exclusion (or any other method to determine cell viability), as the dyes are toxic to a certain extent and some cells might be lost.
11. After the last wash: resuspend cells in F12-K + FCS 10% + L-glutamine 2 m*M* and penicillin/streptomycin solution; dilute to 1×10^5 cells/mL.
12. Put into 6 wells: 1 mL (1×10^5) CTgr-stained cells and 1 mL (1×10^5) CTor-stained cells. For 12 wells, 0.5 mL + 0.5 mL cells per well; for 96 wells, 50 μL + 50 μL cells per well.
13. Allow cells to adhere to the bottom of the wells for 3–6 h. Check adherence.
14. After 3–6 h, remove medium, and add medium ± fusion-inhibiting agent: per 6 wells, 2 mL F12-K + FCS ± 2 μL of 50 m*M* forskolin (in DMSO); per 12 wells, 1 mL F12-K + FCS ± 1 μL of 50 m*M* forskolin (in DMSO) ± inhibiting agent (optional: use cAMP at 10–20 μ*M*, from a 100 m*M* stock in water). Wear gloves when handling forskolin.
15. After 48 h, determine fusion under a fluorescent microscope (10 random pictures phase contrast; fluorescence: green/red/combination per well: calculate nuclei in double fluorescent cytoplasm/all nuclei; *see* **Note 1**).

3.2. HEK239 Fusion Assay and Construct Design

To obtain a better understanding of the molecular players in the fusion process, it might be necessary to investigate the role of certain genes in different cell lines to determine their influences on the fusion process and the physiological changes. This assay can easily be adapted for other applications, for example, interaction and inhibition analyses.

3.2.1. Construction of Syncytin-1 Full-Length and Mutant Expression Constructs

1. Obtain full-length syncytin-1 AF208161 protein coding region through reverse transcriptase polymerase chain reaction (RT-PCR) on RNA isolated from various human choriocarcinoma cell lines using a first strand cDNA synthesis kit and Pwo polymerase (**Table 2**).

Table 2
Primer and siRNA Sequences

Target	Probe	Sequence
sync5′*Bam*HI	Primer 69.5 °C	ATT CTT GGA TCC CCC ATG GCC CTC CCT
sync5′*Hind*III	Primer 66.5 °C	ATT CTT AAG CTT CCC ATG GCC CTC CCT
sync3′stop*Bam*HI	Primer 65 °C	ATT CTT GGA TCC CTA ACT GCT TCC TGC
syncFL3′*Xho*I	Primer 72.6 °C	TCA GAC CTC GAG ACG ACC GCT CTA ACT
sync*Kpn*I5.	Primer 66.5 °C	GTC TGT GGT ACC TCA GCC TAT CGT TGT
siRNAs		
GCM1 201	siRNA	r(CUC CCG CAU CCU CAA GAA G)dTdT
GCM1 815	siRNA	r(CCU ACA GUA GAG ACC U)dTdT

2. Reamplify PCR products for two more cycles using Taq polymerase and clone into the pGEM-T vector (Promega), yielding plasmids pGEM-T–syncytin-1-full length (FL).
3. Reamplify the full-length syncytin-1 open reading frame from the pGEM-T–syncytin-1-FL using Pwo polymerase. Primers match the sequence flanking the pGEM-T polylinker, while the reverse primer contains an *Xho*I linker to facilitate cloning into a mammalian expression vector (*see* **Table 2**).
4. Digest the plasmid pVSV-G (Clontech), encoding the vesicular stomatitis virus glycoprotein (VSV-G) under the control of the cytomegalovirus (CMV) immediate early promoter (pantropic retroviral expression system, BD Clontech) with *Xho*I to remove the VSV-G insert.
5. Ligate the *Sal*I/*Xho*I-digested syncytin-1 insert into the remaining vector, yielding plasmids pCMV–syncytin-1-FL (an expression plasmid encoding the syncytin-1–coding region) and pCMV–syncytin-1-AS (containing the same insert in the antisense [AS] direction as control plasmid).
6. To generate C-terminal truncations of syncytin-1, use plasmid pCMV–syncytin-1-FL as a template for a series of PCRs. A universal 27mer forward primer overlaps with the *Kpn*I site that is found in the central part of the syncytin-1–coding region. Reverse primers match with 12 bases encoding the last four amino acids intend to be encoded and are followed by a TAA stop codon and an *Xho*I linker. The PCR products are digested with *Kpn*I and *Xho*I.
7. Plasmid pCMV–syncytin-1-FL is also to be digested with *Kpn*I and *Xho*I, removing the 3′ part of the syncytin-1–coding region. This part is replaced with the digested PCR products.

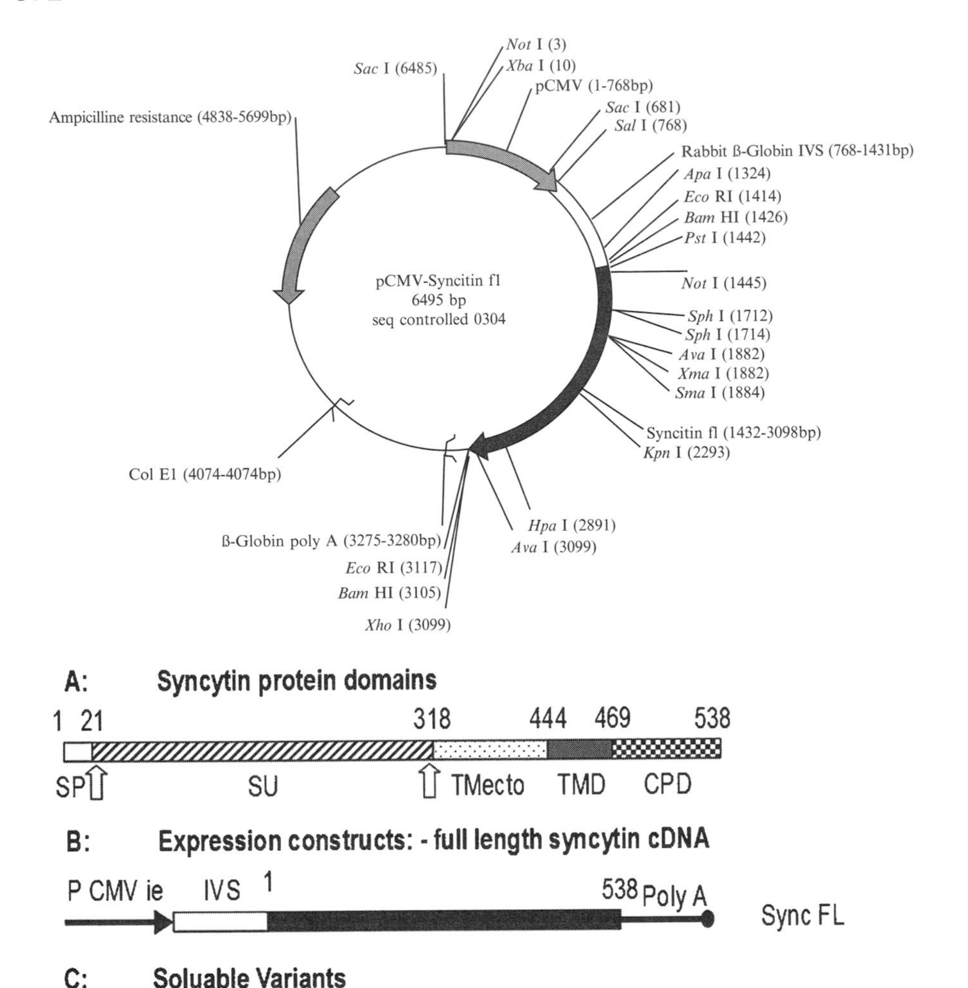

Fig. 3. (I) Schematic plasmid map of the pCMV–syncytin-1 fl *Kursiv* (in black) presents different restriction enzyme recognition sites. pCMV, cytomegalovirus promotor; rabbit β-globin IVS, intron sequence that enhances mRNA stability; syncytin-1 fl, syncytin-1 CDNA; β-globin poly A, polyadenylation signal; Col E1, bacterial recognition site for plasmid replication. **(II)** Schematic overview of the predicted protein domains in syncytin-1 (A) and the constructs encoding the truncated variants of syncytin-1 used in this study (B). **(A)** Syncytin-1 is synthesized as a precursor protein of 538 amino acids. The first 21 amino acids encoding a signal peptide are cleaved off, while there is a furin cleavage site after residue 317 to generate the surface (SU) and the transmembrane (TM) subunits. The TM portion consists of an ectodomain (TMecto), a transmembrane domain (TMD), and a cytoplasmic domain (CPD). **(B)** The expression constructs consist of an immediate early CMV promoter (P CMVie), a β-globin intervening sequence (IVS), the syncytin-1–encoding sequence (black bars), and a β-globin polyadenylation site (Poly A). **(C)** The numbers written above the bars refer to amino acids of the precursor protein. LD shows the area of deleted amino acids.

8. This generates plasmids pCMV–syncytin-1 469, 476, 483, 493, and 515, encoding syncytin-1 proteins truncated after amino acids 469, 476, 483, 493, and 515, respectively (*see* **Fig. 3**). The inserts of these plasmids need to be sequenced
9. The inserts of plasmids pCMV–syncytin-1-FL and of all the truncation variants should be sequenced to check for their sequence integrity.

3.2.2. HEK293 Fusion Assay

1. Transfect subconfluent cultures (70%–80%) of HEK293 cells in 24-well format with pCMV–syncytin-1 plasmid variants and pLEGFP-N1 in a 1:1 (0.5 μg each per well) ratio using Lipofectamine 2000 according to the manufacturer's recommendation (Invitrogen). Medium removal is optional (*see* **Note 2**).
2. Hoechst 33342 dye (Sigma-Aldrich, Taufkirchen, Germany) is added to the culture medium 16–24 h after starting the transfection at 1 μg/mL and cells and are incubated for 15 min.
3. Exchange medium with PBS and view cells by phase contrast and fluorescent microscopy (e.g. Zeiss, Axiovert).
4. Fused cells (syncytia) are recognizable by the presence of multiple nuclei within a large cytoplasm showing a uniform green fluorescence (**Fig. 4**; *see* **Note 3**).
5. For quantitative analysis: take four random pictures comprising at least 2000 nuclei with blue (showing all nuclei) and green (showing all transfected cells) emission filters. Fusion is determined by counting all the nuclei in a field and scoring the number of nuclei within syncytia.
6. Two fusion parameters can be calculated: *(a)* the fusion index (the overall percentage of nuclei within syncytia) and *(b)* the average size of syncytia (the number of nuclei within one single syncytium). Data from these fusion assays can be analyzed using statistical software like Graphpad Prism.

3.3. Inhibition of Induced Fusion in Cell-Based Models

It may be interesting to induce the expression of syncytin-1 and molecules that are involved in the fusion process, without inducing the fusion process itself, to study molecules that are directly and/or indirectly influenced by syncytin-1. This technique is very useful if low-level expression proteins are the targets of investigation, because the fusion process itself can lead to a potential dilution of the protein in the resulting multinuclear syncytia. Additionally, this can be of interest if single cells are needed for certain applications, for example, FACS analysis, in which syncytia are not accessible for analysis as they can clog the machine. Three methods are presented.

3.3.1. BeWo Fusion Interference on the RNA Expression Level

Inhibition of syncytin-1 expression by an antisense or siRNA strategy inhibits fusion on the mRNA level using the following protocol. At the protein level, fusion intervention can be achieved by using specific antibodies, peptides, and

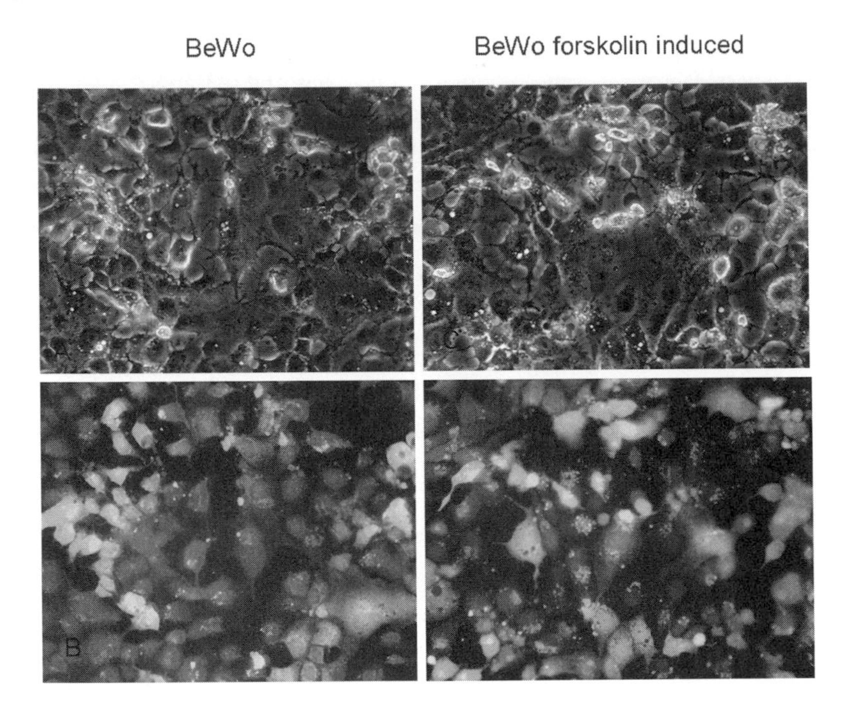

Fig. 4. **(I)** BeWo cells color labeled with green and red fluorescence dye mixed together. **(A)** Nonstimulated BeWos; phase contrast. **(B)** Nonstimulated BeWos; double fluorescent. **(C)** Stim ulated with forskolin; phase contrast. **(D)** Stimulated with forskolin; double fluorescent. Double fluorescent pictures indicate areas of fusion. (*See Color Plates*)

proteins that interfere at different stages ***(9,14)***. Protocols for these topics are presented in the following sections.

1. Color-labeled BeWo cells, as described above (*see* **Subheading 3.1.**), are seeded to 70% confluency and transfected 24 h later with siRNA (approx. 50 n*M*/well) to the gene of interest using Lipofectamine 2000 reagent.
2. Transfection procedure for 1 well of a 12-well plate is as follows. Make Solutions 1 and 2. Solution 1: 6 μL siRNA from 20 m*M* stock in 50 mL OptiMem; Solution 2: 4 μL Lipofectamine 2000 in 50 mL OptiMem.
3. Mix Solutions 1 and 2 and incubate for 30 min at room temperature.
4. Wash target cells with OptiMem media and add 150 mL OptiMem per well to the solution mixture and overlay cells; incubate for 4 h at 37 °C.
5. Add 400 mL of regular media with 20% serum.
6. Transfection efficiency control: use parallel experiment fluorescent-labeled siRNA to ensure at least 90% transfection efficiency, which is determined by fluorescence microscopy. (This control experiment can be used as well for transfection optimization.)
7. Add forskolin (final concentration 25 μ*M*) 24 h later to accelerate syncytial fusion.

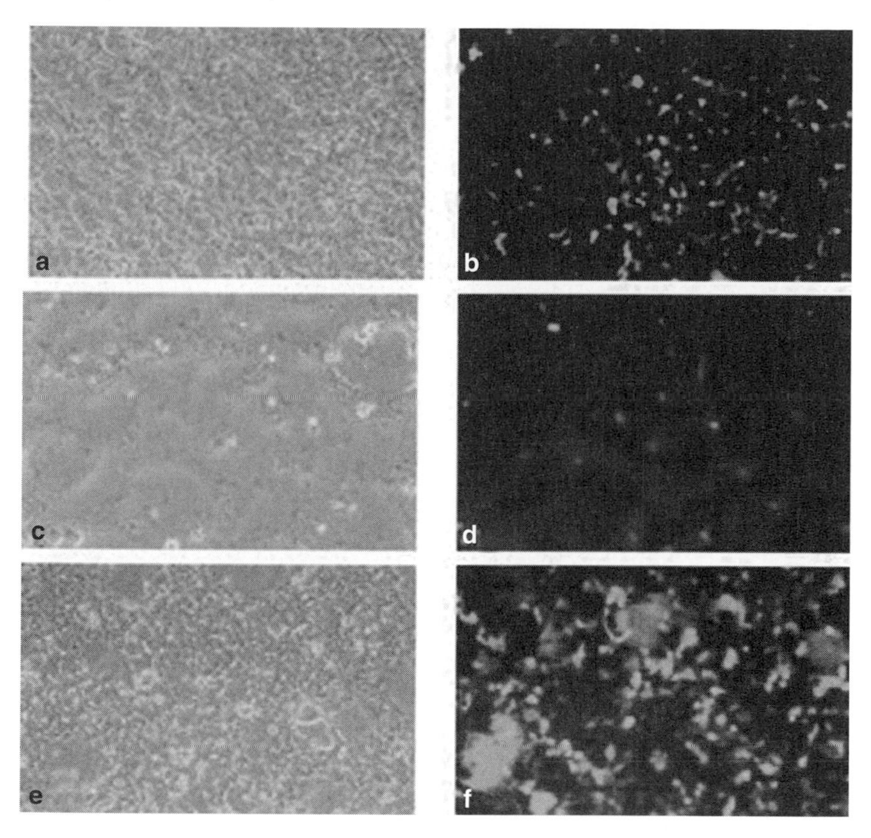

Fig. 4. (continued) **(II)** HEK239 cells were transfected with plasmids encoding syncytin-1 483 + EGFP (1:1) or syncytin-1 antisense + EGFP (1:1). Four hours after transfection, concentrated culture supernatant containing either soluble syncytin-1 (515delTM) or no syncytin (supernatant from syncytin-1-FL or antisense transfected cells) was added to the cultures. Cells were cultured for an additional 18 h and then examined. **(A,B)** HEK293 cells transfected with antisense construct and pLEGFP, no conditioned media added. **(C–F)** Cells transfected with syncytin-1 483 construct and pLEGFP. C and D in conditioned media of antisense-transfected cells and E and F in conditioned media containing soluble syncytin-1. (*See Color Plates*)

8. Assess fusion further 24 and 48 h later using phase contrast and fluorescent microscopy to semiquantitatively monitor the effects of the inhibition of the gene of interest expression.
9. The efficiency of the knock-down can be monitored by different techniques, such as real-time PCR, Northern blot, or ELISA and Western blot analyses, if antibodies are accessible.

10. Extract RNA from the cells using RNeasy kit (Qiagen). (Control experiments include transfection reagent lacking siRNA, nonsilencing siRNA and omission of forskolin.) Monitor the toxicity of siRNA and antisense oligonucleotides with human interferon-α with an ELISA (*see* **Fig. 4**).

3.3.2. Inhibition of Fusion Using Syncytin-1 Soluble Mutants

To investigate receptor interactions (syncytin-1–RDR) we designed syncytin-1 constructs, which result in soluble nonmembrane-bound protein, that can interact with the receptor RDR and inhibit cell fusion. In the HEK293-based model, it is possible to inhibit syncytin-1–mediated fusion with a special designed truncation mutant of the syncytin-1 protein (mutant 515delTM).

3.3.3. Inhibition of Fusion by Creating Soluble Syncytin-1 Mutant Proteins

A soluble variant of syncytin-1 (mutant 515delTM) can be generated as follows.

1. Ligation of two PCR fragments, one from the *Kpn*I site (amino acid 272) to residue 436, with an *Xho*I linker, and one encoding amino acids 470–515, with an *Sal*I and an *Xho*I linker, into plasmid pCMV–syncytin-1-FL (*see* **Fig. 3**) digested with *Kpn*I and *Xho*I.
2. The resulting plasmid, pCMV–syncytin-1–515delTM, encodes a syncytin-1 mutant protein truncated after residue 515, lacking the transmembrane region (residues 437–469; *see* **Fig. 3**).

3.3.4. Production and Concentration of Soluble Syncytin-1 Variant Protein 515delTM

1. Perform transfection of cell line HEK239 in 75-cm surface area flasks, 7.5 mL media, 60%–70% confluency culture dishes with the syncytin-1 variant 515delTM plasmid DNA using the calcium transfection method according to the manufacturer's protocol (Promega; *see* **Note 4**).
2. Cultivate the cells with serum-free media (OptiMem, Invitrogen; *see* **Note 5**).
3. The supernatant is harvested after a cultivation period of 24–42 h. The media needs to be centrifuged at 4000 *g* for 5 min at 4 °C to remove cell debris (*see* **Note 6**).
4. The cell supernatants are then concentrated at least ~10-fold in Amicon spin columns (30-kDa cut off) 3000 *g* for 90 min at 4 °C.
5. The samples are stored at –80 °C until use. Small samples (50 μL) should be kept to analyze protein content by Western blotting (e.g., antibody used: rp69 anti-syncytin; *see* **Table 1**).
6. Transfect cells with syncytin-1 plasmid DNA as described in **Subheading 3.2.1** in 6 or 24 wells.

7. Exchange the media 4h after transfection with the soluble syncytin-1– containing media with 5% v/v fetal bovine serum.
8. Evaluate the cells after 18–24h using standard techniques.

3.3.5. Peptide Inhibition of Fusion

An effective way to inhibit fusion at the protein level in the HEK239 cell-based assay is by using the specific peptide HRB-1 *[11]*. HRB-1 can be synthesized by specialized companies, for example, Aplagen GmBH, using the sequence described in **Subheading 2.4.4**.

1. HEK293 cells are transfected as described in **Subheading 3.2.1**. Nontransfected cells as well as control transfected cells are needed to evaluate the results.
2. Add HRB-1 to a final concentration of 3μg/mL 4–6h after syncytin-1 transfection.
3. Analyze cells after 18–24h with a microscope and manipulate as needed (*see* **Note 7**).

3.4. Biotinylation of Cell Surface Proteins and Determination of Exposed Syncytin-1 Protein

Assuming that fusion relevant proteins need to be exposed at the cell surface, it is necessary to investigate the protein localization on the outer cell membrane to obtain a reliable picture of the fusion site.

1. Twenty-four hours after transfection of HEK293 cells (as described in **Subheading 3.2.1**), wash adherent cells once with PBS at 4°C containing 0.1m*M* CaC12 and 1m*M* MgC12, and incubate in PBS/Ca/Mg with 0.25mg/mL D-biotinoyl-ε-aminocaproic acid-*N*-hydroxysuccinimide ester (biotin-7-NHS; **ref.** ***15***) for 30min at 4°C according to the manufacturer's recommendations.
2. Check viability (should be >90%). Transfected cells can be biotinylated and washed and cell viability determined simultaneously using the Trypan blue exclusion method (*see* below).
3. Give commercial Trypan blue solution 1× (Gibco) to a sample of cells (1 million/mL) and evaluate after 10min incubation number of dead (blue) and living cells (not blue) using a counting grid under the microscope
4. Wash cells twice in PBS/Ca/Mg containing 15m*M* glycine to stop the reaction.
5. Lyse cells with 300μL cold lysis buffer.
6. Incubate cell lysates of surface-biotinylated cells (300μL) with 150μL MACS streptavidin microbeads for 30min at 4°C.
7. Purify biotinylated proteins on MACS columns according to manufacturer's instructions (Miltenyi Biotech).
8. Perform elution with 2 × 35μL hot Laemmli buffer containing β-mercaptoethanol at 95°C.

9. Biotinylated proteins can be electrophoresed, blotted, and immunostained with the rabbit anti-syncytin-1 antibody rp69 (*see* **Table 1** and **Note 8**).

3.5. Fusion in a Placental Tissue-Based System: The Explant Villi Model

3.5.1. Floating Villous Explant Culture

1. Use 11–12-week gestational age placental tissue to eliminate the possibility of trophoblast columns. Alternatively, term/intrauterine growth restriction/preeclamptic placental tissue can be used to, for example, examine different pathologies. Collect tissue samples in Ca/Mg-free PBS on ice and use within 4 h.
2. Use a dissecting microscope with a gooseneck light source, and place tissue in a Petri dish. Dissect the tissue to single stem villi with a good branching structure at the tip for first trimester placental tissue (15–20 mg). For terminal tissue explants (last trimester), use scissors and tweezers to tease the villi apart. Use the tips of the branching structures equivalent to intermediate and terminal villi (15–20 mg), because the stem villi are too big and too differentiated for this kind of experimental setup (*see* **Note 9**). Using fine-tipped tweezers, pick up a villus by the stem and poke it into sterile polystyrene cube (1 cm^3, ultraviolet sterilize for >2 h; *see* **Note 10**). Float in PBS to assess stability of the tissue in the polystyrene cube.
3. Transfer to a 24-well plate in ITS-DMEM/F12 (DMEM/F12 + L-glutamine 2 mm + penicillin/streptomycin/gentamicin/Fungizone + 1% ITS (*see* **Subheading 2.6.1**).
4. Maintain cultures at 37 °C in a 6%–8% ambient oxygen incubator in serum-free media (*see* **Note 11**).

3.5.2. Syncytial Denudation and Syncytia Reformation

To investigate the influence of the syncytium on the fusion process and the ability of the cytotrophoblast to maintain and reconstitute the syncytial layer, the following experimental setup can be used.

1. To assess syncytialization, denude the villi and remove the syncytiotrophoblast using trypsin. Trypsinize (0.125% trypsin in PBS for 11–12-week placentas; prewarmed) for 5 min to remove syncytiotrophoblast (*see* **Note 12**). Control samples are not digested and are used to assess baseline syncytiotrophoblast morphology and the presence of anchoring villi.
2. Neutralize trypsin by washing 3 × 5 min in 10% FBS in PBS. Remove samples for wax embedding at both pre- and postdigestion stages.
3. Place cubes in a 24-well plate in 1.5 mL/well ITS-DMEM/F12 and incubate overnight.
4. The following day treat villi with growth factors, siRNA, or antisense oligos, (*see* **Subheading 3.5.3**), and so forth, and grow for 2–4 days.
5. Removed the explanted villi from the polystyrene cubes after a further 48 or 72 h and fix in 4% fresh paraformaldehyde for 2 h.
6. The specimens were dehydrated and wax embedded prior to paraffin histology and immunohistochemistry.

7. All experiments with cultured villi need to be conducted in triplicate and replicated in at least four separate sets of experiments.
8. Resyncytialization should occur in 48 h (*see* **Note 13**).
9. Fix or extract as wanted.
10. Experiments are conducted in triplicate using individual placentas to reduce any possible bias.
11. Tissue sections are immuno-stained with cytokeratin-7 (CK-7) to assess villous trophoblast integrity or with Ki67 to assess cytotrophoblast proliferation (*see* **Note 12** and **Table 1**).

3.5.3. GCM1 Antisense siRNA/Oligonucleotide Strategy to Inhibit Fusion in Explants for Tissue Use

Two double-stranded siRNA oligonucleotides (21mer) named 201 and 815 against the human GCM1sequence can be purchased from Qiagen (designed using the company algorithm and tested for specificity by BLAST bank analysis; *see* **Table 2**). Phosphorothioate oligonucleotides and controls can be designed and are manufactured by Biognostik (Gottingen, Germany) and others.

1. Mount explants as described above in **Subheading 3.5.1** and incubate in the presence of 1 μM antisense oligonucleotides/oligos (or control siRNAs/ oligos) without any transfection reagent (*see* **Note 12**).
2. Toxicity of siRNA and antisense oligonucleotides can be monitored with the human interferon-α ELISA kit (PBL Biomedical Laboratories).
3. Treat samples as needed and described in **Subheading 3.5.2**.

3.5.4. Histology and Immunohistochemistry

1. Perform immunohistochemistry on rehydrated wax-embedded sections using the standard peroxidase method (***4***).
2. Section paraffin-embedded tissue (thickness 5–10 μm).
3. Deparaffinize sections with xylene, and rehydrate using a graded series of ethanol 70%–100%.
4. Inhibit peroxidase activity by incubating slides in 3% hydrogen peroxide in 10% methanol for 15–30 min.
5. Antigen retrieval is dependent on the antibody used and can include membrane permeabilization by 0.02% Triton X-100, protease digestion using 0.125% trypsin or 5–20 mg proteinase K at 37 °C for 10 min, or microwave heat pretreatment 2–5 min in 10 m*M* sodium citrate at pH 6.0 (acidic) or 1 m*M* EDTA, pH 8 (basic).
6. Use antibodies (source/dilution) as shown in **Table 1**. Antibodies to CK-7 or hCG will distinguish cytotrophoblast from overlying syncytiotrophoblast.
7. Include negative controls by omission of the primary antibody. Slides can be visualized using a Nikon DMRX light microscope and photographed with a Sony PowerHAD 3CCD color video camera DXC-970MD (Sony of Canada Ltd., Willowdale, ON).

4. Notes

1. The fusion index underestimates the real fusion events taking place, because same-color fusion events are not counted.
2. It is not necessary to reach 100% transfection efficiency. We found that 60%–75% confluence gives the best fusion results. The reason is not clear, but probably the expression of receptor and syncytin-1 on the same cell leads to receptor interference, which blocks interaction with corresponding molecules on other cells.
3. It is very easy to see the differences between fused and nonfused cells. The syncytia show a uniform, sometimes cloudlike pattern in the combined cytoplasm. Single cells show more intensive fluorescence.
4. Maximal confluency should be reached after 48 h; maximal confluency and protein expression give highest protein concentration in the supernatant.
5. HEK293 cells can be vital for more than 72 h without FBS; special media conditions help to keep the protein machinery as active as possible.
6. It might be useful to harvest media more than once (e.g., every 8 h) to achieve the highest protein concentrations.
7. Do not wait too long; otherwise cells can get to confluence and/or syncytia break up and are lost.
8. This is a very tricky method and can lead to many problems but gives the best information on what is really happening on the cell surface. The optimum result will be obtained with *(a)* high transfection efficiency (use control plasmid), *(b)* efficient labeling, *(c)* amount of protein, and *(d)* the sensitivity of the downstream analysis (antibody quality). Sometimes it might be wise to pool certain experiments of one sequence together. For example, if Laemmli is used to elute the first sample column, it is possible to perform sequential elution steps on other columns and step by step increase the concentration of the desired protein. Size exclusion columns are used to reduce unwanted byproducts, before binding to the beads, but be aware that biotinylation changes the molecular weight to a certain extent.
9. Explants that are too small will be hard to handle, and those that are too big will fall out of the block.
10. Cut polystyrene blocks from a cuvette box lid; use one made out of small beads with a high density, as the larger, less dense beads do not hold the explants very well.
11. From 6% to 8% oxygen was chosen based on our previous villous explant culture experiments indicating physiologically dissolved oxygen tension (40 mmHg) and optimal syncytiotrophoblast preservation over 5 days. First trimester gestational age was chosen to minimize the likelihood that the villous tips did not contain anchoring columns of extravillous trophoblast, typical of explanted specimens in the early first trimester villi (***16***).
12. In explants it is not necessary to transfect the tissue actively to get efficient uptake of siRNA or antisense, the cytotrophoblast efficiently uptake (***17***).
13. To determine optimal conditions for selective removal of syncytiotrophoblast, explanted clumps of 11–12-week gestation villi need to be transferred individually to tissue culture wells containing PBS plus trypsin (0.05%–0.75%) for varying time periods (30 s to 10 min) at 37 °C. This is important, because time and extent

of trypsinization can vary. It is very important to note that denudation should not be 100%. We found that some residual syncytium is needed to induce syncytiotrophoblast formation in the model.

References

1. Adler, R. R., Ng, A. K., and Rote, N. S. (1995) Monoclonal antiphosphatidylserine antibody inhibits intercellular fusion of the choriocarcinoma line, JAR. *Biol. Reprod.* **53(4)**, 905–910.
2. Lyden, T. W., Ng, A. K., and Rote, N. S. (1993) Modulation of phosphatidylserine epitope expression by BeWo cells during forskolin treatment. *Placenta* **14(2),** 177–186.
3. Drewlo, S., et al. (2006) C-Terminal truncations of syncytin-1 (ERVWE1 envelope) that increase its fusogenicity. *Biol. Chem.* **387(8),** 1113–1120.
4. Baczyk, D., et al. (2004) Complex patterns of GCM1 mRNA and protein in villous and extravillous trophoblast cells of the human placenta. *Placenta* **25(6),** 553–559.
5. Morrish, D. W., et al. (1987) Epidermal growth factor induces differentiation and secretion of human chorionic gonadotropin and placental lactogen in normal human placenta. *J. Clin. Endocrinol. Metab.* **65(6),** 1282–1290.
6. Borges, M., et al. (2003) A two-colour fluorescence assay for the measurement of syncytial fusion between trophoblast-derived cell lines. *Placenta* **24(10),** 959–964.
7. Kudo, Y. and Boyd, C. A. (2002) Changes in expression and function of syncytin and its receptor, amino acid transport system B(0) (ASCT2), in human placental choriocarcinoma BeWo cells during syncytialization. *Placenta* **23(7),** 536–541.
8. Kudo, Y. and Boyd, C. A. (2002) Human placental amino acid transporter genes: expression and function. *Reproduction* **124(5),** 593–600.
9. Mi, S., et al. (2000) Syncytin is a captive retroviral envelope protein involved in human placental morphogenesis. *Nature* **403(6771),** 785–789.
10. Baczyk, D., et al. (2006) Bi-potential behavior of cytotrophoblasts in first trimester chorionic villi. *Placenta* **27(4–5),** 367–374.
11. Blond J. L., Lavillette D., et al. (2000) An envelope glycoprotein of the human endogenous retrovirus HERV-W is expressed in the human placenta and fuses cells expressing the type D mammalian retrovirus receptor. *J Virol.* **74(7),** 3321–3329.
12. Chang, C., et al. (2004) Functional characterization of the placental fusogenic membrane protein syncytin. *Biol. Reprod.* **71(6),** 1956–1962.
13. Morrish, D. W., et al. (1997) In vitro cultured human term cytotrophoblast: a model for normal primary epithelial cells demonstrating a spontaneous differentiation programme that requires EGF for extensive development of syncytium. *Placenta* **18(7),** 577–585.
14. Frendo, J. L., et al. (2003) Direct involvement of HERV-W Env glycoprotein in human trophoblast cell fusion and differentiation. *Mol. Cell Biol.* **23(10),** 3566–3574.

15. Lavillette, D., et al. (1998) A proline-rich motif downstream of the receptor binding domain modulates conformation and fusogenicity of murine retroviral envelopes. *J. Virol.* **72(12),** 9955–9965.
16. Huppertz, B., et al. (2003) Hypoxia favours necrotic versus apoptotic shedding of placental syncytiotrophoblast into the maternal circulation. *Placenta* **24(2–3),** 181–190.
17. Black, S., et al. (2004) Syncytial fusion of human trophoblast depends on caspase 8. *Cell Death Differ.* **11(1),** 90–98.

22

Methods to Fuse Macrophages In Vitro

Agnès Vignery

Summary

Macrophages are mononucleate cells that fuse in rare and specific instances to form osteoclasts in bone or giant cells in chronic inflammatory conditions. Because of the central role these cells play in bone metabolism and in inflammation, respectively, methods to study their formation in vitro are described.

Key Words: Macrophage; fusion; osteoclast; giant cell.

1. Introduction

Macrophages fuse with one another, and perhaps with somatic and tumor cells, to form new cells. Depending on the environment, macrophages differentiate into osteoclasts in bone or giant cells in response to foreign bodies, such as pathogens or implants. This chapter focuses on methods that favor the fusion of macrophages into osteoclasts and giant cells in vitro. Also, because most laboratories use macrophages derived from the bone marrow and from the spleen, and very few use macrophages derived from the lungs and from the peritoneum, if any apart from ours, we will provide more details about assays to fuse macrophages from these tissues.

First, we list the key rules we follow when handling macrophages destined to fuse in vitro:

1. The serum must be of the highest quality, called "U.S. defined fetal bovine serum" by some companies (*see* **Note 1**).
2. The temperature at which macrophages are handled is essential: macrophages tend to adhere very quickly to many different types of plastic, so it is critical to keep the cells at 4 °C at all times.
3. Macrophages are extremely rich in lysosomal enzymes, so, if possible, it is best to lyse them directly in sample buffer for protein or RNA isolation and purification.

From: *Methods in Molecular Biology, vol. 475: Cell Fusion: Overviews and Methods*
Edited by: E. H. Chen © Humana Press, Totowa, NJ

4. Because of the fragility of primary macrophages, it is important not to use trypsin to replate or isolate them. Rather, it is best to wash off the serum, using, for instance, phosphate buffered saline (PBS), and to keep the cells at 4 °C in PBS for 30 min to 1 h and then scrape them in the appropriate buffer. They roll up and detach easily, yet stay intact.

2. Materials

2.1. Fusion of Rat Alveolar Macrophages

1. Any strain of rat, but Fisher 344 females are best; the older and the bigger, the better the yield of macrophages.
2. Butterfly tubing #21 connected to a 5-mL syringe.
3. Lung washing medium: PBS supplemented with 10% fetal calf serum (FCS) or fetal bovine serum (FBS), heat-inactivated, kept at 37 °C.
4. Complete medium: Minimum Essential Medium with Earl's salt (MEME) supplemented with 2 m*M* glutamine, 1× vitamins and 1× penicillin/streptomycin (50 U/mL, 50 µg/mL). Note: Glutamine must be aliquoted and thawed only once (*see* **Note 2**).
5. Heat-inactivated human serum freshly obtained from the blood bank since it is not commercially available. One can never use serum from one's own blood.
6. To prepare serum, use freshly collected human blood (100–1000 mL per donor in red top tubes) kept at room temperature for 1 h to facilitate clotting and then placed in the refrigerator for 2 h to contract the clot. Spin tubes at 300 *g* for 30 min at 4 °C. Collect and pool sera. Heat inactivate serum at 56 °C for 30 min.

2.2. Fusion of Rat Peritoneal Macrophages

1. All strains of rats are acceptable. All reagents necessary to complete the assay are the same as for rat alveolar macrophages. The only differences are the cell collection and the plating density.
2. Tissue culture reagents are the same as for rat alveolar macrophages.
3. 18-gauge needle or bigger connected to a 10-mL syringe.

2.3. Fusion of Mouse Alveolar Macrophages

1. Young adult mice, any strain.

2.4. Fusion of Mouse Bone Marrow Macrophages to Generate Osteoclasts

1. Mice, any strain.
2. Sterile scissors, forceps, scalpels, and blades.
3. 10-mL syringes, 23gauge needles, and a cell strainer.
4. Complete medium: α-MEM supplemented with 2 m*M* glutamine, 1× vitamins, penicillin (100 U/mL), streptomycin sulfate (100 µg/mL), and FBS (*see* **Note 1**).

5. Mouse macrophage-colony stimulating factor (M-CSF) and mouse or human receptor activator of the nuclear factor-κB ligand (RANKL).

2.5. Fusion of Spleen Cell–Derived Macrophages to Generate Osteoclasts

1. Mice (4–12 week old are best), any strain.
2. Two pairs of scissors, two forceps, scalpels, and blades—all sterile.
3. Complete medium: α-MEM supplemented with 2 m*M* glutamine, 1× vitamins, penicillin (100 U/mL), streptomycin sulfate (100 μg/mL), and FBS.
4. 10-cm Petri dishes and sterile glass slides with frosted ends.

2.6. Human Monocyte Fusion Assay

1. Ficoll-Paque Premium (Amersham/GE, catalogue number 17-5442-02).
2. MEME.
3. 20 mL of human blood collected in a 10 mL/EDTA tube (purple top).

3. Methods

3.1. Fusion of Rat Alveolar Macrophages

3.1.1. Macrophage Lavage

1. Sacrifice animals by CO_2 inhalation using a box approved by the animal care and use committee of your institution. Immediately following sacrifice, dip the body of the rat in diluted Lysol to prevent fur contamination. Do not place the heads of rats in Lysol, as it could contaminate the airway and hence the lungs of the animal.
2. To collect alveolar macrophages, lay the rats on their backs and open the skin along the neck to the base of the chest; separate the neck muscles, and expose the trachea.
3. Open a window (at least a 1 cm^2) in the anterior chest wall and make a small incision in the diaphragm before lavage to allow for lung expansion and yet to prevent lung damage. Make sure the lungs are expanding, indicating that the medium gets in there.
4. Make a small opening in the trachea to insert the plastic tubing, which is connected to a 5-mL syringe. The tubing should be inserted not more than 0.5 cm deep inside the trachea to avoid damaging the lungs.
5. Secure the tubing to the trachea by tightening it up with a thread, such as a sewing thread.
6. Inject very slowly 5 mL of PBS/10% heat-inactivated FCS/FBS warmed up to 37 °C.
7. Massage the chest of the animals gently for 3–5 min, and then aspirate the medium.
8. Collect the medium in a 50-mL tissue culture tube placed on ice. The first time, 70% of the medium is recovered, but in subsequent washes, close to 90% of the medium is recovered.

9. Repeat washes 10 times and pool them. Disconnect the needle from the syringe to refill the syringe and leave the needle in place. It should take 30–45 min per rat. Macrophage yield varies between 1×10^6 and 2×10^7 cells per rat, depending on the size of the rat and the efficacy of the lavage.

3.1.2. Macrophage Plating

1. Fill up each 50-mL tube with cold MEME and spin at 300 *g* for 10 min at 4 °C.
2. Discard the supernatant, and pool the cells into a 10- or 50-mL tube, depending on the number of lungs lavaged.
3. Repeat the wash and the spin, and then resuspend the cells at 5×10^6 cells/mL in complete cold MEME supplemented with 10% heat-inactivated human serum. Rat alveolar macrophages are particularly large, about 15 mm in diameter, so they are easily distinguished from red blood cells, which are 7 mm diameter.
4. Plate cells in drops whose size depends on the assay: from 2 µL to 1 mL in 96-well to 10-cm dishes. It is critical *not to cover* the entire surface of the well or dish, but 50% at most (**Fig. 1**). Place drops in the center of each well/dish. To plate 2–10-µL drops, turn off the vent of the hood to avoid dessication of the drop. Also, fill up the outer wells of the dish with PBS to secure moisture.
5. Place dishes in the incubator at 37 °C/5% CO_2. The incubation time varies with the volume plated, but ideally, avoid evaporation, as concentrated salts can kill the cells. The cells settle down and adhere just enough to the plate to contact each other and form a monolayer. Plating is delicate at the beginning, because small drops can quickly dry or escape the center of the well and end up at the periphery of the well. Large drops of 0.5 to 1 mL in 6-well or 10-cm dishes can also escape the center while being moved into the incubator. Incubation time varies between 5 and 20 min.
6. Add complete MEME supplemented with 5% human serum warmed up to 37 °C to cover the cells: from 100 µL in 96-well to 10 mL in 10-cm dishes. Particular attention must be paid when medium is added to the cells. Do not disturb the cells; they are weakly adherent. Macrophages start fusing within hours after plating. Multinucleation reaches 99% within 3–5 days (**Fig. 2**).

3.2. Fusion of Rat Peritoneal Macrophages

3.2.1. Peritoneal Lavage

1. Sacrifice animals by CO_2 inhalation using a box that is approved by the animal care and use committee of your institution. Immediately following sacrifice, dip the body of the rat in diluted Lysol to prevent fur contamination.
2. To collect peritoneal macrophages, lay the rats on their backs and make a 3-cm skin incision along the abdomen to expose the peritoneal muscles.
3. Insert an 18-gauge or larger needle connected to a 10-mL syringe into the peritoneal cavity, and make sure the needle did not enter the intestine and the tip is free in the peritoneal cavity.

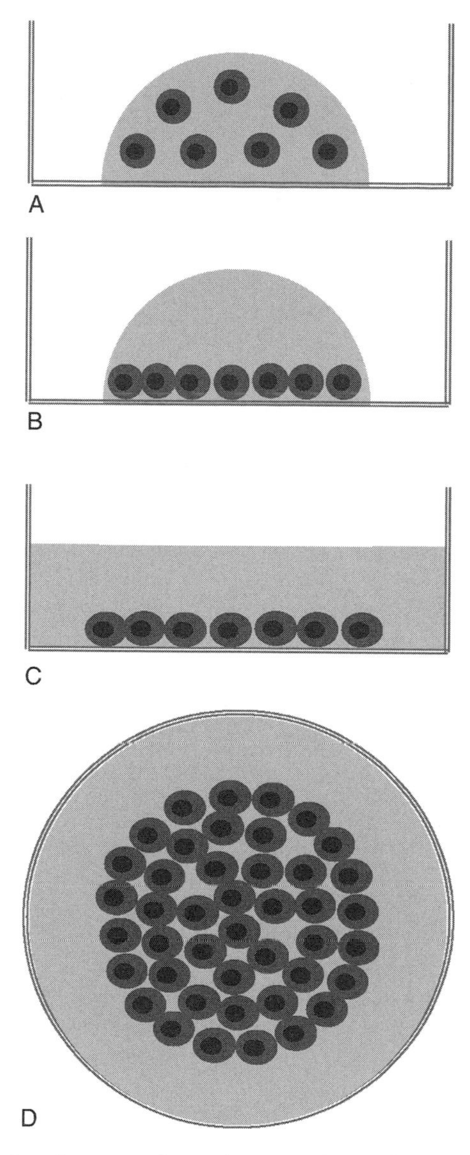

Fig. 1. Plating of alveolar or peritoneal macrophages in fusogenic conditions. The cell density and the size of the drop **(A)** are such that the cells are in contact after they settle down on the dish **(B)** yet cover only half to a third of the bottom of the well or dish. The medium is added to the well or dish only after the cells slightly adhere, so they remain confluent **(C,D)**.

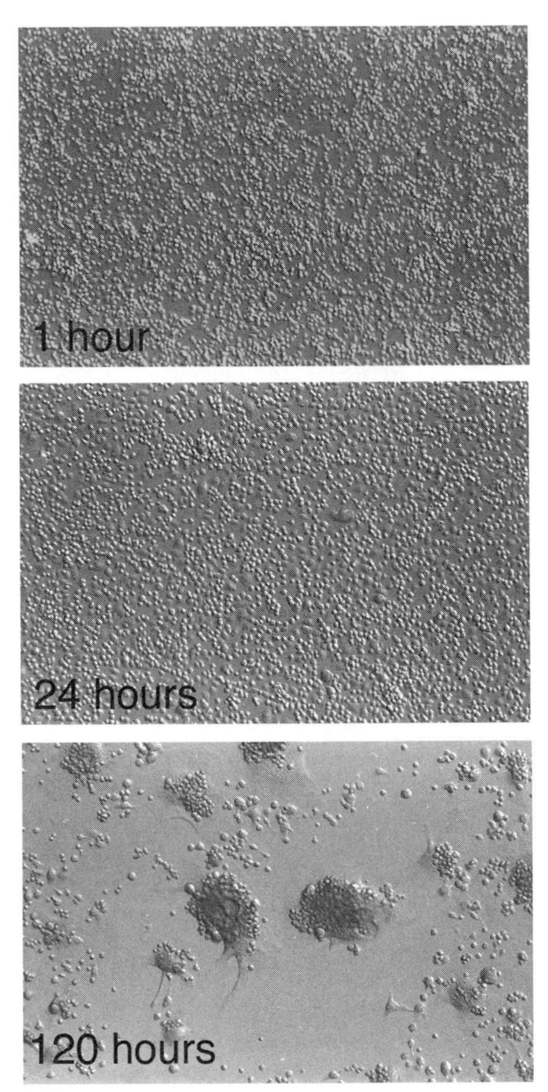

Fig. 2. Plating of alveolar macrophages. Freshly plated macrophages quickly move around and initiate fusion 1 h after plating. Note that by day 5 most of the macrophages have fused into multinucleate cells that contain thousands of nuclei.

4. Fill up the peritoneal cavity with 20–50 mL of 37 °C MEME/10% FBS, massage the abdomen gently for several minutes, and then aspirate the medium. The first time, less than 50% of the medium is recovered. Disconnect the needle from the syringe to refill the syringe and leave the needle in place. Subsequent lavage

recovery reaches nearly 90% of the medium. The yield of macrophages varies with the size of the rat, between 1×10^7 and 5×10^7 macrophages per rat.

3.2.2. Macrophage Plating

1. Pool washes into conical tissue culture tubes placed on ice. It is critical to maintain cells at 4 °C as soon as they are collected to prevent them from adhering to the tube and to aggregate. Keep cells at 4 °C at all times until plated. Once washes are complete, fill up tubes with cold PBS, and spin at 300 *g* for 10 min.
2. Aspirate supernatants, resuspend, and pool pellets into one tube containing complete cold MEME; spin at 300 *g* for 10 min at 4 °C. Resuspend cells at 1×10^7/mL in complete MEME kept at 4 °C supplemented with 10% heat-inactivated human serum.
3. Plate the cells in drops, as for alveolar macrophages, from 2 μL to 1 mL, in 96-well to 10-cm dishes. It is critical *not to cover* the entire surface of the well or dish, but 50% maximum. Place the drops in the center of the well/dish. When plating 2–10-mL drops, turn off the vent of the hood to avoid desiccation of the drop. Also, fill up the outer wells of the dish with PBS to moisten the dish.
4. Place dishes in the incubator at 37 °C/5% CO_2. The incubation time varies according to the cell volume plated, but ideally evaporation should not occur, as concentrated salts will kill the cells. The cells settle and adhere just enough to contact each other and form a monolayer. It is tricky at the beginning, but then you get a sense for the timing. It varies between 5 and 20 min.
5. Add 37 °C complete MEME supplemented with 5% human serum to cover the cells: 100 μL in 96-well to 10 mL in 10-cm dishes. Pay attention when adding the medium: do not disturb the cells. They are weakly adherent. Macrophages should start fusing a few days after plating. Multinucleation reaches 60%–70% within 10 days.

3.3. Fusion of Mouse Alveolar Macrophages

The protocol used to collect alveolar macrophages from mice is similar to that of rats. The main differences with rat alveolar macrophages are the following.

1. The cell size, as mouse alveolar macrophages measure about 10–12 μm diameter and hence are smaller than rat alveolar macrophages.
2. The yield, as mouse macrophage yield varies between 1×10^4 and 2×10^5 cells per mouse, depending on the size of the mouse and the efficacy of the lavage.
3. The plating density, which is higher than that of rat macrophages because mouse macrophages are smaller than those of rats: 1×10^7/mL. Given the yield, only 2–10 μL drops can be plated in 96-well dishes.
4. The fusion rate, as mouse alveolar macrophages reach 30%–50% fusion within 4–6 days.
5. To collect mouse alveolar macrophages, use a JELCO IV catheter with the needle removed for mice and connected to a 1-mL syringe. Use the sheath as tube; depending on the mouse size, use #18 or #20 (for 4-week-old mice). Inject very slowly 1 mL of PBS/10% heat-inactivated FCS/FBS warmed up to 37 °C.
6. Repeat washes 5–10 times.

3.4. Fusion of Mouse Bone Marrow Macrophages to Generate Osteoclasts

3.4.1. Dissection Protocol

1. Autopsy: sacrifice mice by carbon dioxide asphyxiation.
2. Collect femurs and tibiae using forceps and clean adjacent muscle.
3. Place bones in Petri dishes contain α-MEM, on ice.

3.4.2 Bone Marrow Cell Culture

1. Wash bones three times in PBS, once in 70% ethanol, and then once in PBS. Transfer bones to a sterile Petri dish containing α-MEM, and prepare another Petri dish with α-MEM.
2. Cut off both epiphyses to expose the bone marrow cavity. Place all epiphyses in one dish and diaphyses in the other dish. Use sterile scissors to cut epiphyses as small as possible.
3. Wash marrow cavities several times using a 5-mL syringe and a 23-guage needle, until all marrow cells are out of the marrow cavity.
4. Pipette bone marrow cells up and down to break down aggregates until cells are well resuspended.
5. Use a cell strainer to separate particulates from bone marrow cells and wash dishes twice to collect all cells. Pool medium in a 50-mL tube, and spin cells at 300 *g* for 10 min at 4 °C.
6. Resuspend pellets in 5 mL PBS supplemented with 5 mL red blood cell lysis buffer (NH_4CL). Keep cells on ice for 7 min. Add 30 mL of α-MEM to the cells, and spin at 300 *g*, 4 °C, for 10 min.
7. Aspirate medium and resuspend cells in 5 mL α-MEM. Spin and resuspend cells in complete medium (α-MEM, 10% FBS, 10% L929 supernatant, at 10^6/ mL or 1×10^7 cells/dish in 10-mL Corning dishes.
8. Culture cells for 18 h, and then transfer nonadherent bone marrow cells to 50-mL Falcon tube. Wash dishes with 10 mL PBS, and spin cells at 300 *g*, 4 °C, for 10 min.
9. Resuspend and culture cells as in **step 5** for 48 h.
10. Add 5 mL complete medium to each dish, and incubate for 3 days.
11. Aspirate medium, wash twice with PBS, and incubate cells in 10 mL of cold PBS, 4 °C, for 2 h.
12. Scrape the cells into 10 mL PBS, transfer cells to a 50-mL Falcon tube, and spin at 300 *g*, 4 °C, for 10 min. Resuspend cells and plate them at 10^6/ mL in complete α-MEM supplemented 10% FBS.
13. Culture cells for 24 h, change the medium to get rid of dead cells, and replace L929 supernatant with 30 ng/mL recombinant mouse M-CSF and 50 ng/mL recombinant mouse or human RANKL.
14. Change medium every 3 days. Cells fuse within 10–12 days.
15. React multinucleate cells with the substrate for tartrate-resistant acid phosphatase (TRAP).

Of note, some investigators culture bone marrow cells in α-MEM supplemented with 10% FBS, 10 n*M* $1,25(OH)_2$ vitamin D_3, and 100 n*M* dexamethasone instead of RANKL.

3.5. Fusion of Spleen Cell-Derived Macrophages to Generate Osteoclasts

3.5.1. Dissection

1. Euthanize the mice by carbon dioxide asphyxiation or isoflurane inhalation, followed by cervical dislocation.
2. Cut the skin to expose the abdominal musculature.
3. Using a clean set of scissors and forceps, expose the abdominal cavity, remove the spleen, and place it in a Petri dish that contains 10 mL of α-MEM, at room temperature.

3.5.2. Culture

1. Rub the frosted ends of two glass slides against each other to remove excess glass powder, and then rinse the frosted ends in sterile PBS.
2. Place the spleen onto the frosted end of one slide, and then cut the spleen in several pieces with the frosted end of the other slide. Rub the slides against each other to expel the cells from the spleen pieces, and rinse the slide ends with medium.
3. Using a 1-mL pipette, disaggregate the cells and push away the connective tissue pieces on one end of the dish; then collect the cells and place them into a 50-mL tube, on ice.
4. Pool spleen cells into a 50-mL tube, and add α-MEM to fill up the 50-mL tube; spin at 300 *g*, 4 °C, for 10 min.
5. Resuspend cells at 2×10^6/mL in complete medium.
6. Plate cells in 48-well Falcon dishes: 0.5 mL per well (1×10^6 cells per well); adjust the cell number to the surface area of the well if 96-, 24-, or 6-well dishes are used.
7. Add M-CSF (30 ng/mL) and RANKL (50 ng/mL) to the cells.
8. Change the medium on day 3 to get rid of dead cells. Add complete medium as before, supplemented with M-CSF and RANKL.
9. The cells are fused by day 6 or 7.

3.5.3 Fusion of Mouse Bone Marrow and Spleen Cell-Derived Macrophages to Generate Giant Cells

Essentially, the methods used to fuse bone marrow and spleen macrophages to form giant cells are similar to those used to form osteoclasts, the difference being to replace RANKL with interleukin-4 (IL-4; 10 ng/mL) or IL-10 (10 ng/mL). The M-CSF is replaced sometimes by granulocyte-macrophage colony

Table 1
Optional to Test for Best Fusion

No. of cells per milliliter	No. of cells per 50-mL drop	No. of cells per square centimeter
1×10^7	5×10^5	1.3×10^6
5×10^6	2.5×10^5	6.5×10^5
2.5×10^6	1.25×10^5	3.3×10^5
1.25×10	0.625×10^5	1.65×10^5
0.625×10^6	0.3125×10^5	0.82×10^5
0.3125×10^6	0.156×10^5	0.41×10^5

Resuspend cell pellets in complete Minimum Essential Medium with Earl's salt and plate as indicated.

stimulating factor (GM-CSF). It appears that IL-4 is a more potent fusogenic factor than RANKL but leads to the formation of cells that are different from osteoclasts because they are TRAP negative. However, they resorb the substrate onto which they form and adhere. Another difference, which might not affect the differentiation of giant cells, is the use of Iscove's Modified Dulbecco's Medium (IMDM) to culture the cells.

3.6. Human Monocyte Fusion Assay

3.6.1. Isolation of Human Peripheral Blood Monocytes Using Ficoll-Paque Premium

1. Transfer blood to a 50-mL centrifuge tube.
2. Add 10 mL of PBS:MEME (4:1).
3. Underlay with 20 mL of Ficoll-Paque.
4. Centrifuge at 400 *g* for 30 min at room temperature.
5. Remove the lymphocyte layer.
6. Fill up the 50-mL tube with PBS:MEME (4:1).
7. Centrifuge at 300 *g* for 15 min at room temperature.
8. Resuspend cell pellet in 5 mL of complete MEME (10% FCS–1% penicillin/streptomycin–2 m*M* glutamine).
9. Plate cells at 2.5×10^6 cells/mL in a T75 flask, 10 mL per flask, using complete MEME.
10. Incubate cells at 37 °C, 5%CO_2 for 1.5 h.
11. Wash cells twice with plain MEME without serum warmed up to 37 °C.
12. Gently add complete MEME supplemented with 25 ng/mL human M-CSF and 40 ng/mL human RANKL.

The cell density between 2×10^6 cells/mL and 2.5×10^6 cells/mL leads to a plating density of approximately 3×10^5 cells/cm^2.

Table 2
Macrophage Fusion Assays

Species	Cell origin	Yield per animal	Fusogenicity	Time
Rat	Alveolar macrophage	$5–20 \times 10^6$	99% in 10% human serum	3–5 days
	Peritoneal macrophage	$10–50 \times 10^6$	5%–70% in 10% human serum	7–10 days
Mouse	Alveolar macrophage	$1 \times 10^4 – 2 \times 10^5$	30 to 50% in 10% human serum	5–7 days
	Bone marrow–derived macrophage	$1–5 \times 10^6$ after expansion in M-CSF	20 to 50% in M-CSF + RANKL	14–16 days
	Spleen cell–derived macrophage	5×10^6 after expansion in M-CSF	50% in M-CSF + RANKL	6–8 days
	Raw cells: tumor macrophage cell line		50% in RANKL	6–8 days
Human	Monocyte-derived macrophage		50% in M-CSF + RANKL	18–21 days

M-CSF, macrophage-colony stimulating factor; RANKL, receptor activator of the nuclear factor-κB ligand.

4. Conclusion

Whether collected early as monoblasts in the bone marrow, as monocytes in the blood stream, or as peritoneal macrophages in the peritoneal cavity, mononuclear phagocytes have the ability to fuse with themselves in response to a growth factor and a differentiation factor or spontaneously. The yields, the fusion efficacies, and the times required for maximum fusion are listed in **Table 2**.

5. Notes

1. Fetal calf serum and FBS work as well but must be free of endotoxins.
2. Glutamine must be aliquoted and thawed only once because of its short half-life and its instability.

Suggested Reading

Abu-Amer, Y., Ross, F. P., Edwards, J., and Teitelbaum, S.L. (1997) Endotoxin stimulated osteoclastogenesis is mediated by tumor necrosis factor via its P55 receptor. *J. Clin. Invest.* **100,** 1557–1565.

Bi, L. X., Simmons, D. J., and Mainous, E. (1999) Expression of BMP-2 by rat bone marrow stromal cells in culture. *Calcif. Tissue Int.* **64,** 63–68.

Carlo-Stella, C., Di Nicola, M., Milani, R., Longoni, P., Milanesi, M., Bifulco, C., Stucchi, C., Guidetti, A., Cleris, L., Formelli, F., Garotta, G., and Gianni, A. M. (2004) Age- and irradiation-associated loss of bone marrow hematopoietic function in mice is reversed by recombinant human growth hormone. *Exp. Hematol.* **32,** 171–178.

Cui, W., Ke, J. Z., Zhang, Q., Ke, H. Z., Chalouni, C., and Vignery, A. (2006) The intracellular domain of CD44 promotes the fusion of macrophages. *Blood* **107,** 796–805.

Dobson, K. R., Reading, L., Haberey, M., Marine, X., and Scutt, A. (1999) Centrifugal isolation of bone marrow from bone: an improved method for the recovery and quantitation of bone marrow osteoprogenitor cells from rat tibiae and femurae. *Calcif. Tissue Int.* **65,** 411–413.

Gao, Y. H. and Yamaguchi, M. (1999) Suppressive effect of genistein on rat bone osteoclasts: apoptosis is induced through Ca^{2+} signaling. *Biol. Pharm. Bull.* **22,** 805–809.

Grove, J. E., Bruscia, E., and Krause, D. S. (2004) Plasticity of bone marrow-derived stem cells. *Stem Cells* **22,** 487–500.

Harris, R. G., Herzog, E. L., Bruscia, E. M., Grove, J. E., Van Arnam, J. S., and Krause, D. S. (2004) Cells from donor bone marrow can form differentiated epithelial cells in the lung, liver, and skin, without having fused with existing resident cells. *Science* **305,** 90–93.

Helming, L. and Gordon, S. (2007) Macrophage fusion induced by IL-4 alternative activation is a multistage process involving multiple target molecules. *Eur. J. Immunol.* **37,** 33–42.

Kelly, K. A., Tanaka, S., Baron, R., and Gimble, J. M. (1998) Murine bone marrow stromally derived BMS2 adipocytes support differentiation and function of osteoclast-like cells in vitro. *Endocrinology* **139,** 2092–2101.

Kim, M. S., Day, C. J., Selinger, C. I., Magno, C. L., Stephens, S. R., and Morrison, N. A. (2006) MCP-1–induced human osteoclast-like cells are tartrate-resistant acid phosphatase, NFATc1, and calcitonin receptor-positive but require receptor activator of NF-kappaB ligand for bone resorption. *J. Biol. Chem.* **281,** 1274–1285.

Kopen, G. C., Prockop, D. J., and Phinney, D. G. (1999) Marrow stromal cells migrate throughout forebrain and cerebellum, and they differentiate into astrocytes after injection into neonatal mouse brains. *Proc. Natl. Acad. Sci. U. S. A.* **96,** 10711–10716.

Kyriakides, T. R., Foster, M. J., Keeney, G. E., Tsai, A., Giachelli, C. M., Clark-Lewis, I., Rollins, B. J., and Bornstein, P. (2004) The CC chemokine ligand, CCL2/MCP1, participates in macrophage fusion and foreign body giant cell formation. *Am. J. Pathol.* **165,** 2157–2166.

Phinney, D. G., Kopen, G., Isaacson, R. L., and Prockop, D. J. (1999) Plastic adherent stromal cells from the bone marrow of commonly used strains of inbred mice: variations in yield, growth, and differentiation. *J. Cell Biochem.* **72**, 570–585.

Rakopoulos, M., Ikegame, M., Findlay, D. M., Martin, T. J., and Moseley, J. M. (1995) Short treatment of osteoclasts in bone marrow culture with calcitonin causes prolonged suppression of calcitonin receptor mRNA. *Bone* **17,** 447–453.

Saginario, C., Qian, H.-Y., and Vignery, A. (1995) Identification of an inducible surface molecule specific to fusing macrophages. *Proc. Natl. Acad. Sci. U. S. A.* **92,** 12210–12214.

Saginario, C., Sterling, H., Beckers, C., Kobayashi, R.-J., Solimena, M., Ullu, E., and Vignery, A. (1998) MFR, a putative receptor mediating the fusion of macrophages. *Mol. Cell Biol.* **18,** 6213–6223.

Sakai, A., Nishida, S., Okimoto, N., Okazaki, Y., Hirano, T., Norimura, T., Suda, T., and Nakamura, T. (1998) Bone marrow cell development and trabecular bone dynamics after ovariectomy in ddy mice. *Bone* **23,** 443–451.

Seto, H., Aoki, K., Kasugai, K., and Ohya, K. (1999) Trabecular bone turnover, bone marrow cell development, and gene expression of bone matrix proteins after low calcium feeding in rats. *Bone* **25,** 687–695.

Sterling, H., Saginario, C., and Vignery, A. (1998) CD44 occupancy prevents macrophage multinucleation. *J. Cell Biol.* **843,** 837–847.

Strawn, W. B., Richmond, R. S., Tallant, E. A., Gallagher, P. E., and Ferrario, C. M. (2004) Renin–angiotensin system expression in rat bone marrow haematopoietic and stromal cells. *Br. J. Hematol.* **126,** 120–126.

Tropel, P., Noel, D, Platet, N., Legrand, P., Benabid, A. L., and Berger, F. (2004) Isolation and characterisation of mesenchymal stem cells from adult mouse bone marrow. *Exp. Cell Res.* **295,** 395–406.

Uchiyama, S. and Yamaguchi, M. (2004) Inhibitory effect of β-cryptoxanthin on osteoclast-like cell formation in mouse marrow cultures. *Biochem. Pharmacol.* **67,** 1297–1305.

Index

G

H

T

U

V

Z

Printed in the United States of America